CHEMICAL DYNAMICS IN EXTREME ENVIRONMENTS

Advanced Series in Physical Chemistry

Editor-in-Charge

Cheuk-Yiu Ng, *Ames Laboratory USDOE, and Department of Chemistry, Iowa State University, USA*

Associate Editors

Paul F. Barbara, *Department of Chemistry, University of Minnesota, USA*

Sylvia T. Ceyer, *Department of Chemistry, Massachusetts Institute of Technology, USA*

Hai-Lung Dai, *Department of Chemistry, University of Pennsylvania, USA*

Benny Gerber, *The Fritz Haber Research Center and Department of Chemistry, The Hebrew University of Jerusalem, Israel, and Department of Chemistry, University of California at Irvine, USA*

James J. Valentini, *Department of Chemistry, Columbia University, USA*

Advanced Series in Physical Chemistry – Vol. 11

CHEMICAL DYNAMICS IN EXTREME ENVIRONMENTS

Editor

Rainer A. Dressler

Space Vehicles Directorate
Air Force Research Laboratory
USA

World Scientific
Singapore • New Jersey • London • Hong Kong

Published by

World Scientific Publishing Co. Pte. Ltd.

P O Box 128, Farrer Road, Singapore 912805

USA office: Suite 1B, 1060 Main Street, River Edge, NJ 07661

UK office: 57 Shelton Street, Covent Garden, London WC2H 9HE

Library of Congress Cataloging-in-Publication Data
Chemical dynamics in extreme environments / editor Rainer A. Dressler.
    p. cm. -- (Advanced series in physical chemistry ; v. 11)
    Includes bibliographical references and index.
    ISBN 9810241771
    1. Chemical kinetics. 2. Molecular dynamics. 3. Extreme environments. I. Dressler,
Rainer A. II. Series.

QD501.C4627 2000
541.3'94--dc21   00-043482

British Library Cataloguing-in-Publication Data
A catalogue record for this book is available from the British Library.

Printed in Singapore by World Scientific Printers

ADVANCED SERIES IN PHYSICAL CHEMISTRY

INTRODUCTION

Many of us who are involved in teaching a special-topic graduate course may have the experience that it is difficult to find suitable references, especially reference materials put together in a suitable text format. Presently, several excellent book series exist and they have served the scientific community well in reviewing new developments in physical chemistry and chemical physics. However, these existing series publish mostly monographs consisting of review chapters of unrelated subjects. The modern development of theoretical and experimental research has become highly specialized. Even in a small subfield, experimental or theoretical, few reviewers are capable of giving an in-depth review with good balance in various new developments. A thorough and more useful review should consist of chapters written by specialists covering all aspects of the field. This book series is established with these needs in mind. That is, the goal of this series is to publish selected graduate texts and stand-alone review monographs with specific themes, focusing on modern topics and new developments in experimental and theoretical physical chemistry. In review chapters, the authors are encouraged to provide a section on future developments and needs. We hope that the texts and review monographs of this series will be more useful to new researchers about to enter the field. In order to serve a wider graduate student body, the publisher is committed to making available the monographs of the series in a paperbound version as well as the normal hardcover copy.

Cheuk-Yiu Ng

PREFACE

The present volume of the Advanced Series in Physical Chemistry serves to expose an emerging frontier for physical chemists: the elucidation of chemical dynamics in natural and technical environments that are identified as extreme. With the steady increase in affordable computational power, a growing number of macroscopic phenomena are modeled at the molecular level, exerting new demands on the understanding of detailed molecular dynamics in exotic conditions. The development of computational environmental models depends on bidirectional cross-disciplinary collaboration between chemists and technical experts of a number of fields including geophysics, plasma and fluid physics, materials science, and engineering. Not only must researchers attempting to model an extreme environmental system be acquainted with physical chemical expertise and information, but the physical chemist must also be intimately familiar with the parameters of the investigated system in order to identify the relevant processes, and propose a chemical model having a complexity commensurate with the fidelity of the ultimate environmental model. Unfortunately, despite a consensus that the breakthroughs of the future will come from multidisciplinary efforts, interdisciplinary communication has faced considerable resistance. This is mainly attributable to the division of most universities into disciplinary departments, an organizational form that is also reflected in many federal funding agencies. This division ultimately results in a peer structure that is based on disciplinary excellence.[a]

There have been, however, concerted efforts in breaching the barriers between disciplines, a noteworthy example of which is the US Department of Defense Multidisciplinary Research Program of the University Research Initiative that funds cross-departmental efforts to address specific technological objectives, many of which involve exploiting or mitigating the forces

---

[a]N. Metzger and R. N. Zare, *Science* **283**, 642 (1999).

released in systems under extreme conditions. As an additional step in this direction, the present volume attempts to highlight chemical dynamics studies tied to unraveling the mysteries of extreme environments.

This book identifies a number of examples where knowledge of the detailed chemical dynamics is a vital component of the description of a respective extreme environment. A rigid definition of what constitutes an extreme environment does not exist. In selecting the topics, I have chosen environments where the fundamental chemical processes are generally far from thermal and chemical equilibrium, either on short transient time scales, or continuously. At conditions where the energy associated with the various degrees of freedom is not equilibrated, knowledge of the state-to-state dynamics becomes important, and given the limited database on excited state processes, modelers are frequently forced to speculate on the pertinent kinetics. Extreme environments, however, can refer to any condition far from room temperature and atmospheric pressure, a definition adopted by Berman in Chap. 1. In an overview, Berman reverses the general theme of this volume by pointing out how various instances of physical chemical innovation have resulted from the need to understand chemical dynamics under extreme conditions.

Each of the following chapters addresses one or more extreme environments, outlines the associated chemical mechanisms of relevance, and then covers the leading edge science that elucidates the chemical coupling. The chapter topics have been chosen to create a balance between theory and experiment; between gas-phase, solid state, plasma, and interface dynamics; and natural and technical environments. What has resulted is a collection of reviews with overlapping but also highly variable chemical dynamics including timescales ranging from femtoseconds to a billion years. While in one chapter (Chap. 10), the research has not reached the stage at which the molecular structural information is sufficient to commence theoretical dynamical studies, another environment is examined at the level of hyperfine-level transitions (Chap. 4).

In Chap. 2, Raz and Levine investigate a regime of dynamics where the motion along intermolecular coordinates is comparable or faster than that of intramolecular vibrational modes. These conditions exist momentarily when a large cluster impacts a surface at hyperthermal velocities (~ 10 km s^{-1}). In Chap. 3, Boyd describes the challenges facing a direct simulation Monte Carlo modeler of hypersonic flows in a regime intermediate to the continuum and free molecular flow limits. Many of the lessons learned in Chap. 2 find application in this work. In Chap. 4, Heaven takes a close look at the energy transfer and reaction kinetics of chemical laser media. This review is a nice example of how the drive for new technology has fostered new physical chemical techniques as well as the acquisition of kinetic data of fundamental scientific interest. In Chap. 5, Schulz, Volpp and Wolfrum present selected laboratory experiments on gas-phase and surface reactions of importance in combustion systems. The reader is lead through a remarkable progression from the detailed investigation of the H + O$_2$ reaction dynamics to NO$_x$ formation mechanisms in flames, unsteady methanol oxidation under laminar flow conditions, laser diagnostics of surface reactions on a platinum catalyst, to, finally, quantitative laser spectroscopic diagnostics in actual combustion engines. Dressler and Murad describe the phenomenology of meteors in Chap. 6 and review the current understanding of the hyperthermal gas-phase chemical dynamics that are at the origin of various observations. In Chap. 7, Jacobs links the results from detailed studies of the dynamics of hypervelocity surface scattering to physical properties of spacecraft, such as space vehicle drag and charging. Johnson reviews the current understanding of the solid state radiation chemistry of the Galilean moons of Jupiter in Chap. 8. The bombardment by energetic ions and electrons trapped in Jupiter's giant magnetosphere directly affect the appearance of the satellites, and generate volatiles that form the atmosphere and local plasma. In Chap. 9, Minton and Garton review the mechanisms behind atomic oxygen induced degradation of polymers used in low earth orbit. This chapter demonstrates that studies of the appropriate surface interactions are shedding important light on the origin of materials erosion in space. The understanding of the durability of ceramic thermal barrier coatings (TBCs) on the molecular level is the subject of Chap. 10 by Christensen, Jarvis and Carter. TBCs are used to protect turbine blades of aircraft engines, thereby allowing them to run more efficiently at higher temperatures. While Chap. 10 is focused on the structure of ceramic–metal interfaces with the eventual goal of understanding solid state kinetics on a long time scale (years), Chap. 11 by White, Swanson and Robertson examines the ultrafast molecular dynamics associated with shock-induced phase transitions and detonations in solids. The results reported in Chap. 11 show that molecular simulations using reactive potentials provide a powerful probe of the interplay between the continuum properties of shock waves and the atomic scale chemistry.

A better understanding of this interplay is key to the design of safer, more reliable explosives.

Given the broad readership that this volume is intended to address, the authors have taken great care in presenting the material with as general a terminology as possible. Nevertheless, expressions appear that may not be familiar to all readers and would interrupt the flow of the text if explained at the first occurrence. A number of expressions are, therefore, defined in a Glossary at the end of this book.

I thank all of the authors for taking a step outside of their current peer structures and contributing to this volume. I have thoroughly enjoyed interacting with you during the course of the past year. I am particularly grateful to M. Berman for his enthusiastic support of this endeavor and for assisting me in assembling the list of distinguished authors. I am also indebted to C. Kolb, P. Armentrout, E. Murad, G. Pachioni, M. Tagawa, R. Coombe, S. Davis, T. Madden, G. Manke, D. Setser and R. Hardy for donating their precious time in reviewing one or more chapters.

Rainer A. Dressler
Air Force Research Laboratory
Space Vehicles Directorate
AFRL/VSBX, 29 Randolph Rd.
Hanscom AFB, MA 01731-3010

CONTENTS

CHAPTER 1

EXPLORING CHEMISTRY IN EXTREME
ENVIRONMENTS: A DRIVING FORCE
FOR INNOVATION

Michael R. Berman

Air Force Office of Scientific Research
Air Force Research Laboratory, Arlington, VA 22203

Contents

1. Introduction

Our understanding of chemical dynamics has become increasingly more detailed as experimental and theoretical methods have advanced. One of the ultimate goals of these endeavors is to develop reliable, predictive models so that we can foretell the detailed outcome of chemical reactions. The motivation to develop such predictive capabilities is often driven by the need or desire to understand and describe the chemistry in some extreme environments. These extreme environments, with conditions well outside the range of common room temperatures or pressures, can be natural or manmade, and can be part of our everyday experience or far from our terrestrial domain. In many cases, reproducing these extreme environments push experimental and theoretical methods beyond their current limits, necessitating innovation and new developments to describe the chemical dynamics in these regimes. In all cases, a close interplay between experiment and theory is required to develop and validate our descriptions.

In this overview, I will look at a range of examples of extreme environments, and discuss the innovations that are being spawned to address experimental and theoretical challenges. Comprehensive reviews of many of these areas already exist in the literature and are not the intent of this chapter. Nor will I focus on work discussed extensively elsewhere in this volume. Instead, selected examples of current research will be discussed in which innovation in dealing with extreme environments has led to some insights in our understanding of chemical dynamics.

Table 1 lists a variety of environments that can be considered extreme and are the subject of chemical studies. It is difficult to develop general quantitative criteria as to what constitutes an extreme environment. Such categorizations must be viewed from the perspective of the type of system under consideration. For a liquid lubricant in an engine, a temperature of 620 K, above its decomposition temperature, is an extreme environment. For metals exposed to high temperatures, such as titanium and nickel superalloys in aerospace vehicles, temperatures of 1100 and 1400 K, respectively, test their operational limits. In these two cases, thermal degradation of the

Table 1. Conditions in some extreme environments.

Environment	Temperature (K)	Pressure (Pa)
Superfluid helium	< 2.7	
Solid hydrogen	< 4.0	
Earth's atmosphere (100–400 km)	200–2000	
Hydrothermal vents	600	2–3×10^7
Aircraft engine lubricants	620	
Aircraft engine components	2300	1×10^7
Rocket nozzles	2000–4000	1–2.5×10^7
Burning meteoroid	4000	
Detonation of solid explosive	3000–4000	2–4×10^{10}
Earth's core	4000	4×10^{11}

Note: 10^5 Pa = 1 bar $\cong$ 1 atm; 1 GPa = 10 kbar $\cong$ 10^4 atm.

material via bond breakage or chemical reactions such as oxidation can alter the properties of the materials and set the limits for their working environments.

There are a number of natural environments under extreme conditions that have attracted considerable chemical attention. The geochemistry of the earth's interior occurs at temperatures of up to 4000 K and pressures up to 400 GPa.[1,2] In hydrothermal vents on the ocean floor where unique life forms are cultivated, chemistry occurs at temperatures of 600 K and pressures of 20 to 30 MPa.[3,4] The earth's atmosphere also provides a collection of extreme environments. Studies of chemistry in the stratosphere and in polar stratospheric clouds have motivated a wide range of studies at low temperatures and have contributed to our understanding of homogeneous and heterogeneous chemical reactions.[5,6] Meteoroids burning in the atmosphere produce temperatures of 4000 K, and generate a unique high temperature chemistry of metals and metal oxides. (See Chap. 6 of this volume by Dressler and Murad). Charged particles in the upper atmosphere are accelerated by the earth's magnetic field near the poles leading to high energy collisions that produce aurora and interesting chemistry.[7] Considerable attention has also been devoted to the chemistry of other planets and their atmospheres,[8] and the chemistry of their satellites.[9] (Also, see Chap. 8 in this volume by Johnson).

A wide range of manmade circumstances and technological applications also create an important set of extreme environments. Cutting-edge technological applications often stress materials to their limits. For example, combustion processes in rocket engines can occur at temperatures up to 4000 K, and pressures of 25 MPa.[10] In some aircraft engines, efficiency is sacrificed to prevent material failure due to excessive temperatures. Metal parts in aircraft engines are often protected by thermal barrier coatings made of ceramics in order to extend their operating temperature range. (See Chap. 10 in this volume by Christensen, Ashe and Carter). The high temperature stability of the materials on the outer skin of aerospace vehicles also limit performance. The outer skin of a transatmospheric aerospace vehicle would be expected to reach temperatures on the order of 1100 K from the frictional heating of the skin by gas–surface collisions with molecules in the atmosphere.[11] Leading edges of wing surfaces and the walls of the engine inlets could reach temperatures of 1700 K and 2000 K, respectively.[11]

Chemistry at high pressures also has many technological applications. Chemistry in supercritical fluids, for example, is drawing increasing attention for the possibility of replacing organic solvents in synthetic, purification, and separation processes.[12] Extremely high pressure (20–40 GPa)[13] and temperature (3000–4000 K)[14] transients are experienced in the detonation of energetic materials. The study of the chemistry in the detonation front of explosives is an area where experimental research[15] and novel computer simulations (see Chap. 11 of this volume by White) are underway to understand the real time effects on energy transfer and chemical processes. On a microscopic scale, extremely high local pressures can be created at the interface where two materials slide across one another. The pressures produced at the junction where asperities come in contact can reach 1 GPa, exceeding the critical stresses beyond which the materials deform or fracture.[16] A satisfactory understanding and picture of all of these processes will likely rely on the development of reliable simulation methods, developed and validated in close collaboration with experiment.

Another classification of environments that can be considered extreme are highly chemically reactive environments. The sparse but reactive environment of space, the technologically important realm of plasma processing, and chemical laser systems fit into this category. A number of these situations are discussed by Minton (see Chap. 9 in this volume) who looks at the reactions of oxygen atoms with the low-earth orbital translational energy of 5 eV as they collide with spacecraft materials, and Heaven (see Chap. 4 of this volume) who discusses dynamics in chemical laser systems. Again, experiment and theory together must provide a well-grounded, fundamental understanding of the dynamics of these systems if a reliable predictive capability of the macroscopic observables is to be developed.

2. Chemistry at High Temperatures and Pressures

The drive to model the chemistry in flames, in other combustion systems, and in high enthalpy air flows has motivated many studies of chemistry at high temperatures and pressures. Many studies of elementary processes under these conditions have been performed in shock tubes, typically at temperatures in the range 1000 to 2000 K, but temperatures up to 5000 K can be achieved in these devices. In a typical shock tube experiment, a driver gas is stored at high pressure behind a diaphragm, separating it from the sample residing in the length of the tube. When the diaphragm is ruptured, a shock wave travels down the tube, and measurements are often made near the end wall of the tube behind the reflected shock where relatively constant, elevated temperatures can be maintained for hundreds of microseconds. Shock tube work has been reviewed periodically and extensively.[17-19] In fact, much of the existing high temperature rate data frequently used in models of combustion chemistry and other high temperature gas-phase systems has been obtained in shock tubes.[20]

Alternative approaches to measuring reaction rates and dynamical processes at high temperatures have been pursued. Measurements in flames using laser diagnostics[21] and molecular beam mass-spectrometric sampling have been carried out.[22] Direct laser heating of molecules in the gas-phase or heating of an absorbing gas to create a heated channel of gas far from surfaces have been used to obtain mechanistic information and quantitative Arrhenius parameters for homogeneous reactions in the gas-phase.[23,24] The high heating and cooling rates obtained by pulsed-laser-driven heating can produce temperatures in excess of 1000 K that last for only several microseconds. Thus, products of the initial high temperature reaction are often rapidly cooled and thermally stabilized relative to experiments at constant high temperatures. This permits analysis of primary dissociation products that can themselves often thermally dissociate.[25]

Novel methods are now being developed to generate shock waves that rapidly produce extreme environments of high pressure and temperatures in solid materials. Shock wave generation, propagation, and their effects on the chemistry of the material are particularly important in energetic materials, where a shock wave, initiated by a physical impact, could lead to detonation. Studies have been performed in which gas guns are used to propel a large impact plate into a sample, generating pressures of up to 200 GPa. The shocked sample can then be spectroscopically probed in single shot experiments,[14,26] with these spectroscopic studies usually performed in the tens of GPa pressure region. But ultrafast time resolution and precise control of the timing between shock and probe is required to unravel the initial physical and chemical changes and vibrational energy redistribution in condensed-phase materials and to study their evolution following passage of a shock front. New methods have recently been developed to probe these properties in real time, and are providing new insights into the behavior of these materials that directly impact their chemical reactivity.[15]

Dlott and coworkers[27,28] have developed a technique to generate shock waves with fast (< 25 ps) risetimes in specially designed assemblies, that are then probed spectroscopically. In this method, a near-infrared laser pulse is absorbed by a 2-μm thick outer layer of a microfabricated, multilayered target array. A few nanograms of this polymer layer are rapidly ablated and a shock wave that is relatively planar, and of approximately constant velocity throughout the sample is launched into the layers below. A buffer layer delays the arrival of the shock at the sample layer as debris from the ablation is permitted to clear. The shock then impinges on the sample layer, and subsequently travels into a dissipation layer. This layered assembly is supported on a glass substrate which provides optical access for probing by coherent anti-Stokes Raman spectroscopy. This "nanoshock" method has produced transient pressures on the order of 1–5 GPa and temperature rises of several hundreds of degrees in a region about 100 μm in diameter. Since the sample is locally destroyed by the ablation of the shock-generation layer, the sample assembly is translated before the next shot is fired, and reproducible shocks have been generated at a repetition rate of 80 s^{-1}.

The "nanoshock" method allows the collection of vibrational spectra of molecular solids with time resolution of 25 ps. It has been applied to a number of energetic materials.[29] The rapid cooling during unloading of the shock ($\sim 10^{11}$ K s^{-1}) allows little time for thermal decomposition of the sample. This method has recently been used to detect fast mechanical processes in polycrystalline films of the energetic material NTO.[30] The shocks induce partial orientation of the crystals with one another in only a few nanoseconds. This orientation is faster than the time scale for initiating chemical reactions in this system, and may have implications for the mechanism of the initiation of detonation of energetic materials. The importance of orientation in detonation has been previously observed in the directional dependence of shock-wave initiated detonation of single crystals of energetic materials.[31]

The detonation of energetic materials is also critically dependent on energy transfer in molecular solids.[32] Studies of vibrational energy redistribution in polyatomic molecules in the condensed-phase are being carried out on a number of systems.[33,34] Resonant vibrational excitation from ultrashort infrared pulses can drive molecules to effective vibrational temperatures of 4000–5000 K. Vibrational energy redistribution can then be studied by anti-Stokes Raman probing. These IR–Raman studies, with time resolution of about 1 ps, monitor the energy flow throughout almost all of the vibrational states of a polyatomic liquid. A particularly relevant case for energetic materials is nitromethane, a model system for detonation. When the C–H stretching modes of nitromethane are excited, vibrational cooling is observed to occur in three distinctive steps.[34] First, the deposited energy is redistributed to all of the vibrations of the molecule in a few picoseconds. Second, the higher frequency vibrations relax to lower energy modes in a few tens of picoseconds. Finally, these lower energy modes decay to the bath in about 100 ps. The dynamics of coupling of energy into these lowest frequency modes may play a key role in determining the sensitivity of an energetic material to shock initiation.

Other novel methods are being employed to examine chemistry at extremely high temperatures and pressures. For example, Suslick and coworkers use ultrasound to initiate "sonochemistry."[35,36] In this work, ultrasound is directed into a liquid creating bubbles that grow, and then implosively collapse. The rapid collapse, or cavitation, compresses and heats the gas trapped in the bubble producing hot spots with effective temperatures of $\sim$ 5000 K, pressures of $\sim$ 100 MPa, and heating and cooling rates above 10^{10} K s^{-1}. Sonoluminescence, the light that is often produced during cavitation, can provide a probe of the interior of the bubble. The intense temperatures and pressures produced during cavitation can have destructive effects, for example, when it occurs around the propeller of a ship. However, these conditions have been used for some novel chemical synthesis, with particular effect when solid surfaces or powders are in the liquid.[37] In these systems, solid particles are accelerated to high velocities and interparticle collisions can change surface morphology and chemical reactivity. Unique nanostructured materials and biomaterials have been synthesized by these methods.[38,39]

Studies at extremely high static pressure can be carried out in diamond anvil cells.[40] Studies of chemical dynamics in gem anvil cells are beginning to be carried out,[41] and some unique chemical synthesis can also occur at these high pressures. For example, a high pressure polymerized phase of carbon monoxide was recently reported to be synthesized using visible laser light to irradiate a carbon monoxide sample at pressures of over 5 GPa in a diamond anvil cell.[42]

Another class of extreme environments being used for synthetic applications are near critical and supercritical fluids. A number of reviews have looked at reactions in supercritical water.[43,44] Brill and coworkers have developed experimental devices in which to perform infrared and Raman spectroscopic studies of hydrothermal reactions at temperatures up to 700 K and pressures of 35 MPa,[45,46] above the critical point of water (647 K, 22.1 MPa).[47] These studies have elucidated the kinetics and reaction pathways of numerous systems under hydrothermal conditions.

Studies of solvation and energy transfer in supercritical fluids provide an important way to bridge our understanding of dynamics in the gas-phase and liquid phase.[48] Troe and coworkers have performed studies in supercritical fluids to test the applicability of the Isolated Binary Collision (IBC) model of energy transfer over a wide range of density and temperature, spanning the transition from gas-phase to liquid phase.[49,50] In these experiments, subpicosecond lasers were used to excite azulene with about 20 000 cm^{-1} of energy to its first excited singlet state, S_1. This state rapidly underwent internal conversion producing highly vibrationally excited levels of the ground electronic state, S_0^*. Deactivation of the vibrationally excited azulene by collision partners (Xe, CO_2, ethane, propane, n-pentane, n-octane) was monitored by transient absorption in the ultraviolet on the $S_3 \leftarrow S_0$ transition. These experiments, coupled with Monte Carlo simulations, demonstrated that an IBC model of energy transfer proved applicable from densities that spanned the gas-phase to the liquid phase if the collision frequency was expressed in terms of the radial distribution function around the azulene solute molecule, assumed to be a hard sphere in a Lennard–Jones fluid.[50] The radial distribution function accounts for the local cluster formation or solvation that dynamically occurs in supercritical fluids. This clustering increases the local density of the solvent around the solute with respect to the bulk and could be thought of as shielding the excited molecule. This model implies that while the collision frequency varies with density, the average energy transferred per collision $\langle \Delta E \rangle$ remains constant as a function of density, and should be transferable from gases to liquids.[50]

Another probe of the local environment around the azulene solute molecule was the shift of the frequency of the $S_3 \leftarrow S_0$ absorption band, as a function of density.[50] This solvachromic shift, or stabilization of the electronic states of the solute in the presence of the solute, was accurately described using the same radial distribution function used to reproduce the collisional deactivation rates. The applicability of the same radial distribution function to treat collisional deactivation and solvachromic frequency shifts suggests that both have a similar dependence on local density.

Fayer and coworkers have studied the vibrational dynamics of molecules excited in solvents that were in or near supercritical conditions.[51-54] They have observed very interesting behavior in the vibrational lifetime, T_1, as a function of density and temperature near the critical points. In these experiments, a CO stretching vibration in $W(CO)_6$ was excited to $v = 1$, and its lifetime was measured using picosecond infrared lasers. In supercritical CO_2, just above the critical temperature, the vibrational lifetime decreased from over 900 ps to about 700 ps as the density increased from about 2 to 5 mol L^{-1}. The lifetime then reached a plateau, and remained constant in the density range 7 to 12 mol L^{-1}, before beginning to decrease again at higher densities.[51,52] At constant, near-critical density, vibrational lifetimes were studied as a function of temperature.[53,54] Temperature regimes were found in supercritical ethane and fluoroform where the vibrational lifetime increases with increasing temperature. In supercritical CO_2, however, the lifetime decreased approximately linearly with temperature. These interesting and carefully collected data are providing some nonintuitive results that are challenging tests for theories to describe the intermolecular interactions near the critical point.

3. High Temperature Chemistry in the Atmosphere

The earth's upper atmosphere possesses a collection of extreme environments. For example, at altitudes between 100 and 400 km, the temperature of the atmosphere rises from 200 K to 2000 K. In this region of the ionosphere, the dominant ions are O_2^+, NO^+, and O^+, having concentrations between 10^4 and 10^5 cm^{-3}.[55] Until recently, it has not been possible to measure reaction rates for these species throughout the range of temperatures characteristic of the ionosphere. The need to validate rate constants for use in computational models of the ionosphere has spurred the development of instruments to obtain this data.

Viggiano, Morris, and coworkers have developed a high temperature flowing afterglow that they have used for studying ion–molecule reactions at temperatures from 300 K to 1800 K.[56] The high temperature capability enables the observation of the onset of higher energy reaction channels, and combined with data from previous studies, provides insights into the role of internal energy in promoting reactivity. Examples of this are the reactions of O^+ with N_2 and O_2,[57] where prior work in drift tubes had been used to selectively enhance translational energy.[58] Comparing the rate constants obtained using pure thermal excitation with those from nonthermal, translational excitation, it was determined that rotational energy plays a minimal role in controlling reactivity, while vibrational excitation has a much greater, enhancing effect.[57] In the case of the reaction of O^+ with O_2, thermally exciting vibrations increased the reaction rate constant a factor of 5 over the same energy obtained by pure translational excitation. For O^+ with N_2, it was found that thermally exciting N_2 to $v = 2$ increased the reaction rate constant by a factor of 40 relative to the $N_2(v = 0)$ rate constant, in agreement with earlier observations of Schmeltekopf *et al.*[59,60] The role of vibrational excitation in enhancing reactivity has recently also been observed in the reactions of N^+ and N_2^+ with O_2[61] and in the reactions of O^+ with CH_4,[62] where new product channels were seen to open as CH_4 became vibrationally excited due to high thermal excitation. Wodtke and coworkers[63] have also recently reported that vibrational excitation can promote surface chemistry. They observed that NO molecules excited to $v = 13$ and 15 react (via dissociative adsorption) with a copper surface with a reaction probability more than one thousand times greater than for ground state NO.

Ion beam experiments involving high temperature vapors are also being carried out. Levandier *et al.*[64] have developed a high temperature guided-ion beam apparatus in which reaction cross sections can be measured at selected translational energies and target vapor temperatures up to 800 K. While the instrument was originally designed for metal vapor studies, it has also been used to investigate the change in threshold behavior when low-frequency modes of a molecular target are thermally excited.[65] Studies in this system related to meteor chemistry are discussed in more detail in Chap. 6 by Dressler and Murad in this volume.

Another example of a chemically interesting property of the upper atmosphere is nonlocal thermodynamic equilibrium (NLTE). Rotational NLTE has been demonstrated by the observation of pure rotational emission[66–68] from high N levels of OH, NO, and CO in the upper mesosphere/lower thermosphere (85 to 95 km) where pressures are about 5 Pa. For OH ($X\ ^2\Pi$, $v = 0$–3), emission from rotational levels up to $N = 33$ were observed in the airglow in the CIRRIS 1A experiment conducted on the space shuttle.[66] Holtzclaw *et al.*[69] have studied the rotational relaxation of high N states of OH ($X\ ^2\Pi$, $v = 1$–3) by O_2 in a cryogenically-cooled chamber called LABCEDE at 100 K to assess the processes that affect the rotational distribution of OH in the upper atmosphere. They found that rotational relaxation occurred at a gas-kinetic rate, predominantly by single-quantum steps within each vibrational manifold.

4. Low Temperature Chemistry

Moving from the earth's upper atmosphere to the outer planets and inter-stellar space,[70] lower temperature environments become more important. Recent reviews of ion–molecule reaction dynamics[71] and gas-phase reactivity and energy transfer between neutral species at very low temperatures[72] (below 80 K) describe the current status of these fields. An interesting approach developed by Rowe and coworkers is the CRESU (Cinétique de Réactions en Ecoulement Supersonique Uniforme) technique that has been applied to both ion–molecule[73,74] and neutral–neutral chemistry.[75–78] In this method, reaction kinetics are studied in the expansions of Laval nozzles. These isentropic expansions produce gas flows that are relatively high in density (10^{16}–10^{17} molecules cm^{-3}) and have enough collisions to maintain thermal equilibrium. The uniform temperature, density, and pressure of this method makes it well suited for kinetic measurements down to temperatures around 10 K. In studies of neutral–neutral reactions in the CRESU apparatus, laser photolysis production of radicals or other transient species is coupled with laser-induced fluorescence detection. Van Marter and Heaven[79] have also recently used Laval nozzles to study energy transfer between iodine atoms and O_2 ($a^1\Delta$) at temperatures down to 150 K. This regime is of importance to the chemical oxygen iodine laser when operated in a supersonic expansion (see Chap. 4 by Heaven in this volume).

As is often the case, experiments outside the range of previous measurements have provided unexpected results, highlighting the need for caution when extrapolating results beyond the temperature range in which they were obtained. The low temperature studies in the CRESU apparatus have shown that a large number of reactions have rates that increase as the temperature is decreased.[72] This behavior seems characteristic of reactions with no barrier, and are dominated by long-range attractive forces between reacting species that lead to capture. The intermediate complex formed can allow reagents time to find the appropriate orientation for reaction to take place, as has been argued[76] for $CN + C_2H_6$, or can mediate a series of H atom transfers as in the reaction of $CN + NH_3$.[75]

Due to the inverse temperature dependence of the Arrhenius relationship, these very low temperature studies cover a wide range on an Arrhenius plot. Combining CRESU results with higher temperature studies enables the competition of reaction channels with opposing temperature dependencies to be observed in some reactions. This produces an Arrhenius plot that has a minimum at intermediate temperatures, while the rate constant increases with increasing temperature at high temperatures and increases with decreasing temperature at lower temperatures. Such behavior has been argued to imply for the CN + C_2H_6 reaction that a tight transition state dominates at high temperatures, while a loose transition state controls the reaction at low temperatures.[76] This transition between different dynamics in different temperature regimes has previously been observed in some ion–molecule reactions[71,80,81] and radical–molecule reactions that have been studied over wide temperature ranges.[82–85]

Troe and coworkers have been able to study chemical processes over a very wide range of pressures as well as a wide temperature range. For example, they have studied the recombination reaction of O + O_2 to form O_3 at temperatures down to 90 K and pressures up to 100 MPa.[86] Their studies of reactivity and energy transfer have been instrumental in developing a unified understanding of reactivity and energy transfer over an extremely wide range of conditions, including the transition from the gas-phase into the condensed phase.

At even lower temperatures, some unusual properties of matter are displayed. Consequently, new experimental and theoretical methods are being created to explore and describe chemistry in these regimes. In order to account for zero-point energy effects and tunneling in simulations, Voth and coworkers developed a quantum molecular dynamics method that they applied to dynamics in solid hydrogen.[87,88] In liquid helium, superfluidity is displayed in ^{4}He below its lambda point phase transition at 2.17 K. In the superfluid state, helium's thermal conductivity dramatically increases to 1000 times that of copper, and its bulk viscosity drops effectively to zero. Apkarian and coworkers have recently demonstrated the disappearance of viscosity in superfluid helium on a molecular scale by monitoring the damped oscillations of a 10 Å bubble as a function of temperature.[89] These unique properties make superfluid helium an interesting host for chemical dynamics.

The ability to dope impurity atoms or molecules into large helium clusters by a pick-up method, pioneered by the groups of Toennies[90] and Scoles,[91] has helped make studies in superfluid helium clusters more accessible. In this method, an expansion through a nozzle produces a beam of helium clusters. Under appropriate conditions, helium droplets comprising up to 10^6 helium atoms can be formed. These droplets then traverse a collision cell containing a foreign gas at a pressure of 10^{-4}–10^{-3} Pa. Atoms or molecules in the collision cell become embedded in the droplet, typically evaporating several hundred helium atoms from the droplet for each dopant species picked up. The beam of helium droplets doped with a guest species emerging from the collision cell can then be studied by a number of diagnostic methods.

The spectroscopy of doped helium clusters has recently been reviewed from an experimental[92] and theoretical perspective.[93] The inviscid nature of the helium clusters was demonstrated by Toennies and coworkers who, using rovibrational spectroscopy, demonstrated that SF_6 and OCS were nearly free rotors in ^4He clusters.[94,95] Nauta and Miller have utilized the vanishing viscosity and superthermal conductivity of superfluid helium to create unprecedented molecular asssemblies among dopants in helium clusters.[96] In that work, nine HCN molecules were successively picked up by a helium cluster and then self-assembled into a straight-line chain controlled by the long-range dipole–dipole forces among them. The high thermal conductivity of the bath effectively couples energy away before the molecule can rearrange to a lower-energy minimum on the potential energy surface. The potential ramifications to chemical dynamics of such a fast energy-dissipation process competing with intramolecular energy redistribution are intriguing.

5. Conclusions

The need to understand and model chemistry in extreme environments has long been a driving force for innovation in experimental and theoretical chemistry. As has been shown, a wide range of novel methods and systems have been developed to explore chemistry in these regimes. The conditions that can be achieved in some of these laboratory systems are listed in

Table 2. Experimental methods that produce extreme environments.

Experimental method	Temperature (K)	Pressure (Pa)
Helium droplets	< 2.7	
Laval nozzles/CRESU	10–200	
High Temperature Flowing Afterglow	300–1800	
Critical points of water	647	2.2×10^7
Laser-driven nanoshocks	600	4×10^9
Sonochemistry	5000	1×10^8
Diamond anvil cells		5×10^9

Table 2. These methods, which expand the range of conditions under which reactions have been studied, have helped provide new insights into chemical reactivity, the role of internal energy in fostering reactivity, and the transfer of energy in molecular systems.

As simulations play an increasing role in the prediction of the behavior of complex systems, this drive to explore new ranges of experimental conditions is bound to increase rather than decrease. The potential danger of basing simulations on results extrapolated far beyond the range of conditions under which they were obtained has been frequently demonstrated. Thus, experiment and theory must move together to push out their frontiers. New experimental methods must be developed to probe systems in currently uncharted territory providing benchmarks for theory, and validating models that have been developed. And it is imperative that such models be based on a complete and well-tested understanding of the underlying fundamental principles.

The range of extreme environments under study is also bound to increase as more systems are treated by simulations in multidisciplinary efforts. From the chemistry of ignition and detonation of energetic materials at high temperatures and pressures, to chemistry in the cold, sparse interstellar medium, experimental and theoretical challenges abound. Challenges are also posed in the detailed chemistry occurring within materials under stress, as well as the interactions of materials with their environments. For example, simulations of the cracking of molecular materials requires accurate potentials for molecular bonds that are stretched well beyond their equilibrium conditions. Validating potentials in this region will be vital for ensuring confidence in such simulations.

Another increasingly important consideration in many simulations, the linking of information across multiple sets of time and length scales, presents another daunting set of challenges, particularly in nonequilibrium conditions where state-to-state dynamics plays an important role. Simulation methods based on modeling of molecular collisions are emerging that utilize the detailed chemical information that modern chemical physics can provide (see Chap. 3 of this volume by Boyd on Direct Simulation Monte Carlo methods). Given the high dimensionality of the problem, however, even these most detailed particle-based models require considerable simplifying assumptions. Simulations of combustion, or the interaction of objects with reactive high speed flows will have to couple chemistry with fluid dynamics. Particle-based models of such environments must reconcile time scales from femtoseconds for chemistry to seconds for flow effects. It will be critical to develop tractable methods that retain the essence of the important details of chemical dynamics in these coupled models.

Pushing chemistry into new environments has often resulted in some surprising results. While developing methods to explore these new areas presents challenges, they also present opportunities. Already, new synthetic methods have grown out of exploration of chemistry in extreme environments. We must remain poised to take advantage of such new possibilities as they arise from studies of extreme environments.

Acknowledgment

I would like to express my appreciation to V. A. Apkarian, D. D. Dlott, R. A. Dressler, U. Landman and G. A. Voth for useful discussions during the preparation of this chapter.

References

1. J. Verhoogen, in *Physics and Chemistry of the Earth*, Eds. L. H. Ahrens, K. Rankama and S. K. Runcorn (McGraw Hill, New York, 1956), p. 17.
2. K. E. Bullen, *Sci. Am.* **193**, 56 (1955).
3. R. L. Rawls, December 21 issue, *Chem. Eng. News* **76**(51), 35 (1998).
4. V. Tunnicliffe, *Am. Sci.* **80**, 336 (1992).
5. M. J. Molina, L. T. Molina and C. E. Kolb, *Ann. Rev. Phys. Chem.* **47**, 327 (1996).
6. T. Peter, *Ann. Rev. Phys. Chem.* **48**, 785 (1997).
7. H. C. Carlson and A. Egeland in *Introduction to Space Physics*, Eds. M. G. Kivelson and C. T. Russell (Cambridge University Press, New York, 1995), p. 459.
8. J. A. Kaye and D. F. Strobel, *Icarus* **59**, 314 (1983).
9. Y. L. Yung, M. Allen and J. P. Pinto, *Astrophys. J. Suppl.* **55**, 465 (1984).
10. G. P. Sutton, *Rocket Propulsion Elements*, 6th edition (Wiley, New York, 1992).
11. M. A. Steinberg, *Sci. Am.* **225**, 67 (1986).
12. P. E. Savage, *Chem. Rev.* **99**, 603 (1999).
13. T. Urbanski, in *Chemistry and Technology of Explosives*, Vol. 4 (Pergamon Press, Oxford, 1984), p. 6.
14. C. S. Yoo, N. C. Holmes and P. C. Souers, in *Decomposition, Combustion and Detonation Chemistry of Energetic Materials*, Eds. T. B. Brill, T. P. Russell, W. C. Tao and R. B. Wardle (Materials Research Society, Pittsburgh, 1996), p. 397.
15. D. D. Dlott, *Ann. Rev. Phys. Chem.* **50**, 251 (1999).
16. J. Gao, W. D. Luedtke and U. Landman, *Science* **270**, 605 (1995).

17. S. H. Bauer, *Ann. Rev. Phys. Chem.* **16**, 245 (1965).

18. W. Tsang and A. Lifshitz, *Ann. Rev. Phys. Chem.* **41**, 559 (1990).

19. J. V. Michael and K. P. Lim, *Ann. Rev. Phys. Chem.* **44**, 429 (1993).

20. D. L. Baulch, C. J. Cobos, R. A. Cox, C. Esser, P. Frank, T. Just, J. A. Kerr, M. J. Pilling, J. Troe, R. W. Walker and J. Warnatz, *J. Phys. Chem. Ref. Data* **21**, 411 (1992).

21. D. R. Crosley, *ACS Symposium Series* **134**, 3 (1980).

22. J. Vandooren, M. C. Branch and P. J. Van Tiggelen, *Combust. Flame* **90**, 247 (1992).

23. D. F. McMillen, K. E. Lewis, G. P. Smith and D. M. Golden, *J. Phys. Chem.* **86**, 709 (1982).

24. H. L. Dai, E. Specht, M. R. Berman and C. B. Moore, *J. Chem. Phys.* **77**, 4494 (1982).

25. M. R. Berman, P. B. Comita, C. B. Moore and R. G. Bergman, *J. Am. Chem. Soc.* **102**, 5692 (1980); P. B. Comita, M. R. Berman, C. B. Moore and R. G. Bergman, *J. Phys. Chem.* **85**, 3266 (1981).

26. G. I. Pangilinan and Y. M. Gupta, *J. Appl. Phys.* **81**, 6662 (1997).

27. S. A. Hambir, J. Franken, D. E. Hare, E. L. Chronister, B. J. Baer and D. D. Dlott, *J. Appl. Phys.* **81**, 2157 (1997).

28. G. Tas, J. Franken, S. A. Hambir, D. E. Hare and D. D. Dlott, *Phys. Rev. Lett.* **78**, 4585 (1997).

29. D. E. Hare, I.-Y. S. Lee, J. R. Hill, J. Franken, H. Suzuki, B. J. Baer, E. L. Chronister and D. D. Dlott, in *Decomposition, Combustion and Detonation Chemistry of Energetic Materials*, Eds. T. B. Brill, T. P. Russell, W. C. Tao and R. B. Wardle (Materials Research Society, Pittsburgh, 1996), p. 357.

30. J. Franken, S. A. Hambir and D. D. Dlott, *J. Appl. Phys.* **85**, 2068 (1999).

31. J. J. Dick, R. N. Mulford, W. J. Spencer, D. R. Pettit, E. Garcia and D. C. Shaw, *J. Appl. Phys.* **70**, 3572 (1991).

32. A. Tokmakoff, M. D. Fayer and D. D. Dlott, *J. Phys. Chem.* **97**, 1902 (1993).

33. J. C. Deàk, L. K. Iwaki and D. D. Dlott, *Chem. Phys. Lett.* **293**, 405 (1998).

34. J. C. Deàk, L. K. Iwaki and D. D. Dlott, *J. Phys. Chem.* **A103**, 971 (1999).

35. K. S. Suslick, *Science* **247**, 1439 (1990).

36. K. S. Suslick, U. Kidenko, M. M. Fang, R. Hyeon, K. J. Kolbeck, W. B. McNamara, M. M. Mdleleni and M. Wong, *Phil. Trans. Roy. Soc.* **A357**, 335 (1999).

37. S. J. Doktycz and K. S. Suslick, *Science* **247**, 1067 (1990).

38. M. M. Mdleleni, T. Hyeon and K. S. Suslick, *J. Am. Chem. Soc.* **120**, 6189 (1998).

39. K. S. Suslick and M. W. Grinstaff, *J. Am. Chem. Soc.* **112**, 7807 (1990).

40. C. M. Sung, *High Temp. High Press.* **29**, 253 (1997).

41. T. P. Russell, T. M. Allen and Y. M. Gupta, *Chem. Phys. Lett.* **267**, 351 (1997).

42. M. Lipp, W. J. Evans, V. Garcia-Baonza and H. E. Lorenzana, *J. Low Temp. Phys.* **111**, 247 (1998).

43. R. W. Shaw, T. B. Brill, A. A. Clifford, C. A. Eckert and E. U. Franck, December 23 issue, *Chem. Eng. News* **69**(51), 26 (1991).
44. J. W. Tester, H. R. Holgate, F. J. Armellini, P. A. Webley, W. R, Killilea, G. T. Hong and H. E. Barner, *ACS Symposium Series* **518**, 35 (1993).
45. J. W. Schoppelrei, M. L. Kieke, X. Wang, M. T. Klein and T. B. Brill, *J. Phys. Chem.* **100**, 14343 (1996).
46. A. J. Belsky and T. B. Brill, *J. Phys. Chem.* **102**, 4509 (1998).
47. J. M. H. Levelt Sengers, J. Straub, K. Watanabe and P. G. Hill, *J. Phys. Chem. Ref. Data* **14**, 193 (1985).
48. O. Kajimoto, *Chem. Rev.* **99**, 355 (1999).
49. D. Schwarzer, J. Troe, M. Votsmeier and M. Zerezke, *J. Chem. Phys.* **105**, 3121 (1996).
50. D. Schwarzer, J. Troe and M. Zerezke, *J. Chem. Phys.* **107**, 8380 (1997).
51. R. S. Urdahl, K. D. Rector, D. J. Myers, P. H. Davis and M. D. Fayer, *J. Chem. Phys.* **105**, 8973 (1996).
52. R. S. Urdahl, D. J. Myers, K. D. Rector, P. H. Davis, B. J. Cherayil and M. D. Fayer, *J. Chem. Phys.* **107**, 3747 (1997).
53. D. J. Myers, R. S. Urdahl, B. J. Cherayil and M. D. Fayer, *J. Chem. Phys.* **107**, 4741 (1997).
54. D. J. Myers, S. Chen, M. Shigeiwa, B. J. Cherayil and M. D. Fayer, *J. Chem. Phys.* **109**, 5971 (1998).
55. P. M. Banks and G. Kockarts, *Aeronomy*, Part B (Academic Press, New York, 1973).
56. P. M. Hierl, J. F. Friedman, T. M. Miller, I. Dotan, M. Menendez-Barreto, J. Seely, J. S. Williamson, F. Dale, P. L. Mundis, R. A. Morris, J. F. Paulson and A. A. Viggiano, *Rev. Sci. Inst.* **67**, 2142 (1996).
57. P. M. Hierl, I. Dotan, J. V. Seeley, J. M. van Doren, R. A. Morris and A. A. Viggiano, *J. Chem. Phys.* **106**, 3540 (1997).
58. D. L. Albritton, I. Dotan, W. Lindinger, M. McFarland, J. Tellinghuisen and F. C. Fehsenfeld, *J. Chem. Phys.* **66**, 410 (1977).
59. A. L. Schmeltekoph, *Planet. Space Sci.* **15**, 401 (1967).
60. A. L. Schmeltekoph, E. E. Ferguson and F. C. Fehsenfeld, *J. Chem. Phys.* **48**, 2966 (1968).
61. I. Dotan, P. M. Hierl, R. A. Morris and A. A. Viggiano, *Int. J. Mass Spectrom. Ion Phys.* **167/168**, 223 (1997).
62. A. A. Viggiano, I. Dotan and R. A. Morris, *J. Am. Chem. Soc.* **122**, 352 (2000).
63. H. Hou, Y. Huang, S. J. Gulding, C. T. Rettner, D. J. Auerbach and A. M. Wodtke, *Science* **284**, 1647 (1999).
64. D. J. Levandier, R. A. Dressler and E. Murad, *Rev. Sci. Instr.* **68**, 64 (1997).
65. K. Fukuzawa, Y. Osamura, K. Morokuma, D. J. Levandier, Y.-H. Chiu, R. A. Dressler, E. Murad, A. Midey and A. A. Viggiano, in preparation.
66. D. R. Smith, W. A. M. Blumberg, R. M. Nadile, S. J. Lipson, E. R. Huppi, N. B. Wheeler and J. A. Dodd, *Geophys. Res. Lett.* **19**, 593 (1992).

67. J. A. Dodd, S. J. Lipson, J. R. Lowell, P. S. Armstrong, W. A. M. Blumberg, R. M. Nadile, S. M. Adler-Golden, W. J. Marianelli, K. W. Holtzclaw and B. D. Green, *J. Geophys. Res.* **99**, 3559 (1994).
68. P. S. Armstrong, S. J. Lipson, J. A. Dodd, J. R. Lowell, W. A. M. Blumberg and R. M. Nadile, *Geophys. Res. Lett.* **21**, 2425 (1994).
69. K. W. Holtzclaw, B. L. Upschlute, G. E. Caledonia, J. F. Cronin, B. D. Green, S. J. Lipson, W. A. M. Blumberg and J. A. Dodd, *J. Geophys. Res.* **102**, 4521 (1997).
70. E. Herbst, *Ann. Rev. Phys. Chem.* **46**, 27 (1995).
71. M. A. Smith, in *Unimolecular and Bimolecular Reaction Dynamics*, Eds. C. Y. Ng, T. Baer and I. Powis (Wiley, New York, 1994), pp. 183–251.
72. I. R. Sims and I. W. M. Smith, *Ann. Rev. Phys. Chem.* **46**, 109 (1995).
73. G. Dupeyrat, J. B. Marquette and B. R. Rowe, *Phys. Fluids* **28**, 1273 (1985).
74. C. Rebrion, J. B. Marquette and B. R. Rowe, *J. Chem. Phys.* **91**, 6142 (1989).
75. I. R. Sims, J.-L. Queffelec, A. Defrance, C. Rebrion-Rowe, D. Travers, B. R. Rowe and I. W. M. Smith, *J. Chem. Phys.* **97**, 8798 (1992); I. R. Sims, J.-L. Queffelec, A. Defrance, C. Rebrion-Rowe, D. Travers, P. Bocherel, B. R. Rowe and I. W. M. Smith, *J. Chem. Phys.* **100**, 4229 (1994).
76. I. R. Sims, J.-L. Queffelec, D. Travers, B. R. Rowe, L. B. Herbert, J. Karthäuser and I. W. M. Smith, *Chem. Phys. Lett.* **211**, 461 (1993).
77. I. R. Sims, P. Bocherel, A. Defrance, D. Travers, B. R. Rowe and I. W. M. Smith, *J. Chem. Soc. Fraday Trans.* **90**, 1473 (1994); I. R. Sims, I. W. M. Smith, D. C. Clary, P. Bocherel and B. R. Rowe, *J. Chem. Phys.* **101**, 1748 (1994); P. Sharkey, I. R. Sims, I. W. M. Smith, P. Bocherel and B. R. Rowe, *J. Chem. Soc. Fraday Trans.* **90**, 3609 (1994).
78. P. Bocherel, L. B. Herbert, B. R. Rowe, I. R. Sims, I. W. M. Smith and D. Travers, *J. Phys. Chem.* **100**, 3063 (1996); A. Canosa, I. R. Sims, D. Travers, I. W. M. Smith and B. R. Rowe, *Astron. Astrophys.* **323**, 644 (1997); R. A. Brownsword, A. Canosa, B. R. Rowe, I. R. Sims, I. W. M. Smith, D. W. A. Stewart, A. C. Symonds and D. Travers, *J. Chem. Phys.* **106**, 7662 (1997).
79. T. Van Marter and M. C. Heaven, *J. Chem. Phys.* **109**, 9266 (1998).
80. D. Smith, N. G. Adams and T. M. Miller, *J. Chem. Phys.* **69**, 308 (1978).
81. B. R. Rowe, G. Dupeyrat, J. B. Marquette, D. Smith, N. G. Adams and E. E. Ferguson, *J. Chem. Phys.* **80**, 241 (1984).
82. R. A. Perry, R. Atkinson and J. N. Pitts, *J. Phys. Chem.* **81**, 296 (1977); **81**, 1607 (1977).
83. F. P. Tully, A. R. Ravishankara, R. L. Thompson, J. M. Nicovich, R. C. Shah, N. M. Kreutter and P. H. Wine, *J. Phys. Chem.* **85**, 2262 (1981); J. M. Nicovich, R. L. Thompson and A. R. Ravishankara, *ibid.* **85**, 2913 (1981).
84. M. R. Berman and M. C. Lin, *J. Phys. Chem.* **87**, 3933 (1983).
85. M. R. Berman and M. C. Lin, *J. Chem. Phys.* **81**, 5743 (1984).
86. H. Hippler, R, Rahn and J. Troe, *J. Chem. Phys.* **93**, 6560 (1990).
87. J. Cao and G. A. Voth, *J. Chem. Phys.* **99**, 10070 (1993).

88. S. Jang, S. Jang and G. A. Voth, *J. Phys. Chem.* **A103**, 9512 (1999).

89. A. Benderskii, R. Zadoyan and V. A. Apkarian, *J. Chem. Phys.*, submitted.

90. A. Scheidemann, J. P. Toennies and J. A. Northby, *Phys. Rev. Lett.* **64**, 1899 (1990); A. Scheidemann, B. Schilling, J. P. Toennies and J. A. Northby, *Physica* **B165–166**, 135 (1990).

91. S. Goyal, D. L. Schutt and G. Scoles, *Phys. Rev. Lett.* **69**, 933 (1992); S. Goyal, D. L. Schutt and G. Scoles, *J. Phys. Chem.* **97**, 2236 (1993).

92. J. P. Toennies and A. F. Vilesov, *Ann. Rev. Phys. Chem.* **49**, 1 (1998).

93. K. B. Whaley, in *Advances in Molecular Vibrations and Collision Dynamics*, Vol. 3, Ed. J. Bowman (JAI Press, Greenwich, 1998).

94. M. Hartmann, R. E. Miller, J. P. Toennies and A. Vilesov, *Phys. Rev. Lett.* **75**, 1566 (1995).

95. S. Grebnev, J. P. Toennies and A. Vilesov, *Science* **279**, 2083 (1998).

96. K. Nauta and R. E. Miller, *Science* **283**, 1895 (1999).

CHAPTER 2

CHEMISTRY UNDER EXTREME CONDITIONS:
CLUSTER IMPACT ACTIVATION*

T. Raz and R. D. Levine[†]

The Fritz Haber Research Center for Molecular Dynamics
The Hebrew University, Jerusalem 91904, Israel

Contents

*Work supported by the AirForce Office of Scientific Research.
[†]Corresponding author. Fax: 972-2-6513742; E-mail: rafi@fh.huji.ac.il

Introduction

There is a clear need for the understanding of chemical reactivity under extreme conditions.[1] Typical situations where such a need arises in practice is the chemistry in front of a nose cone during reentry of a space vehicle into the atmosphere[2] or the shuttle glow phenomenon.[3] Laboratory techniques for the study of such extreme conditions have, so far, been limited. The prime source of information is from shock tube studies[4-7] where the upper limit of kinetic temperatures that can be conveniently reached is about 5000 K. Fast, focused, laser heating and plasma chemistry are others.[8-10] More specialized techniques include the chemistry of translationally very hot atoms[11] produced via the nuclear recoil technique[12,13] and sonochemistry.[14] The technique discussed in this chapter promises to provide the opportunity of reaching a controllable range of energy density and pressure, and thereby allow a systematic study of both chemical and physical processes which can be driven by extreme conditions.[3,7,15-66] Our work is theoretical and a complete list of our papers in this field is provided as an appendix. We shall also refer to other relevant theoretical and computational studies.[25,54,67-87] Computational studies rely on knowledge of the potential. At the high densities typical of our system, this is very much an unknown.[a] It is therefore worthwhile to point out that all the

---

[a]There are two reasons why so much is unknown. First, at high densities three (and even four) body forces are important. This is particularly so when chemically reactive atoms are present. Then, even for two-body forces, the strongly repulsive regime is not well understood and, in addition, close in, as one approaches the united atom limit, there is considerable promotion of molecular orbitals. This is a "universal" mechanism for electronic excitation which means a breakdown of the Born–Oppenheimer approximation for close collisions.

phenomena that we shall discuss have, at the time of writing, been seen experimentally.[30,33]

The cluster impact technique is presented in detail in Sec. 1. The concept of the cluster impact technique is a simple one — a cold cluster, moving at a supersonic velocity, impacts a hard surface. A rough estimate of the temperature rise within the cluster is provided by the equivalence of temperature to the random part of the kinetic energy. If, upon impact, the entire initially directed velocity of the cluster is rapidly randomized, the cluster temperature will reach a value that is V_0^2 times room temperature where V_0 is the initial velocity in units of the velocity of sound. A cluster impacting the surface at 10 km/s, which is about Mach 30, can therefore be heated to well over 10^5 °K. Our molecular dynamics simulations for a cluster of interacting but otherwise structureless particles verify that this temperature range is accessible even when energy loss to the surface is allowed, cf. Fig. 1 below. In Sec. 4, we will discuss why this thermalization is so rapid. The medium in which the reactions take place is the impact-heated cluster. It is not obvious that cluster induced chemistry is possible because, as we discuss in Sec. 5, the cluster rapidly fragments. An important

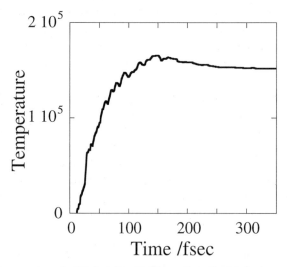

Fig. 1. The temperature in °K (see Eq. (1.1)) vs. time (in fs) during a collision of an initially cold Ar_{125} cluster at an impact velocity of 10 km s^{-1}. Note the sharp rise of the temperature as a result of the impact with the surface and the cooling of the cluster as it expands.

question, therefore, is whether we can achieve activation on a fast enough time scale.

One of our primary observations on cluster impact is that it provides a new regime of dynamics, where the activation process is thermal (due to collisions) but on a very short time scale (ten femtoseconds is typical). This opens a new regime of dynamics, where activation occurs on the same time scale as that of the vibrations of the nuclei. Hitherto, this was commonly possible only by photoactivation. Ordinary chemistry occurs in a regime where collisions are fast on the rotational time scale but slow on the vibrational one.[88] The new regime is described in Sec. 3.

Sections 4, 5 and 6 are arranged according to the consecutive stages of the impact of a cluster with a surface. First, in Sec. 4, we present the new and unique features of the reaction dynamics under the special condition within the impact-heated cluster. Special attention is given to the role of the cluster in determining the efficiency of the studied processes. Section 5 reports molecular dynamics (MD) simulations showing that the finite cluster very rapidly reaches thermal equilibrium and presents a simple mechanism for this facile equilibration. The final stage, that is the fragmentation process of the hot cluster is discussed in Sec. 6. The sudden onset of a shattering regime is shown using both molecular dynamics simulations and an information theory[89,90] analysis. This theoretical prediction was the first new phenomenon to be experimentally verified.[28,33,37] Finally, Sec. 7 presents a comparison between the two theoretical procedures (MD simulations and the maximum entropy information theory procedure) that were used in the study of the burning of air.[81,82,91]

1. Cluster Impact

The essential idea of the cluster impact technique is simple. A cold (glass-like) cluster moving at a supersonic velocity (which experimentally can reach tens of km s^{-1}, 10 km s^{-1} = 0.1 Å fs^{-1}) is incident on a hard surface. The cluster itself is either made up of the reactants (e.g. cluster containing only N_2 and O_2 molecules, see Secs. 3.4 and 6) or the reactants are solvated in a chemically inert medium (e.g. diatomic molecules embedded inside rare gas clusters, see Sec. 3 or, say, in CO_2[63,64]). As the leading edge atoms of the cluster reach and then recede from the surface, the rest of the cluster is still moving forward. Therefore, immediately after the beginning of the impact, high relative kinetic energies, see Fig. 1, and high

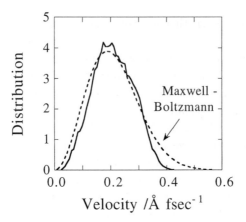

Fig. 2. The three dimensional velocity distribution of all the atoms after impact of a cold cluster of 125 Ar atoms at a surface at a velocity of 25 km/s (= 0.25 Å/fs). Shown for comparison is a Maxwell–Boltzmann functional form for the same mean energy. Even at this high velocity of impact, the velocity distribution after the impact is essentially isotropic. See Sec. 4.

local densities, see Fig. 2, prevail. A short time after the collision with the surface (typically even less than 100 fs), the excess energy is thermalized, see Fig. 3, and a short time later the cluster fragments.

Before the collision with the surface, the direction of the velocity of all the components of the cluster is the same, normal to the surface in most of the figures that we show. Although the velocity of the center of mass is high, the relative kinetic energies of the components of the cold cluster before the collision are very low. As a result of the impact with the surface, the front layers of the cluster change their direction of velocity while the rest of the cluster is still moving forward. High temperature, defined here in terms of the kinetic energy with respect to the center of mass

$$\langle E_{\text{rel}}\rangle \equiv \frac{1}{N}\sum_{i=1}^{N}\frac{1}{2}m_i(\nu_i - \nu_{\text{cm}})^2 \equiv \frac{3}{2}k_BT \quad \text{or}$$

$$T = \frac{1}{3Nk_B}\sum_{i=1}^{N}m_i(\nu_i - \nu_{\text{cm}})^2 \tag{1}$$

are reached rapidly, see Fig. 1.

Figure 1 defines temperature in terms of the mean energy. True temperature is also a measure of the distribution in the speeds of the molecules.

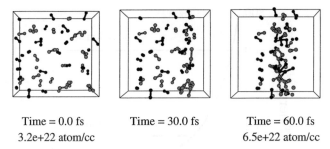

Time = 0.0 fs	Time = 30.0 fs	Time = 60.0 fs
3.2e+22 atom/cc		6.5e+22 atom/cc

Fig. 3. Snapshots during the compression stage when a cluster of 26 N_2 and 25 O_2 molecules impacts a cold surface (not shown in the figure) at 12 km s^{-1}. Computed by molecular dynamics. The impact begins at time zero and the box shown is the smallest cubic enclosure that contains all 51 molecules. The initial density in the box is 3.2 × 10^{22} atoms/cm^3. 60 fs after the impact, the density is 6.5 × 10^{22} atoms/cm^3.

Figure 2 compares the distribution after the impact, computed from molecular dynamics simulations, to the Maxwell–Boltzmann functional form. In Sec. 4, we discuss the reasons why the initially directed energy is so rapidly thermalized.

As a result of the impact with the surface, the cluster is initially quite compressed. At the velocity range of interest (of the order of 10 km s^{-1}), the density can about double, as seen in Fig. 3.

The dynamics under the high densities and high relative velocities, which result from the impact of a cluster with a surface, is the subject of the next section.

2. A New Regime of Dynamics

The important generalization that emerges from our studies is that cluster impact at high velocities provides a new regime of collisional activation, where the intermolecular coupling is comparable or faster than intramolecular motions. At ordinary velocities, only the rotations are in this "sudden" regime. Indeed the rates for rotational–translational energy transfer are rather high.[88] The range of relative velocities in the cluster immediately after the impact is 10–30 km s^{-1}, which means for the example of a rare gas atom–halogen molecule (Rg–X_2) collision at these velocities distances of close approach of less than $\frac{2}{3}$ of the range of the Lennard–Jones Rg–X potential. This is illustrated in Fig. 4.

The duration of the such a hard, close-in, collision is not more than a few fs's, and such collisions are sudden-like with respect to vibrational

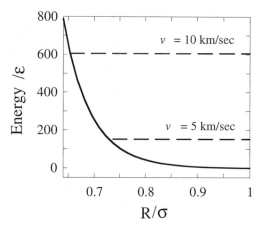

Fig. 4. The Lennard–Jones 12–6 potential between an Ar and an I atom, in units of the well depth ε vs. the atom–atom distance, in units of the range parameter σ. The turning point is shown for two high energy collisions, at the indicated velocities. In units of the well depth, the energies are so high that the turning point is further to the left than one is used to. So much so that the point on the abscissa where $R = \sigma$ is at the end of the scale and the well region is further to the right.

(and, of course with respect to the rotational) motion of the reactants. It follows that vibrotational excitation in a rare gas atom reactant collision will be impulsive like and hence very efficient.[92] At these high velocities, the sequential collisions are quite distinct in time. Despite the high density, the duration of such a close-in collision is shorter than the time between successive collisions and this is also due to the short range of the repulsive force, Fig. 4. The constituents of the cluster are "smaller" than what we are used to as is also seen in Fig. 3.

An important implication is that collisional activation within the cluster can be described as a sequence of binary events. In other words, it is typically one rare gas atom at a time that undergoes a close in collision with a reactant, as illustrated in Fig. 5.

An additional factor that contributes to the validity of the binary picture is the repulsion between the rare gas atoms. It limits the number of atoms that can simultaneously effectively couple to the vibration of the diatomic molecule.

Under the binary approximation we can break the whole history of each cluster impact into a series of elementary steps. Each collision can be treated as an isolated one as is the case in the gas-phase, including the

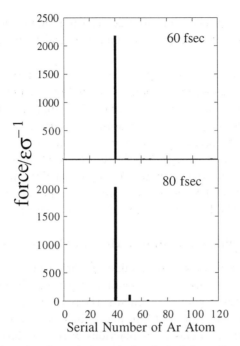

Fig. 5. The force (in reduced units) along the bond of a halogen I_2 molecule, applied by the different rare gas atoms after impact at a velocity of 5 km/s, vs. its (arbitrary) serial number 60 and 80 fs after the impact with the surface. Note how, at any given time, only one or two atoms act to vibrationally excite the molecule.

collision between the reactants. In summary, despite the high densities, gas-phase dynamics can provide useful insights.

The high material and energy density conditions can be established very rapidly, leading to heating on a time scale shorter than even typical intramolecular motions. The important conclusion from our (theoretical) research on cluster–surface collisions, is that the short period while the cluster is hot and compressed is long enough for chemical reactions to take place. In both classes of reactions that we have examined (the simple bond dissociation of diatomic molecules[84,93] and various kinds of four-center reactions,[81,82,91,94,95] see Sec. 3), molecular dynamics simulations show high reactivity. The sudden-like character of the collisions ensures an effective energy transfer between the colliding components of the cluster, including the reactants. This special way of heating, made possible by the impact of a cold cluster on a hard surface (see Sec. 4), is the origin of the sharp behavior of the fragmentation process, as a function of energy, which has the characteristics of a phase transition. This will be further discussed in Sec. 5.

The unique features of our system enable us to use three different theoretical tools — a molecular dynamics simulation, models which focus on the repulsion between atoms and a statistical approach, based on an information theory analysis. What enables us to use a thermodynamic-like language under the seemingly extreme nonequilibrium conditions are the high density, very high energy density and the hard sphere character of the atom–atom collisions, that contribute to an unusually rapid thermalization. These conditions lead to short-range repulsive interactions and therefore enable us to use the kinematic point of view in a useful way.

3. Chemistry within an Impact Heated Cluster

The purpose of our work was to examine whether the compressed hot cluster provides an effective medium for reactions with high barriers. In the simplest case of cluster impact experiments, the reactants are embedded inside a rare gas cluster, and this cold droplet is incident on an inert surface at various velocities. Two classes of reactions were examined in detail using standard molecular dynamics simulations[b]: the dissociation of halogen molecules and four-center reactions.[c]

---

[b]Initial conditions for the cold (50 K) clusters with embedded reactants are chosen by "equilibrating" an initial configuration by a standard molecular dynamics method.[96] All rare gas atom–atom potentials adopt a Lennard–Jones 12–6 functional form with parameters given in Table 1 of Ref. 82. The potential for the diatomic molecule was approximated by a Morse-type functional form, with parameters given in Table 1 of Ref. 97. The potentials for the four-center reactive system were mimicked by a London–Eyring–Polanyi (LEP) four-atom functional form,[98,99] with parameters given in Tables 1 and 2 of Ref. 82. In all cases the potential between the rare gas atoms and the reactive system were taken as a sum of atom–atom potentials, using the Lennard–Jones 12–6 functional form. The initial conditions generated by the equilibration procedure were used as input for the dynamics except that each atom was given the same additional velocity in the direction normal to the surface. The classical equations of motion were integrated for each atom in the cluster. The impact at the surface was described by a hard cube model.[100] We have compared the results of this approximation to results of simulations that used a full mechanical model of the solid[84] and to other approximations.[101] More details on the classical simulations can be found in the original papers.
[c]Concerted four-center reactions are expected to have a high energetic barrier due to an unfavorable orbital correlation.[102] The high barrier for such reactions is due to a curve crossing between the state that originates in the ground state reactants and correlates to the electronically excited products and the state that originates in the electronically excited reactants and correlates to the ground state products.

3.1. *Motivation*

Dissociation of diatomic molecules (and halogens in particular) in shock waves has been extensively studied.[103,104] A lingering intriguing problem has been the unexpectedly low Arrhenius activation energy of such processes.[105,106] Among the factors that have been considered as contributing to the observed kinetic behavior have been the enhancement of the rate of collisional dissociation by internal excitation of the diatomic molecules and the possible role of "multiquantum" transitions in which the molecule gains several vibrational quanta per collision.[103,104,107]

The high relative velocities following impact of a cluster on a surface suggests that such dissociation processes can readily take place when a diatomic molecule embedded inside the cluster is activated by a collision. Molecular dynamics simulations show that beyond a threshold, the yield of dissociation of halogen molecules solvated in a rare gas cluster is a rapidly increasing function of the collision velocity and can reach 100%, see Fig. 6. This, unlike the surface impact induced dissociation of unclustered, cold, halogen molecules where the yield reaches a plateau of below 40%.[108]

The higher yield for dissociation of the clustered molecules correlates with the evidence provided by examination of individual trajectories: molecules can dissociate in one of two pathways — a heterogeneous dissociation

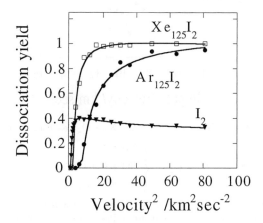

Fig. 6. The yield of dissociation vs. the squared velocity of impact for a bare I_2 molecule and for a molecule embedded in a cluster of 125 atoms of Ar and in a cluster of 125 atoms of Xe. Note how, at a given velocity, the cluster impact dissociation is significantly higher than for the bare molecule.

on the surface or a homogenous one within the cluster, without ever reaching the surface.

The observation of a high dissociation yield encouraged us to examine a much more complicated chemical process — a four-center reaction. Four-center reactions,[98,109–117] involving the switching of covalent bonds, are expected to have a high barrier.[102,118] Beyond any potential barrier to reaction, four-center reactions are also subject to a strong kinematic constraint.[97] However, under the unusual combination of conditions made possible within an impact-heated cluster, the four-center reaction can be driven. A variety of high barrier processes were examined: the bimolecular $N_2 + O_2$ and $H_2 + I_2$ reactions and the unimolecular norbornadiene $\rightarrow$ quadricyclane isomerization. Our primary finding is that the yield of the concerted reaction is quite high because the cluster has a rather essential

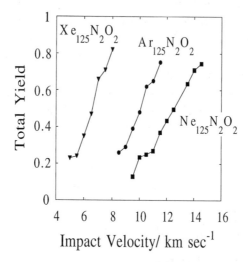

Fig. 7. The reactive possible outcomes in the $N_2 + O_2$ rearrangement collisions are:

$$N_2 + O_2 \rightarrow \begin{cases} NO + NO \\ NO + N + O \\ N + N + O + O \end{cases}.$$

We count the $N_2 + O_2 \rightarrow N + N + O + O$ channel as a reactive one because all the trajectories we examined yielded a four atom final state only when it was preceded by the formation of NO, however briefly. In addition, this channel has a dynamical energy threshold significantly higher than the endoergicity. The total yield is shown vs. the impact velocity for $N_2 + O_2$ molecules embedded in a cluster of 125 rare gas atoms. One can replot this figure in terms of a reduced variable, see Fig. 13.

role to play, as shown in Fig. 7 for $N_2 + O_2$. So much so, that we consider that it is reasonable to speak of "a cluster catalyzed reaction."

In this section we discuss the new features of the dynamics under extreme conditions and the important role of the environment on the high efficiency of the examined process. The ability to break the overall mechanism into series of stages is a result of a clear separation of time scales with which different events take place, as we pointed out in the previous section. This enables us to examine first, for each reaction, the energetic and steric requirements and then to return to those collisions before and after the reaction took place, so as to examine the role of the cluster in enabling such a high reactivity.

3.2. The Reactive System

Both dissociations and four-center reactions have an energy threshold. Since the diatom molecule inside the cluster (before the collision) is cold, the nominal energy threshold for the bond dissociation will be the binding energy. For the four-center reaction, the threshold will be the minimum energy required to surmount the reaction barrier. These energetic requirements can be easily reached by controlled selections of the impact velocity,

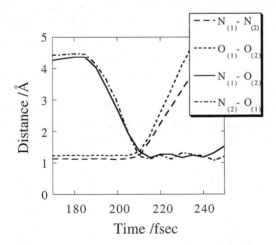

Fig. 8. The bond distance vs. time (in fs) is plotted for the two old bonds and the two new bonds during the $N_2 + O_2 \rightarrow 2NO$ reactive collision in a cluster of 216 Ne atoms for an impact at 10.5 km/s. The bond switching occurs in a concerted fashion in less than 10 fs and in physical units it is clearly faster than the vibrational period of the newly formed NO bonds.

depending on the identity of the rare gas and the energy loss to the surface. However, although these energetic conditions are necessary for the reaction to occur, there are other considerations that are often kinematic in origin. These result in an effective threshold that is higher than the nominal one. On the other hand, it is characteristic of reactions within the cluster that the reactants are preconditioned by vibrational activation. The actual collision that leads to reaction will therefore have a lower threshold than that for cold reactants.

The sudden-like nature of reaction inside the cluster is illustrated in Fig. 8. Preceding the reaction is an activating collision that brings the reactants to a relative velocity of 10–30 km/s. At these energies the entire transversal of the potential energy surface lasts for only 10 fs or so. The reaction is therefore sudden-like with respect to the internal motions of even the lighter reactants (N_2, O_2, H_2). The heavier I_2 molecule is fully standing still during the reaction. Therefore, the bond switching can be described by kinematic considerations.[119–126]

3.2.1. *The Kinematic Model*

The kinematic model that is used to interpret the results of the molecular dynamics simulations has been validated by comparison to exact dynamical calculations for collisions of the isolated reactants, at high energies.[97] The model assumes that the transition from reactants to products occurs in a sudden like manner: up to a given configuration the motion is that of the unperturbed reactants; beyond that configuration it is that of the final, free, products.

The quantitative statement of the model is the kinematic transformation[97,127,128] between the coordinates x_i, $i = 1, 2, 3$ that describe the reactants and the coordinates x'_i, $i = 1, 2, 3$ of the products as given in Eq. (2)

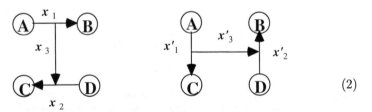

$$\text{(2)}$$

Reactants   Products

The transformation matrix between the two sets of coordinates is:

$$
\begin{pmatrix} x_2' \\ x_2' \\ x_3' \end{pmatrix} = \begin{pmatrix} \dfrac{m_B}{m_{AB}} & \dfrac{m_D}{m_{CD}} & 1 \\[2ex] \dfrac{m_A}{m_{AB}} & \dfrac{m_C}{m_{CD}} & -1 \\[2ex] \dfrac{m_A m_B M}{m_{AB} m_{AC} m_{BD}} & \dfrac{m_C m_D M}{m_{CD} m_{AC} m_{BD}} & \dfrac{m_A m_D - m_B m_C}{m_{AC} m_{BD}} \end{pmatrix} \begin{pmatrix} x_1 \\ x_2 \\ x_3 \end{pmatrix} \qquad (3)
$$

$M = m_A + m_B + m_C + m_D$ and $m_{AB} = m_A + m_B$ where m_X is the mass of atom X. The transformation Eq. (3) is nothing but the identity transformation. The only reason it does not look like one is that the (primed) coordinates used to describe the products, Eq. (2), are not the same as the coordinates used to describe the reactants. That Eq. (3) is really the identity matrix follows from the assumption of the model that the motion of the reactants is unperturbed, up to the transition configuration and that beyond this configuration the products recede without any intermolecular coupling.

In the strict kinematic limit in which no forces act, one takes the time derivative of Eq. (3) to conclude that the transformation Eq. (3) also relates the initial and final velocities. Often however there is a strong repulsion between the products and in the sudden limit this force gives rise to an impulsive change of the velocities. It follows that the velocities transform as

$$
\dot{x}' = (\mathfrak{R} + I)\dot{x} \qquad (4)
$$

where the overdot signifies the time derivative, $\mathfrak{R}$ is the matrix of Eq. (3) and I is a matrix whose components specify the impulses delivered along the directions of the three final velocities in terms of the initial velocities.

The kinematic constraint on four-center reactions follows from Eqs. (3) and (4). For reactions of homonuclear diatomics, the relative velocity of the products $\dot{x}_3'$ is given by

$$
\dot{x}_3' = (m_A/(m_A + m_C))\dot{x}_1 - (m_C/(m_A + m_C))\dot{x}_2 \ \{+I_{3,3} \cdot \dot{x}_3 + \cdots\} \qquad (5)
$$

where the terms in the curly brackets are due to the impulses. Without this contribution it is entirely the internal energies of the reactants (i.e. the velocities $\dot{x}_1$ and $\dot{x}_2$) that can drive the products apart. In the strict kinematic limit, initial translational energy will not be converted into relative motion of the products. If there is repulsion between the products (which is the case for the LEP potential) then the extra terms in Eq. (5) will contribute. The kinematics are however such that the efficiency of the reaction will be considerably enhanced by initial vibrational excitation. An important role of the cluster is not only to accelerate the center of mass motion but also to internally excite the initially cold reactants.

Molecular dynamics simulations have shown[82,97] that for isolated reactants rotational excitation contributes to the enhanced reactivity (cf. Fig. 5, Ref. 97). In the kinematic limit, initial reagent rotational excitation is needed for a finite orbital angular momentum of the relative motion of the products. This is intuitively clear for the $H_2 + I_2 \rightarrow 2$ HI reaction, where there is a large change in the reduced mass.[129] The rather slow separation of the heavy iodine atoms means that rotational excitation of HI is needed if the two product molecules are to separate. This is provided by the initial rotational excitation of the reactants. The extensive HI rotation is evident in Fig. 9 which depicts the bond distances of this four-center reaction on a fs time scale.

The kinematic constraints require that the reactants be energy rich. Even for a high endoergicity as in the $N_2 + O_2$ reaction, much of this

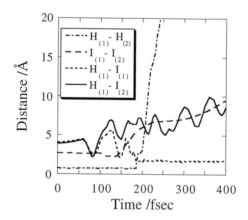

Fig. 9. Bond distances (see inset) vs. time, in fs, for the old and new bonds in the $H_2 + I_2 \rightarrow 2$ HI reaction in an Xe_{125} cluster for an impact velocity of 6 km/s. The large disparity in the vibrational periods of the two reactants make for somewhat more complex dynamics because it takes a rather long time for the very heavy iodine molecule to move. It is therefore mostly the H atoms that move during the bond switching. Also, it is necessary for the reactants to be rotationally excited and the fast rotation of H_2 is very evident in the oscillation of the H–I bond distances before the collision. Similarly, the rotational excitation of the product HI molecules is evident in the oscillation of the H–H bond distance around the I–I bond distance after the collision.

energy must be disposed as excitation of the products. When the reacting diatomics are homonuclear, in the kinematic limit, all of the initial translational energy and some of the initial internal excitation must appear as internal excitation of the products, as expressed by Eq. (3). The newly formed NO molecules are indeed vibrationally hot as can be seen from the amplitude of their motion shown in Fig. 8. We shall shortly note an important post-reaction role of the cluster.

In this section we were concentrated on the reactive system and on the energetic requirements for high reactivity. The kinematic model interprets the energy requirements and energy disposal necessary for higher yield of the four-center reactions. In the next section we discuss the role of the cluster with respect to these points.

3.3. *The Role of the Cluster*

The significant role of the cluster in inducing reaction is illustrated in Figs. 6 and 7 above. In this section we examine the role of the cluster with respect to the requirements presented in the previous section. The T–V effective energy transfer in the heated cluster, and the dependence of this process on the size of the cluster, the identity of the atoms etc. are the primary concepts for understanding the role of the cluster. We will start with the simpler process of bond dissociation and then proceed to the more complicated system — the four-center reaction.

3.3.1. *The Role of the Cluster in the Bond Dissociation Process*

Diatomic molecules embedded in rare gas cluster can dissociate in one of two ways: a heterogeneous dissociation of the molecule on the surface and a homogeneous mechanism where dissociation takes place inside the cluster without the molecule reaching the surface. The sudden collision regime insures that an effective vibrational excitation of the diatomic molecule will occur due to a collision between the diatomic molecule and the rare gas atoms of the cluster. Yet, the size of the cluster and the identity of the rare gas atoms influence the yield of dissociation.

3.3.1.1. *Activation of the reactants — mass effect*

In the initial stage of the impact, the velocity of the embedded diatomic molecule is still the velocity ν of the cluster before impact. Some rare gas atoms have, however, already rebounded from the surface, where they reversed the direction of the velocity and also lost a fraction of the energy. At the velocities in question, the rare gas atom collides with one of the two atoms in the diatomic molecule, while the other atom is essentially a spectator to the energy transfer.[88] The velocity u of the rare gas atom R before it collided with the atom A of the reactants is $u = -a\nu$ where $1 - a^2$ is the fraction of the kinetic energy of atom R which is lost to the surface. The diatomic molecule is initially vibrationally cold so the final velocity of the atom A is

$$\nu' = \left(\frac{\mu - 1}{\mu + 1}\right)\nu + \left(\frac{2}{\mu + 1}\right)u = ((\mu - 1 - 2a)/(\mu + 1))\nu; \quad \mu = M_A/M_R$$

(6)

where ν is the initial velocity of the cluster (and atom A). This simple result for impulsive collisions provides the key to the mass effects in cluster activation. The point is that the dimensionless mass parameter μ, the ratio of the masses of atoms A and R, can span quite a range, from well above unity for He to a small fraction for Xe. Since in such a sudden-like collision the other atom in the diatom molecule is a spectator to the energy transfer, the bigger ν' the higher is the vibrational excitation.

The role of the mass, namely the heavier rare gases are more efficient in inducing vibrational excitation, is very evident in Figs. 6 above and 10.

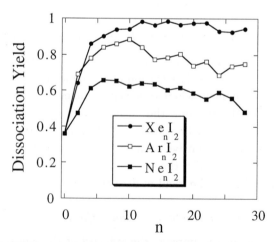

Fig. 10. The yield of dissociation in clusters of different rare gases for an impact velocity of 6 km s^{-1}. Note how the yield is higher for the heavier rare gases.

3.3.1.2. *The role of the size of the cluster*

The qualitative result of the computations conforms to intuitive expectations: the larger the cluster the more important is the homogenous process. The dependence of the dissociation yield on cluster size can be seen in Fig. 11.

Beyond the regime (say fewer than 13 rare gas atoms), where the yield increases with cluster size, there is a decline in the yield, a decline which is more noticeable, the lower is the velocity of impact.

For smaller clusters the primary route for dissociation is the heterogeneous dissociation of the molecule. Just as for isolated collisions in the gas-phase, the probability of dissociation of a diatomic molecule upon impact at the surface is enhanced by its vibrational excitation. As the small cluster impacts the surface, it can be that the halogen molecule reaches the surface immediately, so that it still has the same velocity as that of the cluster center of mass. If that velocity is above the threshold, it will dissociate with about a 40% probability, just as a bare molecule would. The other process that can happen is that the halogen molecule collides first with a cluster atom, an atom that has already hit the surface, and therefore lost a fraction of its translational energy to the surface. Such a collision does two things. It slows the halogen molecule and so reduces its probability to dissociate at the surface. At the same time, at the supersonic velocities of

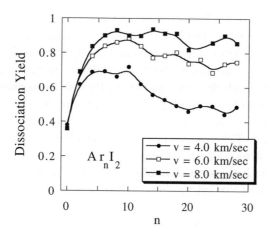

Fig. 11. The yield of dissociation of $I_2(Ar)_n$ clusters at different impact velocities vs. the cluster size. Note that at the lowest velocity, the yield decreases with increasing cluster size.

interest, such a collision efficiently vibrationally excites the halogen molecule. Since vibrational excitation is more efficient than translation in inducing dissociation at the surface, the collision of the halogen molecule with a rare gas atom prior to its hitting the surface enhances the yield of the heterogeneous dissociation.

For larger clusters (where the primary route for dissociation is the homogeneous one), the probability of a molecule to be slowed down by collision with cluster atoms increases, resulting in the yield starting to decrease. At higher velocities of impact this effect is less important, so that the decline in the yield is more moderate (Fig. 11).

The heavier rare gases are larger and more polarizable and so are able to solvate the molecule better and to be in its way as it moves towards the surface. For both the kinematic and steric effects the onset of the homogeneous mechanism of dissociation occurs at a lower velocities for Xe than for the lighter rare gases. Hence in Xe clusters one hardly discerns the transition from the heterogeneous to the homogeneous mechanism as a function of cluster size, cf. Fig. 10.

Collisions that activate the diatomic molecule can occur in succession, but after dissociation no collision induced recombination process is seen in the computations. We will discuss the lack of a "cage" effect in Sec. 3.3.4.

3.3.2. *The Role of the Cluster in Four-Center Reactions*

The efficiency of cluster impact in driving four-center reactions is due to a matching between what the cluster can do to the reactants or products and the very selective energy requirements and the specific energy disposal in a concerted reactive collision, as discussed in details in Sec. 3.2. The cluster serves to provide both the steric and energetic conditions necessary for this reaction. In terms of the impact parameter of the relative motion of the two reactants, their confinement by the cluster keeps it low, so that they do not miss one another. This confinement within the cluster favoring low impact parameter collisions is a key ingredient in why such processes are so efficient. Furthermore, both the activation of the reactants before the reaction and the stabilization of the hot product after it, are due to the cluster atoms.

3.3.2.1. *Activation of the reactants*

The activation process described earlier for the dissociation of the diatomic molecule, is the same for the activation of the reactants before a four-center reaction. After a collision between a rare gas atom and atom A of the reactants, the velocity of atom A changed impulsively. The spectator reactant atom B did not change its velocity. The activating collision will therefore change both the center of mass velocity of the reactants and the relative velocity of the two atoms. The partitioning of the relative velocity between vibrational and rotational excitation of AB depends, of course, on the angle RAB. At a given impact velocity, the heavier the rare gas, the higher is the velocity with which the two reactants will collide after the first activating collision. At higher velocities of impact and for heavier rare gases, the first activating collision can be sufficiently impulsive to transfer large amounts of vibrotational energy that can cause a dissociation of the initially cold diatomic molecule. When this type of collision is followed by a reactive one, the reaction appears to proceed by a three-center, atom–diatom mechanism, which is the first step in the so called "Zeldovich" mechanism.

3.3.2.2. *Stabilization of the products*

After the activation stage, the reactants collide. If reaction occurred, two excited product molecules are rapidly receding from one another. In order

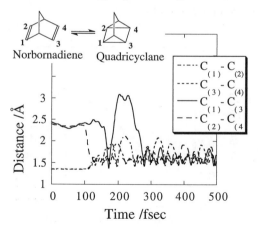

Fig. 12. The role of stabilization by the cluster in the unimolecular isomerization of norbornadiene to quadricyclane in a Ne_{125} cluster at an impact velocity of 6 km/s. Two curves are the bond distances of the C = C double bonds that, during bond switching become single bonds. The two other curves are C to C distances for two carbon atoms that are not regarded as chemically bound in norbornadiene but that form a C–C single bond in quadricyclane. Note the high vibrational energy content of the new bonds and how the molecule is about to dissociate and is prevented from doing so by a deactivating collision.

to achieve high product yield, the role of the cluster atoms is to cool the internal excitation (cf. Fig. 12).

If the rare gas atom is heavier than the atom it collides with $[\mu < 1,$ Eq. (6)], and/or if they are fast, the effect of their colliding with a product molecule is opposite to the desired one; the product warms up further. Other things being equal, a lighter rare gas favors formation of stable molecular products. But a heavier rare gas atom is able to activate the reactants better. Hence, at a given collision velocity, the heavier the rare gas, the higher is the combined yield of all product channels. By specifying as products, not only the bound product molecules but also their dissociated ones (cf. Fig. 7), the yield of the four-center reaction in clusters of all rare gases can be shown on a common reduced energy scale,[82] depending on the mass of the rare gas atoms. This is shown for a 125 rare gas atoms cluster in Fig. 13.

On the other hand, heavier rare gases are not acting as effectively in removing energy from the internally hot, nascent, molecular products. The yield of stable molecular products is favored in lighter rare gases (cf. Fig. 14); where the reactive yield is plotted for Ne and Ar rare gas clusters on a reduced energy scale.

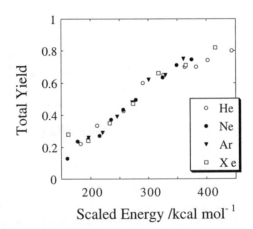

Fig. 13. The yield of all rearrangement processes following a bimolecular $N_2 + O_2$ collision vs. the collision energy in kcal mol^{-1}. (The energy is computed from the velocity of impact using the reduced mass of the reactants.) The results shown are for the reactants embedded in a 125 atoms rare gas cluster where the identity of the rare gas is indicated in the inset (for more details about the energy scaling, see Ref. 82).

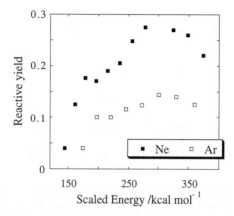

Fig. 14. The yield of molecular products (either one or two NO molecules) in the $N_2 + O_2$ reaction in 125 atoms Ar and Ne clusters plotted vs. the same scaled energy as in Fig. 13. As evident from this plot, the differences between the different rare gases are quite significant. This is primarily due to the heavier rare gases not acting as effectively in removing energy from the internally hot, nascent, NO molecules.

3.3.3. *A Matter of Timing*

The ability to clearly identify the several stages in the overall mechanism and the special role of the cluster in each stage is a result of a very clear separation of the time scales with which different events take place. The duration of a collision is an order of magnitude or shorter than the time between successive collisions. It follows that we are able to examine the contribution of each collision to the total process.

Figure 15 shows the Cl_2 bond distance for two different trajectories for which the initial conditions of the $Ar_{125}Cl_2$ cluster are the same, i.e. the configuration and the center of mass velocity of the clusters at the beginning of each trajectory (before the collision with the surface) are identical. The only differences between the two trajectories are the velocities (randomly chosen from a one-dimensional thermal distribution at 30 K) of the hard cubes that mimic the surface. Despite the rather low temperature of the surface, one of the trajectories results in the dissociation of the diatomic molecule while the other one ends with a vibrationally excited reactant molecule. The effect of the hard cube velocity on the energy of the atom scattering from the surface is negligible but the history of a single trajectory is extremely sensitive to the details of the collisions with the surface, as shown in Fig. 15. This is a characteristic of so called "chaotic" systems. In

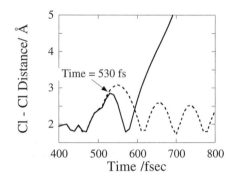

Fig. 15. The Cl_2 bond distance vs. time, in fs, during a collision of a $Ar_{125}Cl_2$ cluster at an impact velocity of 5 km s^{-1} with a surface at 30 K. Shown are the Cl_2 bond distances for two different trajectories for which the initial conditions of the cluster are the same. The only difference between the two trajectories are the (randomly chosen from a thermal distribution at 30 K) conditions for the surface. Despite the rather low temperature of the surface, one of the trajectories is a dissociative one while the other is a nonreactive one.

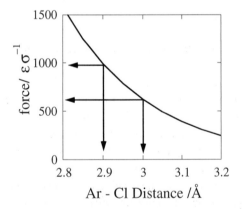

Fig. 16. The Lennard–Jones 12–6 force between Ar and Cl atoms, in reduced units, vs. the atom–atom distance. The magnitude of the force is shown for two Ar–Cl distances. Note how a small change (± 0.1 Å) in the Ar–Cl distance cause a doubling of the magnitude of the force.

Sec. 4, we will have more to say on this. Here, we discuss how small changes are amplified as the cluster evolves in time.

The small changes in the velocities of the cluster atoms recoiling from the surface cause small changes on the position of the atom after the collision, but these small changes can lead to significant changes in the forces acting on the reactants and therefore, the outcome of the trajectory will be very sensitive to the surface conditions. Figure 16 shows the force as a function of the distance between Ar and Cl atoms. Because of the essentially exponential repulsion between atoms, small changes in the Ar–Cl distance result in large changes in the magnitude of the force.

The two trajectories shown in Fig. 15 are seen to be about identical until ~ 530 fs. Figure 17, upper panel, shows the force acting along the Cl_2 bond (the force that will lead to dissociation) for the reactive and the nonreactive trajectories in Fig. 15 at 530 fs. For both trajectories the same Ar atoms exercise the force, but there is a large difference in the magnitude of the forces. This large difference is due to the quite small differences in the positions of the atoms, as shown in the bottom panel of Fig. 17.

The, so called, "timing effect" is the sensitive steric and energetic requirements for the successive collisions inside the heated cluster in order

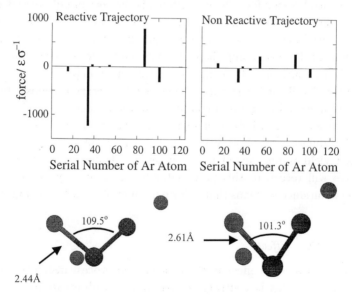

Fig. 17. Top panels: The force (in reduced units) along the bond of the Cl_2 molecule, applied by the different rare gas atoms 150 fs after surface impact at $v = 5$ km s^{-1} vs. the (arbitrary) serial number of the rare gas atoms, for the two trajectories shown in Fig. 15. Bottom panels: The position of the Cl_2 molecule and the rare gas atoms that applied a significant force along the Cl_2 bond, see top panels, for the reactive and nonreactive trajectories shown in Fig. 15. Note how small changes in the position of the atoms (Cl–Ar distance and Cl–Cl–Ar angle) cause a big change in the magnitude of the force applied by the rare gas atoms.

that reaction takes place. For the simplest case of dissociation these require-ments are simpler — a force along the diatomic bond. As we emphasized earlier, it is this simple requirement which means that typically, at a given moment in time, only one rare gas atom is applying an effective force on the molecule.

The high reactivity of the four-center reactions inside the cluster is due to a matching between the energetic and steric requirements of the reaction and the actions of the cluster's atoms on the reactive system. A very essential feature of the four-center reactions in impact heated clusters is the importance of the remarkable timing of the sequence of events that takes place. Unlike the simpler case of dissociation of diatomics embedded in the cluster, the concerted four-center reactions require a fine tuned coordination of many degrees of freedom. The timing is more crucial in the case of the four-center reactions since the whole scenario is richer.

The central reason for a delicate timing is the very impulsive and abrupt changes that are taking place during a typical reactive trajectory. Unless the necessary partner is there at the right time and place, the option for action is over. For every step there is a very narrow window of opportunity.

While the outcome of an individual trajectory depends in a sensitive way on the details at the surface, the ensemble average is robust. In other words, if an ensemble of initial conditions of the cluster is reused, the yield of dissociation is essentially unchanged even though the new run samples a somewhat different set of conditions at the surface. Similarly, the ensemble-averaged yield is unchanged if a new set of cluster initial conditions is used. We will return to this point later but it is an important practical consideration since it means that a finite number of trajectory computations suffices.

3.3.4. *The Cage Effect*

There is no clear cage effect in the dissociation process inside the impact-heated cluster. This is contrary to experimental observations[130-138] and simulations[67,68,137,139-142] of photodissociation of molecules in clusters. This is not unexpected. For caging the dissociation products, the surround-ing medium has to slow them down and to confine them so that they can re-combine. No such strong cage effects are seen in our simulations. Once the halogen molecule has dissociated, the two atoms recede with hardly any noticeable hindrance. We consider that at least two factors, both unique to

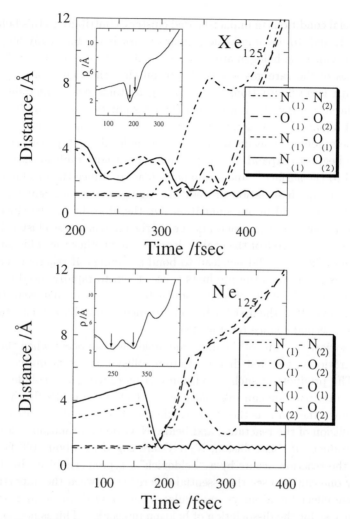

Fig. 18. Bond distances (see legend) vs. time, in fs, for the old and new bonds in the $N_2 + O_2 \rightarrow 2NO$ reaction in a 125 atom cluster. The figure illustrates the higher efficiency of the heavier rare gas atoms in providing a more rigid cage for the outcome of the first bimolecular collision. The inset shows the hyperspherical radius[92] [see Eq. (10)] ρ vs. time in fs. The hyperspherical radius is a measure of how near the four atoms that take part in the reaction are to one another. Top panel: A Xe_{125} cluster at an impact velocity of 7 km/s. Note how the atoms are almost as compressed in their second as in the first bimolecular collisions. (The times of these collisions are indicated by arrows.) Bottom panel: A Ne_{125} cluster at an impact velocity of 12 km/s. The second collision is not very effective in bringing the four atoms together. In both panels only one stable NO molecule is formed.

the special conditions in impact-heated clusters, contribute to this behavior. The first, and, to our mind, the primary factor is the unusually high velocities at which the halogen atoms separate after the dissociation. The high velocities of the rare gas atoms also mean that the cage is a very shaky one, with large fluctuations in the rare gas atoms interatomic distances. Both the size of the atoms is smaller than gas kinetic and the probability of any but a glancing collision is smaller. The second consideration is that by the time a homogenous dissociation of the molecule took place, the cluster already begins to expand. The expansion is most noticeable in the direction parallel to the surface[27] and it too acts so as to reduce the local density.

Like in the dissociation process, no caging of the four-center reaction products was found in our simulation. On the other hand, the cluster can provide a cage for the reactants, i.e. early on when the cluster is very compressed. The role of the caging depends on whether the collision of the two reactants did or did not lead to bond switching. If reaction occurred, the repulsion between the products causes them to rapidly recoil from one another. The role of the cluster atoms is to cool the internal excitation, as shown above. If at the first try bond rearrangement did not occur, the cage insures a second chance (cf. Fig. 18).

Dissociation of one of the diatomic molecules can occur when the steric configuration of the first diatom–diatom collision is not favorable to reaction. The products of such a collision can be caged and recollided, and a three-center reaction can take place as illustrated in Fig. 19. In the trajectory shown therein it is the O_2 molecule that begins to dissociate after the first collision of the reactants, and before N–O bond formation takes place. The products of this collision, $N_2 + O$ are caged for about 100 fs during which the reaction occurs by a "Zeldovich" - type mechanism. In this trajectory one can also see the essential difference between the importance of the cage effect for atom vs. molecules. The O–O distance increase with time just as for the dissociation of halogen molecules. This is not so for the transient N_2O molecule, which is caged for a "long" time.

Because the reactants in a bimolecular collision are typically rotationally excited, whether reaction did or did not occur, the collision products tend to separate in a direction roughly parallel to the surface. They therefore tend to see two "walls" made up of cluster atoms that have a relatively low velocity in the direction parallel to the surface. The efficiency of this cage is therefore quite high and hence it is rare for a bimolecular collision not to be followed by a collision with a cluster atom.

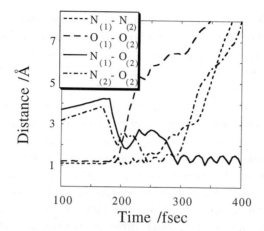

Fig. 19. Bond distances vs. time for the $N_2 + O_2$ reaction in a cluster of 125 Ne atoms at an impact velocity of 12 km/s. Two curves are the two old bonds, N–N and O–O. The other two curves are N to O distances of one oxygen atom from each one of the two nitrogen atoms. Note that dissociation of the O_2 molecule that begins after the first collision of the reactants, and before N–O bond formation takes place. We show the O atom distance to the two N atoms so as to emphasize that a transient hot N_2O molecule was formed. Reaction occurs by the dissociation of this caged molecule.

The role of the cluster in making reactants that failed to react on the first try, try again, is enhanced if it is heavier because it is better able to reverse the velocity of the center-of-mass of the reactive atoms, see Fig. 18. Again there is the correlation that heavier rare gases favor overall reactivity. On the other hand, a heavier rare gas is more likely to induce dissociation of an internally hot nascent product molecule.

The need for precise timing is evident not only before but also after the bimolecular collision. Unless the cage provided by the cluster is there, the deactivation or reactivation, so necessary for a higher yield, will not happen. The cluster catalyses the reaction by being available at the right time and in the right configuration.

3.4. *Rare Gas Clusters Containing Several N_2 and O_2 Molecules*

Our previous molecular dynamics simulations showed that under the unusual combination of conditions made possible within an impact heated cluster, the $N_2 + O_2 \rightarrow 2NO$ reaction can be made to proceed (on the computer) via a four-center mechanism. The high yield was found not only to be due to the high compression and high energy density in the impact heated cluster but also due to the favorable solvation of the reactants by the cluster. Next we examine a more complicated system — rare gas clusters containing several N_2 and O_2 molecules[d] which can exhibit the long sought[145,146] burning of air.

Due to their long range attraction the reactants cluster together and the rare gas atoms surround the reactive molecules. The three roles of the cluster are as in the previous cases, to make sure that the reactions occur through the bulk of the cluster and not only in the layer nearest to the surface, to activate the reactants and to stabilize the products.

Under the high density after the impact, several O_2 and N_2 molecules can interact simultaneously. While the high density conditions prevail, many reactions can occur through different mechanisms: the four-center $N_2 + O_2 \rightarrow 2NO$ reaction and the "Zeldovich" - type atom–diatom reactions, $O + N_2 \rightarrow NO + N$ and $N + O_2 \rightarrow NO + O$, just as in the case of a cluster containing a single N_2/O_2 pair, and also by a new mechanism, a multicenter one, where more than four atoms interact simultaneously. A snapshot of a multiatom configuration is shown in Fig. 20.

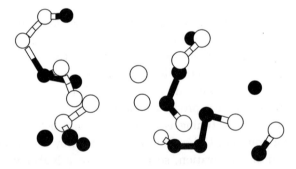

Fig. 20. The location of the reactive atoms (nitrogen: Black, oxygen: White) shortly after a $(O_2)_7(N_2)_7Ne_{97}$ cluster impacted a surface at 12 km/s. A tube connects those atoms that are chemically interacting. Note the 5 and 7 atom configurations.

---

[d]The potential between the reactive atoms that we used is one that has been used before in studies of many-atom systems,[143,144] and which we have examined to insure that it recovers the essential chemical facts that are known to us. (However, one surely does not yet know all that there is to know about the high energy chemistry of the N/O system.) The potential has a functional form that allows for a weakening of a bond between a pair of atoms when one or more other reactive atoms are nearby. In other words, the potential has many-atom terms. It also includes a long range van der Waals attraction which describes the packing in the cold cluster. For more details, see Ref. 95.

N_2O is an important transient species. Examining in detail the history of the MD trajectories indicate that many N_2O molecules formed in the early stages through the new multicenter mechanism and also by a four-center mechanism, as can be seen in Fig. 21 below. The kinematic matrix for the formation of N_2O by the four-center mechanism is given by

$$\begin{pmatrix} z_1 \\ z_2 \\ z_3 \end{pmatrix} = \begin{pmatrix} -m_{CD}/m_{BCD} & m_B M/(m_{AB}m_{BCD}) & 0 \\ 1 & m_A/m_{AB} & m_D/m_{CD} \\ 0 & 0 & 1 \end{pmatrix} \begin{pmatrix} x_1 \\ x_2 \\ x_3 \end{pmatrix}. \quad (7)$$

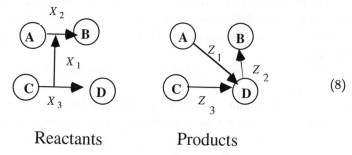

$$(8)$$

Reactants   Products

Here the mass factors are given by $m_{XY} = m_X + m_Y$ etc. and M is the total mass.

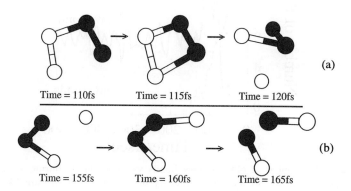

Fig. 21. Two sequences of snapshots of the (a) formation and (b) destruction of N_2O when a cluster of 7 N_2 and 7 O_2 molecules, embedded in 97 Ne atoms impacts on a cold (30 K) surface at (top row) 8 km s^{-1} and (bottom row) 7 km s^{-1}. Note how fast the atoms are rearranging.

It can be seen from Eq. (7) that (1) the CD bond which is sometimes known as the "spectator" bond[127,147] is indeed a spectator in that its velocity is unchanged, (2) unlike the case of a four-center $AB + CD \rightarrow AC + BD$ reaction, here the $A + BCD$ products have a finite relative velocity and so can recede from one another even if the reactants are not internally excited, and (3) the new BD bond is formed with considerable kinetic energy.

N_2O molecules are also formed in the case of a single N_2/O_2 pair (cf. Fig. 19), but are short lived due to a high internal excitation, as the kinematic model predicts. When such a molecule is formed in a cluster containing several N_2/O_2, the reverse reaction, $N_2O + O \rightarrow 2NO$, can take place, during the expansion stage of the cluster, as illustrated in Fig. 21.

The combination of stronger chemical interactions and the high density lead to new possibilities of chemical processes. The high density of chemically interacting atoms allows for a higher reactivity as can be seen in Fig. 22.

Once again we see how cluster impact chemistry provides a new dynamical regime. The unique features of the scheme are the high material and energy density conditions which can be established very rapidly, leading to super heating on a time scale shorter than even typical intramolecular

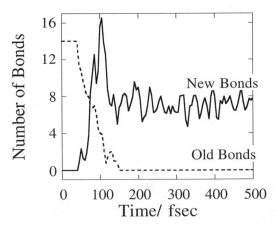

Fig. 22. The number of old and new bonds vs. time (in fs) during a collision of $(O_2)_7(N_2)_7Ne_{97}$ cluster at an impact velocity of 12 km s^{-1}. There are 14 bonds at the beginning. During the compression stage, many atoms interact simultaneously and the total number of bonds reach a maximum value of ~ 20 bonds at 86 fs and at 104 fs. A long period after the impact, 8 hot new bonds are formed. Note the oscillations of the number of new bonds at the end of the trajectory, which is an indication of highly vibrationally excited products, as the kinematic model predicts.

events. In this section we were concerned with the influence of this fast heating on the dynamics of the reactive system embedded inside rare gas clusters. Obviously, the rare gas atoms collide not just with the reactive system but also with themselves, leading to the fast heating of the entire cluster. Due to this effective energy transfer, the localized hot area which is created immediately after the collision with the surface, rapidly dissipates. In the next section we will discuss the mechanism of this fast thermalization.

4. The Rapid Thermalization of the Impact Energy

As a result of the impact with the surface, the leading edge atoms of the cluster reverse their direction of motion and collide with the atoms of the cluster which are still moving forward. Further collisions result in an ultrafast conversion of the initially directed energy to random motion and ultimately to the extensive or complete breakup of the cluster. In this section we show that the finite cluster reaches thermal equilibrium in a very short time that is available before all collisions cease due to the rapid expansion of the cluster. Then we will present a simple mechanism which suggests an interpretation for what causes this fast relaxation.

4.1. *The Translational Thermal Equilibrium*

Figure 23 is a typical output from a MD simulation. It is for an Ar_{125} cluster at an impact velocity of 20 km s^{-1}. It is seen that by 80 fs the velocity distribution of the cluster atoms looks like a three dimensional Maxwell–Boltzmann one.

The quantitative measure of relaxation towards a thermal distribution is the entropy deficiency[89,90] which in our case is given by

$$DS = \int d\nu f(\nu) \ln(f(\nu)/f^0(\nu)). \qquad (9)$$

Here ν is the (scalar) velocity, $f(\nu)$ is the normalized three-dimensional velocity distribution as determined by the molecular dynamics simulation at a given point in time. $f^0(\nu)$ is the three dimensional Maxwell–Boltzmann velocity distribution with a temperature determined by the condition that $f^0(\nu)$ has the same mean energy as the velocity distribution obtained in the simulation after a long propagation time. At thermal equilibrium $DS = 0$ and otherwise it is positive. The larger is DS, the more extreme is the deviation from equilibrium. The results for the entropy deficiency are shown in

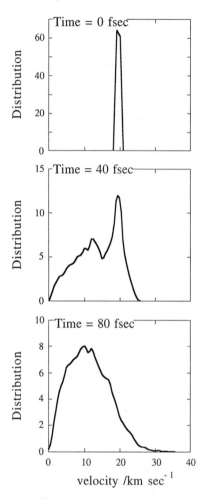

Fig. 23. The distribution of the (scalar) velocity of atoms at different times in a molecular dynamics simulation of the impact of a 125 atom Ar cluster at a surface where the surface is simulated by the hard cube model at the temperature of 30 K. The impact velocity is 20 km s^{-1} or 1 Å per 5 fs where the range parameter of the Ar–Ar potential is 3.41 Å. The mean free path is very roughly of the same magnitude. Thermalization is essentially complete by $\approx$ 80 fs or, after roughly four collisions.

Fig. 24. For the smaller clusters there is extensive but incomplete thermalization. The stabilization of DS at a finite value by about 70 fs is due to the computed velocity distribution not changing any further with time. The cluster shattered before there was time for complete thermal equilibration.

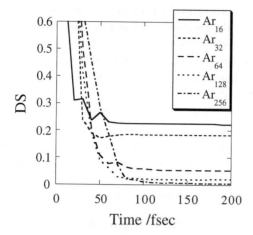

Fig. 24. The entropy deficiency, Eq. (9) vs. time for clusters of different sizes. For thermalization $DS \to 0$ and the value of zero is achieved if and only if the system has equilibrated. When most of the atoms are at or near the surface the relaxation is fairly extensive but is not completed before the cluster shatters.

On the other hand, even for the smaller clusters there is considerable relaxation and for the larger clusters thermalization is complete.

Not discussed herein is very recent experimental work[63,65] in which the thermal distribution of molecules and atoms of a cluster after surface impact has been measured and compared to information theory. There are two reasons why these results should be mentioned. One is that the experiment used a cluster beam incident at an angle ϕ with respect to the normal. It measured the component of velocity parallel to the surface. But in the plane containing the incident beam, there are two such components. One at an angle $\pi - \phi$ with the cluster beam and one at an angle $\pi + \phi$. The observed velocity distribution in both directions was identical. This is a computation-free proof of the equilibration. The other point is the magnitude of the measured temperature which was of the order of 15 000 °K, same as that measured for electrons boiling out of the impact.[38]

What brings the fast relaxation about? In order to answer this question, we examined a simple model of hard sphere collisions in a plane.

4.2. *Model*

A simple mechanical model can exhibit many of the physical features found in the simulation. It is a set of hard spheres (five suffices for all practical

purposes) which can have velocities in a plane, and bounded by two rigid
walls. The rigid walls reverse the x component of the velocity of the end
atoms that collide with them, and each collision between two hard spheres
reverses the components of the two velocities along the line of the center
of the two. Each hard sphere has either two adjacent neighbors or, for the
two end ones, a neighbor on one side and a wall on the other. In short the
spheres are arranged in a sequence but not along a straight line. Knowing
the initial velocities condition of each hard sphere in the chain, one can
analytically calculate[148] the velocity components of all hard sphere at any
given point of time, meaning after any number of collisions. The central
point in this model is that the line joining the centers of any two consecutive
spheres in the chain is randomly oriented — the spheres are not arranged

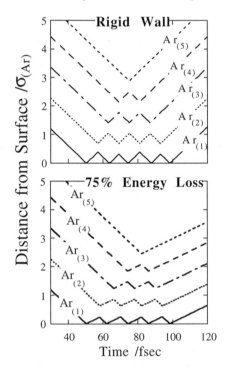

Fig. 25. The distance of spheres from the surface vs. time for a perfect array of 125 hard
spheres colliding with a rigid surface with all atoms of the front face reaching the surface
simultaneously. Computed by an exact dynamical simulation. The five spheres whose
distances are shown are drawn from the five successive layers of the perfect array. Note
that the array is reflected in perfect order. No thermalization takes place and a shock
front propagates without any dispersion, see Sec. 4.3.

exactly along a straight line. Note that in this model when two hard spheres collide, the other spheres are spectators.

The results from the model mimic those obtained by the MD simulations. The main observations are that (1) thermalization is complete in about five collisions, (2) atoms inside the cluster thermalize sooner than the atoms of edges, and (3) thermalization occurs practically just as fast in all directions.

Our conclusion is that it is the lack of structure (or spatial order) of the cluster that is the cause of the rapid thermalization. In all our molecular dynamic simulations, the clusters are glass-like and therefore no spatial order exist. The implication is that a perfectly ordered array of hard spheres will not relax. If a perfect crystal of hard spheres at 0 K is directed towards a rigid surface in such a manner that all the atoms of the front face hit the surface at exactly the same instant then there is no energy dissipation. The cluster rebounds from the inert surface as if it is rigid,[148] see Fig. 25.

4.3. *The Shock Front Propagation*

The constraint of a collision in a given sequence in our simple chain model means that there is a "shock front" propagating through the system, a front which reverses its direction every time an end atom collides with the hard walls. When a perfectly ordered crystal hits a hard wall, one can understand how a dispersion-free propagation of a shock wave is possible. The new feature is that such a shock front was seen in full MD simulations of impact heated clusters, using realistic forces,[101,149] and has been recently studied in more detail.[85]

For a cluster of size n, where one can expect several shells of rare gas atoms (say $n^{1/3} > 4$), the first group of atoms to hit the surface will do so at about the same time. The more rigid the surface, the more will these atoms rebound back into the cluster with comparable velocity vectors. The result is that the downcoming atoms or molecules will, at about the same time, experience collisions with these oppositely moving atoms. The width of such a front, in time, is less than σ/ν where σ is the range of the interatomic potentials. At the velocities used in our simulations on diatomic molecules embedded in rare gas clusters, which are lower than 0.1 Å/fs, this front should last for 50 fs or longer. This time is shorter than the rotational period. Moreover, the range of the anisotropy of the atom–molecule potential means that the translational–rotational coupling will begin to occur at much longer atom–molecule separation than those for effective translational–vibrational coupling. It follows that, of the two internal energies, the rotational motion will be coupled to the shock front before the vibrational motion, and will remain coupled to it after the vibrational motion is already *de facto* isolated. It is therefore the vibrational motion of the halogen molecule that provides the more sensitive test for the "arrival time" of the shock front.

Examination of the trajectories establishes that for both cases shown in Fig. 26, the molecule has not yet reached the surface. We have verified that the inverse relation between velocity and period of the shock front

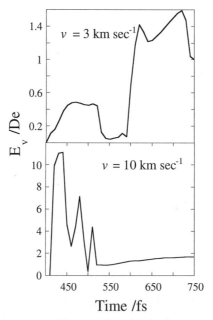

Fig. 26. The vibrational energy, in units of D_e, of Cl_2, embedded in a cluster of 125 Ar atoms vs. time, as a probe for a "shock front." Examination of the trajectories shows that the molecule is moving towards and has not yet reached the surface. It is excited by collisions with rare gas atoms which were at the front of the cluster and have already reflected from the rigid surface, back into the cluster. Upper panel: Low velocity of impact. A layer of rare gas atoms requires ~ 100 fs to move a distance of one σ. In a naive picture, the duration of the front will thus be about 100 fs and so will be the time period between successive fronts. Lower panel: A high velocity where the time scale is expected to be ~ 35 fs. At this high speed there is even some evidence for a third shock front. These results were obtained for impact of the cluster on a rigid wall. A softer surface will cause both a spatial and a temporal dispersion of the shock front.

remains valid up to impact velocities of 15 km/s. In the larger clusters and provided the surface is rigid, a second, often more diffused front can also be discerned due to the second wave of atoms reflected from the surface. Both the duration of the translational–vibrational coupling and the spacing between the fronts are as expected for the two velocities shown in Fig. 26. On the other hand, for a softer surface, the variation in magnitude and direction of the atoms which rebound from the surface tends to average out the sharper front which is characteristic of a hard, rigid surface.

This "micro" shock wave is also seen in Fig. 27 which is for the four-center $N_2 + O_2$ reaction in a $(O_2)_7(N_2)_7Ne_{97}$ cluster impact, where more chemical interactions are possible. It shows the number of atom–atom bonds and the oscillations are due to successive compressions of the reactants by the cluster.

Impact heating of large clusters was shown to be a method for achieving ultrafast heating in the literal sense. Already after two or three collisions the translational distribution is nearly thermal-like and this is particularly so for the atoms "inside" the cluster. At high velocities, energy which is sufficient to fully dissociate the cluster is provided on a time scale comparable to that of a molecular motion. As a result, the cluster completely

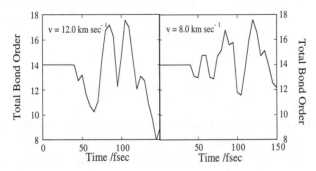

Fig. 27. The bond order between all reactive atoms during the compression stage in a $(O_2)_7(N_2)_7Ne_{97}$ cluster impact at two velocities of impact: 12 km/s (left panel) and 8 km/s (right panel). There are 14 bonds to begin with and during the cluster compression many atoms acquire additional neighbors, cf. Fig. 20. Of course, some molecules dissociate and this makes a negative contribution to the total bond order. The oscillations are due to the "micro shock wave" generated by successive layers of Ne atoms rebounding from the surface. The period of the oscillation (25 and 35 fs for the two panels) is equal to the radius of a Ne atom over the velocity of impact. When more extensive energy dissipation at the surface is allowed, the oscillations are quenched due to the dispersion in both the magnitude and the direction of the velocities of the Ne atoms rebounding from the surface.

fragments. The fragmentation process of the cluster is represented in the next section.

5. The Shattering of Clusters

A short time after the collision with the surface the cluster fragments. An important aspect of the cluster impact induced reactions is that the fragmentation of the cluster prevents the collision-induced dissociation of the energy rich products, by collisions with the atoms of the cluster. Cluster impact heating can induce a tremendous energy deposition in a time much shorter than the rate of expansion of the cluster, as has been shown in the previous section. In this section we discuss the subsequent evolution of this extreme nonequilibrium state, a state that cannot be reached by ordinary heating rates where expansion of the system is faster than the rate of energy deposition. The discussion is not by historical order. The shattering transition that we shall discuss was first seen in a purely theoretical approach, using information theory[150] and was almost immediately verified experimentally.[37] The molecular dynamics simulations[83] and better experiments[28,29,33] followed. The experimental signature of the transition to shattering is the survival of the parent cluster as a function of the impact velocity. Up to a certain velocity of impact the hot cluster rebounds intact from the surface. If one waits long enough, this hot cluster may cool by evaporation but there is a clear time interval during which the rebounding cluster is intact. This will be further discussed below. At a slightly higher impact velocity the parent cluster disappears and many very small fragments appear.

The shattering phenomenon is given much attention because it provides a direct experimental proof of the unusual conditions that are achieved by the impact: The localized energy deposited in the cluster is rapidly distributed before the cluster expands.

5.1. *The Sudden Onset of a Shattering Regime*

Molecular dynamics simulations show a sharp transition, as a function of the impact velocity, between two behaviors: recoil of the intact parent cluster with the possibility of evaporation (i.e. the departure of a small subcluster, one or two atoms) and shattering of the cluster to small fragments, mainly monomers. The hyperradius is a convenient measure for the size of an n

body system, defined by Ref. 151

$$\rho^2 = \left(\prod_{i=1}^{n} m_i\right)^{-1/n} \sum_{i=1}^{n} m_i(\mathbf{r}_i - \mathbf{r}_{cm})^2 \tag{10}$$

$\mathbf{r}_i$ is the position vector of particle i and $\mathbf{r}_{cm}$ is the position of the cluster center of mass. Figure 28 shows the value of ρ at a large distance from the surface for a rebounding cluster of 125 Ar atoms as a function of the velocity of impact. The onset of shattering when the impact velocity exceeds a threshold, is clear from this plot.

The change in ρ at low velocities is due to two processes. Due to the energy deposition, the cluster firstly undergoes a shock wave oscillation, Sec. 4.3, which, upon thermalization, results in shape changes involving the stretching and contraction of the interatomic distances. This causes a slight increase in the value of ρ with superposed oscillations. In addition, after some delay, the warm cluster can cool by the evaporation of one or two atoms, as illustrated in Fig. 29.

At this point there is an experimental[e] upper bound on the duration of shattering of less than 80 ps.[30] The simulations however suggest that shattering occurs in less than a ps, see Fig. 30.

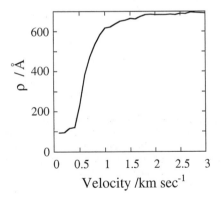

Fig. 28. The hyperradius, ρ, Eq. (10) of a cluster of 125 Ar atoms computed at a point where the center of mass has receded to 75 Å from the surface, vs. the velocity of impact. The positions of the atoms are taken from a molecular dynamics simulation. The onset of shattering is at 0.45 km s^{-1} and the shattering is to individual atoms by about 1.0 km s^{-1}.

---

[e]See also the very recent report.[152]

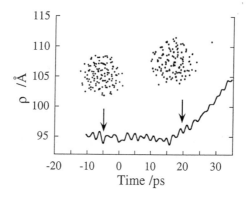

Fig. 29. The hyperradius ρ of a cluster of 125 Ar atoms impacting a surface at a subsonic velocity of 100 ms^{-1}. Results are shown over a long time scale beginning before the impact (at $t = 0$) and for a long time thereafter. It is evident that the slight increase in ρ is due to an evaporation of one Ar atom, see inserts.

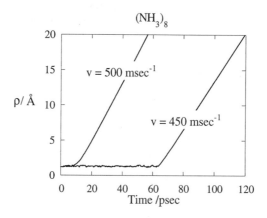

Fig. 30. The hyperradius ρ of the cluster, defined in Eq. (10), in Å vs. time, in ps, for the same cluster impacting a cold surface at two slightly different low supersonic velocities of 450 and 500 ms^{-1} respectively. This figure illustrates how a small increment in the energy of impact changes the outcome dramatically. The cluster (which has identical initial conditions in both runs) evaporates just one monomer at the lower energy but fully shatters at the higher energy.

Finally, the onset of shattering occurs roughly when the cluster is heated to the boiling temperature of an equivalent drop. This temperature is clearly higher, thus deeper is the potential between the constituents of the cluster. This is shown by an example in Fig. 31, where MD simulations of cluster impact are carried out with and without an electrostatic

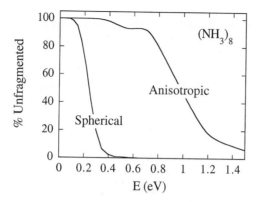

Fig. 31. The fraction of intact parents that recoil from the surface as a function of the impact energy. MD simulations for a cluster of 8 NH_3 molecules interacting only by a central van der Waals potential and by a potential containing in addition a dipolar attraction. The extra attraction shifts the shattering transition to a higher energy.

attraction term between aspherical molecules. The strong long range attraction increases the shattering energy just as hydrogen bonding increases the boiling temperature of bulk ammonia or of water beyond that of spherical molecules of similar polarizabilities.

5.2. *Shattering by Maximal Entropy*

The purpose of this section is to understand why the transition between the two behaviors is such a sharp one. In order to do so, we used information theory analysis[f] and calculated the fragment size distribution as a function of the available energy, the size of the incident cluster and the identity of the constituents of the cluster. By imposing the mean energy as the only energetic constraint, we assume that there are no restrictions on the partitioning of the impact energy. The theory will then imply that there is a thermalization of the translational degrees of freedom of the impact-heated cluster. This is indeed the case for rare gas clusters as we have shown in Sec. 3. When this is not necessarily the case, e.g. for ionic clusters at lower energies, there will not be extensive fragmentation. This section is a technical one and the reader willing to grant the conclusion that there is a shattering phase transition can skip it.

---

[f]The theory we used has been originally applied to the extensive fragmentation induced in polyatomic molecules upon laser multiphoton excitation[90,153] and was more recently applied to the fragmentation[71] of C_{60}.

The distribution of fragment size is computed as one of the maximal entropy subject to two constraints: (1) conservation of matter:

$$N = \sum_{p=1}^{N} pX_p = \sum_{p=1}^{N} \sum_{k=p-1}^{p(p-1)/2} pX_{pk} \qquad (11)$$

where N is the number of atoms in the cluster, and (2) conservation of energy:

$$E = \sum_{p=1}^{N} \sum_{k=p-1}^{p(p-1)/2} \varepsilon_{pk} \,. \qquad (12)$$

Where ε_{pk} is the energy of a species of p atoms and k bonds. X_{pk} be the number of fragments with p atoms ($1 \leq p \leq N$) and k bonds after the impact of a cluster of N atoms with the surface. (The range of k is from $(p-1)$, which is the minimal number of bonds in a connected cluster of p atoms to $p(p-1)/2$, which is the number of all possible pairs.) The distribution of fragment sizes is

$$X_p = \sum_{k} X_{pk} \qquad (13)$$

where one can show[90,153,154] that the number of clusters of a given internal energy is

$$X_{pk} = g_{pk} \cdot \exp(-\beta \cdot \varepsilon_{pk} - \gamma \cdot p) \,. \qquad (14)$$

Here g is the degeneracy, β is the Lagrange multiplier determined by the energy content and γ is the Lagrange multiplier for the conservation of the number of atoms. Summing over all internal energies we have an expression in terms of the partition function Q of the given isomer, Eq. (16) below

$$X_{pk} = g_{pk} Q_{pk} \exp(-\gamma p) \,. \qquad (15)$$

The degeneracy, i.e. the number g_{pk} of different possible isomers of given size p and number of bonds k, is evaluated by graph theory[g] to calculate the number of free connected graphs with p points and k lines[159-162] (for

---

[g]Another possible definition for the number of isomers of a given cluster size is by counting the minima in the potential energy surface of the cluster. Such counts have been reported in the literature,[75,155-158] and an empirical formula that reproduces these results has been suggested. However, later studies[157] suggest that this formula is somewhat of an underestimate of the true value.

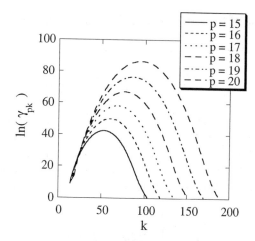

Fig. 32. The number of possible isomers, g_{pk}, for a connected cluster of ρ atoms and k atom–atom bonds, on a logarithmic scale vs. k, for several smaller clusters. Note how steeply the number of possible isomers increases with the number of atoms in the cluster.

details, see Appendix A of Ref. 154). Figure 32 shows the exact value of g_{pk} obtained by the counting procedure described in Appendix A of Ref. 154.

The practical form for the partition function, g_{pk}, [needed in Eq. (15)] for any isomer of a given species is computed[163] as a product of the translational partition function for the center of mass times the partition function of $(3p-3)$ independent isotropic harmonic oscillators (we regard all internal degrees of freedoms as vibrational modes), each truncated at the energy ε (ε is a positive number):

$$Q_{pk} = h^{-3p}\underbrace{(2\pi mpD/\beta)^{3/2}}_{Q_{tr,p}}$$

$$\times \underbrace{(2\pi m/\beta\omega)^{3p-3}(1 - \exp(-\beta\varepsilon))^{3p-3}}_{Q_{vib}^{3p-3}}\exp(\beta k\varepsilon(k)). \qquad (16)$$

Because the vibrotational partition function is referenced with respect to the ground state at zero energy one needs the extra factor, $\exp(\beta k\varepsilon(k))$, which accounts for the energy of the ground state. In real molecules, additional bonds have lower bond energies and therefore we take the energy per bond ($\varepsilon(k)$) in this factor to be a function of the number of bonds.

For this form of the partition function, the mean energy of the isomer is

$$\varepsilon_{pk} = -\partial \ln Q_{pk}/\partial \beta$$

$$= \frac{3}{2\beta} + \frac{3p-3}{\beta} - \frac{(3p-3)\varepsilon \exp(-\beta\varepsilon)}{(1 - \exp(-\beta\varepsilon))} - k\varepsilon(k). \qquad (17)$$

The first two terms in the mean energy will be recognized as the familiar classical equipartition value of $k_B T/2$ per degree of freedom. The remainder is due to the truncation of the atom–atom potential to have a finite dissociation energy. Due to this truncation the contribution of the mean energy per bond does not continue to increase as the temperature increases:

$$\varepsilon_{pk} \rightarrow \begin{cases} -k\varepsilon(k) & \beta \rightarrow \infty \\ (3/2\beta) & \beta \rightarrow 0 \end{cases}. \qquad (18)$$

5.3. *The Results of the Information Theory Analysis*

The primary result of this analysis is that evaporation occurs only at very low levels of excitation (low velocities of impact). Very hot clusters do not evaporate. They shatter into small pieces. The theory does not exhibit an intermediate regime of cluster fission into two (or three, ...) roughly equally sized subclusters, as in a liquid drop model.[164] The transition from the evaporative to the shattering regime is a quite abrupt function of the velocity of impact.

As the available energy increases, the number, X, of possible fragments rapidly increases and reaches its maximal value of N as shown in Fig. 33 top panel. At low velocities of impact one does see an evaporation regime, Fig. 33 bottom panel. The sharp transition from one regime to the other as the energy increases is clear in Fig. 33.

The "sudden" switch over from an intact cluster to independent monomers is due to a competition between two entropic effects. On the one hand, the large entropy of the large cluster is due to the exponentially many possible configurations and on the other hand, the entropy due is to the translational motion of the large number of monomers. Each one of these contributions is exponentially large in cluster size, but they act in opposite direction. The very many possible isomers of a parent cluster of a given size enables it to soak up considerable energy. The translational entropy favors the formation of as many fragments as possible. At the point where the two

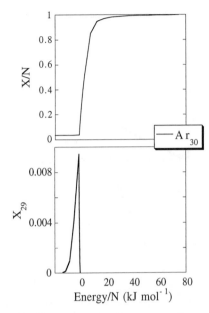

Fig. 33. Results for cluster of 30 Ar atoms. Top panel: The total number, X, of fragments (plotted scaled by the size of the original cluster since $0 < X/N \leq 1$) vs. the energy. The point of sudden increase in the number of fragments is the shattering transition energy. Bottom panel: Plotted is the number of clusters of 29 atoms vs. the energy. When a cluster of 30 atoms evaporates a single atom it necessarily leaves an intact cluster of 29 atoms. At a sharp value of the energy, evaporation gives over to shattering where the cluster fragments to many small fragments without going through the intermediate formation of large subclusters. Note that evaporation is minor route (less than 1% at its maximum) and occurs only over a narrow range in impact velocities. (In our definition of the energy, the total energy will have a negative value as long as the total kinetic energy is below the binding energy of the cluster. The threshold for shattering is thus at zero.)

effects are balanced, a small change in conditions (e.g. in the velocity of impact) results in very large variation in the probabilities of the two outcomes.

5.3.1. *The Shattering Transition as a Phase Transition*

The abrupt change from primarily intact to mainly small clusters as shown in Fig. 33, is accompanied by a discontinuity in the specific heat, Fig. 34.

The phase transition like behavior seen in Fig. 34 occurs when the evaporation gives over to shattering. It is not only the specific heat but

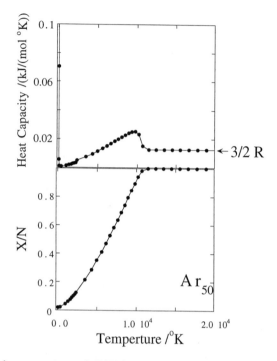

Fig. 34. The heat capacity and X/N (= number of fragments/size of parent cluster) vs. the temperature for a cluster of 50 Ar atoms. Computed by analytically differentiating the energy of the cluster, Eq. (17) with respect to the temperature. Indicated on the drawing is the equipartition value $(3R/2)$ of the heat capacity when the cluster is fully fragmented, so that $X = N$.

also other response functions that manifest a discontinuity at the onset of shattering.[150]

The point about the technique of cluster impact is that it enables one to "heat" the cluster on a time scale short compared to that needed for expansion and *ipso facto* for evaporation. In this way one can prepare "superheated" clusters with enough energy for breaking most or all intermolecular bonds so that the cluster shatters into its constituents. The sharp transition from evaporation regime to the shattering one, shown in this section, indicate a fast translational thermalization of the extreme disequilibrium formed immediately after the impact with the surface. The need for only two constraints in the maximum entropy formalism (conservation of matter and total energy) to predict the experimental results on the fragment size distribution indicates that the ultrafast dissipation is also a feature of the experimental results. We also reiterate that the thermal nature of the final state has by now been directly experimentally verified.

The success of the maximum entropy procedure to predict the shattering of clusters encourages us to use it in more complicated systems, where very little is known about the potential energy surface. In the next section the results from both molecular dynamics simulations and information theory analysis for clusters made up of N_2 and O_2 molecules are presented.

6. The Burning of Air: A Theoretical Study Using Two Complementary Procedures

In the previous section we showed that molecular dynamic simulations and an information theory analysis predicted similar results for the fragmentation process of impact heated rare gas clusters, results which soon thereafter have been confirmed experimentally. In this section we compare the results from both methods on the study of the impact of clusters containing several N_2 and O_2 molecules and clusters containing only N_2 and O_2 molecules. We still have no definitive results for this particular experiment but we hope that this challenge will be taken up soon. Other four-center reactions have, as we write this chapter, been observed.

The reason we employ two rather distinct methods of inquiry is that neither, by itself, is free of open methodological issues. The method of molecular dynamics has been extensively applied, *inter alia*, to cluster impact. However, there are two problems. One is that the results are only as reliable as the potential energy function that is used as input. For a problem containing many open shell reactive atoms, one does not have well tested semiempirical approximations for the potential. We used the many body potential[143,144] which we used for the reactive system in our earlier studies on rare gas clusters containing several N_2/O_2 molecules (see Sec. 3.4). The other limitation of the MD simulation is that it fails to incorporate the possibility of electronic excitation. This will be discussed further below. The second method that we used is, in many ways, complementary to MD. It does not require the potential as an input and it can readily allow for electronically excited as well as for charged products. It seeks to compute that distribution of products which is of maximal entropy subject to the constraints on the system (conservation of chemical elements, charge and

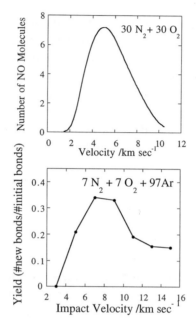

Fig. 35. Two characterizations of the facile formation of new bonds in superheated clusters containing N_2/O_2. Top panel: The yield of NO, computed by the maximum entropy formalism, when a cluster of 30 N_2 and 30 O_2 molecules is superheated. The velocity scale is that which is needed to heat the cluster if there is no energy dissipation to the surface. The experimental "impact velocity" is necessarily equal or higher than this velocity. Bottom panel: The yield of new bonds (as a fraction of the number of initial bonds) computed by a molecular dynamics simulation when a cluster of 7 N_2 and 7 O_2 molecules, embedded in 97 Ar atoms, impacts on a cold (30 K) surface, vs. the impact velocity, previously discussed in Sec. 3.4.

energy).[h] Both methods consider the surface on which the cluster impacts to be chemically inert. It can exchange energy but not atoms or charges with the cluster.[i]

---

[h] At a given volume V, the number of molecules of species j is given by[153,154] $X_j = Q_j(\beta)V\exp(-\beta\Delta E_{0j} - \sum_k \gamma_k a_{kj})$. Here Q_j is the (temperature dependent) partition function[165] of the species j, per unit volume and using the ground state as the zero of energy for each species. ΔE_{0j} is the heat of formation (see the appendix of Ref. 153 for more practical details). β is the Lagrange multiplier imposed by the conservation of energy constraint. The summation over k is overall the distinct chemical elements, (here, two, N and O), and over the charge. a_{kj} is the number of times that element k (or, the total charge, zero for neutral species) contributes to a molecule of species j.

[i] Both restrictions can be relaxed and have been in recent work which allows charge exchange with the surface.

The results from both methods at velocities of impact below 10 km s^{-1} are remarkably consistent: as a result of the ultrafast heating and strong confinement achieved by cluster impact, air burns and forms primarily NO as seen in Fig. 35.

The maximum entropy formalism does predict a higher yield of N_2O molecules, Fig. 36. In our molecular dynamics simulation on this system, many of the N_2O molecules formed during the early stage (i.e. during the compression) react during the subsequent expansion of the cluster (see Sec. 3.4 and Fig. 20 in particular). This may be due to the potential that is used in the simulations not being quite realistic as far as N_2O is concerned. It may also be that had we allowed for an expansion stage in the maximum entropy computations, the number of N_2O molecules would have decreased. Certainly, this is the result expected on the basis of the LeChatelier principle. Allowing for an expansion stage is technically simple, however, this will require the introduction of a rate of expansion, that can only be determined from the MD simulations, and our intent was to keep the two methods as much as possible apart.

A very interesting result of the maximum entropy distribution is that at higher velocities of impact, there is a steep onset of copious production of electronically excited and of charged species as shown in Fig. 37. This behavior is also seen experimentally[32,38] and is the reason why an MD simulation involving only ground state species may be realistic only at impact

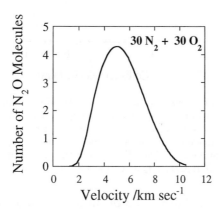

Fig. 36. The yield of N_2O molecules computed by the maximum entropy formalism for a cluster of 30 N_2 and 30 O_2 molecules. The velocity scale is the same as in Fig. 35 top panel. Far fewer N_2O molecules are found in the molecular dynamics simulation, see text.

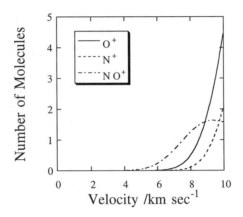

Fig. 37. The yield of charged products computed by the maximum entropy formalism for a cluster of 30 N_2 and 30 O_2 molecules. The velocity scale is the same as in Fig. 35. Note the steep onset for the appearance of charged atoms.

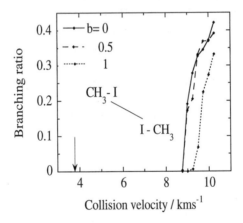

Fig. 38. The probability to exit on the reactive side (of the upper electronic potential energy surface as a function of the relative velocity in a near collinear $CH_3I + CH_3I$ collision, see inset. The results are shown for three impact parameters as indicated. The arrow indicates the nominal energy threshold for accessing the upper electronic surface. Computed by the quantal FMS method.[166] The reactive side includes both the formation of molecular products ($CH_3CH_3 + I_2$ as well as $CH_3 + CH_3 + I_2$ etc.).

velocities below say, 10 km s^{-1}. Since our primary interest is the chemistry under extreme conditions, it appears to put an upper limit on the useful energy range. Even so, the applicable range is quite wide, extending to velocities of impact of about Mach 30.

Full quantum mechanical computations[166] for four-center binary collisions at relative velocities in the 10 km/s range also show a high effective threshold with a steep post threshold rise, Fig. 38. Below the threshold, the probability of a nonadiabatic process is not zero but it is exponentially small. As for ground state four-center reactions, Sec. 3.2.1, here too there is a kinematic constraint. The total energy is conserved. If the ground and excited potential energy surfaces do not intersect then there is a change in the potential energy upon electronic excitation. This is necessarily accompanied by a change in the kinetic energy of the nuclei. Such a change is penalized by a low Franck–Condon factor. (The, so called, exponential gap rule.[88]) Past the effective threshold, the energy is so high that the fractional change in the momenta of the nuclei is small. Hence the small contribution of crossing to the excited state below the effective threshold and the fast rise after.

Concerted four-center reactions are expected to have a high energetic barrier due to an unfavorable orbital correlation. Our original suggestions for cluster impact induced chemistry have been that such nominally "forbidden" reactions could be thermally driven by cluster impact. The high barrier for such reactions is due to a curve crossing between the state that originates in the ground state reactants and correlates to the electronically excited products and the state that originates in the electronically excited reactants and correlates to the ground state products. These reactions offer therefore a good example where there will be a failure of the adiabatic approximation and the dynamics will follow a diabatic correlation.

The computations in Fig. 38 show that for vibrationally cold reactants there is a steep onset for crossing to the upper surface in the velocity range 9–10 km/s. The (avoided) crossing region is "late" and requires extended bonds. Hence, and as also shown by the computations,[166] the crossing is favored by vibrational excitation.

Designing a cluster that, upon impact, favors electronic excitation is of interest because, so far, all the experiments for detecting visible/UV light emission from impact heated clusters have, as far as we know, failed. (Possibly, the phenomenon of shuttle afterglow[3] can be considered an exception to this statement). It could be that any electronically excited states that are formed are very effectively quenched by the coated surfaces that are used as targets. It is however of some concern that the lowest energy indication of nonadiabaticity is electron emission. Of course, this emission could be due to the population of highly promoted (nearly united atom[167]) orbitals in close collisions. In that case there will not be a whole set of electronically excited states that are formed enroute to ionization. The only conflicting result is that we have clear experimental evidence that the velocity distribution of the emitted electrons is characterized by a temperature that is similar to the *translational* temperature of the ions formed upon shattering of the cluster. This suggests an equilibration of the electronic and translational degrees of freedom. Yet, so far, experimental evidence for excitation of bound, excited, electronic states is lacking.

High available energy is only one reason for the new chemistry that is made possible in the hot and compressed cluster. The simulations show that equally important is that collisions in the cluster necessarily occur with rather low impact parameters. (Two molecules moving relative to one another with a high impact parameter will collide with other molecules before they can collide with each other.) The same is true for nonadiabatic transitions. The computed probability of crossing to the upper electronic state decreases rapidly with increasing impact parameter. It is because the cluster favors low impact parameter collisions that the yield of reactive collisions is high.

7. Summary

Cluster – surface collisions at high velocity were shown to induce chemical reactions (on the computer). The special conditions made possible within the rapidly heated cluster opens a window to a new regime of dynamics, where the intermolecular motion is comparable or faster than the intramolecular one. This "sudden" coupling is the key for understanding the special dynamics inside the superhot and superdense cluster. Despite the high density inside the cluster a short time after the collision, the dynamics is more like in the gas-phase than in a condensed one since the duration of each collision is shorter than the time between successive collisions, so that each collision is essentially isolated. That the dynamics is like in the gas-phase does not mean that the surroundings do not play a role. Quite the contrary, the role of the cluster atoms is essential in inducing the reactions. Not just by creating the necessary energetic conditions for the reactions to take place, but also by activating the reactants and deactivating the products. When the environment is part of the reactive system, like in the case of the liquid air clusters, new reaction dynamics can be found through multicenter mechanisms.

A short time after the collision the cluster fragments. Fast translational thermalization precedes this fragmentation process. This ultrafast energy dissipation is shorter than the rate of expansion of the cluster. This phenomenon is demonstrated by the transition from evaporation to shattering as the impact velocity increases. This sharp transition to shattering has the characterization of a phase transition known from macroscopic systems. The system is small enough which enables one to follow each atom during the history of each trajectory, but it is large enough to have a macroscopic characterization. Therefore, impact heated clusters can be an optimal tool to bridge between microscopic to macroscopic systems.

Our research is only the beginning but we believe that the essential conclusions will not qualitatively change.

Epilogue

The existing results obtained from theoretical and computational studies on cluster–surface collisions at various impact velocities are only the beginning of what can be studied using presently available methods. Despite the relatively simple research tools we used until now, our preliminary work has already shown that many unexpected features can arise and opened windows to interesting, unexplored, research areas. We are very gratified that in the short time since our work began there are already published definitive experimental studies that show that cluster impact chemistry is a viable real world possibility. The purpose of this review is not only to survey what we have done but to encourage additional experimental studies and theoretical explorations of this new regime of dynamics.

Acknowledgments

We thank Uzi Even for many discussions and encouragement. We had the benefits of discussions with many colleagues, in particular Michal Ben-Nun, Wolfgang Christen, Christoph Gebhardt, Joshua Jortner, Tamotsu Kondow, Karl Kompa, Israel Schek, Hartmut Schroeder, Akira Terasaki and Hisato Yasumatsu.

Appendix: Complete List of Publications

1. Ballistic and Dissociative Rare-Gas Atom-Halogen Molecule Collisions, M. Ben-Nun, T. Raz and R. D. Levine, *Chem. Phys. Lett.* **220**, 291 (1994).

2. Dissociation Dynamics of Diatomic Molecules Embedded in Impact Heated Rare Gas Clusters, T. Raz, I. Schek, M. Ben-Nun, U. Even, J. Jortner and R. D. Levine, *J. Chem. Phys.* **101**, 8606 (1994).
3. Dissociation of Halogen Molecules in Small and Large Impact Heated Clusters, T. Raz, I. Schek, J. Jortner and R. D. Levine, Yamada Conference XLIII on Structure and Dynamics of Clusters, 1995.
4. Kinematic Model for Four-Center, AB + CD, Reactions, T. Raz and R. D. Levine, *Chem. Phys. Lett.* **226**, 47 (1994).
5. Four-Center Reaction Induced by Cluster Impact, T. Raz and R. D. Levine, *J. Am. Chem. Soc.* **116**, 11167 (1994).
6. Four-Center Reactions: A Computational Study of Collisional Activation, Converted Bond Switching, and Collisional Stabilization in Impact Heated Clusters, T. Raz and R. D. Levine, *J. Phys. Chem.* **99**, 7495 (1995).
7. Concerted vs. Sequential Four-Center Reactions: A Computational Study of High Energy Dynamics, T. Raz and R. D. Levine, *J. Phys. Chem.* **99**, 13713 (1995).
8. Dynamics of Chemical Reactions Induced by Cluster Impact, T. Raz and R. D. Levine, Heidelberg Conference: "Gas-Phase Reaction Systems: Experiments and Models — 100 Years after Max Bodenstein," Springer-Verlag (1995).
9. Fragment Size Distribution in Cluster Impact: Shattering vs. Evaporation by a Statistical Approach, T. Raz, U. Even and R. D. Levine, *J. Chem. Phys.* **103**, 5394 (1995).
10. Shattering of Clusters Upon Surface Impact: An Experimental and Theoretical Study, E. Hendell, U. Even, T. Raz and R. D. Levine, *Phys. Rev. Lett.* **75**, 2670 (1995).
11. On the Shattering of Clusters by Surface Impact Heating, T. Raz and R. D. Levine, *J. Chem. Phys.* **105**, 8097 (1996).
12. On the Burning of Air, T. Raz and R. D. Levine, *Chem. Phys. Lett.* **246**, 405 (1995).
13. Fast Translational Thermalization of Extreme Disequilibrium Induced by Cluster Impact, T. Raz and R. D. Levine, *Chem. Phys.* **213**, 263 (1996).
14. Cluster-Surface Impact Dissociation of Halogen Molecules in Large Inert Gas Clusters, I. Scheck, J. Jortner, T. Raz and R. D. Levine, *Chem. Phys. Lett.* **257**, 273 (1996).

15. The Transition from Recoil to Shattering in Cluster-Surface Impact. An Experimental and Computational Study, W. Christen, U. Even, T. Raz and R. D. Levine, *Int. J. Mass Spectrom. Ion Proc.* **174**, 35 (1998).

16. Collisional Energy Loss in Cluster Surface Impact: Experimental, Model and Simulation Studies of Some Relevant Factors, W. Christen, U. Even, T. Raz and R. D. Levine, *J. Chem. Phys.* **108**, 10202 (1998).

17. The Shattering of Clusters Upon Surface Impact, U. Even, T. Kondow, R. D. Levine and T. Raz, *Comm. Atom. Mol. Phys.* **1D**, 14 (1999).

References

1. G. C. Pimentel, *Opportunities in Chemistry* (National Academy Press, Washington, 1987).
2. C. Park, *J. Thermophys. Heat Trans.* **7**, 385 (1993).
3. E. Murad, *Ann. Rev. Phys. Chem.* **49**, 73 (1998).
4. *Shock Wave in Chemistry*, Ed. A. Lifshitz (Dekker, New York, 1981).
5. W. C. Gardiner, Jr., *Combustion Chemistry* (Springer, Berlin, 1984).
6. W. Tsang and A. Lifshitz, *Ann. Rev. Phys. Chem.* **41**, 559 (1990).
7. J. Sakata and C. A. Wight, *J. Phys. Chem.* **99**, 6584 (1995).
8. E. Grunwald, D. F. Dever and P. M. Keehn, *Megawatt Infrared Laser Chemistry* (Wiley, New York, 1978).
9. V. N. Kondratiev and E. E. Nikitin, *Gas-Phase Reaction* (Springer, Berlin, 1981).
10. *Nonlinear Laser Chemistry with Multiple Photon Excitation*, Ed. V. S. Letokhov (Springer, Berlin, 1983).
11. G. W. Flynn and R. E. J. Weston, *Ann. Rev. Phys. Chem.* **37**, 551 (1986).
12. J. E. Willard, *Ann. Rev. Phys. Chem.* **6**, 141 (1955).
13. R. Wolfgang, *Ann. Rev. Phys. Chem.* **6**, 15 (1965).
14. A. Henglein, *Ultrasonic* **25**, 6 (1987).
15. P. U. Andersson and J. B. C. Pettersson, *Z. Phys.* **D41**, 57 (1997).
16. R. D. Beck, P. S. John, M. L. Homer and R. L. Whetten, *Chem. Phys. Lett.* **187**, 122 (1991).
17. R. D. Beck, P. Weis, G. Brauchle and J. Rockenberger, *Rev. Sci. Instr.* **66**, 4188 (1995).
18. R. D. Beck, J. Rockenberger, P. Weis and M. M. Kappes, *J. Chem. Phys.* **104**, 3638 (1996).
19. M. Benslimane, M. Chatelet, A. D. Martino, F. Pradere and H. Vach, *Chem. Phys. Lett.* **237**, 323 (1995).
20. E. R. Bernstein, *J. Phys. Chem.* **96**, 10105 (1992).
21. R. J. Beuhler, *J. Appl. Phys.* **54**, 4118 (1983).

22. R. J. Beuhler and L. Friedman, *Chem. Rev.* **86**, 521 (1986).
23. A. W. Castleman, Jr. and K. R. Bowen, Jr. *J. Phys. Chem.* **100**, 12911 (1996).
24. M. Chatelet, A. D. Martino, J. Pettersson, F. Pradere and H. Vach, *Chem. Phys. Lett.* **196**, 563 (1992).
25. M. Chatelet, A. D. Martino, J. Petterson, F. Pradere and H. Vach, *J. Chem. Phys.* **103**, 1972 (1995).
26. S. Chen, X. Hong, J. R. Hill and D. D. Dlott, *J. Phys. Chem.* **99**, 4525 (1995).
27. W. Christen, K. L. Kompa, H. Schroder and H. Stulpnagel, *Ber. Bunsenges. Phys. Chem.* **96**, 1197 (1992).
28. W. Christen, U. Even, T. Raz and R. D. Levine, *Int. J. Mass Spectrom. Ion Proc.* **174**, 35 (1998).
29. W. Christen, U. Even, T. Raz and R. D. Levine, *J. Chem. Phys.* **108**, 10202 (1998).
30. W. Christen and U. Even, *J. Phys. Chem.* **102**, 9420 (1998).
31. W. Christen, U. Even, T. Raz and R. D. Levine, *J. Chem. Phys.* **108**, 10202 (1998).
32. U. Even, P. J. d. Lange, H. T. Jonkman and J. Kommandeur, *Phys. Rev. Lett.* **56**, 965 (1986).
33. U. Even, T. Kondow, R. D. Levine and T. Raz, *Com. Atom Mol. Phys.* **1D**, 14 (1999).
34. Y. Fukuda, Ph.D. Thesis, University of Tokyo (1998).
35. J. Gspann and G. Krieg, *J. Chem. Phys.* **61**, 4037 (1974).
36. M. Gupta, E. A. Walters and N. C. Blais, *J. Chem. Phys.* **104**, 100 (1996).
37. E. Hendell, U. Even, T. Raz and R. D. Levine, *Phys. Rev. Lett.* **75**, 2670 (1995).
38. E. Hendell and U. Even, *J. Chem. Phys.* **103**, 9045 (1995).
39. P. M. S. John and R. L. Whetten, *Chem. Phys. Lett.* **196**, 330 (1992).
40. H. Kishi and T. Fujii, *J. Phys. Chem.* **99**, 11153 (1995).
41. C. Lifshitz, in *Cluster Ions*, Eds. C. Y. Ng, T. Baer and I. Powis (Wiley, New York, 1993).
42. T. Lill, H.-G. Busmann, B. Reif and I. V. Hertel, *App. Phys.* **A55**, 461 (1992).
43. T. Lill, F. Lacher, H.-G. Busmann and I. V. Hertel, *Phys. Rev. Lett.* **71**, 3383 (1993).
44. T. Lill, H.-G. Busman, B. Reif and I. V. Hertel, *Surf. Sci.* **312**, 124 (1994).
45. J. A. Niesse, J. N. Beauregard and H. R. Mayne, *J. Phys. Chem.* **98**, 8600 (1994).
46. J. B. C. Pettersson and N. Markovic, *Chem. Phys. Lett.* **201**, 421 (1993).
47. M. Svanberg and J. B. C. Pettersson, *Chem. Phys. Lett.* **263**, 661 (1996).
48. M. Svanberg, N. Markovic and J. B. C. Pettersson, *Chem. Phys.* **220**, 137 (1997).

49. H. Tanaka, T. Hanmura, S. Nonose and T. Kondow, in *Similarities and Difference between Atomic Nuclei and Clusters*, Eds. A. Abe and Y. Lee (The American Institute of Physics, 1998), p. 193.
50. A. Terasaki, T. Tsukuda, H. Yasumatsu, T. Sugai and T. Kondow, *J. Chem. Phys.* **104**, 1387 (1996).
51. A. Terasaki, H. Yamaguchi, H. Yasumatsu and T. Kondow, *Chem. Phys. Lett.* **262**, 269 (1996).
52. A. Terasaki, H. Yasumatsu, Y. Fukuda, H. Yamaguchi and T. Kondow, *Riken Rev.* **17**, 1 (1998).
53. H. Vach, A. D. Martino, M. Benslimane, M. Chatelet and F. Pradere, *J. Chem. Phys.* **100**, 8526 (1994).
54. H. Vach, M. Benslimane, M. Chatelet, A. D. Martino and F. Pradere, *J. Chem. Phys.* **103**, 1 (1995).
55. V. Vorsa, P. J. Campagnola, S. Nandi, M. Larsson and W. C. Lineberger, *J. Chem. Phys.* **105**, 2298 (1996).
56. A. A. Vostrikov, D. Y. Dubovend and M. R. Pretechenskiy, *Chem. Phys. Lett.* **139**, 124 (1987).
57. A. A. Vostrikov and D. Y. Dubov, *Z. Phys.* **D20**, 61 (1991).
58. C. Walther, G. Dietrich, M. Lindinger, K. Luetzenkirchen, L. Schweikhard and J. Ziegler, *Chem. Phys. Lett.* **256**, 77 (1996).
59. H. Weidele, D. Kreisle, E. Recknagel, G. S. Icking-Konert, H. Handschuh, G. Gantefor and W. Eberhardt, *Chem. Phys. Lett.* **237**, 425 (1995).
60. R. Wörgötter, B. Dunser, P. Scheier, T. D. Mark, M. Foltin, C. E. Klots, J. Laskin and C. Lifshitz, *J. Chem. Phys.* **104**, 1225 (1996).
61. R. Wörgötter, J. Kubitsa, J. Zabka, Z. Dolejsek, T. D. Märk and Z. Herman, *Int. J. Mass Spectrom. Ion Proc.* **174**, 53 (1998).
62. G. Q. Xu, R. J. Holland and S. L. Bernasek, *J. Chem. Phys.* **90**, 3831 (1989).
63. H. Yasumatsu, S. Koizumi, A. Terasaki and T. Kondow, *J. Chem. Phys.* **105**, 9509 (1996).
64. H. Yasumatsu, A. Terasaki and T. Kondow, *J. Chem. Phys.* **106**, 3806 (1997).
65. H. Yasumatsu, S. Koizumi, A. Terasaki and T. Kondow, *J. Phys. Chem.* **A102**, 9581 (1998).
66. C. Yeretzian, R. D. Beck and R. L. Whetten, *Int. J. Mass Spectrom. Ion Proc.* **135**, 79 (1994).
67. R. Alimi, R. B. Gerber and V. A. Apkarian, *J. Chem. Phys.* **92**, 3551 (1990).
68. F. Amar and B. J. Berne, *J. Phys. Chem.* **88**, 6720 (1984).
69. M. B. Andersson and J. B. C. Pettersson, *J. Chem. Phys.* **102**, 4239 (1995).
70. R. S. Berry, *J. Phys. Chem.* **98**, 6910 (1994).
71. E. E. B. Campbell, T. Raz and R. D. Levine, *Chem. Phys. Lett.* **253**, 261 (1996).
72. C. L. Cleveland and U. Landman, *Science* **257**, 355 (1992).
73. U. Even, I. Schek and J. Jortner, *Chem. Phys. Lett.* **202**, 303 (1993).
74. F. A. Gianturco, E. Buonomo, G. Delgado-Barrio, S. Miret-Artes and P. Villarreal, *Z. Phys.* **D35**, 115 (1995).

75. M. R. Hoare and J. A. McInnes, *Adv. Phys.* **32**, 791 (1983).
76. C. E. Klots, *Z. Phys.* **D20**, 105 (1991).
77. L. Liu and H. Guo, *Chem. Phys.* **205**, 179 (1995).
78. N. Markovic and J. B. C. Pettersson, *J. Chem. Phys.* **100**, 3911 (1994).
79. L. Ming, N. Markovic, M. Svanberg and J. B. C. Pettersson, *J. Phys. Chem.* **A101**, 4011 (1997).
80. L. Qi and S. B. Sinnott, *J. Phys. Chem.* **B1997**, 6883 (1997).
81. T. Raz and R. D. Levine, *J. Am. Chem. Soc.* **116**, 11167 (1994).
82. T. Raz and R. D. Levine, *J. Phys. Chem.* **99**, 7495 (1995).
83. T. Raz and R. D. Levine, *J. Chem. Phys.* **105**, 8097 (1996).
84. I. Schek, T. Raz, R. D. Levine and J. Jortner, *J. Chem. Phys.* **101**, 8596 (1994).
85. I. Schek and J. Jortner, *J. Chem. Phys.* **104**, 4337 (1996).
86. M. Svanberg, N. Markovic and J. B. C. Pettersson, *Chem. Phys.* **201**, 473 (1995).
87. G. Q. Xu, S. L. Bernasek and J. C. Tully, *J. Chem. Phys.* **88**, 3376 (1988).
88. R. D. Levine and R. B. Bernstein, *Molecular Reaction Dynamics and Chemical Reactivity* (Oxford University Press, New York, 1987).
89. R. D. Levine and A. Ben-Shaul, in *Chemical and Biochemical Applications of Lasers*, Vol. 2, Ed. C. B. Moore (1977).
90. R. D. Levine, in *Theory of Reactive Collisions*, Ed. M. Baer (CRC Press, 1984).
91. T. Raz and R. D. Levine, *Chem. Phys. Lett.* **246**, 405 (1995).
92. M. Ben-Nun, T. Raz and R. D. Levine, *Chem. Phys. Lett.* **220**, 291 (1994).
93. T. Raz, I. Schek, M. Ben-Nun, U. Even, J. Jortner and R. D. Levine, *J. Chem. Phys.* **101**, 8606 (1994).
94. T. Raz and R. D. Levine, *J. Phys. Chem.* **99**, 13713 (1995).
95. T. Raz and R. D. Levine, in *Heidelberg Conference: Gas-Phase Reaction Systems: Experiments and Models — 100 Years after Max Bodenstein* (1995).
96. M. P. Allen and D. J. Tildesley, *Computer Simulations of Liquids* (Clarendon Press, Oxford, 1987).
97. T. Raz and R. D. Levine, *Chem. Phys. Lett.* **226**, 47 (1994).
98. W. Altar and H. Eyring, *J. Chem. Phys.* **4**, 661 (1936).
99. S. Glasstone, K. J. Laidler and H. Eyring, *The Theory of Rate Processes* (McGraw-Hill, New York, 1944).
100. E. K. Grimmelmann, J. C. Tully and M. J. Cardillo, *J. Chem. Phys.* **72**, 1039 (1980).
101. T. Raz, I. Schek, M. Ben-Nun, U. Even, J. Jortner and R. D. Levine, *J. Chem. Phys.* **101** (1994).
102. R. Hoffmann, *J. Chem. Phys.* **49**, 3739 (1968).
103. H. Johnston and J. Birks, *Acc. Chem. Res.* **5**, 327 (1972).
104. H. O. Pritchard, H. Johnston and J. Birks, *Acc. Chem. Res.* **9**, 99 (1976).
105. J. A. Kerr and S. J. Moss, *CRC Handbook of Bimolecular and Termolecular Gas Reactions* (CRC, Boca Raton, 1981).

106. I. W. M. Smith, *Kinetic and Dynamics of Elementary Gas Reactions* (Butterworths, London, 1980).
107. J. H. Kiefer and J. C. Hajduk, *Chem. Phys.* **38**, 329 (1979).
108. R. B. Gerber and A. Amirav, *J. Phys. Chem.* **90**, 4483 (1986).
109. H. C. Andersen, D. Chandler and J. D. Weeks, *Adv. Chem. Phys.* **34**, 105 (1976).
110. J. B. Anderson, *J. Chem. Phys.* **100**, 4253 (1994).
111. S. H. Bauer, *Science* **146**, 1045 (1964).
112. S. H. Bauer, *Ann. Rev. Phys. Chem.* **30**, 271 (1979).
113. J. M. Bowman and G. C. Schatz, *Ann. Rev. Phys. Chem.* **46**, 169 (1995).
114. D. C. Clary, *J. Phys. Chem.* **98**, 10678 (1994).
115. J. Jaffe and S. B. Anderson, *J. Chem. Phys.* **49**, 2859 (1968).
116. L. M. Raff, L. Stivers, R. N. Porter, D. L. Thompson and L. B. Sims, *J. Chem. Phys.* **52**, 3449 (1970).
117. L. M. Raff, D. L. Thompson, L. B. Sims and R. N. Porter, *J. Chem. Phys.* **56**, 5998 (1972).
118. R. J. Pearson, *Symmetry Rules for Chemical Reactions* (Wiley, New York, 1984).
119. P. J. Kuntz, M. H. Mok and J. C. Polanyi, *J. Chem. Phys.* **50**, 4623 (1969).
120. R. D. Levine, *Quantum Mechanics of Molecualar Rate Processes* (Clarendon Press, Oxford, 1969).
121. B. H. Mahan, *J. Chem. Phys.* **52**, 5221 (1970).
122. M. T. Marron, *J. Chem. Phys.* **58**, 153 (1973).
123. M. G. Prisant, C. T. Rettner and R. N. Zare, *J. Chem. Phys.* **81**, 2699 (1984).
124. S. A. Safron, *J. Phys. Chem.* **89**, 5713 (1985).
125. I. Schechter, M. G. Prisant and R. D. Levine, *J. Phys. Chem.* **91**, 5472 (1987).
126. I. Schechter and R. D. Levine, *J. Chem. Soc. Faraday Trans. 2* **85**, 1059 (1989).
127. F. T. Smith, *J. Chem. Phys.* **31**, 1352 (1959).
128. J. O. Hirschfelder, *Int. J. Quant. Chem.* **IIIS**, 17 (1969).
129. S. W. Benson, *Thermochemical Kinetics*, 2nd edition (Wiley, New York, 1976).
130. M. L. Alexander, N. E. Levinger, M. A. Johnson, D. Ray and W. C. Lineberger, *J. Chem. Phys.* **88**, 6200 (1988).
131. M. Gutmann, D. M. Willberg and A. H. Zewail, *J. Chem. Phys.* **97**, 8037 (1992).
132. A. L. Harris, J. K. Brown and C. B. Harris, *Ann. Rev. Phys. Chem.* **39**, 341 (1988).
133. J. M. Papanikolas, J. R. Gord, N. E. Levinger, D. Ray, V. Vorsa and W. C. Lineberger, *J. Phys. Chem.* **95**, 8028 (1991).
134. J.-M. Philippoz, R. Monot and H. V. D. Bergh, *J. Chem. Phys.* **92**, 288 (1990).

135. E. D. Potter, Q. Liu and A. H. Zewail, *Chem. Phys. Lett.* **200**, 605 (1992).

136. J. Schroeder and J. Troe, *Ann. Rev. Phys. Chem.* **38**, 163 (1987).

137. Y. Yan, R. M. Whitnell, K. R. Wilson and A. H. Zewail, *Chem. Phys. Lett.* **193**, 402 (1992).

138. R. Zadoyan and V. A. Apkarian, *Chem. Phys. Lett.* **206**, 475 (1993).

139. A. Garcia-Vela, P. Villareal and G. Delgado-Barrio, *J. Chem. Phys.* **94**, 7868 (1991).

140. D. Scharf, U. Landman and J. Jortner, *J. Chem. Phys.* **88**, 4273 (1988).

141. L. Perera and F. G. Amar, *J. Chem. Phys.* **90**, 7354 (1989).

142. Z. Li, A. Borrmann and C. C. Martens, *J. Phys. Chem.* **97**, 7234 (1992).

143. D. W. Brenner *et al.*, *Phys. Rev. Lett.* **70**, 2174 (1993).

144. J. Tersoff, *Phys. Rev.* **B37**, 6991 (1988).

145. Cavendish, (1785).

146. Priestley, (1779).

147. J. Echave and D. C. Clary, *J. Chem. Phys.* **100**, 402 (1994).

148. T. Raz and R. D. Levine, *Chem. Phys.* **213**, 263 (1996).

149. H. Vach, A. D. Martino, M. Benslimane, M. Chatelet and F. Pradere, *J. Chem. Phys.* **100**, 8526 (1994).

150. T. Raz, U. Even and R. D. Levine, *J. Chem. Phys.* **103**, 5394 (1995).

151. L. M. Delves, *Nucl. Phys.* **8**, 358 (1958).

152. D. G. Schultz and L. Hanley, *J. Chem. Phys.* **109**, 10976 (1998).

153. J. Silberstein and R. D. Levine, *J. Chem. Phys.* **75**, 5735 (1981).

154. T. Raz, U. Even and R. D. Levine, **103**, 5394 (1995).

155. M. R. Hoare, *Adv. Chem. Phys.* **40**, 49 (1979).

156. J. Kostrowicki, L. Piela, B. J. Cherayil and H. A. Scheraga, *J. Phys. Chem.* **95**, 4113 (1991).

157. C. J. Tsai and K. D. Jordan, *J. Phys. Chem.* **97**, 11227 (1993).

158. S. Weerasinghe and F. G. Amar, *J. Chem. Phys.* **98**, 4967 (1993).

159. G. Chartrand, *Introductory to Graph Theory* (Dover, New York, 1977).

160. F. Harary, *Trans. Am. Math. Soc.* **78**, 445 (1955).

161. R. J. Riddel and G. E. Uhlenbeck, *J. Chem. Phys.* **21**, 2056 (1953).

162. G. E. Uhlnbeck and G. W. Ford, *Studies in Statistical Mechanics*, Eds. J. De Boer and G. E. Uhlenbeck (Interscience, New York, 1962).

163. D. R. Herschbach, H. S. Johnston and D. Rapp, *J. Chem. Phys.* **31**, 1652 (1959).

164. H. A. Bethe, *Rev. Mod. Phys.* **9**, 69 (1937).

165. *JANAF Thermochemical Tables*, 2nd edition (NBS, Washington, 1971).

166. M. Chajia and R. D. Levine, *Chem. Phys. Lett.* **304**, 385 (1999).

167. W. Lichten, *Phys. Rev.* **164**, 131 (1967).

CHAPTER 3

NONEQUILIBRIUM CHEMISTRY MODELING
IN RAREFIED HYPERSONIC FLOWS

Iain D. Boyd

*Department of Aerospace Engineering,
University of Michigan, Ann Arbor, Michigan, USA*

Contents

1. Introduction

The hypersonic regime in the Earth's atmosphere is generally characterized
by speeds ranging from 3 km/s, typical of a ballistic missile, to 8 km/s or
higher for a reentering spacecraft. The rarefied flow regime lies between the
continuum and free molecular limits. In continuum conditions, the number of collisions experienced by each molecule or atom in the gas within
a characteristic time is sufficiently large that the velocity distribution is
only perturbed slightly from the equilibrium Maxwellian form. In the free
molecular limit, there are essentially no intermolecular collisions and only
gas–surface interactions and other geometric effects will lead to changes in
macroscopic flow properties. A convenient measure of the collisional state
of a gas flow is the Knudsen number which is the ratio between the average
distance between collisions (the mean free path, λ) and the characteristic
length of the flow (D):

$$\text{Kn} = \frac{\lambda}{D}. \tag{1}$$

It is generally accepted that the rarefied transitional flow regime lies in
the range of $0.01 < \text{Kn} < 10$. In this regime, the role of intermolecular
collisions dictates the overall flow properties significantly. Let us consider
NASA's Space Transportation System (the Space Shuttle) as an example.
In orbit, at 300 km above the Earth's surface, using a characteristic length
scale of 20 m (the vehicle wing span), $\text{Kn} \approx 100$ and this is clearly in
the free molecular limit. Intermolecular collisions have essentially no effect
on the flight of the vehicle in its orbit around the Earth. On the other
hand, at sea level, $\text{Kn} \approx 10^{-9}$, and this is in a regime where a continuum
description of fluid flow will be accurate. During a reentry trajectory, there
will be a portion of the flow that lies in the rarefied regime, and for the
Space Shuttle, this occurs at around 90 to 130 km.

Under hypersonic continuum flow conditions, a strong shock wave forms
in front of the nose of the reentry vehicle. A typical reentry velocity at high
altitude is 7 km/s (about Mach 25). The post-shock temperature for this
Mach number, based on theoretical gas dynamics,[1] is about 15 000 K and
the peak temperature inside the shock is even higher (about 25 000 K).

Under these conditions, both the rotational and vibrational modes of the molecular air species are excited significantly. If we consider a collision between a free stream molecule of air (either N_2 or O_2) traveling at 7 km/s and that of another air molecule in the high temperature part of the shock (with a peak thermal velocity of about 4 km/s) then the resulting collision energy can exceed 11 eV. This is sufficient to dissociate both molecular nitrogen and oxygen. This energy is also enough to produce a small degree of ionization through the formation of ions such as NO^+.

In the rarefied flow regime, all of these collisional mechanisms can also occur. Due to the reduced number of intermolecular collisions, however, all of the energy distributions may be in a state of nonequilibrium. This means that the rates of the kinetic processes cannot be described by the temperature-dependent forms usually employed in continuum models. Instead, models are required that describe these processes at the individual collision level.

Close to the surface of the vehicle, the post-shock temperature decreases rapidly to the surface temperature in a thin layer. Due to the presence of oxygen and nitrogen atoms in the dissociated flow, there is a strong possibility of surface recombination. These processes occur in the rarefied regime also, but here the influence of the gas–surface scattering is significant since the reflected molecules will travel relatively large distances into the flow field before undergoing an intermolecular collision (since the mean free path is large). In other words, the domain of influence of the surface is relatively large in the rarefied flow regime.

In this article, we begin by reviewing the variety of problems associated with laboratory and flight investigations of hypersonic rarefied flows. Then a complete review of computational methodologies is provided. This includes both continuum and particle methods. The focus is on particle methods and details are provided on general concepts. Then, several of the most commonly employed chemistry models are described in detail. There are two components to any chemistry model: the rate of chemical reaction, and the mechanics of chemical reaction. These are discussed separately. Models for gas–surface interaction and hybrid methods that use both continuum and particle descriptions of the gas flow are also briefly reviewed. Results are reviewed for hypersonic conditions in air applicable to the Space Shuttle and to ballistic missiles where both dissociation and exchange reactions are important. The behavior of rarefied flow chemistry models is first considered in a test cell environment. Then, the models are applied to the conditions of a hypersonic flight experiment and comparison is made in detail between the flight measurements and the computational data.

2. Laboratory and Flight Investigations

The investigation of rarefied, hypersonic flows using experimental methods faces several difficulties. In particular, the generation of flight conditions in the laboratory poses two major problems. Firstly, simulation of the low density environment requires special vacuum chambers. Secondly, the production of the large specific enthalpy associated with high speed flow is extremely difficult and expensive. There are a number of facilities around the world that have been used for laboratory studies of rarefied, hypersonic flows. These fall into two classes. Firstly, there are a number of low enthalpy wind tunnels that can produce Mach numbers of interest, but at total temperatures in the range of 300 to 2000 K.[2,3] These facilities are useful for investigating aerodynamics under frozen flow conditions. They are not capable of generating high temperature phenomena of interest such as vibrational excitation and chemical reactions which occur in flight where the total temperature exceeds 10 000 K. Secondly, there are high enthalpy facilities that actually generate chemically reacting flows.[4] However, the densities associated with these facilities place the flow in the continuum regime (Kn < 0.01). It is perhaps surprising to note that there exists a need for development of experimental facilities to generate flows in thermochemical nonequilibrium within the rarefied, hypersonic regime.

When appropriate flow conditions are generated in the laboratory, there are a number of diagnostics methods suitable for diagnosing rarefied, hypersonic flows. Surface measurements such as pressure, shear stress, and heat transfer may readily be made on vehicle models. Intrusive diagnostics are difficult to apply to rarefied flow fields with the main exception being the Patterson probe (see for example in Ref. 3) which is useful for very low density regimes. Flows may be probed nonintrusively using spectroscopic diagnostics such as the electron beam technique[5,6] to obtain density, temperature, and velocity data. However, it should be noted that the electron beam has only been successfully applied in low enthalpy flow conditions. Application to a flow consisting of a chemically reacting mixture significantly increases the difficulties associated with interpretation of the signal detected.

In light of the difficulties associated with ground testing, several flight experiments have been flown to gather data for model and vehicle development. Clearly, such experiments are extremely expensive. A good review of existing data is provided by Park.[7] Some of the more notable flight data include measurement of the aerodynamic characteristics of the Space Shuttle,[8] radiation intensity and spectra,[9-11] vehicle radiative heating,[12] and plasma parameters.[13] Further flight tests are required as new hypersonic regimes become of interest.

3. Computational Modeling

In the absence of any significant body of experimental data for rarefied, hypersonic flows, vehicle design has depended greatly on computational modeling. In this section, we briefly review continuum approaches for computing rarefied gas flows. We then pass on to a more complete description of the methods and models associated with a particle formulation. Hybrid methods that combine continuum and particle methods are reviewed briefly.

3.1. *Continuum Methods*

Under the conditions of interest, it is expected that the velocity distribution functions will be significantly perturbed from the equilibrium Maxwellian form. In principle, the most obvious choice for modeling such flows would be to numerically solve the Boltzmann equation of dilute gas flows.[14] In this integrodifferential equation, the phenomena leading to changes in the velocity distribution function of the gas are analyzed. A primary difficulty in using the Boltzmann equation is that the distribution function is described in phase space (the combination of physical and velocity space). While some progress has been made in the solution of the Boltzmann equation for relatively simple problems,[15,16] there has been no successful computation of chemically reacting flows with this approach.

Using the theory developed by Chapman–Enskog (see Ref. 14), a hierarchy of continuum fluid mechanics formulations may be derived from the Boltzmann equation as perturbations to the Maxwellian velocity distribution function. The first three equation sets are well known: (1) the Euler equations, in which the velocity distribution is exactly the Maxwellian form; (2) the Navier–Stokes equations, which represent a small deviation from Maxwellian and rely on linear expressions for viscosity and thermal conductivity; and (3) the Burnett equations, which include second order derivatives for viscosity and thermal conductivity.

The Euler equations are clearly inappropriate for our flow regime. A significant amount of work has been performed in hypersonic flows using the Navier–Stokes equations.[17,18] It is generally found that these equations can be solved using finite volume methods up to Knudsen numbers of 0.01. At higher Knudsen number, the physical validity of the Navier–Stokes equations is called into question. In addition, as we go further into the rarefied flow regime, it becomes increasingly difficult to obtain converged solutions to the Navier–Stokes equations. At some point, the continuum approach will become prohibitively expensive numerically. The simulation of chemical reactions in the continuum approach relies on temperature dependent rate coefficients. For dissociation reactions, multiple temperature models are employed (for example, the two-temperature model of Park[7]). Since the energy distributions become non-Boltzmann in the rarefied regime, the concept of temperature for any energy mode is ill-defined. Another difficulty for the continuum approach in the rarefied flow regime concerns the removal of activation energy from the internal energy modes. This needs to be performed at the collisional level and this cannot be accomplished in a continuum formulation.

It has been argued that in the higher Knudsen number regime, the Burnett equations will allow continued application of the continuum approach. In practice, many problems have been encountered in the numerical solution and physical properties of the Burnett equations. In particular, it has been demonstrated that these equations violate the second law of thermodynamics.[19] Work on use of the Burnett equations continues,[20] but it appears to be unlikely that this approach will extend our computational capabilities much further into the high Knudsen number regime than that offered by the Navier–Stokes equations.

In addition to the limitations of the continuum approaches in being able to accurately represent transport processes under strongly nonequilibrium conditions, the formulation of physically meaningful boundary conditions may also be problematic. For the Euler equations, the boundary conditions at the vehicle surface must be adiabatic for energy and no slip for momentum. Use of the Navier–Stokes equations allows stipulation of isothermal temperature and slip velocity conditions. However, under strongly nonequilibrium conditions, these boundary conditions will fail to reproduce the physical behavior accurately. The situation for the Burnett equations is even worse since the required boundary conditions must include second order effects.

3.2. *The Direct Simulation Monte Carlo Method*

The direct simulation Monte Carlo method (DSMC) was developed between 1960 and 1980 almost exclusively by Bird.[21] In this method, the large number of real molecules in a gas flow is represented by a much smaller set of model particles. These particles move through physical space and undergo collisions with other particles and with solid boundaries in a manner analogous to the real gas dynamics. The selection of particles that collide is performed statistically and this distinguishes DSMC from the Molecular Dynamics method which handles collisions deterministically. Macroscopic properties such as density and mean velocity are obtained in steady flows by time-averaging particle properties. The general philosophy with the DSMC method is to perform a direct simulation of the gas dynamics rather than to be concerned about whether the method is solving any particular theoretical equation. Nevertheless, it has been demonstrated that the DSMC technique provides solutions that are consistent with the Boltzmann equation.[22]

A key feature of the DSMC technique in comparison to continuum methods is its relatively high computational expense. To allow the decoupling between molecular motion and intermolecular collisions to occur in a physically accurate way, the time-step used in the DSMC technique must be smaller than the mean time between collisions. Similarly, the size of the cells employed in the DSMC computational grid must be of the order of the local mean free path everywhere in the flow domain. These physical restrictions on the size of the numerical parameters results in the time steps and cell sizes employed in DSMC calculations being usually significantly smaller than those employed in continuum computations. For this reason, significant work has been performed in the optimization of the DSMC technique for different types of computer hardware. Examples of specific implementations are described in Refs. 23–26.

Since 1980, a large body of research has been performed using the DSMC technique. Applications include hypersonic flows, spacecraft propulsion systems, materials processing, astrophysics, and flows through micromachines. Recent reviews of the method and applications are provided in Refs. 27–29. It is the purpose of this article to review the status of the DSMC technique specifically in relation to its ability to accurately model the nonequilibrium, chemically reacting flows that are characteristic of rarefied hypersonic conditions.

3.2.1. *Nonreactive Intermolecular Interactions*

The procedures and models employed to simulate intermolecular interactions represent key components in obtaining a physically accurate DSMC computation. Before reviewing in some detail the available models for reacting flows, it is appropriate to briefly discuss the key points of the models employed to simulate nonreacting collision phenomena.

The DSMC method performs collisions by firstly grouping together particles that are likely to collide at each iteration of the method. This is achieved by binning the particles into computational cells that are of the size of a local mean free path. By employing such small cells, it is permissible to ignore the specific locations and trajectories of the particles within the cells in the process of computing collisions. Having binned particles into cells, pairs of particles are selected at random and a probability of collision computed based on the product of their relative velocity and collision cross section. This process is performed for a sufficient number of pairs of particles to ensure that the appropriate number of collisions is simulated. There are a number of different schemes for simulating the collision rate in the DSMC method (for examples, see Refs. 21, 23 and 30). By comparison, there are only a few models for the collision cross section, and the most widely used of these are the variable hard sphere (VHS) of Bird[21] and the variable soft sphere (VSS) of Koura and Matsumoto.[31] If it is decided that two particles collide and that no internal energy exchange or chemical reaction occurs, i.e. the collision is elastic, then the collision mechanics are applied so as to conserve linear momentum and energy in the collision (for details, see Ref. 21).

The inelastic, nonreactive collision events that are included in DSMC computations are energy transfer between the translational collision energy and the internal energy modes due to molecular rotation and vibration, and internal electronic structure. In general, probabilities of these inelastic events are computed based on the energy of each collision. For rotational relaxation, a phenomenological approach is typically employed and general probabilities have been developed by Boyd.[32] In the phenomenological approach, energy is exchanged in the collision using the statistical mechanics scheme of Borgnakke–Larsen[33] that was extended to quantized energy states by Bergemann and Boyd.[34] For vibrational energy exchange, two approaches are adopted. Either an approach analogous to the rotational mode is employed, and a general probability of exchanging energy is computed, e.g. see the works of Boyd[35] and Choquet,[36] or individual state to state transition probabilities are evaluated, e.g. see Gimelshein *et al.*[37] and Koura.[38] By comparison, there has been little work performed on the DSMC modeling of electronic energy exchange. However, this follows a natural extension to the other modes and has been handled both phenomenologically[39] and using detailed kinetics mechanisms.[40] A final important aspect of modeling internal energy relaxation using the DSMC technique concerns the relation between the energy exchange schemes and experimental measurements of relaxation times. This issue has been treated in detail in Refs. 41 and 42.

3.2.2. *Chemistry Modeling*

There are two steps in the computation of chemical reactions using the DSMC technique. In the first step, a reaction probability is evaluated and compared with a random number to determine if the reaction actually occurs. In general, the reaction probability should depend on the translational, rotational, and vibrational energies associated with the collision. In the second step, the mechanics of the reaction must be performed. These tasks are considered separately in the following.

3.2.2.1. *Cross sections*

(a) TCE Model

In the DSMC technique, the probability that a chemical reaction occurs is the ratio of the reaction cross section to the elastic cross section. The most commonly applied chemistry model is the Total Collision Energy (TCE) form employed by Boyd[43] based on a general model proposed by Bird.[21] In this model, the probability of reaction, P, is obtained by integrating the microscopic equilibrium distribution function for the total collision energy, and equating it to a chemical rate coefficient, K_f. Specifically, the mathematical form of the probability is obtained from the following integral:

$$K_f = \langle \sigma g \rangle \int_{\varepsilon_a}^{\infty} P(\varepsilon_c) f_B(\varepsilon_c) d\varepsilon_c, \qquad (2)$$

where σ is the elastic cross section, g is the relative velocity, ε_a is the activation energy of the reaction, and $f_B(\varepsilon_c)$ is the microscopic equilibrium Boltzmann distribution for the total collision energy, ε_c. With application to the reaction of interest here, the total collision energy consists of the translational collision energy, and the sum of the rotational and vibrational energies of the two colliding particles involved in the reaction process. Consider the case where the chemical rate coefficient is given in modified Arrhenius form:

$$K_f = aT^b \exp(-\varepsilon_a/kT),\tag{3}$$

where a and b are constants, and k is the Boltzmann constant. In this case, the reaction probability for the TCE model is:

$$P_{TCE} = A\frac{(\varepsilon_c - \varepsilon_a)^\psi}{(\varepsilon_c)^\chi},\tag{4}$$

where

$$A = \frac{a\epsilon\sqrt{\frac{1}{2}m_r\pi}}{\sigma_{ref}[(2-\omega)kT_{ref}]^\omega k^b}\frac{\Gamma(\zeta+2-\omega)}{\Gamma(2-\omega)\Gamma(\zeta+b+\frac{3}{2})}$$

$$= B\frac{\Gamma(\zeta+2-\omega)}{\Gamma(2-\omega)\Gamma(\zeta+b+\frac{3}{2})}\tag{5a}$$

in which $\epsilon = 1$ for collisions between two particles of the same species, and is $\frac{1}{2}$ otherwise. In addition, ζ is the average number of rotational and vibrational degrees of freedom of the two colliding particles. The parameters ω, σ_{ref}, and T_{ref} are taken from the Variable Hard Sphere collision model,[21] and m_r is the reduced mass of the collision.

The exponents in Eq. (4) are:

$$\psi = b + \frac{1}{2} + \zeta\tag{5b}$$

and

$$\chi = 1 + \zeta - \omega.\tag{5c}$$

The TCE model is highly phenomenological. Under conditions of macroscopic equilibrium, it recovers the rate coefficient, K_f, and maintains equipartition of energy. Its performance at the collision level will be assessed later in this article.

A typical air chemistry mechanism used in DSMC computations is shown in Table 1. Notice that all reactions are described in one direction only. For dissociation reactions, an effective model for recombination has

Table 1. Air chemistry reaction rates (m^3/molecule/s).

	Reaction	Rate coefficient[64]
(1a)	$N_2 + M_D \rightarrow N + N + M_D$	$7.97 \times 10^{-13} T^{-0.5} \exp(-113\ 100/T)$
(1b)	$N_2 + M_A \rightarrow N + N + M_A$	$7.14 \times 10^{-8} T^{-1.5} \exp(-113\ 100/T)$
(2a)	$O_2 + M_D \rightarrow O + O + M_D$	$3.32 \times 10^{-9} T^{-1.5} \exp(-59\ 400/T)$
(2b)	$O_2 + M_A \rightarrow O + O + M_A$	$1.66 \times 10^{-8} T^{-1.5} \exp(-59\ 400/T)$
(3a)	$NO + M_D \rightarrow N + O + M_D$	$8.30 \times 10^{-15} \exp(-75\ 600/T)$
(3b)	$NO + M_A \rightarrow N + O + M_A$	$1.83 \times 10^{-13} \exp(-75\ 600/T)$
(4a)	$O + NO \rightarrow N + O_2$	$1.39 \times 10^{-17} \exp(-19\ 700/T)$
(4b)	$N + O_2 \rightarrow O + NO$	$4.60 \times 10^{-15} T^{-0.55}$
(5a)	$N_2 + O \rightarrow N + NO$	$1.06 \times 10^{-12} T^{-1.0} \exp(-37\ 500/T)$
(5b)	$N + NO \rightarrow N_2 + O$	$4.06 \times 10^{-12} T^{-1.36}$

M_D = diatomic species N_2, O_2, NO; M_A = atomic species N, O.

been described by Boyd.[44] However, because recombination is only an important mechanism at high density, these reactions are generally omitted from DSMC analyses. For exchange reactions (reactions 4 and 5 in Table 1), the backward reaction rates, K_b, are calculated from the forward rate and the equilibrium coefficient, K_{eq}, in the usual way[12]:

$$K_b = \frac{K_{eq}}{K_f}.$$

(6)

For the TCE model, the data for K_b must be curve-fitted to the Arrhenius form of Eq. (3) over the temperature range of interest.

(f) VFD Model

An extension of the TCE model was proposed by Haas and Boyd[45] and Boyd[44] to account for coupling between vibrational energy and collision-induced dissociation. This reaction involves the breaking down of molecular oxygen and nitrogen into atoms, and is the first step in most air chemistry mechanisms. The coupling between a nonequilibrium vibrational energy distribution function and the rate of dissociation is particularly important under rarefied flow conditions. The Vibrationally Favored Dissociation model (VFD) includes an additional dependence of the reaction probability on the vibrational energy of the reactant molecule. This is achieved by including a term in the reaction probability that raises the vibrational energy to some power ϕ. In this case, the macroscopic rate coefficient is obtained as:

$$K_f = \langle \sigma g \rangle \int_{\varepsilon_a}^{\infty} P_c(\varepsilon_c) \int_{\varepsilon_v=0}^{\varepsilon_c} P_v(\varepsilon_v) f_B(\varepsilon_v) f_B(\varepsilon_c - \varepsilon_v) d\varepsilon_v \, d\varepsilon_c \, .$$

For this model, the overall dissociation probability is:

$$P_{VFD} = P_c(\varepsilon_c) P_v(\varepsilon_v) = A'(\varepsilon_v/\varepsilon_c)^{\phi} \frac{(\varepsilon_c - \varepsilon_a)^{\psi}}{(\varepsilon_c)^{\chi}}, \tag{7}$$

where

$$A' = B \frac{\Gamma(\frac{\zeta_v}{2})}{\Gamma(\frac{\zeta_v}{2} + \phi)} \frac{\Gamma(\zeta + 2 - \omega + \phi)}{\Gamma(2 - \omega)\Gamma(\zeta + b + \frac{3}{2})}, \tag{8}$$

where ζ_v is the total number of vibrational degrees of freedom of the dissociating molecule. The parameter ϕ controls the degree of vibration-dissociation coupling such that molecules in higher vibrational energy levels have a higher probability of dissociating. Note that the TCE model is recovered when ϕ is zero. The VFD model has been successfully verified at the macroscopic level through comparison with experimental data both for reacting shock waves of nitrogen,[44] and for homogeneous dissociation of oxygen.[45]

(c) GCE Model

A general form for a DSMC chemistry model was developed by Boyd *et al.*[46] and is termed the Generalized Collision Energy model (GCE). This model affords more control over the dependence of the collision cross section on each of the internal energy modes. Following a similar analytical approach as used in development of the VFD model, the reaction probability is biased separately to the translational, rotational, and vibrational energy modes. The rate coefficient is obtained through the following integration:

$$K_f = \langle \sigma g \rangle \int_{\varepsilon_a}^{\infty} P_c(\varepsilon_c) \int_{\varepsilon_r=0}^{\varepsilon_c} P_r(\varepsilon_r) f_B(\varepsilon_r)$$

$$\times \int_{\varepsilon_{tv}=0}^{\varepsilon_c - \varepsilon_r} P_t(\varepsilon_t) P_v(\varepsilon_v) f_B(\varepsilon_t) f_B(\varepsilon_v) d\varepsilon_{tv} \, d\varepsilon_r \, d\varepsilon_c \, ,$$

where ε_{tv} is the sum of the translational (ε_t) and vibrational (ε_v) energies. The probability of reaction is:

$$P_{GCE} = P_c(\varepsilon_c) P_v(\varepsilon_v) P_r(\varepsilon_r) P_t(\varepsilon_t)$$

$$= A''(\varepsilon_t/\varepsilon_c)^{\alpha}(1 - \varepsilon_r/\varepsilon_c)^{\beta}(\varepsilon_v/\varepsilon_c)^{\gamma} \frac{(\varepsilon_c - \varepsilon_a)^{\psi}}{(\varepsilon_c)^{\chi}}, \tag{9}$$

where

$$A'' = B \frac{\Gamma(\frac{\zeta_v}{2})}{\Gamma(\frac{\zeta_v}{2} + \gamma)} \frac{\Gamma(2 - \omega + \frac{\zeta_v}{2} + \alpha + \gamma)}{\Gamma(2 - \omega + \frac{\zeta_v}{2} + \alpha + \beta + \gamma)} \frac{\Gamma(2 - \omega + \zeta + \alpha + \beta + \gamma)}{\Gamma(2 - \omega + \alpha)\Gamma(\zeta + b + \frac{3}{2})}.$$

(10)

Parameters α, β, and γ control the biasing of the reaction probability to the translational, rotational, and vibrational energy modes respectively. Note that the form of the vibrational energy biasing retains the original VFD model when $\alpha = \beta = 0$ and $\gamma = \phi$. Also, the TCE model is recovered when $\alpha = \beta = \gamma = 0$.

The GCE model is more detailed than the TCE and VFD chemistry models. However, only through the availability of detailed information about the dependence of reaction cross sections on the precollision states of the colliding particles can the parameters α, β, and γ be determined. In the results section, such information for the exchange reaction involving formation of nitric oxide is available from detailed trajectory computations and is used to determine suitable parameters for the GCE model for this reaction.

(d) Threshold Line Model

The threshold line concept was proposed by Macheret and Rich[47] and used to develop a continuum model for dissociation that provides reaction rates as functions of the temperatures associated with the various energy modes. For use in the DSMC technique, Boyd[48] employed the threshold concept to develop a nonequilibrium dissociation model in terms of a dissociation probability based on the translational collision energy and the rotational and vibrational energies of the dissociating molecule.

As described in detail in Ref. 47, the threshold line probability of dissociation has distinct mathematical forms for low and high vibrational levels. At low vibrational levels, there is insufficient velocity along the line of centers of the two atoms that constitute the molecule to match the translational velocity of the colliding atom. This is the requirement that must be met to allow all of the translational energy of the atom to be transferred into the vibrational energy of the molecule and thus cause dissociation. Hence, at low vibrational energies, the threshold energy required to cause dissociation is greater than the difference between the activation energy and the translational collision energy. As proposed in Ref. 47, the boundary between high and low vibrational levels is determined using the following parameter:

$$\alpha = \left(\frac{m}{m+M}\right)^2, \tag{11}$$

where m is the mass of one of the two atoms in the diatomic molecule that is dissociating, and M is the mass of the collision partner. In the case of diatom–diatom collisions, M is the mass of one of the atoms in the second molecule and therefore α is equal to 0.25 for both N_2–N_2 and N_2–N collisions.

In the following expressions, ε_t is the translational collision energy, and ε_r, and ε_v are the rotational and vibrational energies of the reacting molecule. The vibrational energy distribution is modeled as a simple harmonic oscillator using the approach of Bergemann and Boyd.[34] The threshold line model specifies low vibrational energy levels as those for which $\varepsilon_v < \alpha D^*$ where D^* is a modified activation energy that will be defined later. Using these ideas, it may be shown[48] that the dissociation probability of low vibrational levels for diatom–atom collisions is:

$$P_d = A_i \frac{8\sqrt{2}(\varepsilon_t - F)^{\frac{3}{2}}}{3\pi^2 F[(\sqrt{D^*} - \sqrt{\alpha\varepsilon_v})(\sqrt{D^*} - (2 - \sqrt{\alpha})\sqrt{\varepsilon_v})]^{\frac{1}{2}}}, \tag{12}$$

where A_i is a constant of proportionality for reaction i that does not appear in the continuum formulation. It is necessary here in order to calibrate the DSMC model against specified rate coefficients. In the continuum model, A_i has some unknown functional form that ensures a particular Arrhenius rate coefficient is recovered at macroscopic equilibrium. For low vibrational levels, the threshold energy is given as:

$$F = \frac{(\sqrt{D^*} - \sqrt{\alpha\varepsilon_v})^2}{1 - \alpha}. \tag{13}$$

At high vibrational levels ($\varepsilon_v < \alpha D^*$) the threshold energy is simply:

$$F = D^* - \varepsilon_v. \tag{14}$$

Following the procedures outlined in Ref. 47, the dissociation probability at the high vibrational levels for diatom–atom collisions is derived as:

$$P_d = A_i \sqrt{\frac{1 + \sqrt{\alpha}}{1 - \sqrt{\alpha}}} \frac{(\varepsilon_t - F)^2}{4\pi^2 F^{\frac{3}{2}}\sqrt{\varepsilon_v - \alpha D^*}}, \tag{15}$$

where the constant of proportionality A_i is the same as that in Eq. (12).

The participation of rotational energy in the reaction process is included by proposing an effective activation energy of the form[47]:

$$D^* = D - \varepsilon_r + \frac{2\varepsilon_r^{\frac{3}{2}}}{3\sqrt{6D}}, \tag{16}$$

where D is the dissociation energy. At very high rotational energies, the activation energy is reduced due to the strong centrifugal force induced on the molecule. This aspect of the model reduces the dissociation rate in comparison with the equilibrium value when the rotational temperature is less than the translational value. This effect is much smaller than the coupling with the vibrational energy mode.

As with the continuum model, the particle based threshold line probability must at the macroscopic level recover a specified Arrhenius rate coefficient under equilibrium conditions. Unlike the continuum model, however, the Arrhenius constants a and η do not appear in the model. Instead, the constant of proportionality, A_i, must be obtained through calibration of the model against rate data by performing test simulations under equilibrium conditions. Hence, with this approach, it is not possible to guarantee that the model will produce a specific temperature dependence for the dissociation rate. This could perhaps be achieved through inclusion of further dependencies of the dissociation probability on the various energies involved in the collision.

Wadsworth and Wysong[49] made a detailed assessment of the threshold line model (and other dissociation models) for hydrogen dissociation by making comparison with quasiclassical trajectory computations. They found that the original forms of the threshold line models proposed by Macheret and Rich have to be extended to more complete forms in order to avoid singularities at specific values of the collision energies. Some of their findings are shown in the results section.

(e) Alternative Models

The attraction in the GCE-type of chemistry models is their relative ease of use. Provided that rate data is available in an appropriate Arrhenius form, then it is straightforward to model complex reaction mechanisms. There are, however, several problems with these models and this has led in part to the development of alternative schemes. One of the primary concerns with the GCE approach is whether we can expect models developed partly through mathematical convenience to accurately describe complex collision behavior. Several chemistry models have therefore been proposed which are based primarily on the physics of the reaction processes being simulated. The threshold line model is an example of this type of scheme.

The Maximum Entropy approach of Levine and Bernstein[50] has been developed for application in the DSMC technique by Marriott and Harvey,[51] and Gallis and Harvey[52] over several years. The basic idea in the Maximum Entropy approach is that the distribution of energy of the products of a chemical reaction can be described as a deviation away from microscopic equilibrium. The perturbation of the energy distribution away from microscopic equilibrium is described using a series of maximal parameters, λ_i (which are not related to the mean free path). With this approach, the Borgnakke–Larsen scheme is a subset representing zero departure from microscopic equilibrium obtained with all values of λ_i equal to zero. Using the principle of microscopic reversibility, the chemical reaction probability of the forward reaction is obtained from the probability of an energy state appearing in the products of the backward reaction. This concept ensures that detailed balance is satisfied under conditions of macroscopic equilibrium.

The maximal parameters are determined by performing single-cell simulations at constant temperature, and comparing the simulated reaction rates to available data. Maximum Entropy DSMC mechanisms have been developed for high temperature air and for the chemistry of the Martian atmosphere. The approach has mainly been employed only by its originators because it is not clear that the results obtained are significantly different from those achieved with existing models. This may be a consequence of the nature of the reactions important in hypersonic reacting flows.

It is not possible to consider all of the available DSMC chemistry models in detail. Due to its importance in air chemistry, dissociation has received the most attention, and interesting models have been proposed by Koura,[38] Bird,[53] and Lord.[54] Wadsworth and Wysong compared the threshold line model to those of Refs. 38 and 53. It was concluded that the threshold line model and that of Koura[38] gave results consistent with quasiclassical trajectory calculations for hydrogen dissociation. Some of these data will be shown later in the article.

3.2.2.2. *Energy disposal*

Just as important as simulating the rates of chemical reactions is the manner in which energy is distributed across the modes of the products following the reaction event. This topic has received less attention in comparison to the modeling of reaction cross sections. Essentially, only two techniques exist.

(a) Borgnakke–Larsen Scheme

In its original form,[33] the Borgnakke–Larsen scheme is only suited for use with the TCE cross section model. In this scheme, the total post-reaction energy, that includes any change to the prereaction energy due to chemical activation or deactivation, is distributed across the energy modes of the product particles using statistical energy distribution functions. In particular, the post-reaction translational energy fraction, ε_t, is obtained by sampling from the following distribution using an acceptance-rejection technique:

$$f\left(\frac{\varepsilon_t}{\varepsilon_c}\right) = \frac{\Gamma(\frac{5}{2} - \omega + \zeta)}{\Gamma(\frac{5}{2} - \omega)} \left(\frac{\varepsilon_t}{\varepsilon_c}\right)^{\frac{3}{2} - \omega} \left(1 - \frac{\varepsilon_t}{\varepsilon_c}\right)^{\zeta - 1}. \tag{17}$$

The remaining internal energy is then allocated across the internal modes using the appropriate distribution functions. These modes may be quantized as described by Bergemann and Boyd.[34]

A fundamental law that links the reaction cross section with the reaction mechanics is the principle of microscopic reversibility which says that the probability of the forward reaction in a particular energy state should equal the probability of the reverse reaction in the product energy state of the forward reaction. The combination of the Borgnakke–Larsen scheme and the TCE reaction cross section model are consistent with this principle because both the reaction probability and the energy disposal mechanics are based on the total collision energy. In particular, the TCE model does not contain any biasing towards a specific energy mode. Simple modification of the Borgnakke–Larsen scheme has been employed with the VFD model.[44] In this case, the vibrational energy distribution function for the reverse reaction (recombination) is biased in the same way as the biasing on vibrational energy for the forward reaction (dissociation) cross section. In principle, each different chemical reaction model should use a specific form of post-reaction energy disposal. In practice, most of the DSMC models continue to use the original Borgnakke–Larsen scheme for convenience. In the results section, we will assess this approach for an exchange reaction.

(b) Maximum Entropy Scheme

A notable exception to use of the Borgnakke–Larsen model is the Maximum Entropy model. As described in Sec. 3.2.2.1(e), a key feature of this model is the direct coupling of the reaction probability in the forward direction and the energy disposal for the backward reaction. Thus, the maximal parameters λ_i play a role both in the reaction cross sections, and in the post-reaction energy disposal. In the earlier discussions, it was stated that the values of λ_i for a particular reaction are obtained by comparison of simulation results to experimental measurements of rate data. An alternative approach is to determine λ_i through comparison of simulation results to experimental measurements of product energy distributions. The impressive results of two examples of such a study by Gallis and Harvey[55] are shown in Figs. 1 and 2 where the product vibrational energy distributions are provided for the exchange reactions:

$$Cl + HI \rightarrow I + HCl,$$

and

$$O + CS \rightarrow CO + S.$$

The variable "fv" is the fraction of the total energy going into the vibrational mode in terms of post-collision energy disposal. The equilibrium data

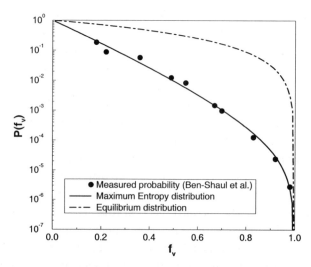

Fig. 1. Vibrational energy distribution of the products of the reaction: Cl+HI → I+HCl (from Gallis and Harvey[55]).

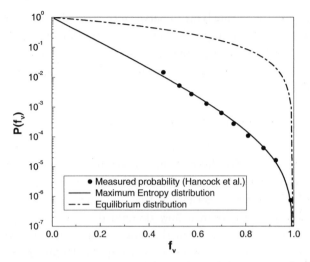

Fig. 2. Vibrational energy distribution of the products of the reaction: $O+CS \rightarrow S+CO$ (from Gallis and Harvey[55]).

shown in Figs. 1 and 2 are the results obtained with the Borgnakke–Larsen scheme. The experimental data are from Refs. 56 and 57, respectively. The Maximum Entropy approach provides an accurate description of the observed distributions since λ comes from the observed deviation from the microscopic equilibrium (maximum entropy) distribution. Unfortunately, experimental data containing this level of detail are very limited.

3.2.3. *Gas–Surface Interactions*

In addition to colliding with other molecules, the particles in a DSMC simulation may interact with solid surfaces. An extensive review of the state of gas–surface interaction modeling is provided by Hurlbut.[58] In almost all DSMC computations, a combination of two ideal limits is employed that may be characterized by surface accommodation coefficients, α. These limits are: (1) specular reflection ($\alpha = 0$) in which reversal of the velocity component normal to the surface is the only change made; and (2) diffuse reflection with full thermal accommodation ($\alpha = 1$), in which new velocity components are sampled from a Maxwellian distribution function at the surface temperature. In reality, the dynamics of particle interaction with real surfaces lies somewhere between these limits and it is common to try and capture such behavior phenomenologically through the use of accommodation coefficients lying in between the specular and diffuse values. In addition, more elaborate models have been developed (e.g. Lord[59]) but these have not yet received widespread use due to the lack of experimental data needed to identify appropriate values of the free parameters required by the model.

3.3. *Hybrid Continuum-Particle Methods*

As a final word on the role of continuum and particle methods for computation of hypersonic rarefied flows, it should be noted that several efforts have been made to combine the optimal aspects of the two techniques.[60-62] The general principle for these methods is that the continuum approach should be used wherever possible in a flow domain due to its superior numerical efficiency. However, regions of the flow field that are in a strong state of nonequilibrium are to be computed using the DSMC method. While ideologically, this is an attractive approach, combining methods based on continuum and particle concepts provides several practical difficulties. For example, it is necessary to be able to determine the physical conditions under which a switch will be made from continuum to particle methods. As an example, the use of a Knudsen number based on local property gradients has been proposed.[63] Another problem involves the communication back and forward between the continuum and particle descriptions of the flow. The data associated with the particle method always contains statistical fluctuations, and these need to be eliminated for communication to the continuum method. The continuum method works with macroscopic quantities such as pressure, temperature, and stream velocity, and these must be related through appropriate distribution functions for use in the DSMC component. All of these difficulties become intensified for a strongly nonequilibrium, chemically reacting flow. This remains an area of ongoing research.

4. Results

To illustrate the present state of chemistry modeling obtainable with the DSMC technique, we consider a particular system in detail. This concerns neutral air chemistry for which a typical kinetics model[64] is shown in Table 1. This neutral species mechanism is appropriate for reentry of the Space Shuttle and for flights of high altitude ballistic missiles.

The underlying goal here is to review the chemistry modeling associated with a particular flight experiment (the second Bow Shock Ultraviolet flight experiment, BSUV-2).[11] This vehicle flew at 5 km/s over an altitude range of about 100 to 60 km and obtained measurements of the ultraviolet emission from nitric oxide and atomic oxygen. The computation of these conditions for prediction of the emission spectra places a great deal of emphasis on the chemistry modeling. At altitudes below about 85 km, atomic oxygen will only be present in significant levels after a degree of dissociation of molecular oxygen has occurred. Above 85 km, photodissociation of oxygen occurs in significant amounts. Over all of the BSUV-2 flight regime, nitric oxide is primarily formed through the first Zeldovich reaction

$$N_2 + O \rightarrow NO + N.$$

So, in order to accurately compute the level of UV emissions measured by BSUV-2 under rarefied flow conditions, it is necessary to employ accurate models for dissociation (primarily of molecular oxygen) and exchange reactions. In addition, it will be found at high altitudes that the modeling of gas–surface interaction becomes important. In summary, the data provided by BSUV-2 tests to a very high level the chemistry models described above for the DSMC technique. The discussions of the DSMC chemistry models begins with consideration of the performance of models in a test cell configuration. Then, consideration is given to computing the conditions relevant to the BSUV-2 experiment.

4.1. *Test Cell Studies*

4.1.1. *Introduction*

A test cell is a convenient, idealized system for investigating in detail the behavior of thermochemistry models employed in the DSMC technique. The system consists of a relatively large number of particles (say 100 000) that are initialized to some state of interest. Without undergoing translational motion or surface collisions, these particles are considered to exist in a single cell and undergo collisions with one another according to the transient conditions. In this way, it is possible to examine macroscopic properties such as overall reaction rate, and microscopic properties such as energy distribution functions. In preparation for the detailed examination of the hypersonic flight experiment, we consider in detail two specific reactions that are important for that system: (1) the dissociation of oxygen; and (2) the first Zeldovich exchange reaction.

In order to evaluate DSMC chemistry models, we require experimental and/or detailed theoretical results. Data of interest that can be measured experimentally include reaction cross sections, and rate coefficients. The most useful type of theoretical data are generated by detailed analysis of the collision and reaction dynamics using potential surfaces obtained from high level quantum chemical methods.

Unfortunately, for oxygen dissociation, the only reliable data is in the form of experimental measurements of the dissociation rate coefficient as a function of temperature. However, for hydrogen dissociation, detailed state-specific data were computed by Martin *et al.*[65] using a quasiclassical trajectory (QCT) approach. These data were compared to three different DSMC chemistry models by Wadsworth and Wysong.[49] One of the conclusions of their study was that the threshold line model gives results that are in very good agreement with the QCT data. Examples of these results are shown in Figs. 3 and 4 for an equilibrium test-cell case where all energy modes have a temperature of 4500 K. In Fig. 3, the vibrational energy distribution of dissociating molecules is shown. There is clearly excellent agreement between the threshold line model and the detailed QCT calculations. In Fig. 4, the rotational energy distribution is shown for two different

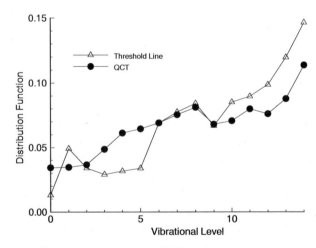

Fig. 3. Vibrational energy distribution of dissociating hydrogen molecules under conditions of thermal equilibrium (from Wadsworth and Wysong[43]).

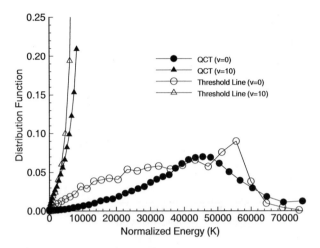

Fig. 4. Rotational energy distribution of dissociating hydrogen molecules under conditions of thermal equilibrium (from Wadsworth and Wysong[43]).

vibrational states. Once again, the DSMC model provides excellent agreement with the theoretical computations. It should be noted that Koura's model[38] was also found to give good agreement with the QCT data in the study of Wadsworth and Wysong.

4.1.2. *Oxygen Dissociation*

Here, we will compare the threshold line model with the TCE chemistry model. These two DSMC chemistry models were used to calculate dissociation rate coefficients under conditions where the translational, rotational, and vibrational modes are in thermal equilibrium with Boltzmann distributions. An isothermal heat bath is simulated that consists of 100 000 particles. Energy is exchanged between the various modes during collision, but chemical reactions are not processed. Instead, the average dissociation probability is evaluated over all collisions and then converted into a rate coefficient. Results for O_2–O_2 dissociation are shown in Fig. 5. The measured rate is that reported by Byron[66]:

$$r_{O_2\text{-}O_2} = 3.32 \times 10^{-3} T^{-1.5} \exp(-59\ 400/T),$$

where the units are cm^3/mol/s. The threshold line model is calibrated at a temperature of 10 000 K giving a leading constant in Eqs. (12) and (15) of

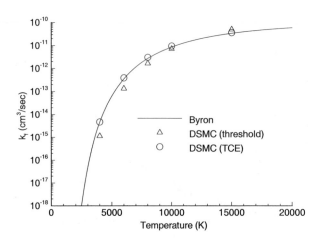

Fig. 5. Dissociation rate coefficient for O_2–O_2 at thermal equilibrium.

$A_{O_2-O_2} = 1.25$. Under these conditions, the TCE model gives excellent behavior. It accurately reproduces the measured data at all temperatures and for all cases considered never gave a reaction probability greater than one. The threshold line model also gives good agreement with the experimental data although it cannot reproduce the temperature dependence exactly. As discussed in Refs. 48 and 49, one of the disconcerting attributes of the threshold line model is the fact that the reaction probabilities under certain conditions can be considerably greater than one. When used in a DSMC computation, only one reaction results from a collision and so the effective maximum collision probability of any kind can only be one. Hence, in the calculation of the rate coefficients, the maximum allowed probability is set to one. It is not clear what the overall effect is in the threshold line model of the larger reaction probabilities.

A key aspect of dissociation models is the manner in which molecules are selected preferentially for reaction based on their vibrational and rotational energies. Further simulations are performed to analyze how this occurs at the collision level in the two DSMC chemistry models. Once again, several million collisions are simulated using 100 000 particles under conditions where the translational and rotational modes are in equilibrium at 10 000 K and different cases are run in which the vibrational temperature is varied from 10 000 K down to 0 K. The rate coefficients computed in this way are shown in Fig. 6. The difference between the two models becomes more significant as the vibrational temperature is reduced. As expected, the TCE model gives larger reaction rates due to the fact that there is no biasing to the vibrational mode with this model. When the vibrational temperature is zero, the TCE dissociation rate is a factor of three higher than that predicted by the threshold line model.

The energy distribution of reacting molecules can be obtained by using an accept-reject approach to identify individual collisions that give rise to

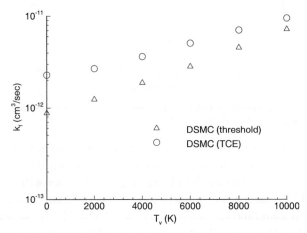

Fig. 6. Dissociation rate coefficient for O_2–O_2 at thermal nonequilibrium ($T_t = T_r = 10\,000$ K).

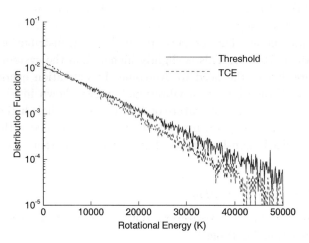

Fig. 7. Rotational energy distribution of dissociating O_2 molecules at nonequilibrium ($T_t = 14\,000$ K, $T_r = 5000$ K, $T_v = 2000$ K).

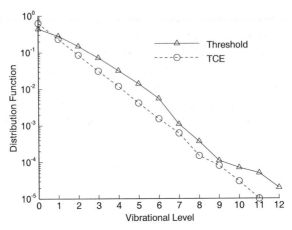

Fig. 8. Vibrational energy distribution of dissociating O_2 molecules at nonequilibrium (T_t = 14 000 K, T_r = 5000 K, T_v = 2000 K).

dissociation. For each of these reactive collisions, the rotational and the vibrational energies of the reacting molecule are recorded. The energy distributions of reactive molecules obtained in this manner with the threshold line and TCE dissociation models for O_2–O_2 collisions are compared in Fig. 7. A nonequilibrium state of T_t = 14 000 K, T_r = 5000 K, and T_v = 2000 K is chosen. These are typical of temperatures found in the BSUV-2 flight experiment that is discussed in the next section. In Fig. 7, it is found that the population of molecules chosen for dissociation by the TCE model has a slightly higher temperature than those selected by the threshold line model. The temperature of the TCE distribution shown in Fig. 7 is about 7000 K. This is slightly higher than the system rotational temperature due to the need to overcome the activation energy of reaction. The TCE distribution of vibrational energy shown in Fig. 8 is also nearly Boltzmann at a temperature of 2500 K, again a little higher than the system vibrational temperature. The distribution obtained with the threshold line model is non-Boltzmann but not very different from the TCE profile.

4.1.3. *Nitric Oxide Formation*

A detailed quasiclassical trajectory (QCT) study of the first Zeldovich reaction (reaction 5a, Table 1)

$$N_2 + O \rightarrow NO + N$$

was reported by Bose and Candler.[67,68] Since it is anticipated that such studies will in the future contribute significantly to the development of DSMC chemistry models, it is appropriate to provide a brief review of the QCT method here. This technique was employed to obtain reaction cross sections and product energy distributions of the reaction at high temperatures and nonequilibrium flow conditions. In the QCT method, the nuclear motion is treated classically, while the electronic part of the problem is solved quantum mechanically. The electronic energy as a function of the nuclear coordinates give the potential energy surfaces, which characterize the reaction path. A classical treatment of the nuclear motion is sufficient at high temperatures where the quantum effects are minimal, and thus, highly averaged reaction attributes are obtained with good accuracy.

It was assumed that the first Zeldovich reaction occurs with the products and the reactants in their ground electronic states,

$$N_2(X^1\Sigma_g^+) + O(^3P_g) \rightarrow N(^4S_u) + NO(X^2\Pi). \qquad (18)$$

This reaction proceeds adiabatically via the two lowest potential energy surfaces, $^3A''$ and $^3A'$, with a degeneracy factor of 1/3 each. These surfaces are usually obtained by analytically fitting the energies obtained from an *ab initio* computational quantum chemistry calculation at various nuclear positions. In Ref. 67, the contracted configuration interaction (CCI) *ab initio* data of Walch and Jaffe[69] was used for both potential energy surfaces. The details of the analytical representation of the $^3A''$ and $^3A'$ surfaces can be found in Refs. 67 and 70, respectively.

To obtain the reaction attributes for a particular set of vibrational, rotational and translational energies, many trajectories were simulated at given values of N_2 vibrational and rotational quantum numbers and N_2–O relative translational energy. The N_2 molecular orientation, vibrational phase and impact parameter were chosen randomly for each trajectory. The reaction attributes were then determined by averaging the outcomes of all collisions. The information obtained is state-specific, so for example, the energy distributions of the reactant and product molecules can be determined. The method used to calculate the vibrational and rotational state of the product molecule is outlined in Ref. 67. With the QCT approach, reaction cross sections were determined solely from the precollision state. The method knows nothing of the fluid flow environment and so the results should be applicable under conditions of both equilibrium and nonequilibrium.

Since, the reaction is highly endothermic, it was extremely difficult to obtain a statistically significant sample of reactive collisions at low energies. Hence, a large number of trajectories were run in the following energy ranges:

$$\varepsilon_t = 1.2 - 18.0 \text{ eV}$$

$$\varepsilon_r = 0 - 3.9975 \text{ eV} \qquad (19)$$

$$\varepsilon_v = 0.146 - 3.62 \text{ eV}.$$

This covers most hypersonic flow regimes with high thermodynamic non-equilibrium. The statistical uncertainties of the rate constants were below 5% for all cases run. It is found that under reentry flow conditions most of the endothermicity is derived from the N_2–O relative translational energy and the NO molecules are formed vibrationally and rotationally very excited. Also, the QCT thermal rate constants showed good agreement with the available experimental data.[67]

(a) Performance of the TCE Model

The QCT reaction cross sections are first compared with the TCE chemistry model. In Figs. 9(a)–9(c), the reaction cross sections are shown as a function of translational collision energy at three vibrational energy levels ($v = 0$, 7, and 13), for rotational levels specified by $J = 0$, 64, and 126 respectively. While the rotational and vibrational quantum numbers J and v refer to the reacting nitrogen molecule, the translational collision energy is that for the N_2–O pair.

These results indicate a number of the basic properties of the QCT data. Firstly, for given values of both J and v, there is variation in the cross sections with translational energy, ε_t. Secondly, for an increase in vibrational quantum number at given values of ε_t and J, there is generally an increase in the cross section. Therefore, the reaction probability varies with vibrational energy. Finally, comparison of Figs. 9(a) through 9(c) clearly demonstrates that there is also a dependence of the reaction cross sections on the rotational energy of the nitrogen molecule. As might be expected, the cross sections increase as the rotational energy of N_2 increases. Taken together, these observations indicate that the reaction cross sections depend on the translational collision energy, and the rotational and

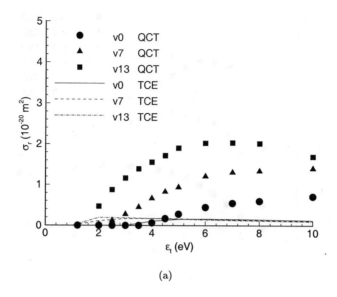

(a)

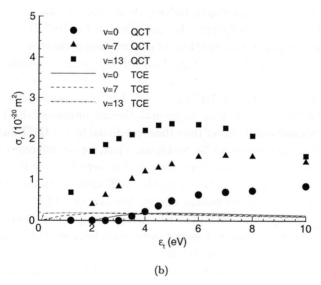

(b)

Fig. 9. (a) Reaction cross sections at $J = 0$: comparison of QCT and TCE results. (b) Reaction cross sections at $J = 64$: comparison of QCT and TCE results. (c) Reaction cross sections at $J = 126$: comparison of QCT and TCE results.

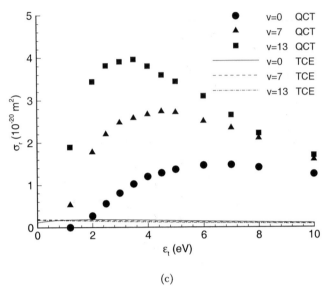

(c)

Fig. 9. (*Continued*).

vibrational energies of nitrogen. Indeed, these observations are reflected in the macroscopic rate coefficients that are calculated in Ref. 67 as a function of the translational, rotational, and vibrational temperatures.

The TCE results shown in Figs. 9(a) through 9(c) are generated using the modified Arrhenius constants recommended by Park *et al.*[71]: $a = 1.07 \times 10^{-12}$ m^3/s, $b = -1.0$, $\varepsilon_a = 5.175 \times 10^{-19}$ J. The TCE results offer very poor agreement with the QCT data. In general, the magnitudes of the TCE cross sections are significantly lower than those predicted by the QCT analysis. In addition, there are three specific problems. First, for several cases, the TCE model gives reaction probabilities greater than zero for $\varepsilon_t = 0$. Obviously, this is physically incorrect. Secondly, the TCE cross sections show almost no dependence on the rotational energy. Finally, the TCE cross sections show almost no dependence on the vibrational energy. The central problem of the TCE model is that it does not distinguish between the translational, rotational, and vibrational energy modes.

These comparisons indicate that an improved DSMC chemistry model must include separate biasing of the reaction cross section to the translational, rotational, and vibrational energies of the collision. This precisely motivated development of the GCE model.

(b) Performance of the GCE Model

In this section, the GCE DSMC chemistry model is calibrated against the QCT database. The parameters now employed in the modified Arrhenius rate coefficient are those derived by Bose and Candler[67] from the QCT data: $a = 9.45 \times 10^{-18}$ m^3/s, $b = 0.42$, $\varepsilon_a = 5.925 \times 10^{-19}$ J. Assuming these values, use of the GCE model requires selection of the parameters α, β, and γ. Through a series of data-fit comparisons, the following values are proposed: $\alpha = 0.2$, $\beta = -0.5$, and $\gamma = 0.3$. This is not a unique set of parameters, but rather one particular set that was found to give satisfactory behavior over the energy range of interest.

In Figs. 10(a)–10(c), comparisons of the QCT data are made with the GCE model. Significant improvements over the TCE model are obtained in terms of the magnitudes of the cross sections. Firstly, the DSMC cross sections now approach zero as the translational collision energy approaches zero for all cases. Secondly, the increases observed in the QCT data as J is increased are now in reasonable agreement with the DSMC model. The mathematical form of the biasing in the GCE model for rotational energy is chosen so as to ensure a nonzero reaction cross section when $J = 0$

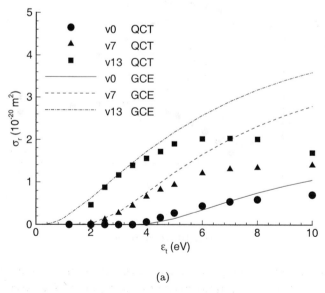

(a)

Fig. 10. (a) Reaction cross sections at $J = 0$: comparison of QCT and GCE results. (b) Reaction cross sections at $J = 64$: comparison of QCT and GCE results. (c) Reaction cross sections at $J = 126$: comparison of QCT and GCE results.

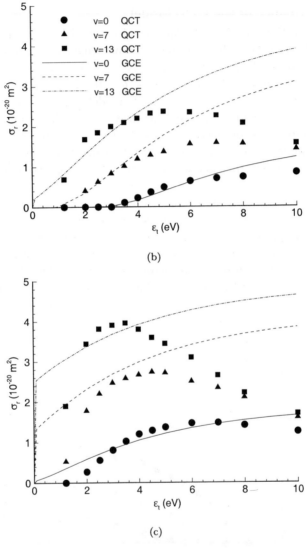

Fig. 10. (*Continued*).

(i.e. zero rotational energy), as found in the QCT analysis and shown in Fig. 10(a). In terms of vibrational energy biasing, the GCE cross section at $v = 0$ is not zero because the ground state vibrational energy is finite. The dependence of the GCE cross sections on vibrational energy matches quite

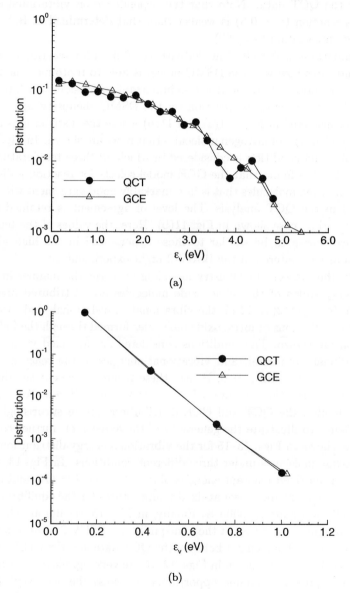

Fig. 11. (a) Vibrational energy distributions of reacting N_2 molecules at $T_t = T_r = T_v = 5000$ K. (b) Vibrational energy distributions of reacting N_2 molecules at $T_t = 14\,000$ K, $T_r = 5000$ K, and $T_v = 1000$ K.

well to the QCT data. Note that the dependence on vibrational energy for this reaction ($\gamma = 0.3$) is weaker than that determined in Ref. 44 for nitrogen dissociation ($\gamma = 2.0$).

Comparison of the GCE model with QCT data for reaction cross sections indicates how well the DSMC model is able to reproduce the rate of chemical reaction. However, it is also important to consider which nitrogen molecules are selected for reaction by the DSMC chemistry model. This issue is considered in Figs. 11(a) and 11(b) where the distributions for the vibrational energy of nitrogen molecules that react are shown. In Fig. 11(a), an equilibrium condition is considered in which all three temperatures are at 5000 K. It is found that the GCE model selects for reaction a distribution of nitrogen molecules that is in remarkably good agreement with those selected in the QCT analysis. The level of agreement is retained in the nonequilibrium case shown in Fig. 11(b). Here, the values of the temperatures are chosen to be similar to those experienced in the high altitude conditions encountered in the BSUV-2 flight experiment.

A further aspect of chemistry modeling concerns the manner in which the energy states of the nitric oxide molecules are distributed upon formation. In Figs. 12(a)–12(c), the vibrational, translational, and rotational energy distributions of nitric oxide molecules formed through the Zeldovich reaction are shown. The conditions considered are for fixed values of the translational, rotational, and vibrational energies of the reactants: $\varepsilon_t = 6.0$ eV, $J = 110$, and $v = 0$. Hence, these results represent the distributions of post-reaction outcomes for one particular collision. For all three energy modes, the QCT and GCE distributions are in surprisingly good agreement. To illustrate the generality of the agreement, further comparisons are shown in Figs. 13–15 for the vibrational energy distribution of the nitric oxide molecules under three different conditions. In Fig. 13, results are shown for fixed reactant energies of $\varepsilon_t = 3.5$ eV, $J = 110$, and $v = 13$. In Fig. 14, results are shown at fixed temperatures for the equilibrium case where all modes are at 5000 K. Finally, in Fig. 15, results are shown for the fixed temperature case at the nonequilibrium BSUV-2 conditions. In all cases, the level of agreement between the QCT data and the GCE results is excellent. The results shown in Figs. 12–15 are very significant. These comparisons represent a unique opportunity to assess the accuracy of using the simple energy partitioning scheme based on the Borgnakke–Larsen model[33] and implemented for a quantized oscillator by Bergemann and Boyd.[34] The results of the present study indicate that these procedures are

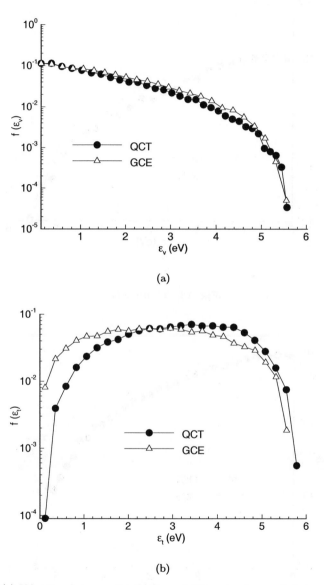

Fig. 12. (a) Vibrational energy distributions of NO molecules created at $\varepsilon_t = 6.0$ eV, $J = 110$, $v = 0$. (b) Translational energy distributions of NO molecules created at $\varepsilon_t = 6.0$ eV, $J = 110$, $v = 0$. (c) Rotational energy distributions of NO molecules created at $\varepsilon_t = 6.0$ eV, $J = 110$, $v = 0$.

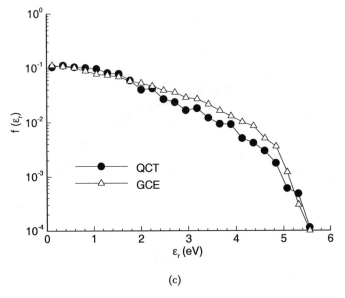

(c)

Fig. 12. (*Continued*).

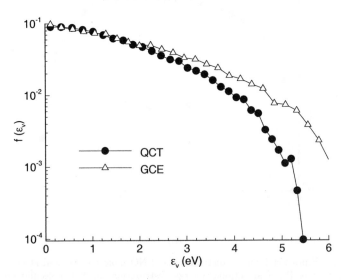

Fig. 13. Vibrational energy distributions of NO molecules created at $\varepsilon_t = 3.5$ eV, $J = 110$, $v = 13$.

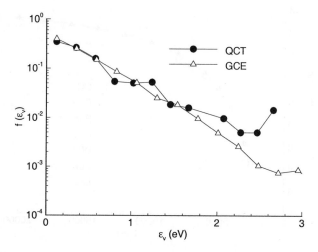

Fig. 14. Vibrational energy distributions of NO molecules created at $T_t = T_r = T_v = 5000$ K.

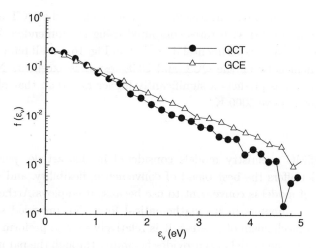

Fig. 15. Vibrational energy distributions of NO molecules created at $T_t = 14\,000$ K, $T_r = 5000$ K, and $T_v = 1000$ K.

capable of accurately reproducing the phenomena observed in the detailed QCT analysis. The good agreement may also be interpreted as indicating that the post-reaction energy for this particular reaction is partitioned in a way that is very close to the statistical distribution assumed in the Borgnakke–Larsen approach.

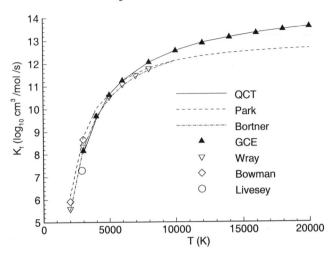

Fig. 16. Overall rate coefficient as a function of temperature.

Finally, the thermally averaged rate coefficients of the QCT and GCE models are compared with rates computed using recommended Arrhenius parameters[71,72] and experimental data[73-75] in Fig. 16. At all temperatures, the agreement between the QCT and GCE results is excellent. Note that the QCT analysis predicts a significantly higher rate than that of Park at temperatures above 5000 K.

4.1.4. *Discussion*

Of the DSMC chemistry models considered in this article, perhaps the GCE model offers the best blend of convenience, flexibility, and accuracy. This type of model is convenient to use because it employs Arrhenius rate coefficient parameters in its mathematical form. So, provided such data exists for a mechanism of interest, it is relatively easy to perform a DSMC computation. The model also provides flexibility through the parameters α, β, and γ that allow individual energy modes to be biased in the selection of reacting particles. These parameters can be identified for a particular reaction if experimental or numerical data exists that describe the variation of reaction cross section with collision energy. Finally, the model provides accuracy in that it will reproduce the measured equilibrium rate coefficients under conditions of equilibrium. While these attributes of the GCE model are very positive, and this model appears to be adequate for nonionizing air reactions, more advanced models such as the Maximum Entropy approach may have an important role to play for extremely nonequilibrium reactions.

The simple test cell configuration allows detailed study of all aspects of DSMC chemistry models. The three important components of any model are the reaction probability, the energy configuration of reactants, and the energy configuration of products. In principle, these data can be measured experimentally. In practice, no data is available for the reactions of interest in hypersonic flows of air. More precise potential energy surfaces will eventually make up for the lack of experimental data. It is hoped that the continued development of these models will eventually provide a database for most of the information needed for many of the reactions of interest. This development may ultimately lead to many of the DSMC chemistry models being discarded, and instead curve-fits to the numerical data being employed directly in computer codes. This could be argued as a requirement in any case, since the model parameters such as α, β, and γ of the GCE model, and the λ_i of the Maximum Entropy model, almost certainly change as a function of temperature due to the complexity of the reaction dynamics.

4.2. *Flow Field Studies*

4.2.1. *Introduction*

Much of the development of the DSMC technique was motivated by the need to model nonequilibrium hypersonic flows. However, very few experiments have been conducted in the laboratory that are suitable for testing the chemical reaction models employed in the DSMC technique. One notable exception resulted from a NATO-AGARD working group on hypersonics. A planetary probe geometry was tested under rarefied, hypersonic flow conditions in several different experimental facilities and a number of research groups generated DSMC results for comparison. Many articles have been published on these studies and details of the experiments and numerical results are provided in Ref. 74. Examples of results obtained using the MONACO DSMC code[26] are shown in Figs. 17 and 18. The shape of the probe can be seen in Fig. 17. The electron beam method was employed in the SR3 wind-tunnel to measure the density field around the geometry.[2] The flow conditions for this case are listed in Table 2. In Fig. 18, the experimental data shown for heat transfer coefficient on the geometry surface was measured in the LENS facility.[4] The flow conditions are again listed in

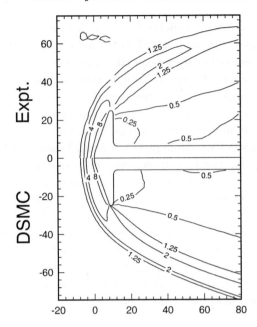

Fig. 17. Comparison of DSMC computation and experimental measurement for the ratio of local to free stream density. Distances are in millimeters.

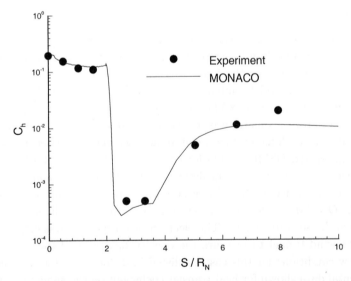

Fig. 18. Comparison of DSMC computation and experimental measurement for coefficient of heat transfer along the surface of the probe.

Table 2. In each case, it is found that the DSMC computations provide excellent agreement with the experimental data. However, it should be noted that these flow conditions gave chemically frozen flow. These comparisons indicate that the DSMC method is an accurate simulation method, but they do not aid in the assessment of chemical reaction models.

Table 2. Free stream conditions for planetary probe studies.

Facility	T_o (K)	ρ_∞ (kg/m^3)	U_∞ (m/s)	Body diameter (cm)	Kn_∞
SR3 (CNRS)	1100	5.2×10^{-5}	1500	5.0	0.01
LENS (Calspan)	4350	13.0×10^{-5}	3250	15.2	0.002

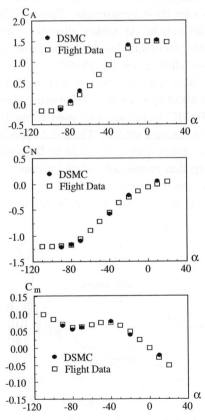

Fig. 19. Comparison of DSMC prediction and reentry flight data for force coefficients as a function of angle of attack, α (from Ivanov *et al.*[78]).

In terms of flight data, for entry into the Earth's atmosphere, the DSMC technique was applied by Rault[77] to predict the aerodynamic and heating coefficients of the Space Shuttle during the rarefied portion of its trajectory. The rarefied portions of trajectories of Russian reentry capsules were analyzed by Ivanov *et al.*[78] using the DSMC technique. Examples of their results are shown in Fig. 19 where the force coefficients are shown as a function of angle of attack (α) at 85 km altitude. Of course, inclusion of chemistry in the DSMC analysis is important for accurate prediction of heat transfer, as pointed out in Ref. 78. However, in general, it is found that the aerodynamic coefficients are relatively insensitive to the particular DSMC chemistry models. A series of studies has developed and applied the DSMC method to investigate the ultraviolet spectra obtained on the second Bow Shock Ultra Violet flight experiment. These studies will be reviewed in detail later in this article. At higher altitudes (over the range of 140 to 300 km) the DSMC technique has been employed to investigate visible emissions of NO_2, so called spacecraft glow.[79]

The DSMC technique has also been applied to predict the aerodynamic and heating coefficients of spacecraft entering other planetary atmospheres. For example, the DSMC technique was used to analyze the Mars entry trajectory of NASA's Viking 1 vehicle. These data were compared to experimental measurements taken during the flight by Blanchard *et al.*[80] as shown in Fig. 20. The Magellan spacecraft orbited around Venus and the later part of its mission was used to perform the first planetary aerobraking maneuvers. In support of these studies, DSMC investigations of the flight of

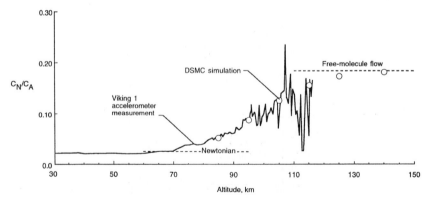

Fig. 20. Comparison of DSMC prediction and Viking 1 flight data for the ratio of force coefficients (from Blanchard *et al.*[80]).

the Magellan spacecraft through the Venusian atmosphere were performed by Rault[81] and by Haas and Schmidt.[82] An analogous study was performed by Haas and Milos[83] for entry of the Galileo probe into the atmosphere of Jupiter.

The Bow Shock Ultraviolet-2 (BSUV-2) hypersonic flight experiment was flown in 1991.[11] The vehicle geometry consisted of a spherically-capped cone with a nose radius of about 10 cm. BSUV-2 reentered the atmosphere at 5.1 km/s and provided data in the altitude range from 110 to 60 km. Experimental measurements of the ultraviolet emission due to nitric oxide and vacuum ultraviolet emission due to atomic oxygen resonance transitions were obtained. Calculation of the radiative emission was performed in a decoupled approach. In initial work, the chemically reacting flow field was computed using continuum (CFD) and direct simulation Monte Carlo (DSMC) methods.[84,85] Then, the emission was predicted from the flow field solutions using the nonequilibrium radiation code NEQAIR.[86] Initial calculations of the NO and O emissions for BSUV-2 were successful for altitudes at 80 km and lower. Above 80 km, the predicted emission was too low by as much as two orders of magnitude at the highest altitude considered. Comparison of the experimental data with the computational results from Candler *et al.*[85] for variation in nitric oxide emission as a function of altitude is shown in Fig. 21.

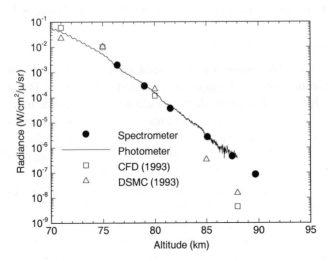

Fig. 21. Comparison of data measured by BSUV-2 and previous numerical predictions for NO emission as a function of altitude.

Broadly speaking, there are three main processes to be modeled in the computation of emission in these flows: (1) fluid mechanics; (2) chemistry; and (3) radiation. The rarefied fluid mechanics should be accurately simulated since both continuum and DSMC methods have been applied and cross checked for consistency. For radiation, the NEQAIR code has been improved and calibrated extensively through comparison with the BSUV data (from both flights 1 and 2) and through comparison with laboratory data. Nevertheless, there remains some uncertainty in the model, particularly as the degree of nonequilibrium increases at high altitude. The degree of uncertainty of the radiation model is estimated to be at least a factor of 2.

As mentioned above, the nonequilibrium radiation code NEQAIR[86] is employed for prediction of ultraviolet emission from the DSMC flow field solutions. The modeling of ultraviolet emission with this code is discussed for nitric oxide in Ref. 84 and for atomic oxygen in Ref. 87. A common assumption made in using the NEQAIR code is that a quasisteady state (QSS) exists for the number densities of the electronically excited species. The assumption requires that the time scale of chemical processes is much smaller than the time scales for diffusion and for changes in overall properties. Under these conditions, the local values of temperatures and ground state species number densities obtained from the DSMC computation may be used to compute the populations of the electronically excited states.

4.2.2. *Nitric Oxide Emission*

Emission for nitric oxide was measured on BSUV-2 along the stagnation streamline using both photometers and spectrometers. Details may be found in Ref. 11. The flight data has already been shown in Fig. 21. In computing new DSMC predictions of ultraviolet emissions, nitric oxide formation will be modeled using the GCE model in place of the TCE model used in Ref. 85, and oxygen dissociation will be modeled using the threshold line model in place of the VFD model ($\phi = 0.5$) used in Ref. 85. Prediction of the magnitude of NO emission using NEQAIR depends primarily on the bulk translational temperature of the gas and the nitric oxide concentration along the stagnation streamline. The variation computed using the DSMC technique along the stagnation streamline of translational and vibrational temperatures, and of nitric oxide concentration, are compared for three different altitudes in Figs. 22(a)–22(c) respectively. The Knudsen numbers for the BSUV-2 vehicle are 0.008, 0.033, and 0.215 for the altitudes of 71, 80, and 90 km, respectively. The results illustrate how the shock wave becomes significantly thicker at higher altitude. It is also clear that a large degree of thermal nonequilibrium exists under all three conditions.

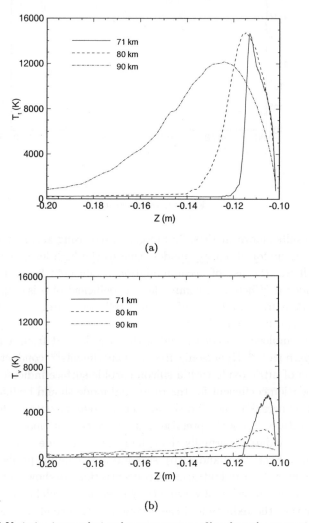

(a)

(b)

Fig. 22. (a) Variation in translational temperature profiles along the stagnation stream-line as a function of altitude. (b) Variation in vibrational temperature profiles along the stagnation streamline as a function of altitude. (c) Variation in nitric oxide concentration profiles along the stagnation streamline as a function of altitude.

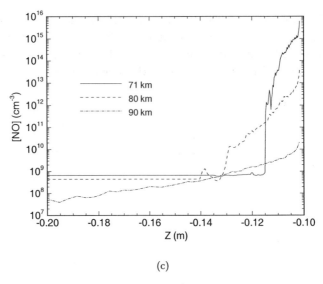

(c)

Fig. 22. (*Continued*).

The results shown in Figs. 22 are generated using accommodation coefficients of one for all energy modes. Due to the high level of rarefaction
in these flows, the use of accommodation coefficients less than unity is
also considered. When an accommodation coefficient of α is employed, this
means that a fraction $1 - \alpha$ of particle reflections from the surface are
treated as specular. For the translational mode, values ranging from 0.7
to 1.0 are employed to correspond with data derived from a hypersonic
flight experiment.[88] Hypersonic free-jet experiments[89] conducted on the
interaction of nitric oxide with a silicon carbide surface indicated that the
accommodation coefficient for the rotational mode should be 0.5, and that
value is used here. For the vibrational energy mode, no data could be found
in the literature. Using the premise that the vibrational mode is less likely
to be accommodated thermally than the rotational mode, a value of 0.1 is
employed. In Figs. 23(a) and 23(b), stagnation streamline profiles of translational temperature and nitric oxide concentration are shown for the 90 km
condition where the effect of varying the gas–surface modeling is considered.
It is clear that the reduction of the accommodation coefficient for translational energy (α_t) leads to an increase in translational temperature and a
small increase in NO concentration. Variation of the other accommodation
coefficients had much less impact on the flow field properties. In addition,

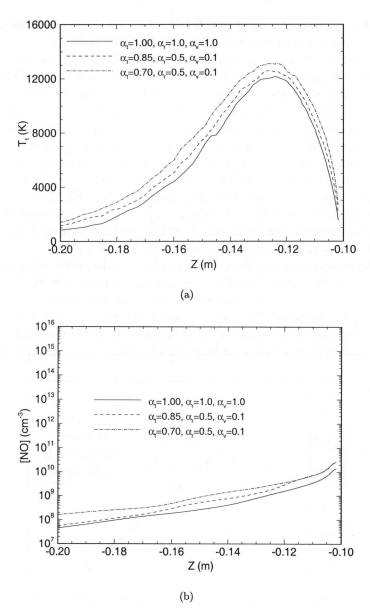

(a)

(b)

Fig. 23. (a) Translational temperature profiles along the stagnation streamline as a function of gas–surface interaction model at 90 km altitude. (b) Nitric oxide concentration profiles along the stagnation streamline as a function of gas–surface interaction model at 90 km altitude.

use of a wall temperature of 1000 K instead of 500 K had almost no effect on the flow except at very small distances from the surface.

Since the NO emission depends on the translational temperature and nitric oxide concentration, it is clearly of interest to consider how the different values of α_t affect the predicted radiation. In Fig. 24, revised emission predictions based on the flow field solutions above are shown. The most important conclusion concerning the data shown in Fig. 24 is that good agreement between measurement and prediction is obtained over the entire altitude range. It is significant to note that the good agreement obtained previously at low altitudes[85] is retained here with all of the advanced chemical models. Another important conclusion from Fig. 24 is that the emission predicted at high altitude is very sensitive to the translational energy accommodation coefficient. At 90 km, there is a factor of 5 difference in NO emission obtained with values of $\alpha_t = 0.7$ and 1.0.

Before considering in detail the dependence of the predicted spectra on gas–surface interaction modeling, it is appropriate to consider the BSUV-2 flight data. In Fig. 25(a), two self-normalized spectra measured at 79 and 90.8 km are compared. The spectral scan at 90.8 km is at an altitude where the surface modeling sensitivity can be best tested. Comparison of this scan with that at 79 km shows the high altitude data to be signal limited. As

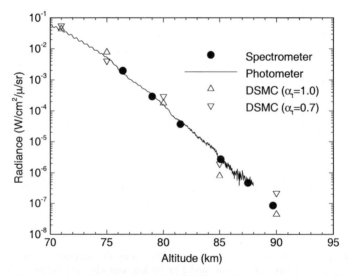

Fig. 24. Comparison of data measured by BSUV-2 and new numerical predictions for NO emission as a function of altitude.

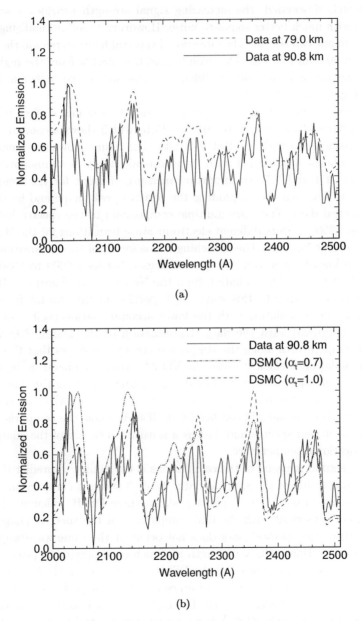

(a)

(b)

Fig. 25. (a) Comparison of normalized NO spectra measured by BSUV-2 at 79 and 90.8 km altitude. (b) Comparison of normalized NO spectra at 90 km altitude.

the vehicle descended, the increasing signal strength provides a scan at 79 km with much better signal-to-noise. However, by self normalizing each spectrum one can see that the vibrational spectral features remain the same between 90.8 and 79 km. The evolution of the spectra from the higher to the lower altitude provides confidence in the use of the higher altitude, noisier spectrum.

The sensitivity of the emission prediction to the gas–surface interaction model is clearly illustrated in Fig. 25(b) which shows a comparison of computed spectra with the BSUV-2 data. The figure shows two computed spectra obtained at 90 km with translational energy accommodation coefficients of 1 and 0.7. An important measure of merit for the computed spectra is how well they reproduce the relative peak shape and heights of the spectral data. There are multiple components to the spectra between 2000 and 2500 Å from different electronic state transitions for the NO and O_2 molecular systems. The branching ratios for all of these transitions are not well known; however, the spectral region between 2000 to 2300 Å is strongly dominated by radiation from the $NO(A \to X)$ transition. Hence, if emphasis is given to this wavelength portion of the spectra it can be seen that the calculation with the lower accommodation coefficient gives better agreement with the data. Examination of the DSMC NO vibrational temperatures along the stagnation streamline shows that the lower accommodation coefficient raises the NO vibrational temperature by about 1000 K. This increase in vibrational temperature increases the number of transitions from the higher vibrational levels of the NO(A) state, thereby increasing the calculated peak height at 2050 Å. Thus, the comparison between calculated spectra and data is a sensitive measure of the computed NO vibrational temperature.

The second measure of the calculations is the ability to predict the absolute magnitude of the spectral data. Table 3 shows a comparison of the absolute magnitude of the highest peak for three DSMC calculations and the data. Consistent with the first comparison of the spectral shape and peak ratios, the spectral peak does not occur at the same wavelength for each spectra. Hence the wavelength at which each maximum peak occurs is also indicated. The data and the DSMC calculations with an accommodation coefficient less than one produce a maximum peak location at approximately the same wavelength. The DSMC calculation with a maximum peak at approximately 2400 Å has a maximum spectral peak occurring at a wavelength that is consistent with a lower NO vibrational temperature.

Table 3. Comparison of maximum spectral peak heights.

Description	Wavelength (Å)	Radiance (W/cm^2/sr/μ)
BSUV-2 data at 90.8 km	2031	0.94×10^{-7}
DSMC, $\alpha_t = 1.00$	2359	1.1×10^{-7}
DSMC, $\alpha_t = 0.85$	2044	2.0×10^{-7}
DSMC, $\alpha_t = 0.70$	2046	4.7×10^{-7}

The table shows that all the calculations and the data are in good agreement, with the worst case being a factor of five. The uncertainty in the absolute magnitude of the spectral radiance data is $\pm 25\%$. The uncertainty in the wavelength calibration of the instrument of about 10 Å is sufficient to explain the discrepancy in the peak wavelength positions for the DSMC calculations with an accommodation coefficient less than one. An accommodation coefficient of 0.85 seems to provide the best overall agreement.

4.2.3. *Atomic Oxygen Emission*

Emission from atomic oxygen at a wavelength of 130.4 nm was measured on BSUV-2 using a photoionization cell along a line of sight that subtended an angle of 36° with the center of the sphere that represents the nose of the vehicle. Details of the experiment and of the radiation model may be found in Ref. 87. Prediction of the atomic oxygen emission depends primarily on the bulk translational temperature and the atomic oxygen concentration along the line of sight. In Figs. 26(a) and 26(b), these flow properties predicted by the DSMC technique are shown for three different altitudes. Qualitatively, these are similar to the data shown in Figs. 22 along the stagnation streamline. The bow-shock becomes thicker and more diffuse at higher altitude. These results are obtained with all accommodation coefficients set to one. When lower values are employed, similar behavior to that obtained along the stagnation streamline is observed. Specifically, the translational temperature is increased at lower values of α_t.

Comparison is made in Fig. 27 of prediction and measurement of atomic oxygen emission as a function of altitude. The oscillations in the measured data are due to precession of the vehicle that occurred during flight. The agreement between the BSUV-2 data and prediction is particularly good at high altitude. As with nitric oxide emission, the radiation at high altitude is found to be sensitive to the gas–surface interaction model. At 90 km, there

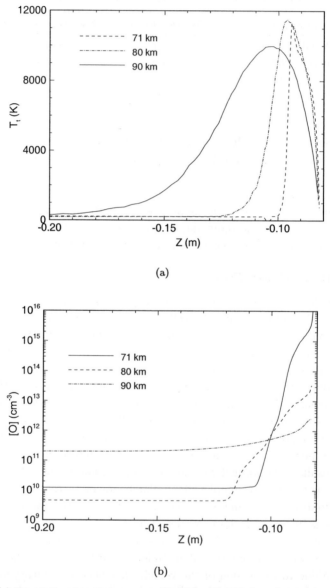

(a)

(b)

Fig. 26. (a) Translational temperature profiles along the 36° line of sight as a function of altitude. (b) Atomic oxygen concentration profiles along the 36° line of sight as a function of altitude.

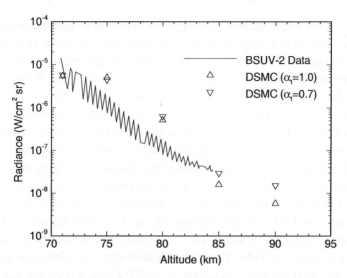

Fig. 27. Comparison of data measured by BSUV-2 and new numerical predictions for O emission as a function of altitude.

is a factor of 3 difference between the emissions predicted using values for α_t of 0.7 and 1.0. The comparison between measurement and computation at lower altitude is not as satisfactory. At 80 km, the atomic oxygen VUV model of Ref. 87 predicts that the flow is optically thin. Hence, the total radiation is governed by the gradient of atomic oxygen concentration along the 36° line-of-sight and the translational temperature. It was found that the O concentration was very sensitive to the $O_2 + N_2$ dissociation rate used and there remains some uncertainty as to its best value. The agreement between the predicted and measured O atom radiance at 70 km seen in Fig. 27 probably occurs due to compensating errors. At that altitude (and lower) the flow *is* optically thicker and the total radiance is influenced by the reabsorption of VUV photons by the cool atomic O in the boundary layer.

4.2.4. *Discussion*

The comparisons between DSMC predictions and experimental measurements of UV emission on the BSUV-2 flight provide strong evidence of the ability of the DSMC technique to accurately model nonequilibrium, hypersonic flows. While a great deal of progress has been made, some words of caution are also appropriate. Clearly, there are a large number of assumptions involved in the computation of the ultraviolet emission in a strongly nonequilibrium flow. It would be much more satisfactory to assess DSMC chemistry models through comparison with simpler, more direct experimental data such as measurements of species densities. So far, no data of this type has ever been taken on a flight experiment. Nevertheless, one of the conclusions of the BSUV-2 studies must be that computer models and codes exist to accurately simulate flows up to a velocity of about 7 km/s.

In the future, there will be considerable interest in studying flows at significantly higher velocity. This is mainly motivated by NASA missions involving the return of vehicles from extraterrestrial bodies such as asteroids, the Moon, and Mars. These vehicles will enter the Earth's atmosphere at velocities as high as 14 km/s. In this regime, the air chemistry kinetics mechanism shown in Table 1 must be considerably extended in comparison to that required for the conditions of BSUV-2 and indeed for the Space Shuttle. In particular, all of the molecules will be dissociated, ionization will become important, and atomic radiation will be responsible for an important fraction of the total vehicle heating. Accurate computation of these flows in the rarefied flow regime will require significant development of the existing DSMC models for the reactions of interest. The progress made in this regard in the BSUV-2 research program makes a strong case for preparing a flight experiment to support the modeling activities.

Acknowledgments

The author gratefully acknowledges the significant contributions to the DSMC Bow Shock research made by Graham V. Candler of the University of Minnesota and Deborah A. Levin of the George Washington University. Thanks are extended to Dean Wadsworth and Ingrid Wysong of the Air Force Research Laboratory, Michael Gallis of Sandia National Laboratories, Mikhail Ivanov of the Institute of Theoretical and Applied Mechanics (Novosibirsk, Russia) and Robert Blanchard of NASA Langley Research Center, for providing data and plots that are included in this article. This work was supported by the Army Research Office under grants DAAH04-93-G-0089 and DAAG55-98-1-0500 with Dr. David Mann as grant monitor, and by the Air Force Office of Scientific Research under grant F49620-96-1-0091 with Dr. Mitat A. Birkan as grant monitor.

References

1. J. D. Anderson, *Hypersonic and High Temperature Gas Dynamics* (Wiley, New York, 1989).
2. J. Allegre and D. Bisch, Report RC 95-2 (CNRS, Meudon, 1995).
3. A. Danckert and H. Legge, *J. Spacecraft* **33**, 476 (1996).
4. M. S. Holden, J. Kolly and K. M. Chadwick, AIAA Paper 95-0291 (1995).
5. E.P. Muntz, *AGARDograph* **132**, 111 (1969).
6. R. J. Cattolica, in *Rarefied Gas Dynamics: Physical Phenomena*, Eds. E. P. Muntz, D. P. Weaver and D. H. Campbell (AIAA, New York, 1989), p. 133.
7. C. Park, *Nonequilibrium Hypersonic Aerothermodynamics* (Wiley, New York, 1990).
8. R.C. Blanchard, R. G. Wilmoth and G. J. LeBeau, *J. Spacecraft* **34**, 8 (1997).
9. A. Seiff, D. E. Reese, S. C. Sommer, D. B. Kirk and E. E. Whiting, *Icarus* **18**, 525 (1973).
10. E. E. Whiting, J. O. Arnold and R. M. Reynolds, *J. Quant. Spectrosc. Radiat. Trans.* **8**, 837 (1973).
11. P. W. Erdman, E. C. Zipf, P. Espy, L. C. Howlett, D. A. Levin, R. J. Collins and G. V. Candler, *J. Thermophys.* **8**, 441 (1994).
12. D. L. Cauchon, NASA TM X-1222 (1966).
13. W. L. Jones and A. E. Cross, NASA SP-252, 109 (1970).
14. W. G. Vincenti and C. H. Kruger, *Introduction to Physical Gas Dynamics* (Krieger, Malabar, 1986).
15. T. Ohwada, *J. Comp. Phys.* **139**, 1 (1998).
16. C. Cercignani, *Comm. Math. Phys.* **197**, 199 (1998).
17. G. V. Candler, *AIAA J.* **31**, 410 (1993).
18. P.A. Gnoffo, *Ann. Rev. Fluid Mech.* **31**, 459 (1999).
19. K. A. Comeaux, D. R. Chapman and MacCormack, AIAA Paper 95-0415, (1995).
20. K. Y. Yun, R. K. Agarwal and R. Balakrishnan, *J. Thermophys.* **12**, 328 (1998).
21. G. A. Bird, *Molecular Gas Dynamics and the Direct Simulation of Gas Flows* (Oxford University Press, Oxford, 1994).
22. W. Wagner, *Math. Comput. Simul.* **38**, 211 (1995).
23. D. Baganoff and J. D. McDonald, *Phys. Fluids* **A2**, 1248 (1990).
24. I. D. Boyd, *J. Comp. Phys.* **96**, 411 (1991).
25. L. Dagum, *Concurrrency-Practice and Experience* **4**, 241 (1992).
26. S. Dietrich and I. D. Boyd, *J. Comp. Phys.* **126**, 328 (1996).
27. G. A. Bird, *Comput. Math. Appl.* **35**, 1 (1998).
28. E. S. Oran, C. K. Oh and B. Z. Cybyk, *Ann. Rev. Fluid Mech.* **30**, 403 (1998).
29. M. S. Ivanov and S. F. Gimelshein, *Ann. Rev. Fluid Mech.* **30**, 469 (1998).
30. K. Nanbu, *J. Phys. Soc. Japan* **45**, 2042 (1980).
31. K. Koura and H. Matsumoto, *Phys. Fluids* **A3**, 2459 (1992).
32. I. D. Boyd, *Phys. Fluids* **A2**, 447 (1990).
33. C. Borgnakke and P. S. Larsen, *J. Comput. Phys.* **18**, 405 (1975).

34. F. Bergemann and I. D. Boyd, *Proc. 18th Int. Symp. Rarefied Gas Dynamics*, Eds. B. Shizgal and D. P. Weaver, Vol. 158 (AIAA, Washington, 1994), p. 174.
35. I. D. Boyd, *J. Thermophys.* **4**, 478 (1990).
36. I. Choquet, *J. Thermophys.* **9**, 446 (1995).
37. S. F. Gimelshein, M. S. Ivanov, G. N. Markelov and Y. E. Gorbachev, *J. Thermophys.* **12**, 489 (1998).
38. K. Koura, *Phys. Fluids* **6**, 3473 (1994).
39. J. D. Anderson, J. D. Foch, M. J. Shaw, R. C. Stern and B. J. Wu, in *Rarefied Gas Dynamics*, Eds. V. Boffi and C. Cercignani (Teubner, Stuttgart, 1986), p. 413.
40. K. K. Kossi and I. D. Boyd, *J. Spacecraft* **35**, 653 (1998).
41. F. E. Lumpkin, B. L. Haas and I. D. Boyd, *Phys. Fluids* **A3**, 2282 (1991).
42. B. L. Haas, D. B. Hash, G. A. Bird, F. E. Lumpkin and H. A. Hassan, *Phys. Fluids* **6**, 2191 (1994).
43. I. D. Boyd, Paper 90-0145 (AIAA, Washington, 1990).
44. I. D. Boyd, *Phys. Fluids* **A4**, 178 (1992).
45. B. L. Haas and I. D. Boyd, *Phys. Fluids* **A5**, 478 (1993).
46. I. D. Boyd, D. Bose and G. V. Candler, *Phys. Fluids* **9**, 1162 (1997).
47. S. O. Macheret and J. W. Rich, *Chem. Phys.* **174**, 25 (1993).
48. I. D. Boyd, *Phys. Fluids* **8**, 1293 (1996).
49. D. C. Wadsworth and I. J. Wysong, *Phys. Fluids* **9**, 3873 (1997).
50. R. D. Levine and R. B. Bernstein, *Molecular Reaction Dynamics* (Oxford University Press, Oxford, 1987).
51. P. M. Marriott and J. K. Harvey, in *Rarefied Gas Dynamics*, Ed. A. E. Beylich (VCH, Weinheim, 1991), p. 784.
52. M. A. Gallis and J. K. Harvey, *J. Fluid Mech.* **312**, 149 (1996).
53. A. B. Carlson and G. A. Bird, in *Rarefied Gas Dynamics*, Eds. J. K. Harvey and R. G. Lord (Oxford University Press, Oxford, 1995), p. 434.
54. R. G. Lord, in *Rarefied Gas Dynamics 20*, Ed. C. Shen (Peking University Press, Beijing, 1997), p. 180.
55. M. A. Gallis and J. K. Harvey, AIAA Paper 96-1849 (June 1996).
56. A. Ben-Shaul, R. D. Levin and R. B. Bernstein, *J. Chem. Phys.* **61**, 4937 (1974).
57. G. Hancock, C. Morley and I. W. M. Smith, *Chem. Phys. Lett.* **12**, 193 (1971).
58. F. L. Hurlbut, in *Rarefied Gas Dynamics 20*, Ed. C. Shen (Peking University Press, Beijing, 1997), p. 355.
59. R. G. Lord, *Phys. Fluids* **7**, 1159 (1995).
60. J. Eggers and A. E. Beylich, in *Rarefied Gas Dynamics*, Eds. J. K. Harvey and R. G. Lord (Oxford University Press, Oxford, 1995), p. 1216.
61. D. B. Hash and H. A. Hassan, *J. Thermophys.* **10**, 242 (1996).
62. T. Lou, D. C. Dahlby and D. Baganoff, *J. Comput. Phys.* **145**, 489 (1998).
63. I. D. Boyd, G. Chen and G. V. Candler, *Phys. Fluids* **7**, 210 (1995).
64. I. D. Boyd and T. Gokcen, *AIAA J.* **32**, 1828 (1994).
65. P. G. Martin, D. H. Schwartz and M. E. Mandy, *Astrophys. J.* **461**, 265 (1996).

66. S. Byron, *J. Chem. Phys.* **30**, 1380 (1959).
67. D. Bose and G. V. Candler, *J. Chem. Phys.* **104**, 2825 (1996).
68. D. Bose and G. V. Candler, *J. Thermophys. Heat Transfer* **10**, 148 (1996).
69. S. P. Walch and R. L. Jaffe, *J. Chem. Phys.* **86**, 6946 (1987).
70. M. Gilibert, A. Aguilar, M. González, F. Mota and R. Sayós, *J. Chem. Phys.* **97**, 5542 (1992).
71. C. Park, J. T. Howe, R. L. Jaffe and G. V. Candler, *J. Thermophys. Heat Transfer* **8**, 9 (1994).
72. M. H. Bortner, *A Review of Rate Constants of Selected Reactions of Interest in Reentry Flow Fields in the Atmosphere* (National Bureau of Standards, Washington, 1969).
73. K. L. Wray and J. D. Teare, *J. Chem. Phys.* **36**, 2582 (1962).
74. C. T. Bowman, *Combust. Sci. Tech.* **3**, 37 (1971).
75. J. B. Livesey, A. L. Roberts and A. Williams, *Combust. Sci. Tech.* **4**, 9 (1971).
76. J. N. Moss and J. M. Price, *J. Thermophys.* **11**, 321 (1997).
77. D. F. G. Rault, *J. Spacecraft* **31**, 944 (1994).
78. M. S. Ivanov, G. N. Markelov, S. F. Gimelshein, L. V. Mishina, A. N. Krylov and N. V. Grechko, *J. Spacecraft* **35**, 16 (1998).
79. D. P. Karipides, I. D. Boyd and G. E. Caledonia, *J. Spacecraft* **36**, 1999.
80. R. C. Blanchard, R. G. Wilmoth and J. N. Moss, *J. Spacecraft* **34**, 687 (1997).
81. D. F. G. Rault, *J. Spacecraft* **31**, 537 (1994).
82. B. L. Haas and D. A. Schmidt, *J. Spacecraft* **31**, 980 (1994).
83. B. L. Haas and F. S. Milos, *J. Spacecraft* **32**, 398 (1995).
84. D. A. Levin, G. V. Candler, R. J. Collins, P. W. Erdmann, E. Zipf and L. C. Howlett, *J. Thermophys.* **8**, 447 (1994).
85. G. V. Candler, I. D. Boyd and D. A. Levin, AIAA Paper 93-0275 (AIAA, Washington, 1993).
86. C. Park, in *Thermal Design of Aero-Assisted Orbital Transfer Vehicles*, Ed. H. F. Nelson (AIAA, Washington, 1985), p. 100.
87. D. A. Levin, G. V. Candler, L. C. Howlett and E. E. Whiting, *J. Thermophys.* **9**, 629 (1995).
88. D. T. Lyons, *Proc. 19th Int. Symp. Rarefied Gas Dynamics*, Eds. J. K. Harvey and R. G. Lord (Oxford University Press, Oxford, 1995), p. 1408.
89. G. Gundlach and C. Dankert, *Proc. 19th Int. Symp. Rarefied Gas Dynamics*, Eds. J. K. Harvey and R. G. Lord (Oxford University Press, Oxford, 1995), p. 974.

CHAPTER 4

CHEMICAL DYNAMICS IN CHEMICAL LASER MEDIA

Michael C. Heaven

Department of Chemistry, Emory University
Atlanta, GA 30322

Contents

1. Introduction

Chemical lasers are of interest because they provide a means for efficient conversion of stored chemical energy into high power laser radiation. Such devices can be self-contained, compact, and suitable for use from a mobile platform. A variety of uses are envisioned for chemical lasers. Industrial applications are mostly associated with cutting and welding processes. Chemical lasers for the decommissioning of nuclear power facilities[1-3] and removal of space debris[4] are applications that are being actively investigated. Military uses include communications, target illumination and directed energy weapons.[3]

The challenge in developing chemical laser systems is to find reactions that will efficiently generate a population inversion in an atom or molecule that can support laser action. It is well known that the excited state pumping conditions become more difficult to achieve as the frequency of the laser transition (ν) increases. This is because the competition between stimulated emission and loss by radiative decay becomes more unfavorable with increasing frequency (the cross section for stimulated emission is frequency independent while the spontaneous decay rate is proportional to ν^3). Consequently, it is easier to create chemical lasers that operate on long wavelength (low frequency) vibrational transitions (e.g. HF chemical lasers) than short wavelength electronic transitions. Despite the difficulties imposed by the pumping requirements, there has been a sustained effort to develop electronic transition lasers that are driven by chemical reactions. The advantages of using short wavelengths are that the photons carry more energy, they couple to targets more effectively, and wavelengths that are readily transmitted by the atmosphere or fiber optics may be obtained.

The best prospect for creating a chemically driven electronic transition laser is to use a transition where the upper laser level is metastable. Chemical methods that generate high yields of certain metastables have been developed, but so far it has not been possible to achieve lasing on the primary products. For example, inverted populations of the metastables $NF(a^1\Delta)$ and $NCl(a^1\Delta)$ can be generated by chemical means, but the $a^1\Delta - X^3\Sigma^-$

transitions are too weak to sustain lasing for reasonable number densities or gain lengths. In fact, due to conflicting requirements it is unlikely that a metastable that can be generated in high yield from a chemical reaction will be suitable for direct lasing. Factors that strongly favor metastable production (e.g. conservation of spin and/or orbital angular momentum) also lead to very small cross sections for stimulated emission back to the ground state. However, the strategy of transferring energy from metastable carriers to less metastable lasing species has proved to be successful. The best known device that uses this approach is the chemical oxygen iodine laser (COIL).[5-7] This relies on near resonant energy transfer from $O_2(a^1\Delta)$ to the upper spin-orbit level $(^2P_{1/2})$ of atomic iodine. Radiative relaxation back to the $I(^2P_{3/2})$ ground state is electric dipole forbidden, but magnetic dipole allowed with a lifetime of 125 ms.[8] This is near optimal for a chemically pumped system. Continuous wave COIL devices with output powers in excess of 40 kW at 1.315 μm have been developed.[7] Energy transfer excitation of $I(^2P_{1/2})$ by $NCl(a)$ is also efficient, and a pulsed iodine laser pumped by this means has been demonstrated.[9] $NF(a)$ carries more energy than $NCl(a)$ or $O_2(a)$, and has been used to pump molecular lasers[10-12] that operate at wavelengths as short as 470 nm. Inverted systems driven by energy transfer from metastable states of silicon or germanium oxide to thallium, gallium, and sodium atoms have been reported by Gole and coworkers.[13,14]

There is a complex interplay of kinetic processes in these energy transfer lasers. The conditions under which they operate are extreme in terms of the concentrations of transient species and the local gas temperatures. Many of the key reactions are difficult to study in isolation as they involve interactions between pairs of radicals or other transient species (e.g. collisions between electronically and vibrationally excited molecules). The impetus to obtain mechanistic and kinetic data for chemical laser systems (potential as well as demonstrated) has stimulated the development of new experimental techniques and produced a substantial body of kinetic data. Beyond the practical value of this work, the kinetic data that have been obtained are of fundamental scientific interest. They provide insights regarding the underlying principles that govern energy transfer processes and rare examples of reactions involving electronically and/or vibrationally excited species.

In this chapter, I will consider chemically pumped electronic transition lasers that are based on energy transfer from $O_2(a)$ or $NX(a)$ metastables, focusing on the energy transfer and reaction kinetics of these devices. As the chemical oxygen iodine laser is at an advanced stage of development, much of the chapter is concerned with this system. Aspects of COIL kinetics that remain to be elucidated are discussed, and recent investigations of COIL kinetics carried out by my group at Emory University are summarized. The remainder of this chapter describes laser schemes that are based on energy transfer from $NF(a)$ and $NCl(a)$. These systems are at the stage where some promise has been shown, but many open questions remain.

2. Brief History and Basic Operating Principles of COIL

The idea that $I^*[= I(^2P_{1/2})]$ could be pumped to lasing by chemical means developed from the initial observations of Arnold *et al.*[15] These investigators detected strong emission from I^* when I_2 was added to a flow of O_2 that had been passed through a microwave discharge. They correctly surmised that I^* was being excited by near resonant energy transfer from $O_2(a^1\Delta)$. Evidently, I_2 was being dissociated by electronically excited O_2, but Arnold *et al.*[15] could not determine the dissociation mechanism. Derwent and Thrush[16-19] studied the $I_2 + O_2(a)$ system in more detail. Among other findings, they concluded that the reaction

$$I + O_2(a) \leftrightarrow I^* + O_2(X) \tag{1}$$

could result in an I^* population inversion, provided the O_2 flow contained significantly more than 17% of $O_2(a)$.[18] The arguments leading to this conclusion are straightforward. The reverse component of reaction (1) is significant, as energy transfer from I^* to O_2 is near resonant (endothermic by 279 cm^{-1}, see Fig. 1). The equilibrium constant is given by

$$K_{eq} = 0.75 \exp\left(\frac{401.4}{T}\right). \tag{2}$$

Due to the ratio of degeneracies, threshold gain is achieved when $[I^*] = [I]/2$. Using the equilibrium constant it is easily shown that the yield of $O_2(a)$ needed to reach threshold gain (Y_{th}) is given by

$$Y_{th} = \frac{1}{2K_{eq} + 1} \tag{3}$$

where the yield is defined by

$$Y = \frac{[O_2(a)]}{[O_2(a)] + [O_2(X)]}. \tag{4}$$

At room temperature the above equations gives $Y_{th} = 0.15$. Such a high yield of metastable is difficult to produce using standard microwave or RF discharge excitation sources. However, chemical methods for generating high concentrations of $O_2(a)$[20] were further developed for laser demonstration experiments. These chemical singlet oxygen generators (SOG) used the solution phase reaction of Cl_2 with H_2O_2 to produce $O_2(a)$. In 1977, Pritt *et al.*[21] measured an I* population inversion when I_2 was added to the flow from a chemical $O_2(a)$ generator. A few months later McDermott *et al.*[5] obtained lasing from I* in a $O_2(a)/I_2$ flow system. This first demonstration, which employed longitudinal flow with a coaxial cavity, provided a cw output power a little above 4 mW. Less than a year later, Benard *et al.*[6] achieved cw output powers in excess of 100 W using a transverse-flow geometry. These early devices used subsonic gas flows. Attempts to

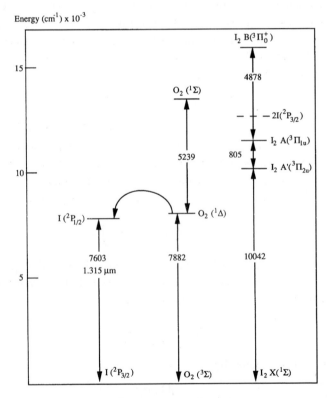

Fig. 1. Energy level diagram showing the low-lying electronic states of O_2, I_2, and I.

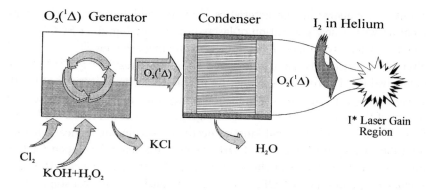

Fig. 2. Schematic diagram of the main components of a typical supersonic COIL system.

scale subsonic devices were moderately successful, and an output power of 4.6 kW was eventually obtained from a 4 m length cavity.[7]

The next critical development was a transition to supersonic flow devices. As studies of the elementary reactions occurring in COIL systems provided a better understanding of the kinetics, it became apparent that there were advantages to be gained from using supersonic flows.[7] Most importantly, the low temperature achieved by supersonic expansion (predicted to be around 150 K for typical devices) displaces the equilibrium of reaction (1) in favor of I* production. The equilibrium constant increases from $K_{eq} = 2.88$ at room temperature to 10.9 at 150 K, and the threshold yield falls to 0.044. Supersonic flow stretches the gain region, which facilitates power extraction from the optical cavity. In addition, the low temperature reduces the Doppler linewidth of the lasing transition, thereby increasing the gain at the line center. These advantages were impressively demonstrated by a supersonic device with a 25 cm length nozzle, which yielded powers comparable to the 4 m subsonic device mentioned above. With further improvements in SOG design, supersonic COIL systems have achieved cw output powers of 40 kW.[7] Figure 2 shows a schematic of a typical supersonic COIL device powered by a rotating disc SOG.

3. Development and Modeling of COIL Systems

The development of COIL systems is still in progress. Detailed characterizations of moderate size COIL systems are being carried out to learn more about device operating characteristics.[22-24] Several laboratories are experimenting with a new generation of lasers powered by liquid jet SOG's.[25-36] The potential advantages of using new gas mixing schemes,[26,27,30-32,37-43] alternative carrier gases and higher operating pressures[26,27,31,39,42,44,45] are being evaluated. From an engineering perspective, the objectives of COIL research programs are to maximize efficiency and minimize the size and weight of the laser. To facilitate the design process, the ability to model the fluid dynamics and chemical kinetics occurring within the device is highly desirable. Computer models of COIL systems are being developed for this purpose, and tested against the performance of prototype devices. Roughly speaking, operation of a COIL device can be considered in two stages; precavity events and intracavity events. The precavity events include generation of $O_2(a)$, injection of I_2 into the primary flow, I_2 dissociation, excitation of I^* and supersonic expansion. The main purpose of modeling the precavity processes is to determine the composition, temperature, flow velocity and homogeneity of the gas mixture delivered to the optical cavity. By the time the gas flow reaches the cavity it contains a substantial concentration of I^* in equilibrium with $O_2(a)$. Intracavity events are strongly influenced by lasing, which depletes the energy stored by $O_2(a)$ through reaction (1), which remains in equilibrium. Hence, the most significant kinetic processes occurring within the cavity are stimulated emission and the energy transfer processes represented by reaction (1). Elastic and inelastic collisions of I^* and I with other species present in the laser cavity are also of importance. Computational models of the intracavity processes describe the conversion of the energy stored within the gas flow to laser output power.

The precavity gas mixing and I_2 dissociation kinetics have proved to be the most difficult processes to characterize and model. In some instances models of COIL systems have been made which consider only the intracavity processes.[46-48] To do this it is assumed that a premixed flow is present in the cavity, with initial concentrations of $O_2(a)$, $O_2(X)$, I^*, and I that are representative of the experimentally observed range. Several groups have developed more elaborate models that consider the precavity and intracavity processes. These models have steadily improved as advances in computational fluid dynamics have been incorporated. Significant progress has been made. Two-[49-51] and three-[52-56] dimensional models that address the complex coupling between fluid dynamics and chemical reactions have replaced early "leaky stream" one-dimensional models of mixing.[57,58] However, it is evident that we still do not have an adequate understanding of the interplay between mixing and chemical kinetics. For example, there are practical advantages associated with operating COIL systems at relatively high pressures. Computer models predict efficient high pressure operation,[51,53,59] but experiments show that the laser power falls rapidly as the pressure is increased beyond the optimal range.[51] Note that high pressure mixing is a difficult process to simulate. Mixing models are more forgiving at low pressures where molecular diffusion is much faster and tends toward the premixed or instantaneous mixing limit.

The kinetic data used in COIL models were mostly determined in the 1980's. A critical review of rate constants available for modeling COIL systems was carried out in 1988. Attempts were made to identify the most important reactions and establish the best estimates for their rate constants. The outcome of this effort was compiled as "The Standard Chemical Oxygen Iodine Laser Kinetics Package,"[60] which is still in use. At this juncture there was a pause in the study of COIL kinetics while attention shifted to modeling of the gas dynamics. It seemed there was little point in refining the kinetic parameters if the local temperatures, pressures, and concentration gradients could not be adequately predicted. Given the advances that have been made in treating the gas dynamics, it is again pertinent to assess and improve the quality of the kinetic data. In the following sections, I will review aspects of the gas-phase kinetics of COIL that are not well established. To organize this material I will divide the discussion into subsections that consider precavity and intracavity kinetics.

4. Precavity Kinetics in COIL Systems

4.1. *The I_2 Dissociation Mechanism*

The mechanism of I_2 dissociation by $O_2(a)$ has been a long-standing problem of central importance for COIL systems.[23,58,61-63] As a single $O_2(a)$ molecule does not carry enough energy to dissociate I_2 (see Fig. 1), the mechanism must involve multiple energy transfer events. The average number (N) of $O_2(a)$ molecules needed to dissociate one I_2 molecule is a quantity that is sensitively dependent on the flow conditions. This parameter constrains the maximum energy that can be extracted from a laser. Typically, models of COIL devices use estimates for N in the range of 4–6.[58] To better understand the factors that influence N there have been several studies of the dissociation process.

In their original work on $I_2 + O_2^*$ Arnold *et al.*[15] proposed two plausible dissociation mechanisms. The first involved O_2 energy pooling to produce $O_2(b)$,

$$O_2(a) + O_2(a) \rightarrow O_2(b) + O_2(X) \tag{5}$$

which then dissociated I_2 by the reaction

$$I_2(X) + O_2(b) \rightarrow I + I + O_2(X). \tag{6}$$

Once an appreciable concentration of I had formed, Arnold *et al.*[15] suggested that $O_2(b)$ formation could be augmented by the sequence

$$O_2(a) + I \rightarrow O_2(X) + I^* \tag{1a}$$

$$O_2(a) + I^* \rightarrow O_2(b) + I. \tag{7}$$

Alternatively, they considered the possibility that $O_2(a)$ could excite a low lying, metastable state of I_2 (denoted by $I_2^\dagger$), and that this excited species was dissociated by subsequent transfer events:

$$O_2(a) + I_2 \rightarrow O_2(X) + I_2^\dagger \tag{8}$$

$$O_2(a) + I_2^\dagger \rightarrow O_2(X) + I + I. \tag{9}$$

Derwent and Thrush[16,19] concluded that reaction (6) could account for the observed dissociation rates. However, their kinetic model required a rather high value for the rate constant ($k_6 = 2 \times 10^{-10}$ cm^3 s^{-1}). Houston and coworkers[64,65] measured the rate constant for removal of $O_2(b)$ by I_2 directly, using pulsed laser excitation. They obtained an upper bound for k_6 that was an order of magnitude smaller than the Derwent and Thrush[16,19] estimate. This result supported the second mechanism proposed by Arnold *et al.*[15] In the next round of studies it was established that an intermediate excited state of I_2 was involved in the dissociation process, and that I^* also participated through a chain branching sequence.

The flow tube studies of Heidner *et al.*[66-68] still constitute the most comprehensive attempts to unravel the dissociation mechanism. In flowing $O_2^*/I_2/H_2O$ mixtures they followed the time dependence of emissions from $O_2(a)$, $O_2(b)$, $I_2(B)$, $I_2(A)$, and I^*. They found that the dissociation kinetics were consistent with a slow initiation sequence, followed by rapid removal of I_2. The overall dissociation rate increased with increasing I_2 concentration, providing a clear indication that an intermediate excited state of I_2 was involved in dissociation. Heidner *et al.*[67] proposed that the initiation process consisted of reactions (8) and (9), with reaction (8) being the rate limiting step. Once atomic I had been liberated, they proposed that the more rapid reaction

$$I^* + I_2(X) \rightarrow I_2^\dagger + I \tag{10}$$

followed by reaction (9) dominates the kinetics [participation of reaction (10) was demonstrated by Hall *et al.*[69] as described below]. Dissociation of $I_2^\dagger$ by I^* was also considered, but as the condition $[I^*] \ll [O_2(a)]$ prevailed it was deemed unlikely to be of importance.

While the fact that dissociation proceeded via an intermediate excited state of I_2 was firmly established, the character of this state could not be determined. The lowest lying electronically excited state $(A'\,^3\Pi_{2u})$ was known to be at too high an energy to be populated by transfer involving a single collision with $O_2(a)$. This appeared to leave assignment of $I_2^\dagger$ to highly excited vibrational levels of $I_2(X)$ as the only possible option, but this did not seem entirely satisfactory. Heidner *et al.*[67] used Ar as a carrier gas in their experiments, and found no dependence of the dissociation rate on Ar pressure. This was surprising, as collisions with Ar would be expected to vibrationally relax I_2. However, if relaxation was dominated by collisions with $O_2(X)$ or H_2O, Heidner *et al.*[67] concluded that the dissociation rate could be insensitive to variations in the Ar pressure over the range investigated.

Heidner *et al.*[67] developed a kinetic model of the dissociation process and attempted to define a subset of critical rate constants by fitting to dissociation rate data. Unfortunately, several of the rate constants of interest could not be uniquely determined as they were strongly correlated. Consequently, two limiting rate constant sets were offered as constraints on a yet-to-be-determined set of final values. The limiting parameter sets (models 1

Table 1. Fitted rate constants from the work of Heidner *et al.*[67]

Reaction	Reaction #	k, Model 1	k, Model 2
$O_2(a) + I_2(X) \rightarrow O_2(X) + I_2^\dagger$	8	7×10^{-15}	7×10^{-15}
$O_2(a) + I_2^\dagger \rightarrow I + I + O_2(X)$	9	3×10^{-10}	3×10^{-11}
$I_2^\dagger + O_2 \rightarrow I_2(X) + O_2$	17	5×10^{-11}	5×10^{-12}
$I_2^\dagger + H_2O \rightarrow I_2(X) + H_2O$	—	3×10^{-10}	3×10^{-10}
$I_2^\dagger + Ar \rightarrow I_2(X) + Ar$	18	4×10^{-12}	1×10^{-14}

and 2) are reproduced in Table 1. Note that three of the rate constants differ by more than an order of magnitude between the two models.

Concurrent with these studies, Hall *et al.*[69] examined energy transfer from I* to $I_2(X)$ [reaction (10)] using laser induced fluorescence (LIF) of the I_2 B–X system to monitor highly excited vibrational levels of the ground state. They observed population of levels in the range $25 < v < 43$ and proposed that reaction (10) was the chain branching step in the dissociation mechanism. They also investigated vibrational relaxation of $I_2^\dagger$ by Ar, He, and I_2. Rate constants for loss of population from $v = 40$ of $k_{He} = (3.1 \pm 0.6) \times 10^{-11}$, $k_{Ar} = (2.3 \pm 0.1) \times 10^{-11}$ and $k_{I_2} = (5.6 \pm 1.2) \times 10^{-11}$ cm^3 s^{-1} were reported. A value for k_{He} that was greater than k_{Ar} was expected on the basis of classical dynamics (see Sec. 4.2). However, the rate constant for relaxation by Ar was surprisingly large, and seemed to be at odds with the data of Heidner *et al.*[67] To resolve this conflict Hall *et al.*[69] suggested that their relaxation rate constant was dominated by $\Delta v = -1$ transfer. Reaction (10) populates levels near $v = 40$, but $I_2(X)$ molecules in levels above $v = 20$ have enough energy to be dissociated by one collision with $O_2(a)$. Hence, multiple vibrational relaxation collisions would be needed to relax the nascent vibrational distribution resulting from reaction (10) to levels below $v = 20$. The effective rate constant for this deactivation process would be much smaller than the $\Delta v = -1$ rate constant. David *et al.*[72] constructed a kinetic model of this sequential relaxation process and found they could reproduce the results of Heidner *et al.*[67] (in particular, the insensitivity of the dissociation rate to Ar pressure) using Hall *et al.*[69] relaxation rate constants.

Van Bentham and Davis[70] demonstrated the presence of vibrationally excited I_2 in a flowing $I_2/O_2(a)$ system using LIF detection. Population of the $33 < v < 44$ levels was observed, and a comparison of the time dependent profiles for $I_2^\dagger$ and I* indicated that $I_2^\dagger$ could be the intermediate from which I* was produced. Van Bentham and Davis[70] briefly discussed the origin of the $I_2^\dagger$ detected in their experiment. Conditions were such that dissociation was probably initiated by reactions (8) and (9), but in the region where spectra for $I_2^\dagger$ were recorded the concentration of I* was high enough for reaction (10) to be the dominant local source of $I_2^\dagger$. Van Bentham and Davis[70] examined the qualitative effect of the He carrier gas on the $I_2^\dagger$ vibrational level population distribution. As expected, vibrational relaxation was observed as the He pressure increased.

The average number of $O_2(a)$ molecules (N) needed to dissociate one I_2 molecule was examined by Alsing and Davis[71] using flow tube techniques. They found that N was critically dependent on the initial I_2 concentration. In a flow containing 60 m Torr of $O_2(a)$, N decreased from 16 to 6 as the initial I_2 pressure was increased from 1 to 2.7 m Torr. There are several energy loss pathways that could account for values of N that are greater than the minimum value of $N = 2$. One obvious mechanism is collisional relaxation of $I_2^\dagger$. Vibrational levels with $v > 20$ carry enough energy to be dissociated by $O_2(a)$, but David *et al.*[72] and Bouvier *et al.*[73] have argued that the probability of dissociating levels with $v < 30$ is low. This prediction was based on a Franck–Condon model, where it was assumed that transfer from $O_2(a)$ can be treated as a vertical excitation of $I_2(X, v)$ to an unstable excited state (e.g. $^1\Pi(1_u)$, 0_u^-, 1_g). Hence, there is a threshold (v_{th}) somewhere in the range $20 < v < 30$ below which $I_2(X, v)$ is not effectively dissociated by $O_2(a)$. In the framework of this interpretation the $I_2^\dagger$ deactivation rate constants reported by Heidner *et al.*[67] correspond to relaxation of the nascent vibrational distributions resulting from E–V transfer [reactions (8) and (10)] to levels below the threshold. High values for N will be seen under conditions where $I_2^\dagger$ deactivation competes with dissociation of $I_2^\dagger$ by $O_2(a)$ [reaction (9)].

Although I_2 vibrational levels in the range $10 \leq v < v_{\text{th}}$ cannot be dissociated by $O_2(a)$ they may still play a role in the overall process. Evidence that transfer from $O_2(a)$ excites $v \geq 10$ levels more rapidly than $v < 10$ levels was obtained by Barnault *et al.*[74] It has been suggested[73,74] that excitation from the mid range vibrational levels populates metastable electronically excited states:

$$I_2(X, \geq 10) + O_2(a) \rightarrow I_2(A'\,^3\Pi(2u)) + O_2(X) \qquad (11)$$

$$I_2(X, \geq 15) + O_2(a) \rightarrow I_2(A\,^3\Pi(1u)) + O_2(X) \qquad (12)$$

$$I_2(A', A) + O_2(a) \rightarrow I_2(B\,^3\Pi(0_u^+)) + O_2(X) \qquad (13a)$$

$$\rightarrow I + I + O_2(X). \qquad (13b)$$

These steps can account for the I_2 B–X and A–X emissions seen in $I_2/O_2(a)$ flames. The $A'\,^3\Pi(2u)$ state does not radiate, but Nota *et al.*[75] used laser excitation of the D'–A' system to show that A' is populated. It was estimated that individual vibrational levels of A' held populations comparable to $I_2(X)$ populations in levels near $v = 35$. The importance of reactions (11)–(13) to the dissociation process has yet to be evaluated. Rate constants for these processes (and the subsequent dissociation steps) are not known.

The I_2 dissociation model in the standard kinetics package for COIL[60] uses Heidner *et al.*[67] model 1 rate constants (cf. Table 1). The standard package was compiled before the work of Bacis and coworkers,[73,74] and the electronic excitation processes defined by reactions (11)–(13) were not considered. Clearly, our understanding of the dissociation kinetics could be improved if the identity of $I_2^\dagger$ was firmly established and reliable kinetic data for this species were available. In recent years we have attempted to address these issues by examining the kinetics of vibrationally excited $I_2(X)$ (in doing this we are implicitly testing the assumption that $I_2^\dagger$ is just vibrationally excited I_2). Our work on this problem is summarized in the following subsections.

4.2. *Rovibrational Relaxation of $I_2(X, v > 20)$*

Rovibrational relaxation of $I_2(X, v = 23, 38,$ and 42) was examined for collision partners that are likely to be present in COIL systems (He, Ar, O_2, N_2, Cl_2, I_2, and H_2O).[76-79] Two pulsed dye lasers (10 ns duration pulses) were used to excite a single rotation–vibration level of $I_2(X)$. For example, the scheme for excitation of $v = 23$ is illustrated in Fig. 3. A third pulsed laser (the probe laser) was used to monitor the population of the initially prepared state and nearby states that were populated by collisional energy transfer. The delay between the excitation and detection pulses could be continuously varied, allowing the time evolution of the population distribution to be followed. Results from this type of measurement are shown in Fig. 4. The upper panel shows a probe laser spectrum taken under collision free conditions. This trace demonstrates clean preparation of the rovibrational level $v = 23$, $J = 57$. The spectrum shows two lines [$P(57)$ and $R(57)$] as the D–X transition is subject to the selection rule $\Delta J = \pm 1$. The middle and lower panels in Fig. 4 illustrate rovibrational energy transfer induced by collisions with H_2O and Ar, respectively. These spectra show regular sequences of strong lines due to rotational energy transfer within the $v = 23$ manifold. In addition, there are several weaker lines (indicated by asterisks) that originate from the $v = 22$ rotational manifold. Careful searches were made for spectral features resulting from upward vibrational transfer ($v \rightarrow v + 1$) and multiquantum downward transfer (single-collision

$v \to v - \Delta v$ events with $\Delta v \geq 2$). These processes were too slow to yield detectable features, and we concluded that relaxation of $I_2^\dagger$ is dominated by $\Delta v = -1$ events.

An aspect of $I_2(X)$ collisional energy transfer that complicated the data analysis is apparent in Fig. 4. While it was obvious that rotational energy transfer was much faster than vibrational transfer (e.g. compare the intensities of the $v = 23$ and $v = 22$ rotational lines in Fig. 4), vibrational energy transfer was fast enough to be detectable well before rotational thermalization of the initially excited vibrational level had been achieved (with the exception of I_2–I_2 collisions). Therefore, in constructing kinetic models of the relaxation processes, rotational and vibrational energy transfer could

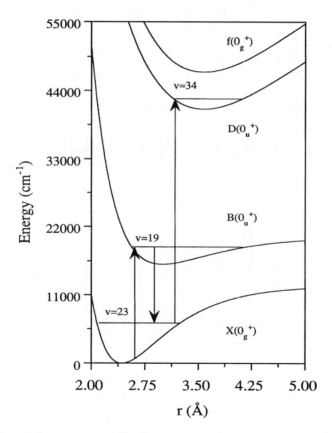

Fig. 3. Potential energy curves of I_2 showing a typical state preparation and detection scheme for studies of rovibrational energy transfer.

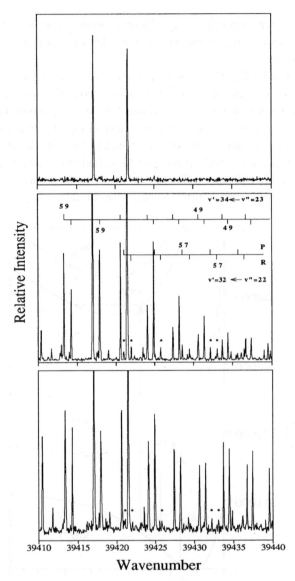

Fig. 4. Spectra showing rovibrational level state preparation and collisional energy transfer. The upper trace demonstrates clean preparation of the $I_2(X)$, $v = 23$, $J = 57$ level under collision-free conditions. The middle and lower traces show the effects of rovibrational energy transfer induced by collisions with H_2O (middle trace) and Ar (lower trace). The peaks marked with asterisks originate from levels populated by vibrational energy transfer.

not be separated. Although we were primarily interested in vibrational relaxation processes, this circumstance necessitated careful concurrent studies of rotational energy transfer.

Rotational and vibrational energy transfer rate constants derived from our measurements are presented in Table 2. Rotational transfer processes could be adequately understood using classical models of angular momentum transfer. For example, total rotational transfer rate constants [$k_{rot} = \sum_{J'} k(vJ \to vJ')$] were proportional to the collision momentum, in accordance with classical predictions. Vibrational energy transfer occurred with little vibration–rotation coupling. The rotational distributions of the $v - 1$ levels were very similar to the distributions in the initial vibrational levels produced by rotationally inelastic collisions. Downward vibrational transfer rate constants were smaller than the rotational transfer rate constants (k_{rot}) by factors in the range of 4–8 (excepting $I_2 + I_2$).

The dependence of the vibrational transfer rate constants on the initially excited v level is a point of interest. Approximate quantum mechanical models predict that rate constants for $\Delta v = -1$ transfer will scale linearly

Table 2. Energy transfer rate constants for $I_2(X)$, $v = 23$ and 38.

M	v	k_{vib}^{M} [a,b]	$k_{\Delta v = -1}^{M}$ [a,c]	k_{rot}^{M} [a,d]
He	23	1.7	1.2	4.7
He	38	1.4	1.0	4.0
Ar	23	1.2	0.8	5.3
Ar	38	0.6	0.4	4.9
O_2	23	1.1	0.8	6.3
O_2	38	1.0	0.7	6.0
Cl_2	23	1.4	1.0	6.4
I_2	23	< 0.3	< 0.3	8.5
I_2	38	< 0.3	< 0.3	6.6
H_2O	23	2.2	1.6	8.7
H_2O	38	2.6	1.9	6.2

[a] Units of 10^{-10} cm^3 s^{-1}.
[b] Rate constant for transfer out of the initial vibrational level to all other vibrational states. Estimated error limits, ±40%.
[c] Estimated error limits, ±40%.
[d] Total rate constant for pure rotational energy tranfer. Estimated error limits, ±25%.

with v.[80] However, it is known that linear v scaling does not work for $I_2(B)$,[81,82] and it is not reflected by classical trajectory calculations for I_2 colliding with a structureless collision partner.[83] For high vibrational levels, classical models predict transfer probabilities that are rather weakly dependent on v. The errors associated with the vibrational transfer rate constants in Table 2 are relatively large, which obscures the vibrational dependence. Nevertheless, the data do suggest a dependence on v that is less than linear, in keeping with semiclassical models.

For the range of collision partners investigated, it appeared that the collision momentum was the most important factor determining the vibrational transfer probabilities. Experimental studies of $I_2(B)$[84] and the classical trajectory calculations of Rubinson *et al.*[83] have shown that there is an optimum value of the collision momentum for efficient transfer. For a fixed temperature this corresponds to an optimum mass for the collision partner. We found that transfer probabilities for $I_2(X)$ increased with increasing collision partner mass for all of the colliders examined, with the exception of I_2. This suggests that the optimum mass for the collider is somewhere between 70 and 254 amu. Note that the models described here considered only vibration to translation $(V–T)$ energy transfer. Vibration to vibration $(V–V)$ transfer is possible for molecular collision partners, but it is improbable for most of the collision partners examined. The large vibrational spacings of O_2, N_2, and H_2O require unreasonably large Δv transitions in $I_2(X)$ to excite the accepting mode. Even the fundamental of $Cl_2(X)$ requires a minimum of $\Delta v = 3$ for $I_2(X, v > 20)$. $V–V$ transfer in $I_2^\dagger + I_2$ collisions can occur with $\Delta v = 2$, but it appears that this is not a significant process.

The temperature dependencies of $I_2(X)$ vibrational transfer rate constants are of relevance to ongoing efforts to improve the performance of supersonic COIL devices. At present, most COIL systems employ subsonic injection of I_2 into the primary $O_2(a)$/He flow. For this situation I_2 dissociation occurs at ambient temperatures. However, mixing schemes that involve transonic and supersonic injection of I_2 are being investigated.[26,27,30–32,38–43] In these devices I_2 may be dissociated in regions where the flow is undergoing rapid cooling or has already reached a low temperature. Some theoretical models predict enhanced vibrational transfer cross sections at low temperatures,[85] which would be detrimental to transonic and supersonic injection devices. Other models predict a more benign decrease with decreasing temperature.

We briefly examined vibrational relaxation of $I_2(X, v > 20)$ induced by collisions with He at low temperatures. These measurements were performed by focusing state preparation and probe laser beams into the downstream region of a He free-jet expansion that had been seeded with I_2. Transfer out of the initial levels $v = 23$ and 42 was examined at 5 K. For these conditions the probe laser spectra were quite simple and uncongested. A typical probe laser spectrum is shown in Fig. 5. The delay used to record this trace was long enough for multiple collisions to occur, so populations are seen in levels corresponding to $\Delta v = -1$ and -2 transitions.

Low temperature vibrational transfer data were analyzed by fitting the time-dependent relative vibrational populations to a kinetic model defined by the rate equations

$$\frac{d[v]}{dt} = \sum_{v'}(-k_{v-v'}[v] + k_{v'-v}[v'])[\text{He}] \tag{14}$$

where $[v]$ and $[v']$ are the relative populations in levels v and v', and $k_{v-v'}$ is the rate constant for transfer from level v to level v'. As the temperature was low, only downward energy transfer was included in the model. Two different assumptions about the dependence of $k_{v-v'}$ on $\Delta v (= v - v')$ were examined. In the first model it was assumed that relaxation occurred by sequential $\Delta v = -1$ steps, so the rate constants for multiquantum jumps were

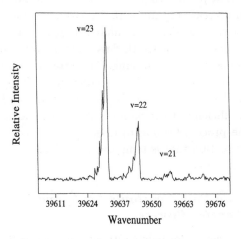

Fig. 5. Vibrationally resolved spectrum showing relaxation of $I_2(X)$, $v = 23$ by collisions with He at a temperature of 5 K.

set to zero. In the second model transfer rate constants were represented by the expression

$$k_{v-v'} = k_{v-(v-1)} \exp[-\alpha(1 - |\Delta v|)] \tag{15}$$

which permits multiquantum transfer.

For low temperature collisions with He, the fit given by the multiquantum jump model was clearly superior to that obtained using the single quantum jump model. Even so, the $\Delta v = -1$ process accounted for more than 70% of the total removal rate constant ($\alpha = 1.1$). For transfer out of $v = 23$, the total removal rate constant was around 1.6×10^{-11} cm^3 s^{-1} at 5 K. This was roughly an order of magnitude smaller than the He vibrational relaxation rate constant at room temperature (1.7×10^{-10} cm^3 s^{-1}). Part of this difference is from the change in the collision frequency. To compensate for this factor, it is helpful to calculate effective vibrational relaxation cross sections from the relationship $\sigma_{v-v'} = k_{v-v'}/\langle v \rangle$, where $\langle v \rangle$ is the average relative speed of the collision partners. Effective cross sections for removal of population from I$_2$, $v = 23$ by collisions with He were 10.1 ± 2.1 Å^2 at 5 K and 9.5 ± 3.0 Å^2 at 300 K. Similarly, for $v = 42$, the vibrational transfer cross sections were 8.7 ± 2.2 Å^2 (5 K) and 7.9 ± 2.2 Å^2 (300 K). These results suggest that the cross section is approximately independent of the temperature and the initial vibrational level over the ranges investigated. This behavior is at variance with simple theories, such as the Schwartz–Slawsky–Herzfeld (SSH) model,[80] that predict that the cross sections should show a linear v-dependence and decrease rapidly with decreasing temperature. Conversely, our data did not show evidence of enhanced cross sections at low collision energies. The implication for COIL devices with transonic and supersonic I$_2$ injection is somewhat encouraging. The rate constants for vibrational relaxation decrease with the collision frequency ($\propto T^{1/2}$) so the dissociation efficiency will not be reduced by enhanced loss of I$_2^\dagger$. Of course, for the dissociation efficiency to remain constant or increase with decreasing temperature, the process leading to dissociation of I$_2^\dagger$ [reaction (9)] must also have a cross section that is temperature independent or that increases as the temperature falls.

4.3. *Deactivation Rate Constants for* I$_2^\dagger$ *and the* I* + I$_2$ *Energy Transfer Process*

Models of the dissociation of I$_2$ in COIL systems and I$_2$/O$_2(a)$ flames must include an adequate description of processes that deactivate I$_2^\dagger$. In principle, models can be constructed that consider the populations in all vibrational levels of $I_2(X)$. While this would be a viable approach for modeling flow tube data, it is not desirable for numerical models of COIL devices. It would add a significant level of complexity (and cost) to track individual quantum states, and is probably an unnecessary embellishment. Heidner *et al.*[67] modeled their flow tube data using effective deactivation rate constants. These may be defined in terms of the overall process (multicollision)

$$I_2^\dagger + M \to I_2(X, v < v_{\text{th}}) + M \tag{16}$$

(in the following deactivation rate constants are indicated by the label $k_M^\dagger$). Deactivation rate constants can be derived from the state-to-state rate constants, provided that the nascent vibrational distribution of $I_2^\dagger$ is known. Prior to our work on this problem, partially relaxed distributions had been observed. Hall *et al.*[69] recorded low resolution spectra for $I_2^\dagger$ generated by reaction (10). They did not attempt to map the population distribution, but noted the range of vibrational levels populated. Van Bentham and Davis,[70] and Barnault *et al.*[74] reported $I_2^\dagger$ vibrational distributions that were recorded in $I_2/O_2(a)$ flames.

The $I_2^\dagger$ distribution of most significance is that resulting from reaction (10), which will become the dominant means of exciting $I_2^\dagger$ once the slow initiation processes have liberated atomic I. It is of interest to note that reaction (8) (E–V transfer from $O_2(a)$, the only energetically possible transfer process) is likely to be slow because near resonant transfer requires a large change in the I_2 vibrational quantum number ($\Delta v \approx 40$). The same restriction does not apply to $I^* + I_2$ transfer as this probably involves passage through a bound I_3 intermediate.

To determine the nascent distribution, we examined $I_2^\dagger$ produced by reaction (10) under low pressure conditions. Pulsed photolysis of I_2 at 496 nm was used to generate I^*. As only a fraction of the I_2 was dissociated, $I_2^\dagger$ then appeared due to $I^* + I_2$ collisions. The excited vibrational levels were probed using LIF of the D–X transition with full rotational resolution.[76,86] The nascent vibrational population distribution extracted from this spectrum is shown in Fig. 6, which reveals a strong preference for near resonant E–V transfer.

With this information in hand we constructed simple models of the $I_2^\dagger$ deactivation process.[76] The first of these allowed only $\Delta v = \pm 1$ transfer steps. Downward transfer rate constants (derived from the experimental data)

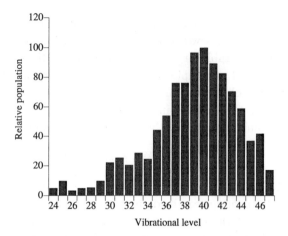

Fig. 6. Nascent vibrational population distribution resulting from the electronic to vibrational energy transfer process $I^* + I_2(X) \rightarrow I + I_2^\dagger$.

were assumed to be either independent of v, or linearly dependent on v. Upward transfer rate constants were calculated using the principle of detailed balance. Collisional relaxation was simulated by numerically integrating the coupled rate equations, with the initial conditions defined by the nascent $I_2^\dagger$ distribution. The concentration of $I_2^\dagger$ was determined by summing the concentrations of all of the $I_2(v > v_{th})$ components. Deactivation rate constants were obtained by computing the lifetime of $I_2^\dagger$ as a function of collider gas pressure. Two limiting values for v_{th} were considered. These were the purely energetic threshold, $v_{th} = 20$, and the threshold estimated from the Franck–Condon model, $v_{th} = 30$. Calculated values for the deactivation rate constants are listed in Table 3. Note that the deactivation rate constants are roughly 20 times smaller than the $\Delta v = -1$ rate constants for the slow deactivation parameters (linear v-scaling with $v_{th} = 20$). This ratio decreases to about 10 for the fast deactivation parameters (rate constants independent of v, with $v_{th} = 30$).

The deactivation model was developed using room temperature energy transfer data that were obtained under conditions where $\Delta v = \pm 2$ events could not be detected. For collisions with He, the low temperature measurements described above did show $\Delta v = -2$ transfer, and provided an estimate for the relative magnitudes of the $\Delta v = -1$ and $\Delta v = -2$ rate constants. To examine the influence of multiquantum transfer on the deactivation model, we recently included $\Delta v = \pm 2$ channels. To estimate the

Table 3. Rate constants for deactivation of $I_2(X)^\dagger$ (298 K).

M	n^a	$k^M(0 \leftarrow 1)^{b,c}$	$k_\dagger^M(v_{th} = 20)^d$	$k_\dagger^M(v_{th} = 30)^d$
He	0	11	5.0	7.8
He	1	0.39	5.1	9.6
Ar	0	6.0	2.8	4.3
Ar	1	0.23	3.0	5.6
O_2	0	8.0	3.7	5.7
O_2	1	0.27	3.5	6.6
N_2	0	13	6.0	9.2
N_2	1	0.34	4.4	8.4
Cl_2	0	10	4.6	7.1
Cl_2	1	0.44	5.7	11
H_2O	0	18	8.3	13
H_2O	1	0.60	7.8	15

[a] Models with no v-dependence ($n = 0$) and linear v-scaling ($n = 1$).
[b] Base rate constants for transfer from $v = 1$ to $v = 0$.
[c] Units of 10^{-11} cm^3 s^{-1}.
[d] Units of 10^{-12} cm^3 s^{-1}.

$\Delta v = -2$ rate constant we assumed that the $k_{\Delta v=-2}/k_{\Delta v=-1}$ ratio determined from the low temperature data (0.3) was also valid for 300 K. With this modification the deactivation rate constants for He were predicted to be a factor of two greater than those listed in Table 3 (for both the slow and fast deactivation parameters).

4.4. *Implications of $I_2^\dagger$ Relaxation Kinetics for COIL Models*

As the deactivation rate constants of Heidner *et al.*[67] are used in most computational simulations of COIL systems, comparisons of deactivation rate constants derived from the energy transfer measurements with Heidner *et al.*[67] results are of particular interest.

In Heidner *et al.*[67] models, the rate constants for dissociation of $I_2^\dagger$ by $O_2(a)$ [reaction (9)] and the processes

$$I_2^\dagger + O_2(X) \rightarrow I_2 + O_2(X) \tag{17}$$

$$I_2^\dagger + Ar \rightarrow I_2 + Ar \tag{18}$$

are strongly correlated. The ratio $k_9/k_{O_2}^\dagger$ was constrained to a value of 6. Model 1 assumed a high rate constant for k_9, which in turn yielded

$k_{O_2}^\dagger = 5 \times 10^{-11}$ cm^3 s^{-1}. For this limit deactivation by O_2 dominated, and the model could tolerate $k_{Ar}^\dagger$ values as high as 4×10^{-12} cm^3 s^{-1}. Model 2 assumed k_6 and $k_{O_2}^\dagger$ values that were an order of magnitude smaller than those of model 1. With slow deactivation by O_2, the model could not permit much deactivation by Ar, and the upper bound for $k_{Ar}^\dagger$ fell to 1×10^{-14} cm^3 s^{-1}. The deactivation rate constants derived from the energy transfer measurements do not fit with either model. The predicted deactivation rate constant for O_2 is in good agreement with the model 2 value, while the Ar deactivation rate constant agrees with model 1. Furthermore, the deactivation rate constants for H_2O are in sharp disagreement. Heidner et al.[67] models use a value for k_{H_2O} of 3×10^{-10} cm^3 s^{-1}. This exceeds the measured $\Delta v = -1$ rate constant, and is a factor of 20 greater than the highest value of $k_{H_2O}^\dagger$ predicted from the energy transfer data.

Data such as those shown in Fig. 4 are visibly in disagreement with the relative values for $k_{Ar}^\downarrow$ and $k_{H_2O}^\downarrow$ from Heidner et al.[67] models. The middle and lower traces were recorded under conditions where the number of collisions with Ar or H_2O were approximately equal. These traces show comparable amounts of $\Delta v = -1$ transfer. The fraction of vibrational transfer induced by H_2O is no more than 4 times greater than that induced by Ar. These data are obviously inconsistent with a ratio of 75 (model 1) or 3×10^4 (model 2). Similarly, the energy transfer data do not support dramatically different values for $k_{Ar}^\dagger$ and $k_{O_2}^\dagger$.

Heidner et al.[67] did not examine deactivation of $I_2^\dagger$ by He. This is an important parameter for COIL, where He is normally used as the carrier gas. The standard kinetics package recommends a value for $k_{He}^\dagger$ of 4×10^{-12} cm^3 s^{-1}, presumably based on model 1 and the notion that $k_{He}^\dagger$ would be roughly the same as $k_{Ar}^\dagger$. The predicted deactivation rate constants for He given in Table 3 are in reasonable agreement with this estimate. Despite this apparent harmony, the value for $k_{He}^\dagger$ was called into question by Carroll,[51] who encountered problems when trying to use the standard package to model the performance of a COIL system operating in the 70–130 Torr pressure range. Carroll[51] treated $k_{He}^\dagger$ as a variable in his model, and found that the behavior of the laser could be reproduced when $k_{He}^\dagger$ was increased to 4.8×10^{-11} cm^3 s^{-1}; a value well outside the range of the rate constants derived from the energy transfer data. This discrepancy could be an artifact caused by other uncertain rate constants or problems with the mixing model at high pressures. Madden[52] has since modeled two-dimensional gain profile maps for COIL using a three-dimensional computational fluid dynamics code. Both normal and high pressure operation was examined. Madden[52] found that the deactivation rate constants derived from vibrational relaxation measurements (the $v_{th} = 30$, $n = 1$ rate constants from Table 3) gave better agreement with experiment than the rate constants from the standard package. These results suggest that the increase in $k_{He}^{\dagger}$ suggested by Carroll[51] may not be needed to model the high pressure data.

Overall the inconsistencies are sufficiently troubling that we do not yet have a secure understanding of the dissociation mechanism. There is no obvious way to reconcile the deactivation rate constants reported by Heidner *et al.*[67] with the vibrational deactivation rate constants presented in Table 3. This is not simply a problem with the absolute magnitudes of the rate constants. The flow tube data were simulated using rate constant ratios that are clearly not reflected in the vibrational relaxation data. It is possible that the inconsistencies arise because we are not comparing equivalent properties. Thus far we have assumed that Heidner *et al.*[67] deactivation rate constants are for vibrational deactivation. However, it is possible that electronically excited states of I_2 also play an important role in the dissociation process [e.g. the neglected reactions (11), (12), and (13)]. If this was the case, the deactivation constants would be trying to represent a combination of vibrational relaxation and electronic quenching effects. Further studies of the dissociation mechanism will be needed to unravel this convoluted problem.

5. Intracavity Kinetics in COIL Systems

5.1. $I^* + O_2(X)$: *Electronic Energy Transfer*

Energy transfer from $O_2(a)$ to I is the most important kinetic processes occurring within the laser cavity. It is easy to see that the efficiency of the laser will be degraded if reaction (1a) is not fast enough to convert the energy stored in $O_2(a)$ before the gas leaves the optical cavity. The rate constant for reaction (1a) at room temperature is well known. As it is technically easier to study the reverse reaction

$$O_2(X) + I^* \rightarrow O_2(a) + I \tag{1b}$$

most determinations of $k_{(1a)}$ were performed by measuring $k_{(1b)}$, and then calculating $k_{(1a)}$ from the equilibrium constant. This procedure is valid provided that processes represented by the reaction

$$O_2(X) + I^* \rightarrow O_2(X) + I \tag{19}$$

are unimportant. This condition was established by Derwent and Thrush,[18] and Young and Houston.[87]

Models of supersonic COIL devices require information about the temperature dependence of $k_{(1a)}$ and $k_{(1b)}$ over the range 150–300 K. The low temperature behavior of reaction (1b) was originally investigated by Deakin and Husain.[88] Their results were fit to a simple Arrhenius expression that yielded a preexponential factor of $(3.6 \pm 1.8) \times 10^{-11}$ cm^3 s^{-1} and an activation energy of 0.2 ± 0.2 kcal mol^{-1}. Deakin and Husain[88] noted that the activation energy was much smaller than the endothermicity of the quenching process. To explain this anomaly, they suggested that the "observed activation energy of 0.2 kcal mol^{-1} may result from an energy barrier of magnitude equal to that of the endothermicity for electronic energy transfer, coupled with a T^{-1} dependence in the preexponential factor." This suggestion is equivalent to the equation

$$k_{(1b)} = \frac{k'}{T} \exp\left(\frac{-401.4}{T}\right) \tag{20}$$

where k' is a constant. Combining Eqs. (2) and (20) gives

$$k_{(1a)} = \frac{0.75k'}{T} = 7.8 \times 10^{-11} \left(\frac{295}{T}\right) \text{ cm}^3 \text{ s}^{-1} \tag{21}$$

where the substitution for k' was made using the room temperature value for $k_{(1b)}$. Equations (20) and (21) are used frequently in models of supersonic COIL devices.

The T^{-1} dependence of $k_{(1b)}$ was unexpected as it implies that there are strong attractive forces acting between $O_2(X)$ and I^* at relatively large internuclear separations. To explore this phenomena in more detail, and to test the reliability of Eqs. (20) and (21), we examined reaction (1b) at a temperature of 150 K.[89,90] These measurements were carried out using a Laval nozzle to provide low temperature gas flows. Traces of I_2 were entrained in He/O_2 mixtures, and pulsed laser photolysis was used to generate I^* by exciting just above the $I_2(B)$ dissociation limit. Near threshold photodissociation was used to ensure that I^* was produced with a translational temperature in equilibrium with the local conditions. I^* decay kinetics were monitored as a function of the O_2 concentration. In one series of measurements time-resolved $I^* \to I$ fluorescence decay curves were recorded. The detector available for the first measurement of this kind was relatively slow, so the decay curves were convoluted with the detector response characteristics.[90] A plot of deconvoluted decay rates versus O_2 concentration yielded a value for $k_{(1b)}(T = 150)$ of $(4.6 \pm 1.0) \times 10^{-12}$ cm^3 s^{-1}, where the error limits did not include systematic deviations that may have been introduced by decovolution. As we were unable to obtain a satisfactory assessment of the data reduction errors, we repeated the measurements when a faster detector became available.[89] Direct analysis of the undistorted decay curves gave a slightly higher estimate for the rate constant of $k_{(1b)}(T = 150) = (6.3 \pm 1.0) \times 10^{-12}$ cm^3 s^{-1}.

Additional experiments were performed using pulsed LIF detection to monitor I^*.[89] Excitation of the $5p^4 6s\,^2 P_{3/2} - 5p^5\,^2 P_{1/2}$ transition near 206.2 nm was used for this purpose. To observe the removal of I^*, the delay between the photolysis and LIF probe pulses was fixed, and the concentration of O_2 varied. Quenching kinetics were extracted from the dependence of the LIF signal on the O_2 concentration. Analysis of these data was complicated by collisional transfer between hyperfine sublevels (discussed in detail below). When the effects of hyperfine transfer were taken into account, the LIF data yielded a value for $k_{(1b)}$ that was in agreement with the estimate based on fluorescence decay kinetics. A simultaneous fit to both data sets established a value of $k_{(1b)}(T = 150) = (7.0 \pm 0.7) \times 10^{-12}$ cm^3 s^{-1}.

Our value for $k_{(1b)}$ was a factor of two smaller than the prediction of Eq. (20). Taken with the room temperature data, our result was consistent with the Arrhenius expression

$$k_{(1b)} = (1.05 \times 10^{-10}) \exp(401.4/T) \text{ cm}^3 \text{ s}^{-1} \qquad (22)$$

with the activation energy equal to the endothermicity of reaction (1b). Combining Eq. (22) with the expression for the equilibrium constant gives a temperature independent value for $k_{(1a)}$ of 7.8×10^{-11} cm^3 s^{-1} for the range $150 \leq T \leq 300$. The consistency of these results with supersonic COIL performance was tested by Madden[52] in CFD calculations. His model yielded predictions that were in good agreement with experimental gain measurements.

The temperature independence of $k_{(1a)}$ implies that there is no barrier for this process. The most probable mechanism would involve surface crossings (seams of intersection) between states that correlate with the $O_2(a) + I$ and $O_2(X) + I^*$ dissociation limits, with the crossings located below the former asymptote. Furthermore, the long-range attractive forces acting between I and $O_2(a)$ must be of moderate strength; just sufficient to compensate for the decrease in the collision frequency as the temperature is reduced.

To gain a deeper understanding of reaction (1), high level *ab initio* calculations were carried out to predict the potential energy surfaces of states that correlate with the $I + O_2(a)$ and $I^* + O_2(X)$ dissociation limits.[91] Given the above considerations, it was of interest to know if there are any deeply bound states that correlate with $I + O_2(a)$, and the energy regions where the surfaces from the two dissociation limits intersect. Technical details of these calculations, including a description of the perturbative method used to treat spin-orbit coupling, are given in Ref. 91.

Spin-coupled calculations revealed crossing seams between potential energy surfaces arising from the $I + O_2(a)$ and $I + O_2(X)$ asymptotes. In a fully adiabatic treatment of this problem the surface crossings were avoided and somewhat difficult to locate. To facilitate interpretation of the adiabatic results, diabatic calculations were preformed with coupling between states from different asymptotes suppressed. Surface intersections were allowed and readily located using the diabatic treatment.

Radial cuts through the potential surfaces were calculated for several different angles. These cuts showed their deepest minima and lowest energy curve crossings for Jacobi angles around $50°$. Figure 7 shows diabatic cuts for $\theta = 50°$. Note that the states are bound by 250 cm^{-1} or less, indicative of weak dispersive interactions. Several curve crossings occur below the $O_2(a) + I$ dissociation limit. This supported the notion that electronic energy transfer between I and O_2 is mediated by surface intersections, and that quenching of I^* by $O_2(X)$ should be characterized by an activation energy of $E_{act} = \Delta U = 279$ cm^{-1}. A rough value for $k_{(1a)}$ was obtained by considering the curves shown in Fig. 7. The avoided crossings occur in the $R = 6.5$–7 au range, so it was assumed that collisions with impact parameters of $b \leq 6.5$ au could result in transfer. As only two of the four states that correlate with $I + O_2(a)$ are crossed by lower states, only half of the collisions will occur on suitable surfaces. It was further assumed that collisions that traverse the crossing region sample the states statistically. As two upper surfaces are crossed by three lower surfaces, this yields a probability of $\frac{3}{5}$ for exit on one of the lower surfaces. These considerations lead

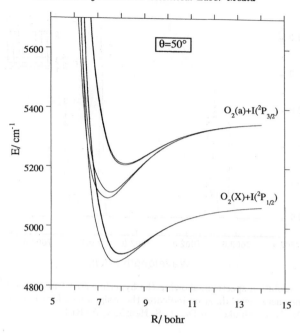

$\theta=50°$

5600

5400

$O_2(a)+I(^2P_{3/2})$

E/ cm^{-1}

5200

$O_2(X)+I(^2P_{1/2})$

5000

4800

5   7   9   11   13   15

R/ bohr

Fig. 7. Diabatic *ab initio* potential energy curves for $I + O_2$. These calculations correspond to a fixed O–O distance of 2.287 bohr and a Jacobi angle of 50°.

to a transfer cross section of $\sigma_f = \frac{1}{2} \cdot \frac{3}{5}\pi b_{max}^2 = 40$ au^2. Taken with the average collision velocity, this yields a 295 K rate constant of $k_f = 5.5 \times 10^{-11}$ cm^3 s^{-1}. Given the severity of the approximations, this is in reasonable agreement with the experimental value of $(7.8 \pm 0.6) \times 10^{-11}$ cm^3 s^{-1}.

5.2. *Collisional Transfer between the Hyperfine Sublevels of $I(^2P)$*

Continuous wave operation of COIL is facilitated by the hyperfine structure of the atom. Iodine has a nuclear spin of $\frac{5}{2}$, so the $^2P_{1/2}$ and $^2P_{3/2}$ levels are split by hyperfine interactions. Figure 8 shows the allowed transitions between the hyperfine sublevels and a high resolution emission spectrum. The $F' = 3 \rightarrow F'' = 4$ transition is most intense, and this is the laser line under normal conditions. Collisional relaxation between the hyperfine sublevels of $^2P_{3/2}$ maintains the population inversion, while transfer between the $^2P_{1/2}$ levels extracts energy stored in the $F' = 2$ level. Hence, if it is not sufficiently rapid, hyperfine relaxation can limit power extraction.

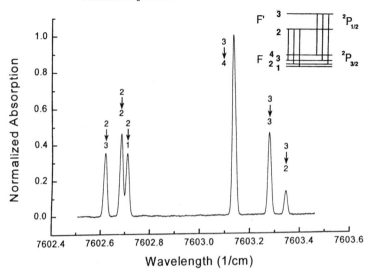

Fig. 8. High resolution spectrum showing the hyperfine structure of the I $^2P_{1/2}$–$^2P_{3/2}$ transition. The energy level diagram indicates the assignments for the transitions (figure supplied by Dr. G. C. Manke and Dr. T. S. Henshaw, AFRL).

Theoretical models predict that hyperfine relaxation of I* and I occurs by distinctly different mechanisms.[92,93] Electrostatic interactions between I $^2P_{3/2}$ and a collision partner can reorient the electronic angular momentum vector (**J**) and thereby change the magnitude of the total angular momentum, F (the nuclear spin, **I**, is a spectator during these events). However, electrostatic interactions cannot induce transitions from M_J to $-M_J$ when J is half-integer (M_J being the space-fixed projection of **J**).[94] This is not unduly restrictive for a $^2P_{3/2}$ state, so any collision partner will be able to induce hyperfine transfer. In contrast, this selection rule will not permit electrostatically induced transfer between the hyperfine sublevels of $^2P_{1/2}$. Magnetic and/or exchange interactions are required to cause transfer in this case. For COIL systems this theory predicts that all species present in the device will contribute to hyperfine relaxation of I $^2P_{3/2}$, while only I ^2P and $O_2(X)$ will cause significant hyperfine relaxation of I*. Furthermore, as the pressure of $O_2(X)$ will be much greater than that of atomic iodine, collisions with $O_2(X)$ should dominate I* hyperfine relaxation.[47,61,95]

Rate constants for transfer between the hyperfine levels of I $^2P_{3/2}$ have not been measured. Yukov[92] provided a theoretical estimate of 2.3 × 10^{-9} cm^3 s^{-1} for transfer induced by collisions with He, while Okunevich and Perel[96] suggested a general value of 4×10^{-10} cm^3 s^{-1} for any collision partner. Based on these predictions, it is usually assumed that equilibrium is maintained among the ground state hyperfine levels in laser systems (both photolytic and COIL devices).

Hyperfine relaxation between the sublevels of I* was investigated by Thieme and Fill[97,98] using pulsed photolytic iodine lasers. Photolysis of C$_3$F$_7$I was used to generate inverted I* in an amplifier cell. Pulses from a high power iodine laser were used to disturb the equilibrium between the I* $F = 3$ and $F = 2$ levels by saturating the $F' = 3 \leftrightarrow F'' = 4$ transition. Delayed pulses from a low power iodine laser were then used to monitor gain on the 3–4 and 2–2 transitions, and thereby monitor relaxation back to equilibrium. In the analysis of these data it was assumed that relaxation between the ground state levels was rapid, and that the observed kinetics reflected transfer between the I* $F = 2$ and 3 levels. Thieme and Fill[97,98] concluded that hyperfine relaxation of I* was primarily induced by collisions with ground state iodine atoms. A rate constant of 1.01×10^{-9} cm^3 s^{-1} was reported for $F = 2 \rightarrow F = 3$. They also suggested that I* + I* collisions make a small contribution to hyperfine relaxation. Transfer by the other open-shell species in the system, C$_3$F$_7$, appeared to be unimportant.

Hyperfine transfer induced by collisions with O$_2$ was observed during our recent studies of I* + O$_2$(X) quenching when LIF detection was employed.[89,99] The linewidth of the laser used to excite LIF was narrow enough to permit partial resolution of the hyperfine structure. Scans over the atomic transition are shown in Fig. 9. Transitions arising from $^2P_{1/2}$ $F = 2$ and 3 were resolved as the separation between these levels is relatively large (0.66 cm^{-1}). Hyperfine splitting of $5p^46s\,^2P_{3/2}$ is on a much smaller scale, and was not resolved. Photolysis of I$_2$ via the B state continuum was used to generate I*. In principle this should yield statistical population of the $F = 2$ and $F = 3$ sublevels ($[F = 2]/[F = 3] = 5/7$). Unexpectedly, we found that near threshold dissociation resulted in nonstatistical hyperfine level populations. The initial distributions depended on the photolysis wavelength and temperature. Photolysis of a 150 K sample at 498 nm produced a distribution with $[F = 2]/[F = 3] = 1.3$ (this is apparent in the upper trace of Fig. 9). Reducing the temperature to 10 K and photodissociating at 499 nm increased the ratio to $[F = 2]/[F = 3] \approx 8$. Electronic state mixings near the dissociation limit of I$_2$(B) are the most probable cause of these anomalous population distributions.

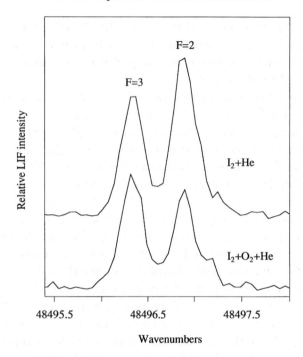

Fig. 9. LIF scans over the I* $5p^46s(^2P_{3/2})$–$5p^5(^2P_{1/2})$ transition. I* was generated by 498 nm photolysis of I_2 in a Laval nozzle expansion at 150 K. The upper trace shows peak intensities that reflect the nascent hyperfine level populations. The lower trace shows the effect of relaxing this distribution by collisions with $O_2(X)$.

Collisions between I* and closed-shell collision partners (He, Ar, and N_2) did not relax the nonequilibrium hyperfine distribution to any measurable degree. But, in accord with the theoretical picture outlined above, hyperfine relaxation was observed for collisions with $O_2(X)$. For example, the shift towards equilibrium can be seen in Fig. 9 by comparing traces recorded in the absence and presence of O_2. Hyperfine transfer induced by collisions with O_2 was examined at the temperatures $T = 295$, 150, and 10 K. For $T \geq$ 150 K, I* quenching and hyperfine transfer occurred at similar rates. Hence, both processes were considered in the kinetic model used to extract the rate constants. At the lowest temperature (10 K) quenching was negligible. The hyperfine transfer rate constants were found to be independent of temperature, within the experimental errors. Values of $k(F = 2 \rightarrow F = 3) =$ $(2.3 \pm 0.5) \times 10^{-11}$ and $k(F = 3 \rightarrow F = 2) = (1.6 \pm 0.5) \times 10^{-11}$ cm^3 s^{-1}

were obtained. As described in the previous section for $k_{(1a)}$, the temperature independence implies that the hyperfine transfer cross sections increase with decreasing temperature due to long range attractive forces.

The measured hyperfine transfer rate constants were smaller than previous estimates. Cerny *et al.*[100] reported a value for $k(F = 3 \to F = 2)$ of 1.5×10^{-10} cm^3 s^{-1} ($T = 300$ K), derived from an analysis of spectral line pressure broadening data. Alekseev *et al.*[93] obtained a theoretical estimate for the transfer rate constants of 10^{-10} cm^3 s^{-1}. Churassy *et al.*[61] obtained good agreement between the performance of their COIL device and a computer model that included hyperfine relaxation with $k(F = 2 \to F = 3) = 10^{-10}$ cm^3 s^{-1}. Copeland and Bauer,[48] and Zagidullin and Nicolaev[47] considered hyperfine relaxation in models of COIL gain saturation. In these studies the sum of the rate constants, $k(F = 2 \to F = 3) + k(F = 3 \to F = 2)$, was set equal to 10^{-10} cm^3 s^{-1}. Copeland and Bauer[48] assumed that the constants were temperature independent, while Zagidullin and Nicolaev[47] assumed a temperature dependence of the form $k(F = 2 \to F = 3) + k(F = 3 \to F = 2) = 10^{-10}\sqrt{T/300}$. At 150 K these estimates are factors of two to five greater than the present results.

5.3. *Pressure Broadening and Velocity Changing Collisions*

Optical power extraction is influenced by homogeneous and heterogenous broadening of the I* $\to$ I 3–4 transition. As the cavity modes are much narrower than the atomic line, the gain depends on the width and shape of the line. Any process that broadens the line, such as pressure broadening or increases in the Doppler width due to heating, will reduce the gain at the line center. The relative magnitudes of the homogenous and heterogeneous linewidths also have important implications for optical power extraction. Consider a situation where homogenous broadening is small compared with the Doppler width. Stimulated emission selectively depletes a specific subgroup of the I* velocity distribution and the energy stored in the remaining velocity groups will not be available, unless velocity changing collisions can rapidly replenish the hole in the distribution. If relaxation of the velocity distribution is slow compared to the stimulated emission rate, the laser will exhibit heterogeneous optical saturation characteristics. This is, of course, undesirable for optimal power extraction. Conversely, if relaxation of the velocity distribution is fast, and/or the homogeneous width of the line is significant compared to the Doppler width, all of the energy stored by I* will be available to the radiation field.

Pressure broadening coefficients have been measured by three different techniques. The results of these studies are collected in Table 4. Padrick and Palmer[101] used laser gain measurements to determine broadening coefficients. Pulsed photolysis of CF_3I was used to generate an inverted population of I^* in an amplifier cell. The small signal gain was measured using low power pulses from an independent I^* laser. Broadening coefficients were derived from the dependence of the gain on the pressure of buffer gas added to the amplifier cell. The collision partners He, Ne, Ar, Kr, Xe, N_2, and CO_2 were investigated. For the rare gases the coefficients were in agreement with the predictions of simple models.

High resolution Fourier Transform spectrometers have been used to examine the 3–4 transition lineshape in both absorption and emission. Engleman et al.[8] performed absorption measurements using thermal dissociation of I_2 to generate atoms. Their experiments were conducted at relatively high temperatures (> 900 K), so temperature scaling schemes were used to extrapolate the results to room temperature. Neumann et al.[102] obtained the broadening coefficient for O_2 at room temperature by resolving the atomic fluorescence from the $I_2 + O_2(a)$ reaction. Cerny et al.[100] also examined resolved fluorescence spectra. They used 496.5 nm radiation from a cw argon ion laser to generate I^* by photolysis of I_2. Most recently, Davis et al.[103] have used tunable diode lasers to examine the absorption lineshapes. Their measurements were made at room temperature, with cw photolytic generation of I.

Table 4. Pressure broadening coefficients for the I $^2P_{1/2}$–$^2P_{3/2}$ transition.

Collision Partner	Broadening Coefficient (MHz/Torr)[a]	Reference
He	3.2	103
He	3.6 ± 0.3	101
O_2	5.0	103
O_2	7.5 ± 1.5	102
O_2	4.8 ± 0.3	100
O_2	1.7 ($T = 1000$ K)	8
N_2	5.5	103
N_2	6.2 ± 0.8	101

[a]Coefficients for 295 K, unless otherwise specified.

As can be seen in Table 4, agreement between the various determinations of the pressure broadening coefficients is good, with the exception of the O_2 result of Neumann *et al.*[102] Davis *et al.*[103] note that this measurement may have been biased by instrumental distortion.

While the data base for pressure broadening is well established, far less is known about the rate constants for velocity changing collisions. Zagidullin *et al.*[47,104] and Copeland and Bauer[48] used computer models to examine the influence of velocity changing collisions on gain and power extraction. Rate constants for velocity relaxation were estimated from diffusion coefficients and theoretical models. For devices operated at low pressures, the calculations of Copeland and Bauer[48] show that the extraction efficiency is very sensitive to the estimates used for the velocity relaxation rate constants. Velocity relaxation of I* has been examined using photolysis of CF_3I or $n\text{-}C_3F_7I$ to produce translationally hot I*.[105,106] Pulsed laser photolysis was employed, which generates fragments with spatially anisotropic velocity distributions. This situation permits the study of two collisional relaxation processes; relaxation of the speed distribution (speed changing collisions) and randomization of the spatial anisotropy (direction changing collisions). Cline *et al.*[105] probed I* stimulated emission lineshapes using a tunable diode laser. From the temporal evolution of the lineshape they derived anisotropy (k_b) and speed distribution (k_v) relaxation rate constants resulting from collisions with $n\text{-}C_3F_7I$. For this heavy collision partner they found that the anisotropy relaxed about 2.5 times faster than the speed distribution. Collisions with He were also investigated. Although the relaxation rate constants were not quantified, Cline *et al.*[105] noted that He was an inefficient moderator of fast I*, and that the speed distribution and anisotropy relaxed at similar rates. Nicholson *et al.*[106] used the temporal characteristics of a photolytic I* laser to probe velocity relaxation. A 248 nm excimer laser was used to photodissociate CF_3I in an optical cavity where lasing on the 3–4 transition was observed. Both the build-up time and the temporal profile of the laser pulse were dependent on the presence of inert buffer gas in the photolysis cell. Kinetic modeling was used to evaluate k_b and k_v for the collision partners Ne, Kr, Xe, N_2, and CO_2. As expected, the rate constants increased with increasing mass of the collision partner, with $k_b > k_v$. Velocity relaxation by He proved to be too slow to measure. Nicholson *et al.* gave an upper bound for k_b and k_v of 10^{-11} cm^3 s^{-1}. Overall, the measured velocity relaxation rate constants were appreciably smaller than estimates from theory and diffusion coefficients. The small rate constants for He suggest that neglect of velocity relaxation processes in models of low pressure COIL devices may be a poor approximation.

5.4. *Diagnostics for COIL Systems*

For development and evaluation of COIL models it is essential to have reliable data from characterizations of prototype laser devices. Information about the local concentrations of key species such as I, I*, I_2, $O_2(X)$, $O_2(a)$, Cl_2, and H_2O are required, along with temperature, pressure and flow velocity data. For determinations of species concentrations and temperatures it is desirable to have measurement techniques that do not interfere with the gas flow and mixing. Consequently, spectroscopic diagnostics are frequently employed. The species most easily quantified by spectroscopic means are $Cl_2(X)$ (absorption at 350 nm) and $I_2(X)$ (absorption at 488 nm). The O_2 $a \rightarrow X$ emission has been used to determine $O_2(a)$ concentrations and SOG yields. Concentrations are derived from absolute measurements of the emission intensity, or relative intensities are calibrated against ESR measurements. The former method relies on accurate knowledge of the Einstein spontaneous emission coefficient for the $a-X$ transition. For many years the value of $A = 2.58 \times 10^{-4}$ s^{-1} derived by Badger et $al.$[107] from absorption measurements was assumed to be correct. Recently, Mlynczak and Nesbitt[108] reanalyzed absorption data for the $a-X$ system, and proposed an A coefficient of 1.47×10^{-4} s^{-1}. If this is correct, determinations based on Badger et $al.$[107] rate for spontaneous emission would have underestimated $O_2(a)$ concentrations by a factor of 1.76. As such a large change in the emission rate has substantial implications for COIL diagnostics, Kodymová and Spalek[109] have carefully calibrated absolute emission intensities against ESR concentration measurements. From these data they obtained an A coefficient of $2.24 \pm 0.40 \times 10^{-4}$ s^{-1}, in good agreement with the generally accepted value of Badger et $al.$[107]

Tunable diode lasers have been used to probe the gain in COIL devices.[22,27,39,110-114] The dependence of the centerline gain on various flow parameters has been examined. Two-dimensional gain maps have been constructed that provide information about the mixing of iodine in the primary flow. During the past few years, Davis and coworkers[103,115-118] have developed diode laser diagnostics to the point where they can be used to monitor H_2O, $O_2(X)$, small signal gain and local temperature. They have achieved very high sensitivities using a dual-beam approach with balanced radiometric detection. In principle this apparatus is capable of making absorption measurements that are close to the shot noise limit. H_2O concentrations have been monitored by scanning over the $3_{03} \leftarrow 2_{02}$ line of the $\nu_3 + \nu_2$ combination band at 1392.5 nm. $O_2(X)$ is detected via the rotational lines of the b–X 0–0 electronic transition near 760 nm. Local temperatures have been determined from the I or H_2O lineshapes. Keating *et al.*[112] used this system to create two-dimensional gain and temperature maps for a small-scale supersonic COIL device. As expected, the gain was seen to decrease in regions where the temperature increased. However, local temperatures were notably higher than the predictions of a two-dimensional model of the fluid dynamics. The reasons for this discrepancy are currently being investigated. Furman *et al.*[39,113] have recently used diode laser measurements of $O_2(X)$ and H_2O concentrations, in combination with standard methods for determining $Cl_2(X)$ and temperature, to characterize the performance of a jet-type singlet oxygen generator. The concentration of $O_2(a)$ was inferred from $O_2(X)$. This approach avoids the above mentioned uncertainties associated with emission measurements.

6. Present Status of COIL Kinetics

After more than twenty years of research and development, a great deal is known about the kinetics of COIL systems. There are still, however, significant questions that remain to be resolved. One of the more critical issues, the kinetics of I_2 dissociation, is still obscure. Efforts to characterize elementary reactions, such as vibrational relaxation of $I_2(X)$, have not removed the ambiguities. Instead, the results highlight the uncertain nature of several rate constants used in the standard kinetic model for COIL systems. The nature of the intermediate state (or states) of I_2 involved in dissociation is called into question. Currently accepted rate constants for deactivation of the intermediate appear to be incompatible with the assumption that it is vibrationally excited $I_2(X)$ alone.

Temperature dependent studies of reaction (1) show that the energy transfer rate constant for the pump reaction (1a) is approximately independent of temperature, and does not follow the T^{-1} dependence suggested by earlier studies. For supersonic devices operating at temperatures near 150 K the pump rate estimated from the T^{-1} dependence is too high by a factor of two. Recent measurements of rate constants for I* hyperfine level relaxation and velocity changing collisions also yielded values that were smaller than had previously been assumed. These processes limit the power extraction in COIL resonators if they are not sufficiently rapid. For most simulations it is assumed that relaxation is rapid, to the extent that equilibrium is maintained between the hyperfine levels and velocity subgroups when intense radiation fields are present in the cavity. These assumptions should be reexamined now that experimentally determined rate constants are available.

At this point the standard kinetics package for COIL should be viewed as an intelligent parameterization of our accumulated knowledge of devices. It is probable that some of the rate constants are seriously in error, and significant mechanisms may have been omitted. The consequence of this situation is that the models are successful in reproducing the behavior of devices operating under typical conditions, but extrapolations to predict the properties of devices outside this domain (e.g. high pressure operation) cannot be made with confidence. This is not to say that the present generation of computer models are deficient. The problems now stem, primarily, from the reliability of the input data.

To improve our understanding of COIL, both kinetic measurements and careful characterizations of prototype devices should be pursued in concert. Accurate diagnostic studies can provide species concentration and temperature data on operating COIL devices that should be used to test and extend models of the chemical kinetics and fluid dynamics. Combining this approach with laboratory studies of key reactions will provide the means for establishing a unique set of kinetic parameters, and a more complete understanding of energy flow. Ultimately, such advances could lead to fully predictive computational models and substantive changes in COIL design concepts.

7. Laser Concepts Based on $NF(a^1\Delta)$ and $NCl(a^1\Delta)$ Metastables

7.1. *General Considerations*

The success of the COIL program generated interest in the possibilities for creating analogous chemically driven laser systems. It was quickly recognized that the metastable $a^1\Delta$ states of NF and NCl may be suitable energy carriers for lasers. These molecules are isovalent with O_2, and have the same pattern of low lying electronic states. In ascending energy order these are $X^3\Sigma^-$, $a^1\Delta$, and $b^1\Sigma^+$. Radiative relaxation of the $a^1\Delta$ states is electric dipole forbidden by spin and orbital angular momentum selection rules. Both NF(a) and NCl(a) can be generated by chemical means. These species carry more energy than $O_2(a)$ (see Table 5), providing opportunities for constructing lasers that operate at wavelengths shorter than the COIL 1.3 μm radiation. Initially the potential for creating shorter wavelength lasers was one of the primary motivations for efforts to develop devices driven by NF(a). More recently work on NF(a) and NCl(a) has been motivated by the fact that high concentrations of these metastables can be generated using purely gas phase processes. This is of practical value as the two phase chemistry used to produce high concentrations of $O_2(a)$ in COIL poses significant technical challenges for construction of compact lightweight generators. Metastable generators suitable for use in air-borne or space-based lasers are more easily constructed using gas phase chemistry.

Table 5. Selected properties of $O_2(a^1\Delta_g)$, NF($a^1\Delta$) and NCl($a^1\Delta$) metastables.

	$O_2(a^1\Delta_g)$	NF($a^1\Delta$)	NCl($a^1\Delta$)
Radiative lifetime	74 min. [a]	5.6 s [b]	3 s [c]
Self-annihilation[d] rate constant (cm^3 s^{-1})	4.4×10^{-17} [e]	$(5 \pm 2) \times 10^{-12}$ [f]	$(7.2 \pm 0.9) \times 10^{-12}$ [g]
Energy pooling[h] rate constant (cm^3 s^{-1})	2.7×10^{-17} [i]	5.7×10^{-15} [f]	1.5×10^{-13} [j]
T_0(cm^{-1})	7882	11 435	9280
Energy defect for transfer to I* (cm^{-1})[k]	279	3832	1677

[a]References 107 and 109.
[b]Reference 131.
[c]References 195 and 196.
[d]$M(a) + M(a) \rightarrow$ products.
[e]S. A. Lawton, S. E. Novic and H. P. Broida, *J. Chem. Phys.* **66**, 1381 (1977).
[f]Reference 146.
[g]Reference 186.
[h]$M(a) + M(a) \rightarrow M(b) + M(X)$.
[i]R. F. Heidner, C. E. Gardner, T. M. El-Sayed, G. I. Segal and J. V. V. Kasper, *J. Chem. Phys.* **74**, 5618 (1981).
[j]Reference 124.
[k]$M(a, v = 0) + I \rightarrow M(X, v = 0) + I^*$.

Reactions that provide high yields of NX(a) have been identified. Among the most promising for use in lasers is the sequence[119–125]

$$X + HN_3 \rightarrow HX + N_3 \tag{23}$$

$$X + N_3 \rightarrow NX(a) + N_2; \quad X = F \text{ or } Cl. \tag{24}$$

To initiate and sustain this sequence, high flow rates of halogen atoms may be derived from the types of combustor used in chemical HF lasers. Another promising reaction for generation of NF(a) is[126–131]

$$H + NF_2 \rightarrow NF(a) + HF. \tag{25}$$

Once the NX metastables have been generated they must be transferred to the laser cavity and mixed with the species that will lase (or a precursor for that species). The properties of NX(a) metastables are such that transfer and mixing will pose more difficult problems for NX based systems than they do for COIL. To illustrate this point relevant properties for $O_2(a)$, NF(a) and NCl(a) are presented in Table 5. One of the most important factors governing the ability to generate and transport high concentrations of metastables is the rate constant for bimolecular self-annihilation (M(a) + M(a) $\rightarrow$ products). The rate constants for self-annihilation of NF(a) and NCl(a) are more than five orders of magnitude greater than the corresponding rate constant for $O_2(a)$. Destruction processes include energy pooling events of the type M(a) + M(a) $\rightarrow$ M(b) + M(X). Pooling accounts for nearly $\frac{2}{3}$ of the $O_2(a)$ self-annihilation process, but it is of marginal significance for NF(a) and NCl(a). For the conditions within laser systems, NX(a) metastables are more readily removed by quenching or chemical reaction than $O_2(a)$. A consequence of the relatively fragile nature of NX(a) metastables is that mixing and transport times must be kept to an absolute minimum.

Using NX(a) to replace $O_2(a)$ in chemical iodine lasers is a topic of current interest. It has been established that transfer from NCl(a) can excite I efficiently, at rates that are high enough to sustain a population inversion.[9,132,133] There is also evidence that NF(a) can excite I.[134,135] It is probable that iodine lasers driven by NX(a) will operate at much higher temperatures than COIL. If combustors are used to generate F or H atoms, the primary gas flow will be hot. In addition, as NF(a) or NCl(a) carry more energy than is needed to excite I, the pump process will also liberate heat.

Current models predict operating temperatures for subsonic NCl(a)/I* devices well above 600 K. Supersonic cooling may bring the temperature into the 400–500 K range. A strongly inverted I* population will have to be maintained to offset the effect of Doppler broadening on the gain at line center. Another implication is that knowledge of the temperature dependencies of the more important reactions and energy transfer steps will be needed.

Although the constraints associated with using NF(a) or NCl(a) as energy carriers and transfer agents are challenging, pulsed lasers driven by these metastables have been demonstrated. In the following, I will review the development of these devices and related studies of elementary reaction kinetics.

7.2. Lasers Driven by Transfer from NF(a)

Production of NF(a) from reaction (25) was originally observed by Clyne and White.[129] Further studies of this reaction showed that the branching ratio for formation of NF(a) was in excess of 85%,[126,136] and that the HF product was formed in vibrationally excited states.[131] Emission from NF(b) was also observed when NF$_2$ was added to a flow of H atoms. Herbelin and Cohen[130] showed that the b state was populated by the energy pooling process

$$\text{NF}(a) + \text{HF}(v \geq 2) \rightarrow \text{NF}(b) + \text{HF}(v - 2). \qquad (26)$$

Herbelin *et al.*[137] realized that this system could be scaled to provide high densities of NF(a), and were able to generate concentrations near 6×10^{15} cm^{-3} in a subsonic flow facility. Direct lasing of the a–X transition is not feasible owing to the very small cross section for stimulated emission. Hence, species that could readily accept energy from NF(a) or upconvert the stored energy were sought. The energy pooling reaction

$$\text{NF}(a) + \text{I}^* \rightarrow \text{NF}(b) + \text{I} \qquad (27)$$

was examined as a means to generate high concentrations of NF(b).[138,139] The motivation for this effort was the possibility of constructing a laser operating at 528.8 nm on the NF b–X transition. High yields of NF(b) were achieved, but gain was not demonstrated.

Energy transfer from NF(a) to Bismuth atoms

$$\text{NF}(a) + \text{Bi}(^4S_{3/2}) \rightarrow \text{NF}(X) + \text{Bi}(^2D_{3/2}) \qquad (28)$$

looked promising as this process is near resonant (17 cm^{-1} exothermic) and the upper level has a suitable radiative decay rate ($A = 25$ s^{-1}, fast enough to give a reasonable stimulated emission cross section, but not so fast that pumping by collisional energy transfer becomes impractical). Flow tube kinetic studies showed that transfer from NF(a) to Bi* was rapid and efficient.[140,141] Attempts to achieve lasing on the Bi $^2D_{3/2} \rightarrow {}^4S_{3/2}$ transition were not successful, but during these experiments an intense blue emission from the BiF A–X bands was routinely observed. Herbelin and Klingberg[141] speculated that BiF was being formed by the reaction of NF(a) with Bi($^2D_{3/2}$). Excitation of the BiF $A0^+$ state was attributed to sequential transfer events involving $b^1\Sigma^-$ as an intermediate state. Although pumping of BiF(A) by NF(a) appeared to be efficient, the short radiative lifetime of the A state (1.4 μs)[142] and difficulties in producing high concentrations of Bi vapor prevented lasing of this system under flow mixing conditions.

Benard and coworkers[11,12,143] eventually demonstrated lasing of the NF(a)/BiF system using pulsed methods to initiate both the chemistry and energy transfer processes in premixed gas samples. Their methods for pulsed generation of NF(a) evolved from studies of the reactions of FN$_3$ in flows that contained D and F atoms.[143] In these experiments F atoms were obtained by discharging F$_2$, while D atoms were generated by adding D$_2$ to the F atom stream. The F + D$_2$ reaction also produced vibrationally excited DF and released heat. Emission from the NF a–X system was observed when FN$_3$ was added to this flow. Benard *et al.*[143] argued that NF(a) was being produced from FN$_3$ by either thermal dissociation or vibrational energy transfer from DF(v). The notion that FN$_3$ could be dissociated by vibrationally excited HF or DF was tested by exposing FN$_3$/HF or DF mixtures to pulses from an HF/DF laser. NF(a) was produced by this means, but the temporal profiles indicated that heating, rather than vibrational energy transfer was responsible for the decomposition to NF(a) and N$_2$ (interestingly, it appeared that vibrational transfer caused dissociation of FN$_3$, but NF(a) was not formed by this process).

Once they had established that thermal decomposition of FN$_3$ gives NF(a), Benard *et al.*[143] used this approach to generate high instantaneous concentrations of NF(a). The 10.6 μm output from a CO$_2$ laser was used to pulse heat FN$_3$/SF$_6$/He mixtures. This radiation was only absorbed by the SF$_6$, which transferred the energy to the surrounding bath gas. The temperature increased rapidly to around 1000 K, causing complete dissociation of the FN_3. This laser pyrolysis technique yielded instantaneous $NF(a)$ concentrations of about 3×10^{16} cm^{-3}.

Gain on the BiF $A–X$ transition was subsequently demonstrated using laser pyrolysis of $FN_3/(CH_3)_3Bi/SF_6$/He mixtures. Benard and Winkler[12] devised a sensitive cavity ring-down technique to measure gain in this system. The probe radiation was provided by an etalon-narrowed pulsed dye laser. To achieve the required sensitivity, shot-to-shot intensity fluctuations of the probe laser were cancelled out by a scheme that involved multimode operation. The near vertical character of the BiF $A–X$ transition had to be considered in selecting the transition to be used for the gain measurement. Transitions to the lowest vibrational levels of the ground state were not suitable, as high concentrations of $BiF(X)$ $v'' = 0$ and 1 were present immediately after pyrolysis. The 1–4 band was identified as having the best combination of population ratio and transition probability. Probing the overlapping P(43) and R(70) lines of this band near 471 nm, Benard and Winkler[12] measured a gain of 3.6×10^{-4} cm^{-1}. Factors that limited the gain were radiative loss of $BiF(A)$, $NF(a)$ self-annihilation, and quenching by $(CH_3)_3Bi$ (and its decomposition products). The latter problem constrained the amount of Bi that could be present.

Encouraged by the gain measurement, Benard[11] went on to demonstrate lasing of the BiF $A–X$ transition. As the CO_2 laser pyrolysis technique was not optimal for longitudinal pumping of a laser cavity, Benard[11] experimented with an alternative method for pulsed pyrolysis of FN_3. He constructed a shock tube reactor where the shock wave was generated by discharge ignition of $F_2/H_2/O_2$ mixtures. In this apparatus $FN_3/(CH_3)_3Bi/$He mixtures were shock heated, resulting in intense emission from the BiF $A–X$ system. When the windows of the reactor were replaced by aligned high reflectance mirrors, the intensity and temporal behavior of the emission at 470 nm was consistent with unsaturated lasing. However, as in the CO_2 laser pyrolysis experiments it appeared that quenching by $(CH_3)_3Bi$ and/or its dissociation products was a limiting factor.

The next species considered for transfer pumping by $NF(a)$ was BH.[10,144,145] This molecule was an attractive candidate for several reasons. BH can be sequentially excited to the $A^1\Pi$ state ($T_0 = 23073.9$ cm^{-1}) by two near resonant transfer steps. It may be obtained by decomposition of volatile inorganic precursors, which was thought to be an advantage because these materials and their decomposition products would probably be less efficient quenchers of electronically excited states than organic compounds and their fragments.[144] Lastly, the rovibrational levels of BH(A) are widely spaced. The level spacing issue is of importance in any molecular laser. Partitioning of the excited state population among the vibrational and rotational levels reduces the gain that can be achieved on a single rovibronic line. Consequently, higher densities of the upper laser level can be achieved if the partitioning is limited by low temperatures or (more conveniently) by widely spaced energy levels. For this reason, diatomic hydrides are generally considered to be good lasing species.

Given the short radiative lifetime of BH(A) ($\tau = 0.17$ μs),[145] it was evident that attempts to lase this species by transfer excitation would require pulsed initiation of the chemistry in a premixed system. To be compatible with pulsed pyrolysis generation of NF(a), a volatile precursor that would decompose to give BH on heating was needed. Benard *et al.*[144] used tetraazidodiborane (TADB), an energetic compound that they synthesized for the first time from the reaction of B_2H_6 with HN_3. Mixtures of TADB, FN_3, and SF_6 were pulse-heated using CO_2 laser excitation. Resonance lamps operating on the BH A–X transition were used to monitor relative populations in the A and X states. Benard *et al.*[144] were able to show that thermal decomposition of TADB did, indeed give BH(X), and that BH(A) could be excited by rapid sequential transfer from NF(a). inversion was obtained on the 0–0 transition, and gain of 10%/cm was reported. Subsequently, Benard and Boehmer[10] achieved low power lasing on the A–X 0–0 band using pulsed pyrolysis of B_2H_6/FN_3/SF_6 mixtures.

The difficulties in constructing short wavelength NF(a) transfer lasers can be traced back to the relatively large rate constant for NF(a) self-annihilation,[146] and problems with obtaining sufficient concentrations of the lasing species. As BiF(A) and BH(A) have short radiative lifetimes, rapid pumping was needed to overcome radiative losses. This demanded high concentrations of NF(a), such that the time scale for self-annihilation was short compared to typical flow mixing times. Hence, generation of NF(a) in premixed systems was essential. Efficient extraction of the energy from NF(a) at rates that would dominate over loss by self-annihilation required high concentrations of the lasing species. The required emitter concentrations were barely attained in the BiF and BH laser demonstrations. Further progress with these systems was prevented as compounds that would yield high concentrations of BiF or BH without introducing efficient quenchers could not be found. The long lifetime of Bi($^2D_{3/2}$) was better suited for excitation by chemical means, but unfortunately the reaction between $Bi(^2D_{3/2})$ and $NF(a)$ was too fast to permit pumping to inversion.

As noted above, the possibility of using $NF(a)$ to replace $O_2(a)$ in iodine lasers is a topic of current interest. In a system where $NF(a)$ is obtained from reactions (23) and (24), I atoms are readily introduced by the reaction

$$F + DI \rightarrow DF + I \tag{29}$$

(use of DI rather than HI is favored as HF quenches I* by near resonant E–V transfer. This process is much slower for the DF product). The key question for an $NF(a)/I$ laser system is whether transfer from $NF(a)$ to I is sufficiently efficient and rapid. This process is far off resonance (exothermic by 3832 cm^{-1}) so a physical transfer mechanism would result in a very small transfer rate constant. However, it is possible that collisions between $I(^2P_{3/2})$ and $NF(a)$ will sample a chemically bound intermediate state. Were this to happen the transfer rate constant could be large, provided the branching fraction to ground state products was small.

Indirect evidence for transfer from $NF(a)$ to I was obtained by Du and Setser[135] in a study of $NF(a)$ quenching kinetics. Quenching by I_2 was found to be fast ($k = 1.5 \times 10^{-10}$ cm^3 s^{-1}) and dominated by the physical process

$$NF(a) + I_2 \rightarrow NF(X) + I_2 . \tag{30}$$

Du and Setser[135] monitored the production of $NF(X)$ by using energy transfer from $N_2(A)$ to excite NF b–X emission. Du and Setser[135] noted that addition of I_2 to a flow of $NF(a)$ enhanced the NF b–X emission. To explain this result they suggested the sequence

$$2NF(a) + I_2 \rightarrow 2NF(X) + 2I \tag{31}$$

$$NF(a) + I \rightarrow NF(X) + I^* \tag{32}$$

followed by excitation of $NF(b)$ via reaction (27). Addition of ICl to a flow of $NF(a)$ produced an even stronger emission from $NF(b)$. This could be explained by assuming that $NF(a)$ is more efficient in dissociating ICl than I_2, and/or that I* is less rapidly quenched by ICl than I_2. Du and Setser[135] also examined the effect of adding I atoms to a flow of $NF(a)$. Greatly enhanced $NF(b)$ emission was observed, consistent with reactions (32) and (27). A rate constant for quenching of $NF(a)$ by I of $(1.8 \pm 0.4) \times 10^{-11}$ cm^3 s^{-1} was obtained, but the branching fraction for reaction (32) was not determined.

Evidence that NF(a) can efficiently excite I was reported by Benard,[134] who generated high concentrations of NF(a) and I by shock heating FN$_3$/I$_2$/He mixtures. Intense emissions from NF(b), NF(a) and I* were observed, but the system did not reach threshold gain for the I* transition. The temporal profiles of the emissions were analyzed by fitting to kinetic models. Benard[134] found the data were consistent with a large value for the rate constant for reaction (32) [$(8.5 \pm 0.5) \times 10^{-11}$ cm^3 s^{-1} at 1000 K]. Failure to reach threshold gain was attributed to rapid quenching of I* in the shock heated mixture.

In contrast to studies that indicate reasonably efficient transfer from NF(a) to I, work by Ray–Hunter and Coombe[147] suggests that the room temperature branching fraction may be too small to attain an inversion. These investigators examined the effect of adding Cl atoms to flowing F/HN$_3$/I$_2$ mixtures. NF(a) was generated by reactions (23) and (24), while I atoms were derived from

$$I_2 + F \rightarrow I + IF. \qquad (33)$$

With no Cl present, emission of I* from reaction (32) was detected. When Cl atoms were introduced via the reaction

$$F + H/DCl \rightarrow H/DF + Cl \qquad (34)$$

strong enhancement of the I* emission was observed. The proposed mechanism for enhancement was the sequence

$$Cl + N_3 \rightarrow NCl(a) + N_2 \qquad (35)$$

$$NCl(a) + I \rightarrow NCl(X) + I^*. \qquad (36)$$

Based on kinetic models of this system, Ray–Hunter and Coombe[147] argued that the branching fraction for reaction (32) was less than 5%. However, as their kinetic model contained estimates for several unknown rate constants, the conclusions of this study are uncertain.

7.3. Studies of Elementary Reaction Kinetics of Relevance to NF(a) Laser Systems

Kinetic studies of NF(a) generation and removal processes were conducted in parallel with efforts to demonstrate transfer lasers. Most kinetic measurements have been carried out in discharge flow reactors, using reactions

(23) and (24) (with $X = F$) or reaction (25) to generate NF(a). The rate constant for the first step in the sequence for generating NF from HN$_3$,

$$F + HN_3 \rightarrow HF + N_3 \tag{37}$$

was measured by David and Coombe[148] [$(1.6 \pm 0.2) \times 10^{-10}$ cm^3 s^{-1}] and Habdas *et al.* [$(1.1 \pm 0.2) \times 10^{-10}$ cm^3 s^{-1}].[123] In the latter study the vibrational state distribution of the product HF was determined from infrared chemiluminescence. Hewett and Setser[149] examined the competing path

$$F + HN_3 \rightarrow HNF + N_2 \tag{38}$$

using LIF to detect ground state HNF. This was found to be a minor channel, proceeding with a rate constant of $(6.3 \pm 3.5) \times 10^{-12}$ cm^3 s^{-1}. For the secondary reaction

$$F + N_3 \rightarrow NF(a) + N_2. \tag{39}$$

Habdas *et al.*[123] obtained a rate constant of $(5 \pm 2) \times 10^{-11}$ cm^3 s^{-1} and a branching fraction in excess of 85%. Du and Setser[146] narrowed the uncertainty range for the rate constant to $(4.0 \pm 1.5) \times 10^{-11}$ cm^3 s^{-1}, and Liu *et al.*[119] determined a rate constant of $(5.8 \pm 0.6) \times 10^{-11}$ cm^3 s^{-1} for removal of N$_3$ by F atoms. Collectively, these studies establish the F + HN$_3$ system as a useful source of NF(a). Setser and coworkers[135,146,150–152] went on to use this source for measurements of the rate constant for NF(a) self-annihilation and rate constants for quenching of NF(a) by more than sixty different quenching partners. Of particular relevance to possible laser schemes, they found that quenching rate constants for H$_2$, HF, HCl, N$_2$ and O$_2$ were all relatively small. Du and Setser[146] suggested that removal of NF(a) by reaction was the more common quenching mechanism. Wategaonkar *et al.*[153] demonstrated the presence of the reactive channel in quenching by CO by observing the NCO product. Habdas and Setser[154] used the F + HN$_3$ generation method to examine energy pooling between NF(a) and vibrationally excited HF [reaction (26)]. Self-pooling of NF(a),

$$NF(a) + NF(a) \rightarrow NF(b) + NF(X) \tag{40}$$

was found to be inefficient,[146] with a rate constant of $(5.7 \pm 1.0) \times 10^{-15}$ cm^3 s^{-1}.

The reaction of H with NF_2 is another good source of $NF(a)$, but has been used less frequently because N_2F_4, the precursor of NF_2, is not readily available. The rate constant for reaction (25) was measured by Cheah *et al.*[127] [$(1.5 \pm 0.2) \times 10^{-11}$ cm^3 s^{-1}] and Malins and Setser (1.3×10^{-11} cm^3 s^{-1}).[131] Vibrational level population distributions for the HF product have also been characterized.[123] Secondary reactions in the H+NF_2 system were examined by Cheah and Clyne,[126] who derived estimates for rate constants for the processes

$$H + NF(a) \rightarrow N(^2D) + HF \qquad (41)$$

$$H + NF(X) \rightarrow N(^4S) + HF \qquad (42)$$

and

$$NF + NF \rightarrow N_2 + 2F. \qquad (43)$$

Improved rate constants for reactions (41) and (42) were subsequently reported by Davis *et al.*[155,156]

Heidner and coworkers[136,157] developed pulsed photolysis of NF_2 as a convenient source of both $NF(a)$ and $NF(X)$. They used time-resolved techniques to examine the kinetics of $NF(a)$ (detected by a–X emission)[158] and $NF(X)$ (detected by LIF of the b–X transition).[136,158–162] Photolysis of NF_2 yields vibrationally excited $NF(X)$, which facilitated studies of vibrational relaxation. Heidner *et al.*[162] determined rate constants for relaxation by CO_2 and SF_6. They also measured rate constants for the reaction of $NF(X)$ with NF_2 [$(2.4 \pm 0.2) \times 10^{-12}$ cm^3 s^{-1}] and the disproportionation reaction (43) [$(3.5 \pm 2) \times 10^{-12}$ cm^3 s^{-1}].[162]

Dissociation of NF_2 by 193 nm light was found to be very inefficient. Heidner *et al.*[136] used this property to advantage in a study of the H + NF_2 branching ratio. The reaction was initiated by pulsed photolysis of mixtures of NF_2 and HBr (which is readily dissociated by 193 nm light). LIF signals from $NF(X)$ could not be detected, while $NF(a)$ from reaction (25) was easily observed. These results confirmed that the branching fraction to $NF(a)$ was greater than 90%, as had been deduced from the flow tube experiments. Photolysis generation of $NF(a)$ was used to examine quenching by the species NF_3, I_2, and O_2.[158] The rate constants obtained were in good agreement with flow tube results.

Koffend *et al.*[139] used pulsed photolysis of NF_2/HI mixtures at 248 nm to characterize the energy pooling processes defined by reaction (27). A novel feature of this experiment was the method used to fix the I^*/I concentration ratio. Following the work of Herbelin *et al.*[138] they used the output from a pulsed I^* laser to saturate the atomic transition in the reaction cell. A pooling rate constant of $(5 \pm 2) \times 10^{-11}$ cm^3 s^{-1} was obtained.

As energy pooling processes excite $NF(b)$, quenching of this species is also relevant to the kinetics of energy transfer lasers. For rate constant determinations Setser and coworkers[163–166] used collisions of Ar* metastables with NF_2 or dc discharges through Ar/NF_2 mixtures to generate $NF(b)$ (the latter was found to be more effective[166]). After studying a range of collision partners, Setser *et al.*[163,165] concluded that quenching typically occurred by E–V transfer processes of the kind

$$NF(b) + M \rightarrow NF(a) + M(v > 0). \tag{44}$$

Accurate values for the radiative lifetimes of $NF(b)$ and $NF(a)$ are needed for quantitative interpretation of b–X and a–X emission signals. The radiative lifetime of $NF(b)$ has been measured by Tennyson *et al.*[167] (22.6 $\pm$ 1.7 ms) and Cha and Setser[164,165] (19 $\pm$ 2 ms). A radiative lifetime for $NF(a)$ of 5.6 s was reported by Malins and Setser.[131]

A considerable data base of quenching and energy transfer rate constants for $NF(a)$ and $NF(b)$ has been established. Many of the reactions of importance for chemical laser systems have been determined, although the greater majority of these measurements have been made at room temperature. As noted in Sec. 7.1, it will be of interest in future studies of NF kinetics to examine the temperature dependencies of the rate constants.

7.4. *Lasers Driven by Transfer from NCl(a)*

The potential for using $NCl(a)$ in a chemically pumped laser was first examined by Benard *et al.*[168] These investigators used pulsed CO_2 laser pyrolysis of $ClN_3/SF_6/Ar$ mixtures to generate high concentrations ($> 10^{15}$ cm^{-3}) of $NCl(a)$. From the rate of $NCl(a)$ decay in this system they estimated an upper limit for the $NCl(a)$ self-annihilation rate constant of $> 8 \times 10^{-12}$ cm^3 s^{-1}. Energy pooling between $NCl(a)$ and $NF(a)$

$$NCl(a) + NF(a) \rightarrow NCl(X) + NF(b) \tag{45}$$

was examined using pyrolysis of $ClN_3/FN_3/SF_6/Ar$ mixtures. A modest enhancement of the $NF(b)$ emission was seen when $NCl(a)$ was present, indicating a small but significant rate constant for reaction (45). Much stronger NF(b) signals were obtained when a small quantity of I_2 was added to the pyrolysis cell. This effect was attributed to pumping of I* by NCl(a) [reaction (36)] followed by pooling between I* and NF(a) [reaction (27)]. Benard et al.[168] briefly explored the possibility that NCl(a) and NF(a) might participate in cooperative pumping of IF(B) (an optically pumped IF laser operating on the B–X transition had been demonstrated previously by Davis and coworkers[169–171]). CF_3I was added to the pyrolysis cell to provide a source of IF. Emission from IF(B) was detected, but it was not particularly strong. Benard et al.[168] proposed that either slow pumping of IF or poor production of IF from CF_3I was limiting the concentration of IF(B) formed.

Interest in the use of NCl(a) in chemical iodine laser systems was greatly increased by the Bower and Yang[172] report of a gas kinetic rate constant for reaction (36). NCl(a) was generated in a flow tube by reactions (35) and (37). Further downstream I atoms were produced by the reaction of I_2 or ICl with F atoms. Emission from the NCl a–X system was observed in the absence of the I atom precursor. With I atoms present the a state emission was dramatically reduced and intense emission from I* was detected. From an analysis of the time required to reach steady state conditions, Bower and Yang[172] estimated that the removal process

$$NCl(a) + I \rightarrow products \qquad (46)$$

occurred with a rate constant that exceeded 8×10^{-11} cm^3 s^{-1}. Furthermore, they argued that the dominant quenching mechanism was the E–E transfer process represented by reaction (36). This implied that the rate constant for reaction (36) was unusually large, given that the energy defect for transfer resulting in NCl($X, v = 0$) is unfavorable (1677 cm^{-1}). Bower and Yang[172] suggested that a near resonant process involving formation of NCl($X, v = 2$) might account for the large rate constant. Subsequently, Yang et al.[132] went on to show that transfer from NCl(a) could pump I to inversion in a flow reactor. To demonstrate the presence of an inversion, they monitored the concentration of I* using 206 nm resonance absorption. Saturating pulses from an I* laser were used to modulate the absorption. If an inversion was present Yang et al.[132] expected to see an increase in the transmitted 206 nm light as the saturating pulses would reduce the I* population through stimulated emission. This expectation was confirmed, and they reported gain of 10^{-4} cm^{-1}.

To determine the minimum conditions needed for demonstration of an NCl(a)/I* laser, the value of the rate constant for reaction (36) was required. Ray and Coombe[173] measured this parameter using pulsed 193 nm photolysis of ClN$_3$/CH$_2$I$_2$ mixtures to simultaneously produce NCl(a) and I. The I* temporal profiles showed clear evidence of pumping by reaction (36). By applying the steady state approximation to the maximum of the I* profile, and estimating the concentrations of the reactants, Ray and Coombe[173] obtained a rate constant for reaction (36) of $(1.8 \pm 0.3) \times 10^{-11}$ cm^3 s^{-1}. The error bounds reflected the statistics of the measurements. Systematic errors resulting from the assumptions used in estimating the reagent concentrations were also considered. It was assumed that the photolysis quantum yield for production of NCl(a) was unity, and that I* accounted for only 5% of the I atoms from CH$_2$I$_2$ dissociation. Ray and Coombe[173] concluded that reasonable deviations from these assumptions could have caused their rate constant to be underestimated, but by no more than a factor of two. Their upper limit for the rate constant was considerably smaller than the magnitude suggested by Bower and Yang,[172] but still large enough to be of practical interest. In addition, Ray and Coombe[173] observed peak I* concentrations that corresponded to excitation of an appreciable fraction ($\approx 25\%$) of the I atoms present in the photolysis volume.

The absorption of 193 nm light by ClN$_3$ is not particularly strong [values for σ_{193} of 1.38×10^{-18} (Clark and Clyne[174]) and 2.6×10^{-18} cm^2 (Henshaw *et al.*[175]) have been reported]. For the photolysis intensity used by Ray and Coombe[173] (76 mJ cm^{-2}) only a small fraction of the ClN$_3$ was dissociated. Quenching of NCl(a) by undissociated ClN$_3$ was found to be rapid, but I* signals were seen to persist for several milliseconds. To explain this behavior Ray and Coombe[173] proposed that I* quenching generates NCl(a) via the reaction

$$I^* + ClN_3 \rightarrow I + NCl(a) + N_2 \,. \qquad (47)$$

Repumping of I* then ensues and ClN$_3$ is depleted by chain decomposition.

Ray and Coombe[9] achieved lasing of the NCl(a)/I* system by scaling-up their quenching experiment. The apparatus used for this demonstration is shown in Fig. 10. An excimer laser capable of providing 700 mJ pulses at 193 nm was used to generate high densities of NCl(a) metastables. The I* laser cavity consisted of mirrors coated for $> 99.5\%$ reflectivity at 1.315 μm. Representative I* emission signals from this device are shown in Fig. 11.

Note that the duration of the photolysis pulse was nominally 10 ns. All three traces show long lived I* fluorescence that was attributed to chain decomposition of ClN$_3$ and repumping of the I atoms. Ray and Coombe[9] speculated that I*, NCl(a) and vibrationally excited N$_2$ may participate in this process as chain carriers. Trace a in Fig. 11 was recorded under subthreshold conditions and shows no evidence of lasing. Traces b and c were recorded with above threshold conditions, and lasing spikes can be seen superimposed on the I* fluorescence. The long delay between photolysis and lasing was an intriguing property of this system. A kinetic model indicated that the density of NCl(a) resulting from the initial photolysis event was not enough to get above threshold while an appreciable concentration of undissociated N$_3$Cl (an efficient quencher of I*) remained. Chain decomposition reduced the N$_3$Cl concentration to levels that permitted brief bursts of lasing. The model was successful in predicting delayed lasing, but the

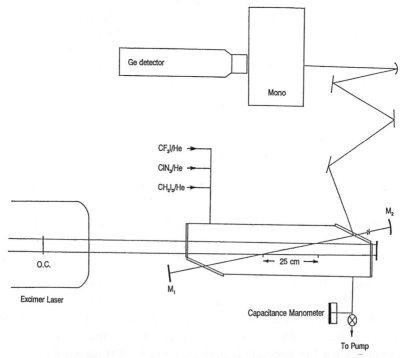

Fig. 10. Diagram of the apparatus used to demonstrate NCl(a) pumping of an I* laser. Mixtures of ClN$_3$ and CH$_2$I$_2$ in He were photodissociated by pulses from an excimer laser operating at 193 nm. Reproduced with permission from Ref. 9.

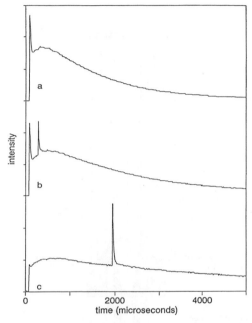

Fig. 11. Time resolved I* emission signals obtained from the apparatus shown in Fig. 10. Trace *a* was recorded under subthreshold conditions. Above threshold conditions were used to record traces *b* and *c*. The sharp spikes in these traces indicate bursts of lasing. Reproduced with permission from Ref. 9.

delays were not as large as the observed intervals. Ray and Coombe[9] suspected that a transient absorber of 1.3 μm radiation, such as vibrationally excited $I_2(X)$, prevented lasing at earlier times.

The gain demonstration of Yang *et al.*[132] and the NCl(a)/I* photolysis laser[9] operated under conditions that were marginally above threshold. The signals were subject to large shot-to-shot fluctuations, making it difficult to extract reliable quantitative information from the data. To better characterize the gain in NCl(a)/I* systems, Herbelin *et al.*[133] developed a subsonic flow reactor that could be used in conjunction with a sensitive diode laser diagnostic for measuring gain on the iodine transition. The essential features of the flow reactor are shown in Fig. 12. Fluorine atoms were generated by a dc discharge through a F_2/He mixture. A flow of DCl was injected just beyond the discharge, and the flow rate adjusted to convert all of the F atoms to Cl by reaction (34). Further downstream in the reactor, at a point where the conversion to Cl atoms was complete, HI was

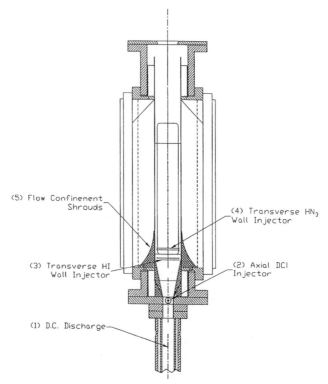

(5) Flow Confinement Shrouds

(4) Transverse HN$_3$ Wall Injector

(3) Transverse HI Wall Injector

(2) Axial DCl Injector

(1) D.C. Discharge

Fig. 12. Subsonic flow device used to measure gain for the NCl(a) + I system. Reproduced with permission from Ref. 133.

injected into the flow. This provided I atoms from the reaction

$$Cl + HI \rightarrow HCl + I. \tag{48}$$

Lastly, HN$_3$ was injected to provide NCl(a) from

$$Cl + HN_3 \rightarrow HCl + N_3 \tag{49}$$

followed by reaction (35). Representative absorption and gain measurements from this flow reactor are shown in Fig. 13. Trace 1 shows a scan over the I atom 3–4 line taken with no HN$_3$ added to the flow. The absorbance indicated a ground state I atom concentration of 6.0×10^{13} cm^{-3} from reaction (48). Trace 2 shows a scan over the atomic line with HN$_3$ added to the flow. A peak gain of 0.02% per cm was observed, indicating that $\frac{2}{3}$ of the I atoms had been transferred to the excited state. Efforts are

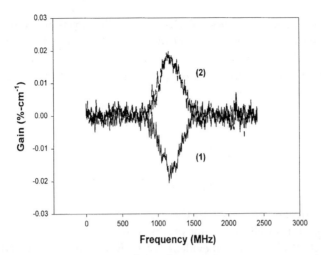

Fig. 13. Diode laser scans over the I $^2P_{1/2}$–$^2P_{3/2}$ 3–4 line from the flow reactor shown in Fig. 12. Trace (1), which shows absorption by ground state I, was recorded with no HN_3 present in the system. The flow conditions were 0.7 mmole/s F_2, 2.0 mmole/s DCl, and 0.03 mmole/s HI. Trace (2) shows the effect of adding 3.3 mmole/s of HN_3 to the flow specified for trace (1). Gain on the 3–4 line is clearly demonstrated. Note that the x axis of this plot is a relative scale. Reproduced with permission from Ref. 133.

currently underway to increase the reagent flows in this device, and see if the gain can be raised to a level suitable for demonstration of cw lasing.

7.5. Studies of Elementary Reaction Kinetics of Relevance to NCl(a) Laser Systems

Clyne and Clark[176,177] carried out some of the earliest work on reactions that produce electronically excited NCl. They observed emission from NCl(b) when Cl atoms reacted with ClN_3. N_3 radicals were detected and they proposed that production of NCl(b) was linked to the reaction of Cl with N_3. Overall, their results were consistent with the sequence

$$Cl + ClN_3 \rightarrow Cl_2 + N_3 \tag{50}$$

$$Cl + N_3 \rightarrow NCl(a) + N_2 \tag{35}$$

$$NCl(a) + NCl(a) \rightarrow NCl(b) + NCl(X). \tag{51}$$

Clyne and Clark[176] assumed that NCl(b) was an alternative product of reaction (35), but it was later shown to originate from the pooling reaction.

Mass spectrometric studies of the $Cl + ClN_3$ system were made by Combourieu et al.[178] and Clyne et al.[179-181] Data from the former group were analyzed by Jourdain et al.[182] who reported a rate constant of $(3.7 \pm 0.6) \times 10^{-12}$ cm^3 s^{-1} for reaction (50). Clyne et al.[180,181] examined the disproportionation reaction

$$NCl(X) + NCl(X) \rightarrow 2Cl + N_2 \quad k = (8.1 \pm 1.8) \times 10^{-12} \text{ cm}^3 \text{ s}^{-1} \quad (52)$$

and removal of $NCl(a)$ by Cl_2

$$NCl(a) + Cl_2 \rightarrow NCl_2 + Cl. \quad (53)$$

By adding quenching agents to the $NCl(a)$ flow, Clyne and MacRobert[180] demonstrated that NCl_2 was not produced by reaction of $NCl(X)$ with Cl_2. Clyne et al.[181] also obtained evidence of E–E transfer between $NCl(a)$ and O_2,

$$NCl(a) + O_2(X) \rightarrow NCl(X) + O_2(a). \quad (54)$$

The room temperature rate constant for the primary reaction of Cl with HN_3 [reaction (49)] has been determined by Yamasaki et al. [$(8.9 \pm 1.0) \times 10^{-13}$ cm^3 s^{-1}][183] and Manke and Setser [$(1.1 \pm 0.3) \times 10^{-12}$ cm^3 s^{-1}].[125] Herbelin et al.[133] used reaction (49) in their gain measurement experiments. They noted that the reaction was too slow for use in an ambient temperature chemical laser device, but their gain measurement (performed at 475 K) implied that the rate constant must increase substantially with increasing temperature. This was confirmed by Manke et al.[184] who found that the rate constant increased by a factor of 10 over the range from 300 to 470 K.

The reaction of H atoms with NCl_2,

$$H + NCl_2 \rightarrow HCl + NCl(a) \quad (55)$$

provides another source of $NCl(a)$. NCl_2 radicals may be obtained from the reaction of NCl_3 with H atoms,

$$H + NCl_3 \rightarrow HCl + NCl_2. \quad (56)$$

So far, this method for generating $NCl(a)$ has not received much attention. Schwenz et al.[185] used these reactions in a qualitative study of $NCl(a)/I$ transfer.

Flow tube studies of NCl(a) kinetics require chemical generation of the metastables in a prereactor before the reactants or quenching agents are added to the flow.[133,185-187] The dimensions of the prereactor are determined, in part, by the time needed for the generation reactions to go to completion. As reactions (49) and (50) have small rate constants at room temperature, they are not optimal for generation of NCl(a). Slow generation reactions require the use of inconveniently long prereactors where there may be significant wall losses. Pritt and Coombe[120] circumvented these problems by using reaction (37) to generate N_3, followed by reaction (35) to give NCl(a). By this means they could generate higher concentrations of the metastable, permitting the first observation of the $a-X$ emission bands.[188] Subsequently, mixed F/Cl/HN$_3$ flows have been used in several studies of NCl(a) kinetics.[119,124,186,187] A typical flow reactor for such measurements is shown in Fig. 14. Manke and Setser[124] used this apparatus to characterize the reaction of Cl with N_3. They reported a rate constant of $(3 \pm 1) \times 10^{-11}$ cm^3 s^{-1} for removal of N_3 by Cl, and a lower limit of 0.5 for the yield of NCl(a). As part of this study they also determined rate constants for removal of NCl(a) by F atoms $[(2.2 \pm 0.7) \times 10^{-11}]$ and Cl atoms $[(1.0 + 1.0/-0.5) \times 10^{-12}$ cm^3 s$^{-1}]$, and the self-pooling reaction (51) ($\approx 1.5 \times 10^{-13}$ cm^3 s^{-1}). Hewett *et al.*[187] used F/Cl/HN$_3$ generation in an extensive study of NCl(a) removal rate constants. Measurements were reported for forty different collision partners. In most instances they found that the NCl(a) removal rate constants were smaller than or comparable to those for NF(a). As with NF(a), they suggested that many of the collision partners removed NCl(a) by reactive channels. Quenching of NCl(a) by O_2 was noteworthy in that the rate constant was larger than for O_2 quenching of NF(a). As this is probably an $E-E$ transfer process in both instances, it is expected that the smaller energy gap for NCl(a) would result in a larger rate constant. Hewett *et al.*[187] examined energy pooling between NCl(a) and NF(a) [reaction (45)] and found that it was more efficient than NF(a) self-pooling [reaction (40)], but less so than NCl(a) self-pooling (the rate constant was estimated to be in the (3–5) $\times 10^{-14}$ cm^3 s^{-1} range). Henshaw *et al.*[186] used a flow tube apparatus with provisions for heating to study the temperature dependence of the removal of NCl(a) by I atoms [reaction (46)], and the room temperature rate constant for reaction (36). The data for reaction (46) were well represented by the Arrhenius expression

$$k(T) = 1.1 \times 10^{-10} \exp(-519/T) \text{ cm}^3 \text{ s}^{-1} \tag{57}$$

over the range from 300 to 482 K. The rate constant of Henshaw *et al.*[186] for reaction (36) at 300 K, $(1.5 \pm 0.7) \times 10^{-11}$ cm^3 s^{-1} was in good agreement with the results of Ray and Coombe,[173] and when divided by the total removal rate constant $[(2.1 \pm 0.4) \times 10^{-11}$ cm^3 s$^{-1}]$ defined an approximate branching fraction for reaction (36) of 0.7. This was in accord with an earlier measurement by Schwenz *et al.*[185] that provided a lower limit for the branching fraction of > 0.6. Henshaw *et al.*[186] also reported a rate constant for reaction (35) $[(1.6 \pm 0.4) \times 10^{-11}$ cm^3 s$^{-1}]$ in good agreement with the results of Manke *et al.*[125] Computer modeling of the cw NCl(a)/I* gain experiment also supports a high branching fraction for reaction (36). Madden[189] found that a value of 0.75 yielded results that were in good agreement with the observed gain.

Quenching constants for NCl(a) have also been measured using UV photolysis of ClN$_3$ to generate the metastable. Pulsed excimer laser photolysis of ClN$_3$ at 193 or 248 mn produces high yields of electronically excited NCl (mostly in the a state),[175,190-192] but there are some subtle complications associated with this source. Ray and Coombe[193] observed unusual time profiles for the NCl a–X emission resulting from 248 nm photolysis. Biexponential decays were observed, and for some conditions the signal exhibited secondary maxima. Ray and Coombe[193] reported that 248 nm photolysis

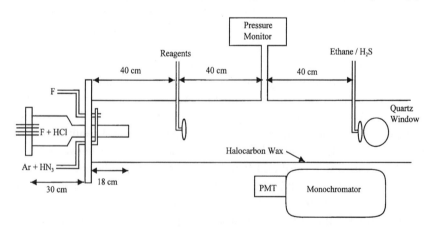

Fig. 14. Flow tube reactor used to study NCl(a) formation and removal kinetics. Note that this apparatus has a two stage prereactor. The first stage is used to generate Cl atoms from F + HCl. The second stage (the first 40 cm of the main reactor tube) is used to generate N$_3$ from the F + HN$_3$ reaction. The inlet for ethane or H$_2$S near the observation window is used for calibration of F or Cl atom concentrations. Reproduced with permission from Ref. 124.

can initiate chain decomposition of ClN_3, and that the chain might be carried by vibrationally excited N_2. With this assignment the chain branching step would be of the form

$$ClN_3 + N_2(v) \rightarrow NCl(a) + N_2 + N_2(v') \tag{58}$$

where $N_2(v')$ is vibrationally excited N_2 resulting from the decomposition of the ClN_3. This is an interesting process, but it does not result in a source of $NCl(a)$ that is suitable for controlled studies of reaction kinetics. Ray and Coombe[193] found that 193 nm photolysis of ClN_3 yielded $NCl(a)$ emission signals that were better behaved. Conditions that produced near single exponential decays were established, and used to study $NCl(a)$ removal processes. Ray and Coombe[193] used this technique to examine energy transfer from $NCl(a)$ to I^*, as described in the previous section. They also examined quenching of $NCl(a)$ by O_2, Cl_2, H_2, HCl, DCl, HF, Ar, and He. Quenching by He and Ar was too slow to be detected. For the diatomic species, quenching (or reaction) rate constants in the range of 6×10^{-13}–1.8×10^{-11} cm^3 s^{-1} were obtained.[193] Surprisingly, flow tube measurements[187] have yielded very different values for the rate constants for all but one of these quenchers (there is good agreement between the results for O_2). The ratios $k(\text{photolysis})/k(\text{flow tube})$ are 45, > 680, 327, and 164 for Cl_2, H_2, HCl, and HF, respectively. Henshaw *et al.*[175] used 193 nm photolysis of ClN_3 to measure rate constants for quenching by F_2, Cl_2, and Br_2. Their results exceed the flow tube values by factors of 7, 73, and 10, respectively. The discrepancies between the flow tube and photolysis cell rate constants appear to be repeatable. Both experiments involve complex chemical environments and there may be unidentified interfering processes. Hewett *et al.*[187] made systematic tests for the presence of processes that could generate additional $NCl(a)$, and thereby lead to artificially low removal rate constants. Their system appeared to be free of such complications. Conversely, Ray and Coombe[193] and Henshaw *et al.*[175] were careful to ensure that the photolysis experiments were not biased by interfering emissions from excited species other than $NCl(a)$, and that the kinetics were not distorted by chain processes. At present the source of the discrepancies has not been identified. Komissarov *et al.*[194] are currently using time-resolved laser absorption measurements to examine the kinetics of $NCl(X)$ in ClN_3 photolysis experiments. These experiments will shed light on the secondary reactions associated with photolysis at 193 and 248 nm.

To date, the self-annihilation rate constant for NCl(a) has only been determined using photolysis techniques. Henshaw *et al.*[175] were able to generate NCl(a) concentrations of 10^{15} cm^{-3}, which was high enough to reveal the second order decay characteristics of self-removal. They reported a rate constant of $(7.2 \pm 0.9) \times 10^{-12}$ cm^3 s^{-1}.

The radiative lifetime of NCl(a) has not been determined in the gas phase, but it has been calculated using high level theoretical methods[195] and measured for NCl isolated in a solid Ar matrix at 15 K.[196] After correction for the effects of the matrix environment the latter was consistent with a gas phase lifetime of 3.7 s. This was within the range of 2.5–5.9 s obtained from *ab initio* theory.[195] The corrected lifetime for NCl(b) obtained from the matrix measurements (1.9 ms) was also in good agreement with the theoretical estimate (1.8 ms),[195] and a later gas phase determination $(2.0 \pm 0.4$ ms).[197] The quenching kinetics of NCl(b) have been examined by Pritt *et al.*[190] and Setser and Zhao.[197] The former group used pulsed photolyis of ClN$_3$ at 193 nm to generate NCl(b), while the latter used a dc discharge through a dilute flow of NFCl$_2$ in He. For the b state, photolysis and flow tube techniques yielded rate constants that were in moderately good agreement. For collisions with diatomic and small polyatomic molecules the quenching rate constants were relatively small, in the 10^{-14}–10^{-13} cm^3 s^{-1} range.

8. Prospects for NX(a) Driven Laser Systems

Chemical methods for producing high density flows of NF(a) have been established and methods for generating similar flows of NCl(a) are being developed. The difficulties in using these species in energy transfer lasers stem from the relatively high rate constants for self-annihilation. To extract the energy from these metastables the transfer and mixing times need to be short, while the rate of transfer to the lasing species must be fast. Ideally, high transfer rates should be achieved using processes that have near gas kinetic rate constants, so that the concentration of the lasing species can be relatively low.

Experiments with lasers driven by transfer from NF(a) were constrained because optimal energy acceptors have not yet been found. The molecules that were pumped to inversion (BiF and BH) were excited by multiple energy transfer events, and could not be generated without introducing troublesome quenching agents. To make further progress with NF(a) driven systems it will be necessary to identify suitable energy recipients. Energy transfer excitation of I is an attractive possibility, but the available kinetic data for this process are ambiguous. It appears that the rate constant for this process may be too small for a device operating near room temperature, but it may become sufficiently large at elevated temperatures. Clearly, further studies of $NF(a) + I$ kinetics are required.

While the idea of using $NCl(a)$ to pump laser systems has been considered for many years, efforts to examine the potential of this system are relatively recent. Energy transfer to I is the only option that has been examined in any detail. Initial studies show that this system can support lasing, and recent cw gain measurements are encouraging. Questions regarding the scaling of this scheme are now being examined. In parallel with efforts to develop the $NCl(a)/I$ system it will be of value to examine transfer between $NCl(a)$ and other possible laser species.

The use of $NBr(a)$ in chemical laser systems is a topic that may be worthy of investigation. Coombe and coworkers[121,198] and Hewett and Setser[199] have shown that $NBr(a)$ is produced by the rapid reaction of Br atoms with N_3, although the branching fraction may favor $NBr(X)$ production. The radiative lifetime of $NBr(a)$ is estimated to be around 200 ms.[199] This is somewhat short for energy storage purposes, but the stimulated emission cross section for the a–X transition may be large enough for direct lasing of NBr. In this context it is of interest to note that the a–X energy separations of NCl and NBr are resonant, so transfer between these species may be facile.

Further studies of $NX(a)$ kinetics are needed to explore different chemical means for generation, to identify new energy acceptors, and to examine potentially detrimental quenching processes. It would also be of value, for both practical and fundamental reasons, to determine the products resulting from $NX(a)$ removal processes.

Note Added in Proof

Since this chapter was compiled, cw chemical lasing of I* in an $NCl(a)/I$ flow reactor has been demonstrated by Henshaw, Manke II, Madden, Berman and Hager (work submitted for publication). Komissarov and Heaven have reexamined the rate constant for self-annihilation of $NCl(a)$. They obtained a room temperature value of $(7 \pm 2) \times 10^{-13}$ cm^3 s^{-1}, which is significantly smaller than previous estimates.

Acknowledgments

A great deal of the research reviewed in this chapter, including work from my own laboratory, has been supported by the Air Force Office of Scientific Research. I am particularly grateful to Dr. Michael Berman of AFOSR for his sustained interest in chemical laser systems and fundamental aspects of energy transfer dynamics. Over the years I have had many helpful discussions with members of the chemical laser community. In this context I would like to acknowledge Dr. D. J. Benard (Rockwell International), Prof. R. D. Coombe* (Denver University), Dr. P. G. Crowell* (Logicon RDA), Dr. S. J. Davis* (PSI), Dr. E. A. Dorko (AFRL), Dr. G. D. Hager (AFRL), Dr. T. J. Madden* (AFRL), Dr. G. C. Manke* (AFRL), Lt. Col. G. P. Perram* (AFIT), Dr. D. N. Plummer (Logicon RDA), Prof. S. Rosenwaks (Ben–Gurion University, Israel) and Prof. D. W. Setser* (Kansas State University). I have also received helpful comments from Dr. J. Kodymova (Academy of Sciences of the Czech Republic, Prague), Dr. R. A. Dressler (AFRL, Hanscom) and Mr. T. van Marter (Emory University). This manuscript was prepared with support from grant AFOSR F49620-98-1-0054.

References

1. J. Adachi, N. Takahashi, K. Yasuda and T. Atsuta, *Prog. Nucl. Energy* **32**, 517 (1997).
2. J. Vetrovec, *Proc. SPIE-Int. Soc. Opt. Eng.* **3574**, 461 (1998).
3. W. P. Latham and B. Quillen, *Laser Inst. Am.* [*Publ.*] **85**, A187 (1998).
4. W. L. Bohn, *Proc. SPIE-Int. Soc. Opt. Eng.* **3343**, 119 (1998).
5. W. E. McDermott, N. R. Pchelkin, D. J. Benard and R. R. Bousek, *Appl. Phys. Lett.* **32**, 469 (1978).
6. D. J. Benard, W. E. McDermott, N. R. Pchelkin and R. R. Bousek, *Appl. Phys. Lett.* **34**, 40 (1979).
7. K. A. Truesdell, C. A. Helms and G. D. Hager, *Proc. SPIE-Int. Soc. Opt. Eng.* **2502**, 217 (1995).
8. R. Engleman, Jr., B. A. Palmer and S. J. Davis, *J. Opt. Soc. Am.* **73**, 1585 (1983).
9. A. J. Ray and R. D. Coombe, *J. Phys. Chem.* **99**, 7849 (1995).
10. D. J. Benard and E. Boehmer, *Appl. Phys. Lett.* **65**, 1340 (1994).
11. D. J. Benard, *J. Appl. Phys.* **74**, 2900 (1993).
12. D. J. Benard and B. K. Winker, *J. Appl. Phys.* **69**, 2805 (1991).
13. J. R. Woodward, S. H. Cobb, K. Shen and J. L. Gole, *IEEE J. Quant. Electron.* **26**, 1574 (1990).

---

*These people were kind enough to read this manuscript and provided helpful comments.

14. K. K. Shen, H. Wang and J. L. Gole, *IEEE J. Quant. Electron.* **29**, 2346 (1993).
15. S. J. Arnold, N. Finlayson and E. A. Ogryzlo, *J. Chem. Phys.* **44**, 2529 (1966).
16. R. G. Derwent and B. A. Thrush, *J. Chem. Soc. Faraday Trans. 2*, **68**, 720 (1972).
17. R. G. Derwent, D. R. Kearns and B. A. Thrush, *Chem. Phys. Lett.* **6**, 115 (1970).
18. R. G. Derwent and B. A. Thrush, *Chem. Phys. Lett.* **9**, 591 (1971).
19. R. G. Derwent and B. A. Thrush, *Faraday Discuss. Chem. Soc.* **53**, 162 (1972).
20. D. R. Kearns, *Chem. Rev.* **71**, 385 (1971).
21. A. T. Pritt, Jr., R. D. Coombe, D. Pilipovich, R. I. Wagner, D. Benard and C. Dymek, *Appl. Phys. Lett.* **31**, 745 (1977).
22. R. F. Tate, B. S. Hunt, C. A. Helms, K. A. Truesdell and G. D. Hager, *IEEE J. Quant. Electron.* **31**, 1632 (1995).
23. C. A. Helms, J. Shaw, G. D. Hager, K. A. Truesdell, D. Plummer and J. Copland, *Proc. SPIE-Int. Soc. Opt. Eng.* **2502**, 250 (1995).
24. T. L. Rittenhouse, S. P. Phipps and C. A. Helms, *IEEE J. Quant. Electron.* **35**, 857 (1999).
25. F. Kitatani and K.-I. Fujii, *Reza Kenkyu* **27**, 127 (1999).
26. M. V. Zagidullin and V. D. Nikolaev, *Proc. SPIE-Int. Soc. Opt. Eng.* **3688**, 54 (1999).
27. B. D. Barmashenko, D. Furman and S. Rosenwaks, *Proc. SPIE-Int. Soc. Opt. Eng.* **3574**, 273 (1998).
28. H. Tanaka, H. Shimada, M. Endo, S. Takeda, K. Nanri, T. Fujioka and F. Wani, *J. Adv. Sci.* **10**, 152 (1998).
29. M. V. Zagidullin, *Proc. SPIE-Int. Soc. Opt. Eng.* **3574**, 569 (1998).
30. M. V. Zagidullin, V. D. Nikolaev, N. A. Khvatov and M. I. Svistun, *Proc. SPIE-Int. Soc. Opt. Eng.* **3574**, 246 (1998).
31. D. Furman, B. D. Barmashenko and S. Rosenwaks, *IEEE J. Quant. Electron.* **34**, 1068 (1998).
32. D. Furman, B. D. Barmashenko and S. Rosenwaks, *Proc. SPIE-Int. Soc. Opt. Eng.* **3268**, 146 (1998).
33. J. Kodymova and O. Spalek, *Japan J. Appl. Phys.* Part 1 **37**, 117 (1998).
34. M. Endo, F. Wani, S. Nagatomo, D. Sugimoto, K. Nanri, S. Takeda and T. Fujioka, *Proc. SPIE-Int. Soc. Opt. Eng.* **3574**, 253 (1998).
35. S. Rosenwaks, I. Blayvas, B. D. Barmashenko, D. Furman and M. V. Zagidullin, *Proc. SPIE-Int. Soc. Opt. Eng.* **3092**, 690 (1997).
36. W. E. McDermott, J. C. Stephens, J. Vetrovec and R. A. Dickerson, *Proc. SPIE-Int. Soc. Opt. Eng.* **2987**, 146 (1997).
37. K. M. Grunewald, J. Handke, W. O. Schall and L. V. Entress-Fursteneck, *Proc. SPIE-Int. Soc. Opt. Eng.* **3574**, 315 (1998).
38. D. Furman, B. D. Barmashenko and S. Rosenwaks, *Appl. Phys. Lett.* **70**, 2341 (1997).

39. D. Furman, E. Bruins, B. D. Barmashenko and S. Rosenwaks, *Appl. Phys. Lett.* **74**, 3093 (1999).
40. M. V. Zagidullin, V. D. Nikolaev, M. I. Svistun and N. A. Khvatov, *Quant. Electron.* **29**, 114 (1999).
41. M. V. Zagidullin, V. D. Nikolaev, M. I. Svistun and N. A. Khvatov, *Kvantovaya Elektron.* (*Moscow*) **25**, 413 (1998).
42. V. N. Azyazov, M. V. Zagidullin, V. D. Nikolaev and V. S. Safonov, *Kvantovaya Elektron.* (*Moscow*) **24**, 491 (1997).
43. M. V. Zagidullin, V. D. Nikolaev, M. I. Svistun, N. A. Khvatov and N. I. Ufimtsev, *Kvantovaya Elektron.* (*Moscow*) **24**, 201 (1997).
44. M. Endo, S. Nagatomo, S. Takeda, M. V. Zagidullin, V. D. Nikolaev, H. Fujii, F. Wani, D. Sugimoto, K. Kunako, K. Nanri and T. Fujioka, *IEEE J. Quant. Electron.* **34**, 393 (1998).
45. D. L. Carroll, D. M. King, L. Fockler, D. Stromberg, W. C. Solomon, L. H. Sentman and C. H. Fisher, *Proc. Int. Conf. Lasers* **21st**, 257 (1999).
46. G. D. Hager, C. A. Helms, K. A. Truesdell, D. Plummer, J. Erkkila and P. Crowell, *IEEE J. Quant. Electron.* **32**, 1525 (1996).
47. M. V. Zagidullin and V. D. Nikolaev, *Kvantovaya Elektron.* (*Moscow*) **24**, 423 (1997).
48. D. A. Copeland and A. H. Bauer, *IEEE J. Quant. Electron.* **29**, 2525 (1993).
49. T. T. Yang, R. A. Cover, V. Quan, D. M. Smith, A. H. Bauer, W. E. McDermott and D. A. Copeland, *Proc. SPIE-Int. Soc. Opt. Eng.* **2989**, 126 (1997).
50. D. A. Copeland and T. T. Yang, *Proc. SPIE-Int. Soc. Opt. Eng.* **2989**, 90 (1997).
51. D. L. Carroll, *AIAA J.* **33**, 1454 (1995).
52. T. J. Madden, Ph.D. Dissertation, University of Illinois, Urbana, 1997.
53. T. J. Madden, D. L. Carroll and W. C. Solomon, *Proc. Int. Conf. Lasers* **18th**, 232 (1996).
54. T. J. Madden and W. C. Solomon, *AIAA* **97-2387**, 23 (1997).
55. W. Masuda, M. Hishida, S. Hirooka, N. Azami and H. Yamada, *JSME Int. J., Ser.* **B40**, 209 (1997).
56. W. Masuda, M. Satoh, H. Fujii and T. Atsuta, *JSME Int. J. Ser.* **B40**, 87 (1997).
57. B. D. Barmashenko, A. Elior, E. Lebiush and S. Rosenwaks, *J. Appl. Phys.* **75**, 7653 (1994).
58. B. D. Barmashenko and S. Rosenwaks, *Aiaa J.* **34**, 2569 (1996).
59. D. L. Carroll, *Proc. Int. Conf. Lasers* **18th**, 225 (1996).
60. G. P. Perram and G. D. Hager (Air Force Weapons Laboratory, Kirkland AFB, NM, USA, 1988), p. 33.
61. S. Churassy, R. Bacis, A. J. Bouvier, C. P. Dit Mery, B. Erba, J. Bachar and S. Rosenwaks, *J. Appl. Phys.* **62**, 31 (1987).
62. B. D. Barmashenko, A. Elior, E. Lebiush and S. Rosenwaks, *Proc. SPIE-Int. Soc. Opt. Eng.* **1810**, 513 (1993).
63. G. P. Perram, *Int. J. Chem. Kinet.* **27**, 817 (1995).

64. R. G. Aviles, D. R. Muller and P. L. Houston, *Appl. Phys. Lett.* **37**, 358 (1980).
65. D. F. Muller, R. H. Young, P. L. Houston and J. R. Wiesenfeld, *Appl. Phys. Lett.* **38**, 404 (1981).
66. R. F. Heidner, III, C. E. Gardner, T. M. El-Sayed and G. I. Segal, *Chem. Phys. Lett.* **81**, 142 (1981).
67. R. F. Heidner, III, C. E. Gardner, G. I. Segal and T. M. El-Sayed, *J. Phys. Chem.* **87**, 2348 (1983).
68. R. F. Heidner, III, *J. Photochem.* **25**, 449 (1984).
69. G. E. Hall, W. J. Marinelli and P. L. Houston, *J. Phys. Chem.* **87**, 2153 (1983).
70. M. H. Van Benthem and S. J. Davis, *J. Phys. Chem.* **90**, 902 (1986).
71. P. M. Alsing, S. J. Davis and G. L. Simmons, private communication (1999).
72. D. David, V. Joly and Fausse, *Proc. 7th Int. Symp. Gas Flow Chem. Lasers* (Springer, Berlin, 1987), p. 156.
73. A. J. Bouvier, R. Bacis, A. Bouvier, D. Cerny, S. Churassy, P. Crozet and M. Nota, *J. Quant. Spectrosc. Radiat. Transfer* **49**, 311 (1993).
74. B. Barnault, A. J. Bouvier, D. Pigache and R. Bacis, *J. Phys. IV*, **1**, C7/647-C7/650 (1991).
75. M. Nota, A. J. Bouvier, R. Bacis, A. Bouvier, P. Crozet, S. Churassy and J. B. Koffend, *J. Chem. Phys.* **91**, 1938 (1989).
76. W. G. Lawrence, T. A. Van Marter, M. L. Nowlin and M. C. Heaven, *J. Chem. Phys.* **106**, 127 (1997).
77. W. G. Lawrence, T. A. VanMarter, M. L. Nowlin and M. C. Heaven, *Proc. SPIE-Int. Soc. Opt. Eng.* **2702**, 214 (1996).
78. M. L. Nowlin and M. C. Heaven, *J. Phys. IV*, **4**, C4/729-C4/737 (1994).
79. M. L. Nowlin and M. C. Heaven, *J. Chem. Phys.* **99**, 5654 (1993).
80. J. T. Yardley, *Introduction to Molecular Energy Transfer* (Academic, New York, 1980).
81. J. I. Steinfeld, *J. Phys. Chem. Ref. Data* **16**, 903 (1987).
82. J. I. Steinfeld, *J. Phys. Chem. Ref. Data* **13**, 445 (1984).
83. M. Rubinson, B. Garetz and J. I. Steinfeld, *J. Chem. Phys.* **60**, 3082 (1974).
84. J. I. Steinfeld and W. Klemperer, *J. Chem. Phys.* **42**, 3475 (1965).
85. S. K. Gray and S. A. Rice, *J. Chem. Phys.* **83**, 2818 (1985).
86. M. L. Nowlin, Ph.D. Thesis, Emory University, 1994.
87. A. T. Young and P. L. Houston, **78**, 2317 (1983).
88. J. J. Deakin and D. J. Husain, *J. Chem. Soc. Faraday Trans. II*, **68**, 1603 (1972).
89. T. Van Marter and M. C. Heaven, *J. Chem. Phys.* **109**, 9266 (1998).
90. T. Van Marter, M. C. Heaven and D. Plummer, *Chem. Phys. Lett.* **260**, 201 (1996).
91. A. L. Kaledin, M. C. Heaven and K. Morokuma, *Chem. Phys. Lett.* **289**, 110 (1998).
92. E. A. Yukov, *Soviet J. Quant. Electron.* **3**, 117 (1973).

93. V. A. Alekseev, T. L. Andreeva, V. N. Volkov and E. A. Yukov, *Soviet Phys. JETP* **36**, 238 (1973).
94. A. Gallagher, *Phys. Rev.* **157**, 68 (1967).
95. D. A. Copeland and A. H. Bauer, *Proc. Int. Conf. Lasers* **14th**, 138 (1992).
96. A. N. Okunevich and V. I. Perel, *Soviet Phys. JETP* **31**, 356 (1970).
97. W. Thieme and E. Fill, *J. Phys. D: Appl. Phys.* **12**, 2037 (1979).
98. W. Thieme and E. Fill, *Opt. Comm.* **36**, 361 (1981).
99. T. A. Van Marter and M. C. Heaven, *Proc. SPIE-Int. Soc. Opt. Eng.* **3612**, 125 (1999).
100. D. Cerny, R. Bacis, B. Bussery, M. Nota and J. Verges, *J. Chem. Phys.* **95**, 5790 (1991).
101. T. D. Padrick and R. E. Palmer, *J. Chem. Phys.* **62**, 3350 (1975).
102. D. K. Neumann, P. K. Clark, R. F. Shea and S. J. Davis, *J. Chem. Phys.* **79**, 4680 (1983).
103. S. J. Davis, W. J. Kessler and M. Bachmann, *Proc. SPIE-Int. Soc. Opt. Eng.* **3612**, 157 (1999).
104. M. V. Zagidullin, V. I. Igoshin and N. L. Kupriyanov, *Kvantovaya Elektron.* (*Moscow*) **11**, 382 (1984).
105. J. I. Cline, C. A. Taatjes and S. R. Leone, *J. Chem. Phys.* **93**, 6543 (1990).
106. J. W. Nicholson, W. Rudolph and G. Hager, *J. Chem. Phys.* **104**, 3537 (1996).
107. R. B. Badger, A. C. Wright and R. T. Whitloch, *J. Chem. Phys.* **43**, 4345 (1965).
108. M. G. Mlynczak and D. J. Nesbitt, *Geophys. Res. Lett.* **22**, 1381 (1995).
109. J. Kodymová and O. Spalek, *SPIE Proc.* **3612**, 92 (1999).
110. R. F. Tate, B. S. Hunt, G. D. Hager, C. A. Helms and K. A. Truesdell, *Proc. Int. Conf. Lasers* **16th**, 609 (1994).
111. R. F. Tate, B. S. Hunt, G. D. Hager, C. A. Helms and K. A. Truesdell, *Proc. SPIE-Int. Soc. Opt. Eng.* **2502**, 272 (1995).
112. P. B. Keating, C. A. Helms, B. T. Anderson, T. L. Rittenhouse, K. A. Truesdell and G. D. Hager, *Proc. Int. Conf. Lasers* **19th**, 194 (1997).
113. D. Furman, B. D. Barmashenko and S. Rosenwaks, *IEEE J. Quant. Electron.* **35**, 540 (1999).
114. E. Lebiush, B. D. Barmashenko, A. Elior and S. Rosenwaks, *IEEE J. Quant. Electron.* **31**, 903 (1995).
115. S. J. Davis, M. G. Allen, W. J. Kessler, K. R. McManus, M. F. Miller and P. A. Mulhall, *Proc. SPIE-Int. Soc. Opt. Eng.* **2702**, 195 (1996).
116. S. J. Davis, K. W. Holtzclaw, W. J. Kessler and C. E. Otis, *Proc. SPIE-Int. Soc. Opt. Eng.* **2702**, 202 (1996).
117. M. G. Allen, K. L. Carleton, S. J. Davis and D. Palombo, *Proc. SPIE-Int. Soc. Opt. Eng.* **2122**, 2 (1994).
118. M. G. Allen, K. L. Carleton, S. J. Davis, W. J. Kessler, C. E. Otis, D. A. Palombo and D. M. Sonnenfroh, *Appl. Opt.* **34**, 3240 (1995).
119. X. Liu, M. A. MacDonald and R. D. Coombe, *J. Phys. Chem.* **96**, 4907

(1992).

120. A. T. Pritt, Jr. and R. D. Coombe, *Int. J. Chem. Kinet.* **12**, 741 (1980).

121. A. T. Pritt, Jr., D. Patel and R. D. Coombe, *Int. J. Chem. Kinet.* **16**, 977 (1984).

122. R. D. Coombe and A. T. Pritt, Jr., *Chem. Phys. Lett.* **58**, 606 (1978).

123. J. Habdas, S. Wategaonkar and D. W. Setser, *J. Phys. Chem.* **91**, 451 (1987).

124. G. C. Manke, II and D. W. Setser, *J. Phys. Chem.* **A102**, 7257 (1998).

125. G. C. Manke, II and D. W. Setser, *J. Phys. Chem.* **A102**, 153 (1998).

126. C. T. Cheah and M. A. A. Clyne, *J. Chem. Soc. Faraday Trans. 2*, **76**, 1543 (1980).

127. C. T. Cheah, M. A. A. Clyne and P. D. Whitefield, *J. Chem. Soc. Faraday Trans. 2*, **76**, 711 (1980).

128. C. T. Cheah and M. A. Clyne, *J. Photochem.* **15**, 21 (1981).

129. M. A. A. Clyne and I. F. White, *Chem. Phys. Lett.* **6**, 465 (1970).

130. J. M. Herbelin and N. Cohen, *Chem. Phys. Lett.* **20**, 605 (1973).

131. R. J. Malins and D. W. Setser, *J. Phys. Chem.* **85**, 1342 (1981).

132. T. T. Yang, V. T. Gylys, R. D. Bower and L. F. Rubin, *Opt. Lett.* **17**, 1803 (1992).

133. J. M. Herbelin, T. L. Henshaw, B. D. Rafferty, B. T. Anderson, R. F. Tate, T. J. Madden, G. C. Manke, II and G. D. Hager, *Chem. Phys. Lett.* **299**, 583 (1999).

134. D. J. Benard, *J. Phys. Chem.* **100**, 8316 (1996).

135. K. Y. Du and D. W. Setser, *J. Phys. Chem.* **96**, 2553 (1992).

136. R. F. Heidner, III, H. Helvajian, J. S. Holloway and J. B. Koffend, *J. Phys. Chem.* **93**, 7818 (1989).

137. J. M. Herbelin, R. R. Giedt and H. A. Bixler, *J. Appl. Phys.* **54**, 28 (1983).

138. J. M. Herbelin, M. A. Kwok and D. J. Spencer, *J. Appl. Phys.* **49**, 3750 (1978).

139. J. B. Koffend, B. H. Weiller and R. F. Heidner, III, *J. Phys. Chem.* **96**, 9315 (1992).

140. G. A. Capelle, D. G. Sutton and J. I. Steinfeld, *J. Chem. Phys.* **69**, 5140 (1978).

141. J. M. Herbelin and R. A. Klingberg, *Int. J. Chem. Kinet.* **16**, 849 (1984).

142. R. F. Heidner, III, H. Helvajian, J. S. Holloway and J. B. Koffend, *J. Chem. Phys.* **84**, 2137 (1986).

143. D. J. Benard, B. K. Winker, T. A. Seder and R. H. Cohn, *J. Phys. Chem.* **93**, 4790 (1989).

144. D. J. Benard, E. Boehmer, H. H. Michels and J. A. Montgomery, Jr., *J. Phys. Chem.* **98**, 8952 (1994).

145. E. Boehmer and D. J. Benard, *J. Phys. Chem.* **99**, 1969 (1995).

146. K. Y. Du and D. W. Setser, *J. Phys. Chem.* **94**, 2425 (1990).

147. A. Ray-Hunter and R. D. Coombe, unpublished results (1998).

148. S. J. David and R. D. Coombe, *J. Phys. Chem.* **89**, 5206 (1985).

149. K. B. Hewett and D. W. Setser, *J. Phys. Chem.* **A101**, 9125 (1997).

150. K. Y. Du and D. W. Setser, *J. Phys. Chem.* **95**, 4728 (1991).
151. K. Du and D. W. Setser, *J. Phys. Chem.* **97**, 5266 (1993).
152. E. Quinones, J. Habdas and D. W. Setser, *J. Phys. Chem.* **91**, 5155 (1987).
153. S. Wategaonkar, K. Y. Du and D. W. Setser, *Chem. Phys. Lett.* **189**, 586 (1992).
154. J. Habdas and D. W. Setser, *J. Phys. Chem.* **93**, 229 (1989).
155. S. J. Davis, W. T. Rawlins and L. G. Piper, *J. Phys. Chem.* **93**, 1078 (1989).
156. S. J. Davis and L. G. Piper, *J. Phys. Chem.* **94**, 4515 (1990).
157. R. F. Heidner, III, H. Helvajian and J. B. Koffend, *J. Chem. Phys.* **87**, 1520 (1987).
158. B. H. Weiller, R. F. Heidner, J. S. Holloway and J. B. Koffend, *J. Phys. Chem.* **96**, 9321 (1992).
159. B. H. Weiller, R. F. Heidner, III, J. S. Holloway and J. B. Koffend, *Proc. Int. Conf. Lasers*, 710 (1991).
160. H. Helvajian, R. F. Heidner, III, J. S. Holloway and J. B. Koffend, *Proc. SPIE-Int. Soc. Opt. Eng.* **1031**, 661 (1989).
161. R. F. Heidner, J. S. Holloway, H. Helvajian and J. B. Koffend, *J. Phys. Chem.* **93**, 7818 (1989).
162. R. F. Heidner, III, H. Helvajian, J. S. Holloway and J. B. Koffend, *J. Phys. Chem.* **93**, 7813 (1989).
163. X. Y. Bao and D. W. Setser, *J. Phys. Chem.* **93**, 8162 (1989).
164. H. Cha and D. W. Setser, *J. Phys. Chem.* **91**, 3758 (1987).
165. H. Cha and D. W. Setser, *J. Phys. Chem.* **93**, 235 (1989).
166. K. Du and D. W. Setser, *J. Phys. Chem.* **95**, 9352 (1991).
167. P. H. Tennyson, A. Fontijn and M. A. A. Clyne, *Chem. Phys.* **62**, 171 (1981).
168. D. J. Benard, M. A. Chowdhury, B. K. Winker, T. A. Seder and H. H. Michels, *J. Phys. Chem.* **94**, 7507 (1990).
169. S. J. Davis and L. Hanko, *Appl. Phys. Lett.* **37**, 692 (1980).
170. S. J. Davis, L. Hanko and R. F. Shea, *J. Chem. Phys.* **78**, 172 (1983).
171. S. J. Davis, L. Hanko and P. J. Wolf, *J. Chem. Phys.* **82**, 4831 (1985).
172. R. D. Bower and T. T. Yang, *J. Opt. Soc. Am. B: Opt. Phys.* **8**, 1583 (1991).
173. A. J. Ray and R. D. Coombe, *J. Phys. Chem.* **97**, 3475 (1993).
174. T. C. Clark and M. A. A. Clyne, *Trans. Faraday Soc.* **65**, 2994 (1969).
175. T. L. Henshaw, S. D. Herrera, G. W. Haggquist and V. A. Schlie, *J. Phys. Chem.* **A101**, 4048 (1997).
176. T. C. Clark and M. A. A. Clyne, *Trans. Faraday Soc.* **66**, 877 (1970).
177. T. C. Clark and M. A. A. Clyne, *Trans. Faraday Soc.* **66**, 372 (1970).
178. J. Combourieu, G. Le Bras, G. Poulet and J. L. Jourdain, *16th Int. Combust. Symp.*, Massachusetts Institute of Technology (1976).
179. M. A. A. Clyne, A. J. MacRobert and L. J. Stief, *J. Chem. Soc. Faraday Trans. 2*, **81**, 159 (1985).
180. M. A. A. Clyne and A. J. MacRobert, *J. Chem. Soc. Faraday Trans. 2*, **79**, 283 (1983).
181. M. A. A. Clyne, A. J. MacRobert, J. Brunning and C. T. Cheah, *J. Chem. Soc. Faraday Trans. 2*, **79**, 1515 (1983).

182. J. L. Jourdain, G. Le Bras, G. Poulet and J. Combourieu, *J. Combust. Flame* **34**, 13 (1979).

183. K. Yamasaki, T. Fueno and O. Kajimoto, *Chem. Phys. Lett.* **94**, 425 (1983).

184. G. C. Manke, T. L. Henshaw, T. J. Madden and G. D. Hager, *Chem. Phys. Lett.* **310**, 111 (1999).

185. R. W. Schwenz, J. V. Gilbert and R. D. Coombe, *Chem. Phys. Lett.* **207**, 52 (1993).

186. T. L. Henshaw, S. D. Herrera and L. A. Schlie, *J. Phys. Chem.* **A102**, 6239 (1998).

187. K. B. Hewett, G. C. Manke, D. W. Setser and G. Brewood, *J. Phys. Chem.* **A104**, 539 (2000).

188. A. T. Pritt, Jr., D. Patel and R. D. Coombe, *J. Mol. Spectrosc.* **87**, 401 (1981).

189. T. J. Madden, private communication (1999).

190. A. T. Pritt, Jr., D. Patel and R. D. Coombe, *J. Chem. Phys.* **75**, 5720 (1981).

191. R. D. Coombe, D. Patel, A. T. Pritt, Jr. and F. J. Wodarczyk, *J. Chem. Phys.* **75**, 2177 (1981).

192. R. D. Coombe and M. H. Van Benthem, *J. Chem. Phys.* **81**, 2984 (1984).

193. A. J. Ray and R. D. Coombe, *J. Phys. Chem.* **98**, 8940 (1994).

194. A. V. Komissarov, G. C. Manke, S. J. Davis and M. C. Heaven, *Proc. SPIE-Int. Soc. Opt. Eng.* **3931**, 138 (2000).

195. D. R. Yarkony, *J. Chem. Phys.* **86**, 1642 (1987).

196. A. C. Becker and U. Schurath, *Chem. Phys. Lett.* **160**, 586 (1989).

197. Y. Zhao and D. W. Setser, *J. Chem. Soc. Faraday Trans.* **91**, 2979 (1995).

198. R. D. Coombe and C. H. T. Lam, *J. Chem. Phys.* **79**, 3746 (1983).

199. K. B. Hewett and D. W. Setser, *Chem. Phys. Lett.* **297**, 335 (1998).

CHAPTER 5

FROM ELEMENTARY REACTIONS TO COMPLEX COMBUSTION SYSTEMS

Christof Schulz, Hans-Robert Volpp* and Jürgen Wolfrum

Physikalisch-Chemisches Institut, Universität Heidelberg
Im Neuenheimer Feld 253, D-69120 Heidelberg, Germany

Contents

*Author to whom correspondence should be addressed: Tel.: 0049-6221-545012 (new), Fax: 0049-6221-545050, E-mail: aw2@ix.urz.uni-heidelberg.de

1. Introduction

Although combustion has been used by mankind for already more than one million years,[1] it is still the most important technology providing the energy supply for our modern day civilizations.[2] Utilization of combustion leads to the release of unwanted pollutants such as carbon monoxide, unburned hydrocarbons, nitric oxides, and soot[3] which affect our environment.[4] The rapidly increasing world population with the drastically increasing energy consumption demands substantial efforts to optimize combustion processes used for propulsion, heating, manufacturing and energy conversion in order to reduce pollutant formation while, at the same time, maintaining a high efficiency for the conversion of chemical to mechanical, thermal, and electrical energy. Until now, however, construction and optimization of technical combustion devices in general are essentially empirical processes, calling to a large degree on the experience of engineers and technicians employing methods based on trial and error methods and global performance measurements. With the growing number of performance and environmental protection requirements that must be met, this kind of approach is reaching its limits. Because combustion processes usually consist of a complex interaction of homogeneous and heterogeneous chemistry and transport processes,[5] the development of numerical simulation tools that can accurately predict combustion and its characteristics for practical applications has to be based on a detailed understanding of the underlying chemical reaction systems and physical processes.[6]

In the past, combustion modeling was directed towards fluid mechanics that included global heat release by chemical reaction. The latter was often described simply with the help of thermodynamics, assuming that the chemical reactions are much faster than the other processes like diffusion, heat conduction, and flow. However, in most cases chemistry occurs on time scales which are comparable with those of flow and molecular transport. As a consequence, detailed information about the individual elementary reactions is required if transient processes like ignition and flame quenching or pollutant formation shall be successfully modeled.[3] The fundamental concept of using elementary reactions to describe a macroscopic

chemical net reaction can be traced back to the work of Max Bodenstein at Heidelberg University who studied in 1894 the kinetics of the
photo-induced decomposition of hydrogen iodide using the sun as a natural
photolysis-light-source.[7] Since then, the development of flash photolysis,[8]
chemiluminescence,[9] and molecular beam techniques[10] have dramatically
expanded the experimental possibilities for the investigation of elementary
reactions. Laser-based techniques which combine pulsed laser photolysis for
reactive species generation with time- and state-resolved laser-based reaction product detection paved the way for detailed studies of the molecular
dynamics of elementary reactions.[11-13] Today, femtosecond laser "pump-
and-probe" techniques pioneered by Ahmed Zewail and his group[14] can
provide detailed information about reactive systems on their way through
the transition state.[15] Nanosecond laser "pump-and-probe" techniques, on
the other hand, can be used to determine asymptotic scalar and vectorial
quantities of the reactive collision.[16] In addition, laser diagnostics methods
which allow nonintrusive measurements of multidimensional temperature
and species distributions in technical systems play an important role in
the validation of the existing mathematical combustion models for reactive
flows.[17] Most recently, new nonlinear laser spectroscopic methods made it
possible to monitor bonding and location of reactive surface species during
heterogeneous combustion at catalytic surfaces under realistic pressure
conditions.[18-20]

In this chapter we will present selected examples from ongoing studies
in our laboratory in which reaction systems of increasing order of complexity are investigated. We will start with new results from chemical
dynamics studies of the $H + O_2$ gas-phase chain branching combustion reaction, which can be used to validate dynamical simulations carried out
on *ab initio* potential energy surfaces. The second example describes recent experiments in which absolute radical concentrations were measured
in a low pressure premixed stoichiometric NO seeded CH_4/O_2 flame to
quantify uncertainties of different kinetic mechanisms currently used in
NO_x formation models. Thereafter, we will describe laser-induced ignition
experiments that allow for comparison with results from direct numerical simulations of unsteady CH_3OH oxidation under laminar flow conditions. After this, we report on results from laser diagnostics experiments in which a novel nonlinear surface vibrational spectroscopy technique employing infrared visible sum frequency generation allows *in situ*
investigations of surface reactions on a polycrystalline platinum catalyst

under technical realistic pressure conditions. After this we will present different laser spectroscopic approaches for quantitative measurements of fuel, temperature and NO distributions in internal combustion engines.

2. Dynamics Studies of the H + O₂ → OH + O Elementary Combustion Reaction

The bimolecular reaction of hydrogen atoms with molecular oxygen

$$H(^2S) + O_2(^3\Sigma_g^-) \rightarrow OH(^2\Pi) + O(^3P) \quad \Delta H_0 = 69.5 \text{ kJ/mol}, \quad (1)$$

plays a central role in combustion chemistry. As one of the most important chain branching steps, reaction (1) is of fundamental importance in the ignition of H_2/O_2-mixtures, as well as in the oxidation of hydrocarbons. As a consequence, the rate of reaction (1) strongly influences important parameters of the combustion process, e.g. burning velocities.[21]

In 1928, Haber and coworkers investigated the role H atoms play in various combustion processes[22] and suggested a first reaction mechanism[23] for the oxidation of hydrogen, in which reaction (1) was considered as a main chain branching step. Since then, many experimental studies on the kinetics of reaction (1)[24] have been carried out including theoretical investigations using statistical,[25] quasiclassical trajectory (QCT),[26] and quantum mechanical (QM) methods. The QM studies can be divided into calculations for total angular momentum $J = 0$[27] and $J > 0$,[28] where the latter ones were performed using approximate methods. Only recently have rigorous QM calculations for $J = 1, 2, 5$ and $J = 10$ been reported.[29] In the QCT and QM studies of reaction (1) mainly three different potential energy surfaces (PESs) were used: the Melius–Blint (MB) PES,[30] the Double Many Body Expansion (DMBE) IV PES of Varandas and coworkers,[31] and a Diatomics-in-Molecules (DIM) PES recently derived by Kendrik and Pack.[32]

Chemical dynamics experiments in which OH product quantum state distributions and an absolute reaction cross section for reaction (1) could be measured were reported in 1984.[33] Subsequent experiments revealed additional details about the reaction dynamics, including nascent $OH(^2\Pi)$ spin-orbit and Λ-doublet rotational fine structure state distributions,[34] $O(^3P)$ product fine structure state distributions,[35] and OH angular momentum polarization distributions,[34(a),36] as well as differential cross sections.[37] The experimental results indicate that depending on the reagent collision energy

($E_{c.m.} < 1.2$ eV), the reaction goes through a HO_2^* reaction intermediate over a deep well in the $HO_2(\tilde{X}^2A'')$ PES leading to statistical OH rotational distributions and differential cross sections which show a forward-backward symmetry as expected from long lived HO_2^* complexes. At higher energies the mechanism changes and becomes a direct one, favoring formation of highly rotationally excited $OH(v'' = 0)$ reaction products that are strongly forward scattered with respect to the initial H atom velocity direction. The dependence of the OH product rotational excitation on the collision energy of the reagents could be well reproduced in QCT calculations on the ground state $HO_2(\tilde{X}^2A'')$ MB-PES.[26(b)] Significant discrepancies between measured[37] and calculated[26(b)] differential cross sections were found at high collision energies which suggest that the multisurface nature of the $H + O_2$ reaction has to be taken into account at higher energies.

The combination of the symmetries of the $H(^2S)$ and $O_2(^3\Sigma_g^-)$ reagents leads to two PESs with $^2A''$ and $^4A''$ symmetry (in C_s point group), for which the lowest energy $HO_2(\tilde{X}^2A'')$ PES correlates adiabatically with $OH(^2\Pi) + O(^3P)$ products along a minimum energy path with no barriers apart from that imposed by the reaction endothermicity.[38(a)] For linear and T-shaped $H + O_2$ approach geometries however, barriers do exist which result from the intersecting $^2\Sigma^-$ and $^2\Pi$ surfaces (in $C_{\infty h}$ point group) and the intersecting $^2A_2$ and $^2B_1$ surfaces (in C_{2v} point group),[38(b)] respectively. The height of the barrier due to the C_{2v} conical intersection has been calculated to be 1.7 eV with respect to the reagent zero-point energy.[38(b)] In Ref. 37, it was suggested that the experimentally observed sharp maximum in the absolute reaction cross section around $E_{c.m.} = 1.8$ eV[39] is a consequence of an enhanced reaction probability due to the presence of an additional sideways component in reactivity. The subsequent decrease of the experimental reaction cross section at $E_{c.m.} > 1.8$ eV[40] was attributed to a change in the underlying collision dynamics from an adiabatic to a diabatic one. It was suggested that at high reagent velocities collisions tend to follow the repulsive diabatic $^2B_1$ curve for approach angles close to 90°, thus closing the window on the sideways approach to reactivity.[37]

All absolute reaction cross section measurements reported in Refs. 33, 36, 40 were based on OH product detection via laser-induced fluorescence (LIF) employing the diagonal bands of the $OH(A^2\Sigma^+ - X^2\Pi)$ UV transition. However, due to predissociation of the $A^2\Sigma^+$ state[41] only the $OH(v'' = 0)$ and $OH(v'' = 1)$ product states could be probed in these experiments. As a consequence, in the reaction cross section measurements at $E_{c.m.} > 1.6$ eV where significant amounts of $OH(v'' > 1)$ can be produced, the contribution of the higher OH product vibrational states to the total reactive cross section had to be estimated by surprisal extrapolation.[42] Recently, we have developed an improved approach for the measurement of absolute reactive cross sections for reaction (1). This method, in which pulsed laser photolysis (LP) for generation of translationally energetic H atoms is combined with vacuum ultraviolet (VUV) LIF detection of $O(^3P)$ atom reaction products, allows the measurement of the complete $O(^3P_{j=2,1,0})$ fine structure state distribution even at high reagent collision energies. The new technique and the associated results are described in the following.

2.1. *Experimental Technique*

The experimental setup used for the reaction dynamics studies is schematically depicted in Fig. 1. The principle of the photolytic calibration method utilizing the LP/VUV-LIF technique for absolute reactive cross section measurements has been described in detail elsewhere.[43] Therefore, only a brief summary of the experimental method will be given in the following. The $H + O_2$ reaction dynamics studies were carried out in flowing mixtures of H_2S/O_2 and HBr/O_2 at room temperature. $[H_2S]:[O_2]$ and $[HBr]:[O_2]$ ratios were between 1:2 and 1:6, at a total pressure of typically 120–180 mTorr. Gases had the following purities: H_2S (UCAR, 99.99%), HBr (Messer Griesheim, 99.8%), O_2 (Messer Griesheim, 99.995%). For photolytic calibration measurements, NO_2 (Messer Griesheim, 99.98%) and SO_2 (Messer Griesheim, 99.98%) could be pumped through the reaction cell at pressures of typically 1–14 mTorr. The flow rates (regulated by calibrated mass flow controllers) were maintained high enough to ensure gas renewal between successive laser shots at the laser repetition rate of 6 Hz. The cell pressure was monitored by an MKS-Baratron.

A schematic description of the pulsed LP/VUV-LIF "pump-probe" method for the single collision studies of translationally energetic H atoms with O_2 is given in Fig. 2. H atoms with average H–O_2 center-of-mass collision energies of $E_{c.m.} = 1.0$ eV and $E_{c.m.} = 2.5$ eV were generated by pulsed laser photolysis of H_2S and HBr using KrF and ArF excimer lasers, which provided photodissociation wavelengths of $\lambda_{pump} = 248$ nm and $\lambda_{pump} = 193$ nm, respectively. The duration of the photolysis laser pulses were 15–20 ns. Typically, 100–250 ns after the "pump" laser pulse,

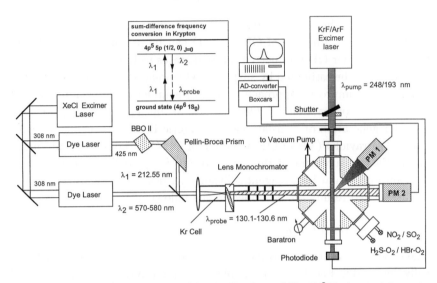

Fig. 1. Experimental apparatus used for the $H + O_2 \rightarrow OH + O(^3P)$ chemical dynamics studies. The Kr four-wave mixing scheme for the generation of tunable vacuum-UV laser radiation for $O(^3P)$ atom LIF detection is shown as an inset.

the $O(^3P)$ atom reaction products were detected by a second "probe" laser pulse (pulse duration 10–15 ns) *via* vacuum-ultraviolet (VUV) one-photon LIF. $O(3s\,^3S^0 \leftarrow 2p\,^3P_j)$ transitions at $\lambda_{\text{probe}} = 130.22$ nm $(j = 2)$, 130.48 nm $(j = 1)$, and 130.60 nm $(j = 0)$ were used to determine nascent $O(^3P_j)$ fine structure distributions. The tunable VUV probe laser radiation was generated by resonant sum difference frequency mixing, $\omega_{\text{probe}} = 2\omega_1 - \omega_2$, in a Krypton gas cell[44] using the output of two dye lasers simultaneously pumped by a XeCl excimer laser. In the four-wave mixing process the frequency ω_1 $(\lambda_1 = 212.55$ nm) is two-photon resonant with the Kr $4p$–$5p\,(\frac{1}{2}, 0)$ transition (see Fig. 1). The second frequency ω_2 could be tuned in the wavelength range 570–580 nm to cover the $O(3s\,^3S^0 \leftarrow 2p\,^3P_j)$ VUV transitions. A bandwidth of ~ 0.45 cm^{-1} was determined for the VUV radiation. VUV-LIF signals were detected through a band pass filter (ARC model 130-B-1D) by a solar blind photomultiplier (THORN EMI model 9413 B, denoted as PM 1 in Fig. 1). Each point of the measured $O(^3P_{j=2,1,0})$ Doppler profiles shown in Fig. 3 was averaged over 30 laser shots. VUV probe beam and photolysis laser intensities were recorded during the measurements by a second solar blind photomultiplier (denoted

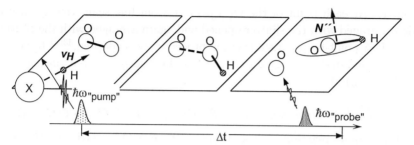

Fig. 2. Schematic description of the pulsed laser photolysis/laser-induced fluorescence (LP/LIF) "pump-probe" method for $H + O_2$ chemical dynamics studies. Translationally energetic H atoms are generated by pulsed laser photolysis of appropriate precursor molecules HX and the nascent $O(^3P_{j=2,1,0})$ atoms produced in the reaction $H + O_2$ are detected under collision free conditions via LIF.

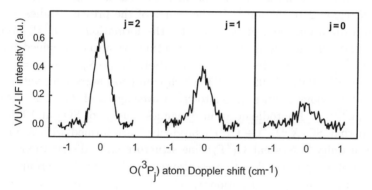

Fig. 3. Doppler profiles for the three fine-structure states, $j = 2, 1, 0$, of $O(^3P)$ atoms formed in the reaction $H + O_2$. Translationally excited H atoms with an average collision energy of $E_{c.m.} = 1.0$ eV were generated by pulsed laser photolysis of 50 mTorr H_2S at 248 nm in the presence of 100 mTorr of O_2. The $O(^3P)$ atom reaction products were detected 180 ns after the photolysis laser pulse by pulsed vacuum-UV laser-induced fluorescence.

as PM 2 in Fig. 1) and by a photodiode, respectively, and used to normalize the LIF signals.

It was found that the VUV probe beam itself produced appreciable $O(^3P_{j=2,1,0})$ atom LIF signals *via* the photolysis of O_2 and NO_2/SO_2, respectively. In order to subtract these "background" $O(^3P_{j=2,1,0})$ atoms from the $O(^3P_{j=2,1,0})$ atoms produced in the $H + O_2$ reaction and in the 248/193 nm photolysis of NO_2/SO_2, respectively, an electronically controlled mechanical shutter was inserted into the photolysis beam path (see

Fig. 1). At each point of the $O(^3P_{j=2,1,0})$ atom line scan, the signal was first averaged with the shutter opened and again averaged with the shutter closed. A point-by-point subtraction procedure was adopted,[43(a)] to obtain directly and on-line a signal free from "probe-laser-generated" background $O(^3P_{j=2,1,0})$ atoms.

2.2. *Absolute Reaction Cross Section for* $H + O_2 \rightarrow OH + O$

Total absolute reaction cross sections for the $H + O_2 \rightarrow OH + O(^3P)$ reaction were obtained by means of a photolytic calibration method using 248 nm laserphotolysis of NO_2 and 193 nm laser photolysis of SO_2 as reference sources of well-defined $O(^3P)$ atom concentrations. The UV photochemistry of NO_2 and SO_2 has been studied in detail.[45] Formation of $O(^1D)$ atoms in the UV photolysis of NO_2 was investigated in the wavelength region 210–255 nm.[46] In the course of the present studies we measured a quantum yield of $\phi_0 = 0.8$ for the formation of $O(^3P)$ products in the 248 nm photolysis of NO_2. In addition, we also determined the nascent $O(^3P_j)$ fine structure state distribution, $N_{j=2}/N_{j=1}/N_{j=0} = (0.64 \pm 0.05)/(0.27 \pm 0.01)/(0.09 \pm 0.01)$, which is in good agreement with the results of previous studies at photolysis wavelengths of 337, 266, 226 and 212 nm.[35(b),47] Photodissociation dynamics studies of SO_2 at 193 nm revealed that $SO(X\,^3\Sigma^-) + O(^3P)$ products[48] are formed with a quantum yield of unity.[49] Nascent $O(^3P_j)$ fine structure state distributions for the 193 nm photolysis wavelength were also reported by different groups.[43(b),50]

After the $O(^3P_j)$ distribution $n_{j=2}/n_{j=1}/n_{j=0}$ for the $H+O_2 \rightarrow OH+O$ reaction has been measured for a given translational energy (see Fig. 3), the total absolute reaction cross section σ_R can be obtained by comparing, e.g. the $O(^3P_{j=2})$ atom signal, $S_R(j = 2)$ produced in the reaction, with the $O(^3P_{j=2})$ atom signal, $S_{Cal}(j = 2)$ produced in the NO_2 photolysis (see Fig. 4) using the following formula[43(b)]:

$$\sigma_R = \gamma \frac{S_R(j = 2)N_{j=2}}{S_{Cal}(j = 2)n_{j=2}} \times \frac{\phi_0\sigma_{NO_2}[NO_2]}{\sigma_{H_2S}[H_2S][O_2]v_{rel}\Delta t}. \qquad (2)$$

Here v_{rel} is the relative velocity which corresponds in the present case (H_2S photolysis at 248 nm in the presence of O_2) to the average translational energy of $E_{c.m.} = 1.0$ eV of the $H + O_2$ reactant pair. The average translational energy was calculated using the $H + SH$ energy partitioning, data obtained in the 248 nm photodissociation dynamics studies of H_2S.[51] [...]

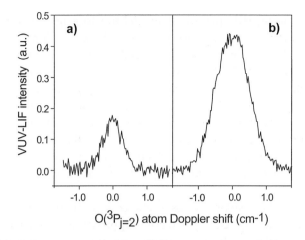

Fig. 4. Comparison of Doppler profiles of $O(^3P_{j=2})$ atoms produced (a) in the reaction $H + O_2$ at $E_{c.m.} = 1.0$ eV and (b) produced in the 248 nm photolysis of 13 mTorr of NO_2.

denote the gas-phase concentrations which can be calculated from the corresponding partial pressures. σ_{NO_2} is the optical absorption cross section of NO_2 at the wavelength of 248 nm (3.3×10^{-20} cm^2)[52] and σ_{H_2S} is the corresponding absorption cross section of H_2S (2.6×10^{-20} cm^2).[53] Δt is the time delay between pump and probe laser pulses. Values of $n_{j=2}$ for the relative population of the $O(^3P_{j=2})$ reaction product state at $E_{c.m.} = 1.0$ eV were taken from the $O(^3P_j)$ distribution, $n_{j=2}/n_{j=1}/n_{j=0} = (0.61 \pm 0.07)/(0.29 \pm 0.06)/(0.10 \pm 0.04)$, measured in the present study. The factor γ in Eq. (2) is a correction accounting for the different degrees of absorption of the VUV probe laser radiation by the H_2S/O_2 mixture and NO_2 respectively. The absorption correction could be directly determined from the known distances inside the flow reactor and the relative difference in the VUV probe laser attenuation measured in the reaction and calibration runs. Under typical experimental conditions the absorption correction factor was $\gamma = 0.79$. In order to determine the total absolute reactive cross section for the $H + O_2$ reaction in different experimental runs, integrated areas of the $O(^3P_{j=2})$ fluorescence excitation curves (see Fig. 4) were determined under identical experimental conditions and evaluated using Eq. (2). For $E_{c.m.} = 1.0$ eV, seven calibration measurements yielded an average value for the reaction cross section of $\sigma_R(1.0 \text{ eV}) = (0.17 \pm 0.04)$Å^2. A second absolute reaction cross section, $\sigma_R(2.5 \text{ eV}) = (0.12 \pm 0.04)$Å^2, was measured by using HBr photolysis at 193 nm to generate translationally energetic H atoms with an average value of $E_{c.m.} = 2.5$ eV.[54] The latter value was calculated by weighted averaging over the bimodal collision energy distribution, 2.1 eV (15%) and 2.6 eV (85%), originating from the two possible HBr dissociation channels: $H + Br(^2P_{1/2})$ and $H + Br(^2P_{3/2})$.[55] In order to obtain $\sigma_R(2.5$ eV$)$, 10 calibration runs were performed using the procedure described in Ref. 43(b), which is based on the 193 nm laser photolysis of SO_2. Experimental errors were calculated from the errors of the entries of Eq. (2) on the basis of simple error propagation. Uncertainties of the values of the optical absorption cross sections were estimated to be 10%.

In Fig. 5, the reaction cross sections $\sigma_R(1.0$ eV$)$ and $\sigma_R(2.5$ eV$)$ obtained with the new $O(^3P)$ detection approach are depicted (filled circles) together with results from earlier absolute measurements from this laboratory[40,43(b)] and relative[39] (solid line) reaction cross section measurements. The most recent results are, within the combined error limits, in agreement with the results of the previous measurements carried out at comparable collision energies. The value obtained in the present study for

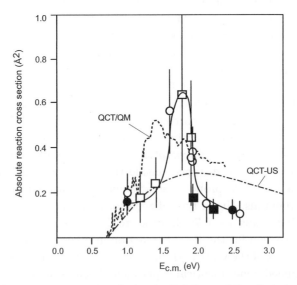

Fig. 5. Absolute reaction cross section for the $H + O_2 \rightarrow OH + O$ elmentary reaction as a function of collision energy. (o) from Ref. 40; ($\square$) and solid line from Ref. 39; ($\blacksquare$) from Ref. 43(b); ($\bullet$) $O(^3P)$ VUV(LIF) work; dashed lines represent theoretical results from QCT/QM (Ref. 56) and QCT-US [Ref. 26(e)] calculations.

$\sigma_R(1.0\text{ eV})$ confirms the value, $\sigma_R(1.0\text{ eV}) = (0.20 \pm 0.08)\text{Å}^2$, which was measured using OH-LIF product detection in combination with a calibration method based on H_2O_2 photolysis.[40(b)] While in the latter experiments all energetically accessible OH product states, $OH(v'' = 0, N'')$ could be directly detected via LIF, the values of the reaction cross sections measured for translational energies 1.2–2.6 eV (open circles and squares in Fig. 5) were obtained using a linear surprisal extrapolation[42] based on the measured $OH(v'' = 0)/OH(v'' = 1)$ population ratio to determine the contribution of higher OH vibrational states to the total reaction cross section and a numerical simulation procedure to deconvolute the measured $OH(v'' = 0, 1; N'')$ rotational product state distributions.[39] Only the reaction cross sections $\sigma_R(1.9\text{ eV})$ and $\sigma_R(2.2\text{ eV})$ (depicted as filled squares in Fig. 5) were recently obtained using a $O(^3P_j)$ VUV-LIF detection and calibration method similar to the one employed in the present study.

The results of two recent dynamical calculations of the $H + O_2(v = 0, j^* = 1)$ reaction on the DMBE IV ($\tilde{X}^2A''$) PES are also shown in Fig. 5.[56] The dashed lines represent results of a QCT study in which a unified statistical (US) approach was employed to account for the problem of zero-point energy conservation[26(e)] and a calculation in which a QCT/QM hybrid model[56] was employed to estimate total reaction cross sections from quantum mechanical ($J = 0$) reaction probabilities, respectively.[27] While the QCT-US calculations cannot reproduce the experimentally observed pronounced maximum in the reaction cross section, the QCT/QM approach predicts a maximum of the reaction cross section, which is in qualitative agreement with the experiments. However, for $E_{\text{c.m.}} > 2.0$ eV both types of single-PES calculations lead to reaction cross sections which are considerably higher than the experimental results. In this energy regime, where the translational energy becomes comparable to the energy of the ($^2A_2 - ^2B_1$) conical intersection, nonadiabatic coupling to the repulsive $^2B_1$ diabatic PES is expected to influence the collision dynamics significantly.[37] As a consequence, additional calculations which include all relevant PESs and their nonadiabatic couplings are clearly needed for a more detailed comparison between experiment and theory at high collision energies. In addition, rigorous QM calculations of reactive cross sections at intermediate energies would be desirable to get more insight into the origin of the peak in the reaction cross section suggested by the experiments.[39] Further experimental studies are currently under way in which the method presented in this article is used to measure cross sections at collision energies between 1.0 and 1.9 eV in order to verify the presence of the pronounced maximum in the excitation function of reaction (1).

3. Absolute Radical Concentration Measurements and Modeling of Low Pressure Flames

Formation and destruction of nitrogen oxides in flames have been extensively studied both experimentally[57] and theoretically.[58-60] These studies have provided an understanding of the interaction of hydrocarbon radicals and nitrogen containing species involved in NO formation and destruction chemistry. Different reaction mechanisms have been developed and successfully applied to the modeling of spatial profiles of intermediate species. However, significant uncertainties remain regarding rate coefficients and reaction product branching ratios of a number of key reactions responsible for the formation and removal of free radicals. Relative radical concentration measurements have the potential to facilitate the development of detailed reaction mechanisms capable of reproducing qualitative trends for a wide range of fuels. However, the lack of absolute concentration measurements of key radical species makes it difficult to resolve uncertainties associated with the predicted absolute concentration levels. This situation has been improved by recent experiments using laser absorption, degenerate four-wave mixing and LIF which allowed quantitative measurements of CH_3, OH and CH radicals in low pressure flames.[61-63] In particular, the situation with respect to the CH radical is encouraging because of its predominant role in the "prompt" NO formation.[3] In the following, absolute CH and CN concentrations determined by calibrated LIF measurements will be compared to results of flame structure calculations in order to quantify the uncertainties of current kinetic models regarding the formation and destruction chemistry of the CH and CN radicals.

3.1. *Experimental Technique*

The experiments were performed in a low pressure (10 Torr) stoichiometric CH_4/O_2 flame stabilized on a McKenna type burner. The burner facility and the optical setup is schematically depicted in Fig. 6. Seeded flames were studied through the controlled addition of a small amount of NO (2% of the total gas flow) through a separate mass flow controller. For species

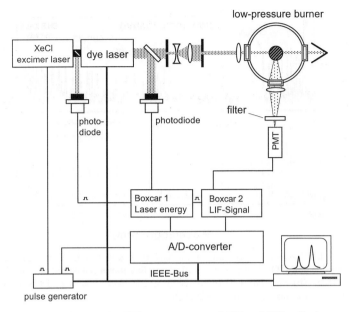

Fig. 6. Experimental setup for LIF measurement of CH and CN radical concentration distributions in a NO seeded stoichiometric CH_4/O_2 low pressure flame.

concentrations determination, LIF detection of $CH(X^2\Pi)$ and $CN(X^2\Sigma^+)$ using the $(B^2\Sigma^- - X^2\Pi)$ and $(B^2\Sigma^+ - X^2\Sigma^+)$ optical transitions, respectively, as well as N_2 Rayleigh calibration measurements[63] were performed. The experimental setup for LIF measurements featured tunable radiation around 387 nm originating from a dye laser pumped by a XeCl excimer laser which was passed unfocused (1 mm diameter) through the flame. Using Quinolon dye dissolved in dioxane the laser-beam energy was varied by neutral density filters, dichroic mirrors and an iris in front of the entrance window of the burner housing. The pulse energy was monitored continuously with a pyroelectric joulemeter (Laser Probe) behind the exit window. The laser pulses had a spectral bandwidth of 0.25 ± 0.03 cm^{-1} at 388 nm as measured with a monitor etalon. LIF and Rayleigh signals from the probe volume were imaged onto the entrance slit of a 0.5 m monochromator (CVI, Digikrom 480) and detected by a photomultiplier (Hamamatsu, R4332) behind the exit slit. To capture all broadband fluorescence light after excitation, the grating was used in 0th order with a Schott filter combination (WG 335/BG 3) in front of the entrance slit, for most of the measurements. For time integrated intensity measurements and the determination

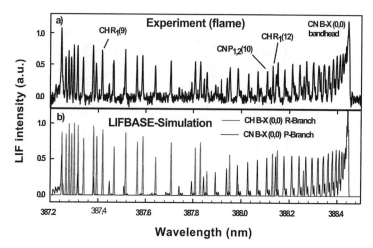

Fig. 7. (a) LIF excitation spectrum of the CH B–X and CN B–X optical transitions in a NO seeded CH_4/O_2 low pressure (10 Torr) flat flame measured 5 mm above the burner surface. (b) Simulation (LIFBASE[64]) of CH B–X (black) and CN B–X (gray) LIF spectra for a temperature of $T = 2000$ K.

of fluorescence lifetimes, the signal from the photomultiplier (PMT) was fed into a gated integrator and a digital storage oscilloscope, respectively. A measured LIF excitation spectrum of CH and CN radicals is shown in Fig. 7(a) and compared with a spectrum [Fig. 7(b)] calculated using the LIFBASE program.[64]

For comparison with the flame simulations, the relative CH and CN concentration profiles along the centerline of the burner were recorded followed by a calibration of the relative concentration profiles using a N_2 Rayleigh calibration method.[63] Linear LIF was used to determine the CH and CN signal intensities as a function of height above the burner. In this case the relationship between the detected LIF intensity I_{LIF} and, e.g. the CH number density N_{CH} is given by

$$I_{LIF} = N_{CH} f_B \frac{B}{4\pi c} \frac{\Gamma(\nu)}{\Delta\nu_{Laser}} \frac{\tau_{eff}}{\tau_0} E_{Laser} \alpha, \qquad (3)$$

where f_B is the Boltzmann factor, B the Einstein coefficient of absorption for the excited rovibronic transition, E_{Laser} is the laser energy, $\Gamma(\nu)$ the line shape overlap, $\Delta\nu_{Laser}$ the laser bandwidth and α is the optical collection and transmission efficiency.[63] To determine absolute number densities from the LIF signal the effective lifetime τ_{eff}, i.e. the fluorescence quantum yield

τ_{eff}/τ_0, has to be determined. This was done by direct measurement of the total fluorescence decay time of the excited transitions at different positions in the flame to account for the specific collisional quenching environment. The remaining unknown factor α was determined from the slope of the Rayleigh scattering signal intensity I_{Rayleigh} (see Fig. 8) as a function of the product of number density N_{N_2} of nitrogen and laser energy E_{Laser}[63]:

$$I_{\text{Rayleigh}} = \frac{1}{hc\nu}\frac{\partial\sigma}{\partial\Omega}N_{\text{N}_2}E_{\text{Laser}}\alpha\,. \tag{4}$$

Excitation and detection geometry, filter selection and electronic settings of the PMT/gated integrator are kept the same as in the LIF measurements. For absolute calibrations, the $R_1(9)$ and $R_1(12)$ transitions of CH at 387.42 and 388.15 nm were selected [see Fig. 7(a)]. The CN calibration was performed using the $P_{1,2}(10)$ transition at 388.11 nm. For the determination of concentration profiles of CH, the $R_1(9)$ line was used. For CN, relative LIF intensity profiles taken from transitions of the unresolved $P(0,0)$-bandhead at 388.44 nm were compared with profiles taken with the $P_{1,2}(10)$ line. This comparison showed no difference within the error limits and therefore the bandhead was chosen in order to obtain a better signal to noise ratio.

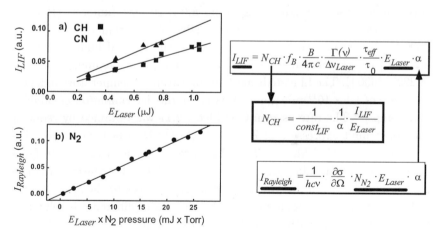

Fig. 8. (a) LIF signal intensity *versus* laser energy from the excited CH B–X $R_1(12)$ and the excited CN B–X $P_{1,2}(10)$ lines. (b) Calibration of the detection system by N_2 Rayleigh scattering. Right side: Schematic description of the Rayleigh scattering calibration procedure for the measurement of absolute CH radical concentration N_{CH}. Details are described in the text.

3.2. Flame Modeling Calculations

The flame structure is modeled by solving the conservation equations for a laminar premixed burner-stabilized flame[65] with the experimental temperature profile determined in previous work using OH-LIF.[66] Three different detailed chemical kinetic reaction mechanisms are compared in the present work. The first one, denoted in the following as Lindstedt mechanism, is identical to the one reported in Ref. 67 where it was applied to model NO formation and destruction in counterflow diffusion flames. This mechanism is based on earlier work of Lindstedt and coworkers and it has subsequently been updated to include more recent kinetic data.[68] In addition, the GRI-Mech. 2.11 (Ref. 59) and the reaction mechanism of Warnatz[69] are applied to model the present flame.

An analysis of the three mechanisms reveals significant differences with respect to key reaction pathways. In the Lindstedt mechanism, a rate for the $^3CH_2 + O_2$ reaction based on the work of Dombrowsky and Wagner is used.[70] This rate is in excess of one order of magnitude slower at a temperature of 1800 K than the rates used in the two other mechanisms. Further differences can be found in the rate constants for the $^3CH_2 + H \rightarrow CH + H_2$ reaction. There are substantial uncertainties with respect to this reaction and estimates range from 8.0×10^9 m^3/(kmol s) (Ref. 71) to more than 5.0×10^{11} m^3/(kmol s) (Ref. 72) at temperatures above 2000 K. The recent work of Röhrig *et al.*[73] has deduced limiting rates of 3.2×10^{10} m^3/(kmol s) and 2.3×10^{11} m^3/(kmol s) in the temperature range of 2200–2600 K. The Lindstedt mechanism features a temperature independent rate of 1.1×10^{11} m^3/(kmol s) based on the determination of Böhland and Temps.[74] The rate used in GRI-Mech. 2.11 was obtained by reducing the upper limit expression of Zabarnick *et al.*[72] by a factor of three. The rates adopted in GRI-Mech. 2.11 and the present mechanism are thus fairly close to the median value suggested by Röhrig *et al.*[73] By contrast, the rate expression adopted in the Warnatz mechanism is close to the lower limit of Ref. 71 at combustion temperatures. A third major difference is related to the reactions of CH with O_2 and H_2O. GRI-Mech. 2.11 and the Warnatz mechanism adopt the CEC recommendation of 3.3×10^{10} m^3/(kmol s) for the $CH + O_2$ reaction.[75] This rate is close to the room temperature determination of Berman *et al.*[76] However, Markus *et al.*[77] investigated the same reaction at high temperatures (2500–3500 K) and obtained a rate constant of 7.5×10^{10} m^3/(kmol s) which is used in the Lindstedt mechanism. This higher value has recently been given support by the study of Röhrig *et al.* who obtained a value of 9.7×10^{10} m^3/(kmol s) (2200–2600 K).[73] The reaction between CH and H$_2$O is also a potentially major sink for the CH radical. The mechanism of Lindstedt features a rate expression adopted from Baulch *et al.*[75] with the product specified as an adduct (CH$_2$OH). The same total rate expression is used by Warnatz. However, two product channels (CH$_2$O + H and ^{3}CH$_2$ + OH) are specified with a branching ratio of 4:1. GRI-Mech. 2.11 is based on the same source assuming a single CH$_2$O + H reaction product channel. However, the rate constant is assigned a value three times higher than the CEC recommendation.[75]

Reactions between C$_1$ hydrocarbon radicals and NO are mainly responsible for "reburn" in the present flame and the computed CN concentration is sensitive to the rates and product branching of these reactions. GRI-Mech. 2.11 and the Lindstedt mechanism feature rate constants and product distributions for the reactions of NO with C and CH obtained from the work of Dean *et al.*[78] and Dean and Bozzelli.[79] The mechanism of Warnatz only considers CN and HCN products for the reactions of NO with C and CH. However, both the present and Warnatz mechanisms adopt the rate constant and product distribution for the NO+^{3}CH$_2$ reaction from the work of Bauerle *et al.*[80] The total rate suggested in Ref. 80 is about ten times slower at 1800 K than that used in GRI-Mech. 2.11. HCN is a major product of the NO reburn chemistry and its subsequent reactions with H, O atoms and OH radicals lead to the formation of CN, NCO and NH radicals. Glarborg and Miller[81] have performed a combined experimental and theoretical study of lean hydrogen cyanide oxidation in the temperature range 900–1400 K. The main oxidation route of HCN was found to follow the sequence HCN + OH → CN + H$_2$O and CN + O$_2$ → NCO + O. This oxidation sequence is considered by all three mechanisms along with reactions featuring the O and OH radicals. The rate constants in all three mechanisms are similar for these reactions. There are, however, two major differences. The Lindstedt mechanism features the rate for the HCN + O → CN + OH reaction determined by Dean and Bozzelli.[79] This rate is more than one order of magnitude slower than those adopted in the other two mechanisms. It may also be noted that the rate constant for the CN + O reaction used in the mechanism of Warnatz is about seven times slower than the rates adopted in the Lindstedt mechanism and GRI-Mech. 2.11.

3.3. *Results and Discussion*

The concentrations of the CH and CN radicals were determined to have maximum values of $(1.3 \pm 0.5) \times 10^{12}$ cm^{-3} and $(3.9 \pm 1.5) \times 10^{10}$ cm^{-3} respectively. CH and CN-LIF measurements were carried out under linear excitation conditions. This was verified for the strongest optical transitions of both radicals by plotting the LIF signal intensity versus laser energy, as shown in Part (a) of Fig. 8. For each data point the laser was scanned over the complete spectral line to account for possible background signals. N$_2$ Rayleigh calibration curves are shown in Part (b) of Fig. 8.

In addition, CH$_3$ and OH radical concentration profiles are depicted in Fig. 9 which were measured using the *in situ* long path absorption technique described in Ref. 61(a). The OH radical profile was determined from LIF measurements using the $(A^2\Sigma^+; v' = 1 - X^2\Pi; v'' = 0)$-transition with the absolute calibration being obtained via absorption measurements at the R-bandhead of the $(A^2\Sigma^+; v' = 0 - X^2\Pi; v'' = 0)$-transition around 307 nm.[61(a)] The relative NO concentration profile shown in Fig. 9 was determined by LIF measurements at the $(A^2\Sigma^+; v' = 0 - X^2\Pi; v'' = 0)$-transition using the $R_2(21)$ spectral line and long path absorption measurements around 225 nm at a height of 38 mm above the burner were performed for the absolute calibration.[61(b)]

Computed concentration profiles of OH, CH$_3$, NO, CH and CN using the three different mechanisms are compared with the measured absolute data in Fig. 9. It can be seen that the shape and location of the OH profile is adequately predicted by the three mechanisms in the main reaction zone. There is also excellent agreement between the computed and measured CH$_3$ profiles. However, the peak CH$_3$ concentration predicted by GRI-Mech 2.11 is higher than the measurements by about 30%. The initial reduction of NO is well reproduced by the mechanisms. In agreement with the current measurements the introduction of the NO dopant results in only a minor (about 10%) reduction in computed CH levels. The CH profile obtained with the mechanism of Warnatz agrees exceptionally well with the experimental results. The predictions using the Lindstedt mechanism and GRI-Mech 2.11 are higher than the measurements by factors of 1.7 and 2.1, respectively. All three mechanisms overpredict the CN concentrations substantially.

A reaction path analysis reveals both similarities and differences in the modeling of the formation and destruction chemistry of the CH and CN

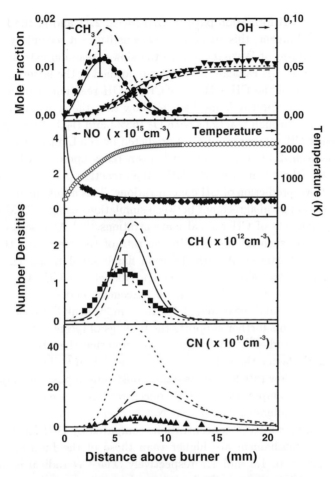

Fig. 9. Measured OH (▼), CH₃ (•), NO (◆), CH (■) and CN (▲) absolute concentration profiles in the 10 Torr $CH_4/O_2/NO$ flame. Comparison with results from chemical kinetics modeling using the Lindstedt reaction mechanism (—) Ref. 67, the GRI 2.11 (– – –) Ref. 59, and the Warnatz mechanism (...) Ref. 69. The experimental errors for the absolute CH and CN concentrations represent 2σ of several independent calibration measurements.

radicals in the three mechanisms. Because of the large amount of molecular oxygen available in the present flame, the consumption of hydrocarbon radicals proceeds to a significant extent via reactions with O_2. The computed CH concentrations are particularly sensitive to the rates of the $^3CH_2 + O_2$, $^3CH_2 + H$, $CH + O_2$ and $CH + H_2O$ reactions. The $^3CH_2 + H$ reaction is the dominant CH formation path in GRI-Mech. 2.11 and in the Lindstedt mechanism. However, this reaction only constitutes a secondary pathway in the mechanism of Warnatz due to the slower reaction rate adopted as discussed above. CH radicals are primarily formed in the latter mechanism via the reverse of the $CH + H_2O \rightarrow {}^3CH_2 + OH$ reaction. Computational results using the Lindstedt mechanism show that the consumption of CH occurs mainly through reactions with O_2 (65%) and H_2O (12%). Although similar peak CH concentrations are predicted by the Lindstedt mechanism and GRI-Mech. 2.11, the latter mechanism relies upon an arbitrarily increased rate (three times) for the $CH + H_2O$ reaction in order to prevent a significant overprediction of CH concentrations. This adjustment was introduced to obtain good agreement between simulations using GRI-Mech. 2.11 and measurements of CH radical concentrations.[63(a)] As discussed in detail in Ref. 67, such an increase in the rate of the reaction $CH + H_2O$ would appear inconsistent with diffusion flame simulations. The Lindstedt mechanism has also been applied to model the $CH_4/O_2/N_2$ flame studied by Luque *et al.*[63(a)] and good agreement between the computed $(1.7 \times 10^{12} \text{ cm}^{-3})$ and measured $(1.5 \times 10^{12} \text{ cm}^{-3})$ peak CH concentrations was obtained. This further supports the findings of recent experimental reaction kinetics studies[73,77] in which it was found that the $CH + O_2$ reaction is significantly faster than the CEC recommendation.[75] The present work indicates that a low rate for this reaction constitutes a possible major cause of errors in the computations of CH concentrations in premixed flames.

The major differences in the HCN/CN chemistry have been outlined above and the peak HCN concentrations computed by GRI-Mech 2.11 and the Warnatz mechanism are higher than those of the Lindstedt mechanism by factors of 1.5 and 1.7 respectively. The CN radical is primarily formed from HCN through the hydrogen abstraction reactions and subsequently consumed by the reactions with O_2 and the O and OH radicals. Therefore, predictions of CN radical concentrations are sensitive to the predictions of the CH and ${}^3CH_2$ radicals. The measured CN:CH ratio of 1:34 can be compared with the computed values using the Lindstedt mechanism (1:17), GRI-Mech 2.11 (1:12) and the Warnatz mechanism (1:2.7) respectively. The significant over-prediction of CN concentrations obtained with the mechanism of Warnatz is predominantly due to uncertainties in the C atom chemistry and the branching of the $CH + NO$ reaction.

The modeling calculations using three different reaction mechanisms have been shown to produce good agreement for the OH and CH_3 radicals.

The amount of NO reduction in the flame is also well predicted. It is further shown that a reasonable agreement for the CH and CN concentration profiles can be obtained with the use of recent kinetic data for the dominant reaction channels. However, significant differences in the CH radical formation and destruction chemistry featured in the three mechanisms were found. The key uncertainties regarding the accurate prediction of absolute CH concentrations in the present flame could be related to the existing uncertainties in the kinetic data of the $^3CH_2 + O_2$, $^3CH_2 + H$ and $CH + H_2O$ reactions. It is evident that the rates and product distributions for these reactions need to be determined at combustion temperatures.

4. Laser-Induced Ignition Processes

Optimal control of ignition processes is one of the key factors in improving the performance of many combustion devices. To this end, a detailed understanding of unsteady combustion phenomena is required. In order to develop quantitative mathematical models for complete simulation of ignition processes that include detailed chemistry, experimental studies of simple systems are particularly useful. The experimental techniques should allow visualization of the ignition process in time and space. This can be done by two-dimensional imaging of OH radicals with planar laser-induced fluorescence (PLIF).[82,83] The application of tunable excimer lasers with their narrow bandwidth, high pulse energies and high repetition rates allows the effective excitation of short lived predissociative states of the OH-radicals.[84]

An experimental setup[85] using PLIF to investigate the laser-induced ignition process is shown in Fig. 10(a). A quartz reactor with suprasil cell walls and SrF_2 windows was used to study the CO_2-laser-induced ignition of CH_3OH/O_2-mixtures. The coincidence of the $9P(12)$ CO_2-laser line in the $(001)–(020)$ band with the $R(12)$-CO stretch fundamental band of the methanol molecule at 9.6 μm allows controlled heating and subsequent ignition of the mixture. For temperature measurements, OH radicals formed during flame propagation can be excited in two rotational states of the $(v' = 3, v'' = 0)$ vibrational band in the $(A^2\Sigma^+ - X^2\Pi)$-transition at 248 nm using either two separate KrF-excimer lasers or a single two-wavelength KrF excimer laser.[86]

Each laser beam was formed into a light sheet 30 mm in height and 0.4 mm thick using quartz lenses. The light sheets are spatially overlapped

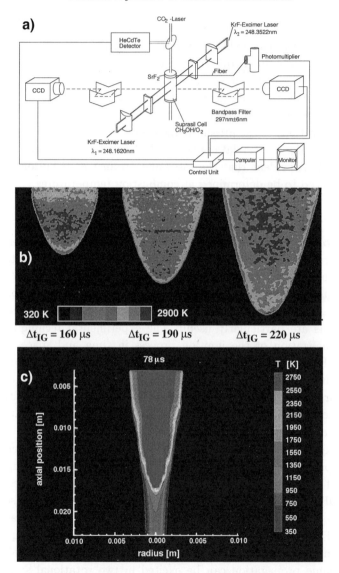

Fig. 10. Thermal ignition of methanol/oxygen mixtures. (a) Experimental setup. Mixtures were ignited by a CO_2 laser beam. Flame propagation is visualized by 2D-LIF of OH radicals. For 2D-temperature measurements, different OH rotational levels are probed simultaneously using two tunable KrF excimer lasers. (b) Development of the 2D-temperature field during the ignition of a CH_3OH/O_2 mixture ($\Phi = 0.9$, $p = 300$ mbar). (c) Numerical simulation of the temperature field 78 μs after CO_2 laser ignition (data shown with kind permission from U. Maas and I. Gran).

with a temporal delay of 100 ns to separate the signals excited by the different pulses. Bandpass reflection filters consisting of four narrowband dielectric mirrors (297 ± 6 nm, transmission $> 90\%$, blocking 5×10^4) were used to isolate the ($v' = 3$, $v'' = 2$) fluorescence bands of OH. The fluorescence was detected by gated image intensified CCD cameras. The excitation of two different rotational transitions of OH radicals, starting from $N'' = 8$ and $N'' = 11$, allowed the measurement of spatially corresponding LIF image pairs.

Using the Boltzmann distribution, the ratio of these images can be converted into temperature fields shown in Fig. 10(b). Calculating temperatures from a two-line LIF measurement requires a careful consideration of a number of effects. The influence of fluorescence quenching, usually the dominating deactivation process for excited OH radicals under atmospheric pressure conditions, is reduced in the experiments by the predissociating nature of the excited state[87,88] to less than 3%. The tunable excimer laser emits polarized light, which in turn induces LIF signals with a spatial preference, depending on the transition that had been excited.[89] Collisions can redistribute the spatial alignment; this means that the ratio of the LIF images might depend on the nature of the collider gas composition. This effect, as well as the variation of the rotational energy transfer rates in the ground state,[90] can be accounted for with a calibration procedure, where temperatures measured by the two-line LIF method are compared to pointwise coherent anti-Stokes Raman spectroscopy (CARS) temperature measurements.[91]

For the direct numerical simulation, the two-dimensional system of coupled ordinary differential and algebraic conservation equations was solved numerically by spatial discretization using finite differences.[92] Two-dimensional imaging of the temperature as well as the numerical simulation[93] show a conical flame front [Fig. 10(b) and Fig. 10(c)]. This can be explained by the fact that a channel within the reaction cell is preheated by the CO_2 laser-beam. Therefore, the flame propagates faster in the axial than in the radial direction. The fast axial flame propagation is caused mainly by successive ignition along the cell axis, due to different induction times that follow the axial temperature gradient. Typical propagation speeds of the ignited CH_3OH/O_2-mixtures investigated are 30 m/s in the radial direction and 130 m/s in the axial direction. This observation gives an important clue for the understanding of knock phenomena in spark ignition engines. Unwanted self-ignition occurs here due to local temperature fluctuations as small as 20 to 30 K ("hot spots") during the adiabatic compression phase. This local ignition, which can be monitored, e.g. by PLIF of formaldehyde,[94] forms pressure waves that produce temperature jumps that can accelerate the flame propagation in the same way as shown in Fig. 10(c). Later on, the combustion wave reaches the pressure wave, and the system is subject to a transition to detonation.[95]

5. Laser Diagnostics of Catalytic Combustion

Catalytic combustion is a promising alternative combustion technology for burning fuel in lean mixtures which can result in a significant reduction of pollutant formation, improved ignition, and enhanced stability of flames.[96] Heterogeneous combustion processes are determined by interaction of diffusion processes from and to the gas phase, by adsorption and desorption on the surface, and by surface diffusion and surface reaction kinetics. Concentrations of reactants and products on the surface are connected with adsorption and desorption equilibria depending on the gas-phase concentrations of reactants and products, and by transport processes. Thus, depending on the operating conditions, different partial processes can become rate determining leading to a completely different behavior of the global surface reaction. As a consequence, the development of appropriate mathematical models for the numerical simulation of surface reactions and their coupling to the surrounding gas phase is essential for the detailed understanding of heterogeneous catalysis under technical relevant conditions.

Computational tools for the description of heterogeneous reaction systems have been developed recently, which include detailed surface chemistry as well as detailed models for molecular multispecies transport.[97] In contrast to flame chemistry, only a few complete surface reaction mechanisms have been derived, which are mainly based on studies of elementary surface reaction steps carried out under ultrahigh vacuum (UHV) conditions and on well-defined single crystal surfaces.[98] The use of this kind of surface kinetics data in the modeling of technical processes, which usually take place at high pressure (pressure gap) and on polycrystalline catalyst materials (materials gap), emphasizes the importance to develop *in situ* diagnostic techniques for molecular level studies of adsorbed species under practical catalytic combustion conditions. While laser spectroscopic techniques are now a normal working tool in gas-phase combustion diagnostics,[17] quantitative laser-based *in situ* diagnostics methods for the investigation of heterogeneous combustion processes are still under development.

The use of optical methods which probe interface electronic and vibrational resonances offers significant advantages over conventional surface spectroscopic methods in which, e.g. beams of charged particles are used as a probe, or charged particles emitted from the surface/interface after photon absorption are detected.[99] Recently, three-wave mixing techniques such as second-harmonic generation (SHG) have become important tools to study reaction processes at interfaces.[100] SHG is potentially surface-sensitive at nondestructive power densities, and its application is not restricted to ultrahigh vacuum (UHV) conditions.[101] However, SHG suffers from a serious drawback, namely from its lack of molecular selectivity.[102] As a consequence, SHG cannot be used for the identification of unknown surface-species.

This drawback can be overcome with the use of infrared visible (IR-VIS) sum frequency generation (SFG), which allows surface vibrational spectroscopic measurements with submonolayer sensitivity.[103] In addition to molecular specificity, SFG allows a direct probe of the effect of the interface on the bond frequencies of the adsorbed molecules.[20,104,105] Recent experiments, in which SFG was used to monitor bonding and location of CO surface species during heterogeneous CO oxidation on a Pt(111) single crystal over a wide pressure range (10^{-6}–10^3 mbar), showed that SFG can be utilized to bridge the "pressure gap" in the CO/Pt-system.[106,107] The work described in the following focussed on the extension of these studies to assess the SFG techniques' potential for "bridging the pressure and the materials gap" in the Pt/CO-system by investigating the possibility for quantitative *in situ* CO coverage measurements during CO oxidation on a polycrystalline Pt foil catalyst under laminar flow conditions.

5.1. *Experimental Technique*

The surface species diagnostics experiments were carried out in a reaction chamber, which allows studies over a wide pressure range from UHV-conditions (3×10^{-10} mbar) up to atmospheric pressure. In the high pressure regime this experimental arrangement can be used to investigate adsorption/desorption and reactive processes of well-defined stagnation point flows of reactant mixtures on the catalyst surface. The reaction chamber schematically depicted in Fig. 11 was equipped with a quadrupole mass spectrometer for Thermal Programmed Desorption (TPD) measurements, a Ar^+ sputter source, a retarding field analyzer (RFA) for Auger Electron

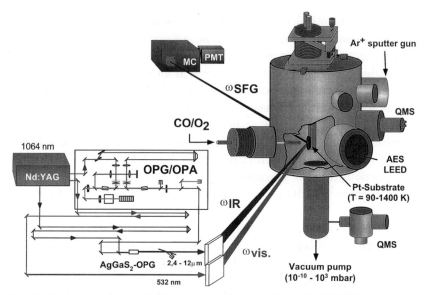

Fig. 11. Experimental setup for the *in situ* detection of chemisorbed CO during catalytic combustion of CO on Pt using optical infrared-visible sum frequency generation (SFG) and mass spectrometry. A mode-locked Nd:YAG laser system is used to provide the visible laser beam (second harmonic: 532 nm) and to pump an optical parametric system to generate infrared radiation (ω_{IR}) tunable with a pulse duration of 25 ps. MC: monochromator, PMT: Photomultiplier, AES: Auger Electron Spectrometer, LEED: Low Energy Electron Diffraction Spectrometer, QMS: Quadrupole Mass Spectrometers for CO Thermal Desorption (TD) and CO_2 production rate measurements.

Spectroscopy (AES) and Low Energy Electron Diffraction (LEED) studies and CaF_2 and quartz windows serving as entrance and exit ports for the laser beams and the sum frequency signal, respectively. A second quadrupole mass spectrometer was connected to the vacuum line behind the reaction chamber which could be used for on-line monitoring of stable reaction products, e.g. CO_2, in the exhaust gas.

The Pt catalyst was mounted on a copper block and could be translated, tilted and rotated by means of a manipulator fitted with a differentially pumped rotary feedthrough. The Pt foil (Advent, purity > 99.99%) could be resistively heated. The mounting allowed to work in the temperature range 300–1600 K using direct sample heating with a proportional-integral-derivative (PID) control unit. The temperature of the catalyst was measured by a Ni-NiCr thermocouple spot welded to the Pt-foil. Clean platinum surfaces could be obtained by applying several cycles of Ar^+ ion sputtering followed by oxidation at 1000 K. After the last sputter cycle, at 300 K with 3 keV Ar^+ ions for 45 min, the sample was heated to 750 K for 5 min. Then the sample was cooled down to 300 K. During the CO oxidation measurements the premixed flows were 30 sccm O_2 (Messer Griesheim, 99.995%), 15 sccm CO (Messer Griesheim, 99.994%), and 105 sccm Ar (Messer Griesheim, 99.998%) at a total pressure of 20 mbar.

For the detection of chemisorbed CO, a 40 ps mode-locked Nd:YAG laser system was used which has been described in detail elsewhere.[20] A

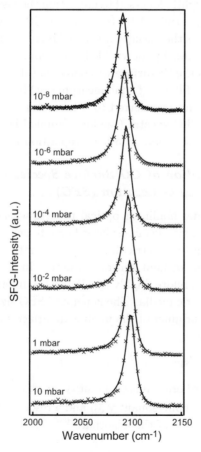

Fig. 12. Pressure dependence (p_{CO} = 10^{-8}–10 mbar) of the SFG spectra of CO chemisorbed on a polycrystalline Pt catalyst at a substrate temperature of 300 K. Experimental data points are represented by crosses, the solid lines represent results of least-square fits. Details of the fitting procedure are described in the text.

part of its output was frequency doubled to 532 nm and used as the visible input frequency for the SFG process. The other part was used to pump an optical parametric system to generate infrared (IR) radiation tunable in the frequency range (1800–2200 cm^{-1}) with a pulse duration of 25 ps and a bandwidth of 7 ± 1 cm^{-1}. The visible and the IR laser-beams were p-polarized and overlapped at the surface. Incident angles were 55° for the visible and 35° for the IR beam with energies of 400 μJ and 30 μJ per pulse, respectively. The spot size of the visible beam was 5 mm diameter and the infrared beam was slightly focussed to fall within the visible. Baselines of the SFG spectra were determined by blocking the IR-laser-beam. Therefore, the baseline represents the background signal due to stray light generated in the reaction cell windows by the visible laser-beam. The CO sum frequency signal reflected from the Pt surface was detected (after filtering of scattered light with a dielectric filter and a monochromator) by a photomultiplier and a gated integrator, and transferred to a laboratory computer. Each point of the recorded CO-SFG spectra (see for example Fig. 12) was obtained by averaging over 120 laser shots at a laser repetition rate of 10 Hz.

5.2. *In Situ Detection of CO Surface Species Using Sum Frequency Generation (SFG)*

SFG is a second order nonlinear optical process where a tunable infrared (ω_{IR}) is mixed with a visible (ω_{VIS}) laser-beam to generate a sum frequency output (ω_{SFG}).[101] In the electric dipole approximation, this process is only allowed in a medium without centrosymmetry. In the CO gas/Pt-system the SFG signal is, therefore, highly specific to the interface region bounded by the centrosymmetric media. The generated SFG signal intensity I_{SFG} is proportional to the product of the nonlinear surface susceptibility $\chi_S^{(2)}$ and its complex conjugate:

$$I_{\text{SFG}} \propto \chi_S^{(2)} \cdot \text{Conj}(\chi_S^{(2)}) = |\chi_S^{(2)}|^2. \tag{5}$$

$\chi_S^{(2)}$ represents a third order tensor quantity which can be modeled in lowest order as the sum of a nonresonant $\chi_{NR}^{(2)}$ and a resonant term $\chi_R^{(2)}$ associated with a vibrational mode of the adsorbate[108]:

$$\chi_S^{(2)} = \chi_{NR}^{(2)} + \chi_R^{(2)}. \tag{6}$$

The resonant term $\chi_R^{(2)}$, which is associated with a vibrational mode of a surface layer of adsorbates, can be expressed, following the coupled-wave approach, as[109]:

$$\chi_R^{(2)} = \frac{N T_X M_X \Delta\rho}{\hbar(\omega_{IR} - \omega_X + i\Gamma)} = \frac{A_R}{(\omega_{IR} - \omega_X + i\Gamma)}, \tag{7}$$

where N denotes the surface density of adsorbed molecules, T_X is the infrared transition moment, M_X is a term proportional to the Raman transition moment of the vibrational mode of the adsorbed molecule X. ω_X is the corresponding molecular vibrational frequency, Γ is the homogeneous Lorentzian half-width of the vibrational mode, and $\Delta\rho$ is the population difference between the ground and vibrational excited state. If the adsorbate vibration is both Raman and infrared active, the resonant contribution becomes significant as the IR laser is tuned through the vibrational transition. As a result, the SFG spectrum provides similar information as can be obtained by conventional vibrational spectroscopy with the difference that the SF signal originates predominately from surface species. In addition, the SF output is coherent and highly directional. A_R defined via Eq. (7) denotes the amplitude of the resonant contribution. The nonresonant term $\chi_{NR}^{(2)}$ can in principle consist of contributions originating from the adsorbate, the surface, or from cross terms resulting from adsorbate surface interaction. The nonresonant term $\chi_{NR}^{(2)}$ can be expressed as

$$\chi_{NR}^{(2)} = A_{NR} e^{i\Phi}, \tag{8}$$

where A_{NR} is the magnitude of the vibrationally nonresonant contribution due to electronic excitations of the Pt surface and the adsorbate, Φ is its phase relative to the vibrational resonance. Inserting (7) and (8) into (6) yields the following expression ($X = CO$):

$$\chi_S^{(2)} = \frac{A_R}{(\omega_{IR} - \omega_{CO} + i\Gamma)} + A_{NR} e^{i\Phi}, \tag{9}$$

which was used together with Eq. (5) for the numerical simulation of the SFG spectra of chemisorbed CO molecules to determine the parameters ω_{CO}, Γ, A_R, A_{NR} and Φ. The CO SFG spectra depicted in Fig. 12 were recorded on polycrystalline Pt during room temperature CO adsorption experiments in the CO pressure range (10^{-8}–10 mbar). Therefore the IR laser was tuned over the frequency range $\omega_{IR} = 1950$–2150 cm^{-1} in which the stretching vibration of terminally adsorbed CO molecules can be excited. The solid lines in Fig. 12 are the results of a numerical least squares fit of the experimental data points (crosses) using Eq. (9) together with Eq. (5). The numerical analysis revealed a frequency of the CO vibrationally resonant contribution of $\omega_{CO} \approx 2095$ cm^{-1}, which is characteristic of the stretching vibration of CO bound "on top" at Pt surface atoms.

In Fig. 13, CO-SFG spectra recorded in the CO pressure range (1–1000 mbar) are depicted, which show at higher pressure ($p_{CO} \geq 50$ mbar) the appearance of a new CO surface species which dominates the measured SFG spectra at the highest CO pressures. The solid lines in Fig. 13 are

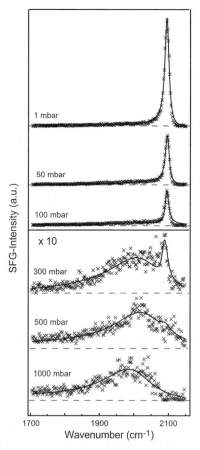

Fig. 13. Pressure dependence ($p_{CO} = 1$–1000 mbar) of the SFG spectra of CO chemisorbed on a polycrystalline Pt catalyst at a substrate temperature of 300 K. The changes in the SFG spectra indicate the presence of a new CO surface species at higher CO pressures.

the results of a numerical least squares fit of the experimental data points (crosses) using the following "two species expression,"

$$\chi_S^{(2)} = \frac{A_R}{(\omega_{IR} - \omega_{CO} + i\Gamma)} + \frac{A_R^*}{(\omega_{IR} - \omega_{CO^*} + i\Gamma^*)} + A_{NR}e^{i\Phi}, \qquad (10)$$

together with Eq. (5). The (*) in Eq. (10) is used to denote the second CO surface species. The SFG spectra were completely reversible and reproducible with variation of the CO gas pressure. The numerical least squares fit analysis of the SFG spectra of Fig. 13 yielded a frequency value of $\omega_{CO^*} \approx 2045$ cm^{-1}. A similar observation was reported by Somorjai and coworkers[110] who investigated CO chemisorption on a Pt(111) single crystal face up to atmospheric CO pressures. In these studies the peak at 2045 cm^{-1} was attributed to the reversible formation of an incommensurable overlayer of CO and to the formation of multiply bonded platinum carbonyl binary complexes [Pt–(CO)$_n$, $n = 1$–4] at heavily reconstructed Pt surface structures.

5.3. *CO Detection During Catalytic CO Oxidation*

The SFG spectra shown in Fig. 14 were obtained during CO oxidation at different substrate temperatures in the range $T = 300$–700 K at a total pressure of 20 mbar ($p_{CO} = 2$ mbar, $p_{O_2} = 4$ mbar, $p_{Ar} = 14$ mbar). The CO and O$_2$ partial pressures were close to the values present in the exhaust gas of an spark ignition (SI) engine. The SFG spectra of Fig. 14 were recorded at fixed surface temperatures by tuning the IR laser over the frequency region 1950–2150 cm^{-1} in which stretching vibrations of terminally adsorbed CO can be excited. Due to the low partial pressure of CO, the SFG spectra recorded during CO oxidation could be well described by the "single CO species" expression of Eq. (9). Solid lines in Fig. 14 are the results of a numerical least squares fit of the experimental data points (crosses) using Eq. (9) together with Eq. (5). The numerical analysis was used to determine the parameters ω_{CO}, Γ, A_R, A_{NR} and Φ as a function of the substrate temperature. According to Eq. (8), the nonresonant part of the SFG spectra can be described by the amplitude of the vibrationally nonresonant contribution, A_{NR}, and the phase between the nonresonant and the vibrationally resonant contribution, Φ. In the present study only a weak dependence of $\chi_{NR}^{(2)}$ on the substrate temperature was observed. Over the temperature range $T = 300$–620 K where CO could be detected, the nonresonant amplitude was considerably smaller than the resonant amplitude (for example: $A_{NR}/A_R \approx 7 \times 10^{-2}$ at a substrate temperature of 620 K). The values of the vibrational frequency ω_{CO}, the resonant amplitude A_R, and the Lorentzian width (FWHM) 2Γ, which determine the resonant contribution of the adsorbed CO, are plotted versus substrate temperature in Fig. 15.

In CO adsorption studies on a polycrystalline platinum foil, a pronounced change of the resonance frequency ω_{CO} was observed.[105] In the latter experiments, at $T = 300$ K a value of (2096 ± 4) cm^{-1} was obtained while at $T = 660$ K, the highest temperature where adsorbed CO was observed, a value of (2057 ± 5) cm^{-1} was found. The latter frequency is lower

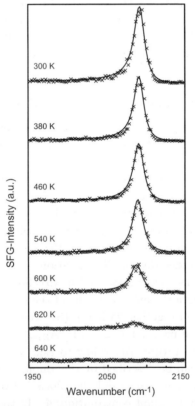

Fig. 14. SFG spectra measured during the catalytic oxidation of CO on a polycrystalline Pt foil at different substrate temperatures at a total pressure of 20 mbar ($p_{CO} = 2$ mbar, $p_{O_2} = 4$ mbar, $p_{Ar} = 14$ mbar).

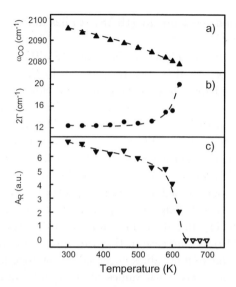

Fig. 15. (a) CO vibrational resonance frequency ω_{CO}; (b) Width (FWHM $= 2\Gamma$) of the resonant contribution of the SFG spectra; (c) Resonant amplitude A_R as defined in the text. Values are plotted against the Pt substrate temperature. Dashed lines are drawn just to guide the eye. The open triangles at substrate temperatures $T > 620$ K in (c) indicate values where within the detection limit no chemisorbed CO could be observed ($A_R \approx 0$).

than the frequencies observed on the (111), (110)[104] and (100)[111] low index platinum surfaces in the low coverage limit and it was attributed to CO molecules strongly bound to defect sites. Following Ref. 112, a low frequency of 2057 cm^{-1} is predicted for CO molecules bound at kinked-site Pt atoms with a coordination number of 6. In the present CO oxidation studies over the whole temperature range ($T = 300$–620 K) where chemisorbed CO could be detected, the resonance frequency ω_{CO} remained at values 2095–2078 cm^{-1} typical of CO molecules molecules "on top"-bound at Pt terrace sites [see Fig. 15(a)]. Hence, the main difference between the earlier CO chemisorption[105] and the present CO oxidation data is the absence of the low frequency CO contribution from defect sites in the latter case. This observation could be explained by a higher sticking probability of O_2 at the defect sites which blocks CO defect site adsorption when both CO and O_2 are present in the gas-phase. Preferential adsorption of O_2 on step sites has been observed in electron stimulated desorption ion angular distribution (ESDIAD) studies on Pt[3(111) $\times$ (100)][113] and during infrared reflection absorption spectroscopy (IRAS) studies during CO oxidation on Pt[4(111) × (100)].[114]

The temperature dependence of the width (FWHM = 2Γ) of the resonant contribution of the SFG spectra is shown in Fig. 15(b). Deconvolution of the measured spectra with the laser bandwidth was not performed. Equation (9) was directly used together with Eq. (5) for the numerical simulation to obtain the necessary parameters. The same method was applied in Ref. 104 to analyze SFG spectra of CO terminally adsorbed on a disordered Pt(111) surface where a value of $2\Gamma = 10$ cm^{-1} was reported at $T = 300$ K for saturation coverage. In the same study, a value of $2\Gamma = 12$ cm^{-1} was determined for CO SFG spectra measured on Pt(110) at $T = 300$ K. The corresponding CO frequencies were found to be 2093 cm^{-1} for disordered Pt(111) and 2094 cm^{-1} for Pt(110), respectively. For both Pt(111) and Pt(110), an increase in the width was observed with decreasing coverage. The maximum values observed at low coverage were 17 cm^{-1} and 14 cm^{-1}, respectively, with values of 2076 cm^{-1} and 2079 cm^{-1} for the corresponding frequencies.[104] In the CO oxidation studies a value of $2\Gamma = (12 \pm 1)$ cm^{-1} was determined at room temperature which increases to a value of $2\Gamma = (20 \pm 2)$ cm^{-1} at $T = 620$ K where the CO coverage is the lowest. Hence the line widths and frequencies measured in the present CO oxidation studies on polycrystalline Pt are comparable with those observed in room temperature Pt(111) and Pt(110) UHV single crystal CO exposure experiments[104] as well as with those measured in high pressure CO chemisorption studies on polycrystalline Pt, in which a value of $2\Gamma = (17 \pm 1)$ cm^{-1} was obtained at $T = 620$ K.[107]

Although a decrease of the resonant amplitude A_R versus substrate temperature as shown in Fig. 15(c) clearly indicates that the equilibrium concentration of chemisorbed CO decreases with increasing surface temperature, care has to be taken in using A_R as a measure of the CO surface coverage. Prior to deducing — for a given adsorbate/surface system — the relative CO coverage from SFG measurements, it is indispensable to carry out calibration measurements with another independent surface sensitive method.[115] In the present work the SFG signals were calibrated against Thermal Programmed Desorption (TPD) measurements. In these calibration measurements SFG spectra of CO adsorbed on the polycrystalline Pt foil were recorded in the temperature range $T = 300$–480 K at a CO pressure of 10^{-8} mbar under adsorption/desorption equilibrium conditions (at $T \geq 480$ K, no SFG signal from chemisorbed CO could be observed). SFG

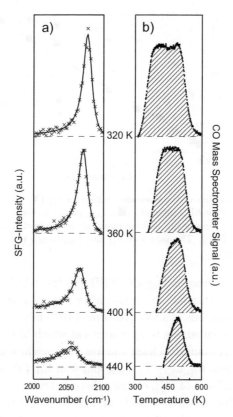

Fig. 16. (a) SFG spectra of CO adsorbed on the polycrystalline Pt foil recorded at different substrate temperatures at a CO pressure of 10^{-8} mbar. SFG intensity is plotted versus the frequency of the tunable IR laser. Experimental data points are represented by crosses, the solid lines represent results of least-squares fits. Details of the fitting procedure are described in the text. The depicted SFG spectra are on the same vertical scale. (b) TPD spectra of CO obtained under molecular flow conditions recorded after the corresponding SFG measurements shown in (a). Desorption was started at different substrate temperatures at the same CO pressure of 10^{-8} mbar. All TPD spectra are also on the same vertical scale.

spectra [Fig. 16(a)] were recorded at fixed surface temperatures by tuning the IR laser over the frequency region 1950–2150 cm^{-1} in which stretching vibrations of terminally adsorbed CO can be excited. The surface coverage was derived from the integrated area of the TPD spectra [Fig. 16(b)] measured after each SFG experiment and normalized to the value at the saturation coverage at 300 K. A set of SFG spectra are shown in Fig. 16

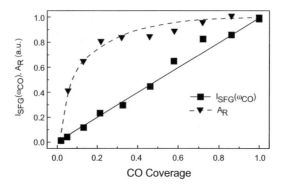

Fig. 17. Plot of the values of the resonant amplitude A_R (▼) and the maximum value of the resonant part of the SFG intensity (■), $I_{SFG}(\omega_{CO})$, derived from the SFG spectra depicted in Fig. 16(a), versus the relative coverage determined from the corresponding TPD spectra of Fig. 16(b). The values of A_R and $I_{SFG}(\omega_{CO})$ were normalized to unity at saturation coverage.

together with the corresponding CO-TPD spectra, which were obtained — in order to avoid isothermal desorption after pumping off the gas — under molecular flow conditions at a CO pressure of 10^{-8} mbar.[116] Desorption was started at the same substrate temperatures at which SFG spectra were previously recorded. A numerical least squares fit procedure was used to determine the parameters A_R, ω_{CO}, Γ, A_{NR} and Φ of Eq. (9). In this evaluation only a weak dependence of $\chi_{NR}^{(2)}$ on the substrate temperature was observed. In Fig. 17, the values of the resonant amplitude A_R and the maximum value of the resonant part of the SFG intensity, $I_{SFG}(\omega_{CO})$, obtained from the corresponding parameters A_R, ω_{CO} and Γ, are plotted versus the relative CO coverage of the Pt foil. As can be seen in Fig. 17, $I_{SFG}(\omega_{CO})$ is the spectroscopic quantity which correlates linearly with CO surface coverage.

5.4. *Measurements of CO Coverage During Catalytic CO Oxidation*

The SFG spectra shown in Fig. 14 obtained during CO oxidation were evaluated in order to determine $I_{SFG}(\omega_{CO})$ and hence the relative CO coverage as a function of substrate temperature as depicted in Fig. 18. Comparison with the CO coverage measured during CO chemisorption studies[105] showed that if CO and O_2 is present in the stagnation flow, the CO surface coverage at $T = 300$ K is reduced. However, simultaneous mass-spectrometric measurements of the CO_2 formation (see Fig. 18) revealed

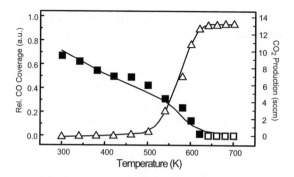

Fig. 18. CO coverage (■) as a function of the Pt substrate temperature measured during CO oxidation at a pressure of 20 mbar. (□) indicate temperatures where equilibrium CO surface coverage was too low to be detected. CO_2 production (△) simultaneously measured by mass spectrometry in the exhaust gas. The measurement were carried out under laminar flow conditions in a stagnation point flow onto the Pt catalyst surface. Solid lines are results of a numerical reactive flow simulation for a $CO/O_2/Ar$-stagnation point flow onto the Pt foil corresponding to the experimental flow conditions (CO: 15 sccm; O_2: 30 sccm; Ar: 105 sccm).

that the observed decrease of the CO coverage at $T = 300$ K is not due to reaction but is the result of the competitive adsorption between CO and O_2. Also, the following slight decrease in the CO surface coverage with temperature in the range 300 K $< T <$ 500 K is due to CO desorption rather than reaction. After the onset of reaction at $T >$ 500 K, on the other hand, the CO coverage decreases rapidly with increasing substrate temperature. At $T >$ 620 K, no chemisorbed CO could detected, indicating very low equilibrium CO surface concentrations in the temperature range $T = 640$–740 K.

Detailed CO oxidation experiments in which CO was monitored via SFG up to atmospheric were carried out in the group of Somorjai.[106] However, while the present CO oxidation studies were performed at a low CO partial pressure ($p_{CO} = 2$ mbar) as typically present in the exhaust gas of a SI engine and on a more realistic Pt catalyst surface, the CO oxidation studies of Ref. 106 were carried out at a much higher CO partial pressure ($p_{CO} = 53$–133 mbar) over a Pt(111) single crystal. In the latter studies, structural changes of the Pt(111) surface induced by the high pressures of CO were observed which gave rise to a variety of new CO surface species leading to "multiresonance" SFG spectra. Under the low CO partial pressure conditions of the present CO oxidation studies, such spectral features were not observed (see Fig. 14), suggesting that no major reorganization of the polycrystalline Pt surface takes place up to temperatures where CO is almost completely converted into CO_2. The solid lines shown in Fig. 18 represent results of numerical simulations of a stagnation point flow of a $CO/O_2/Ar$-mixture onto the Pt foil corresponding to the conditions of the experiment.[117] In the calculations, detailed models for molecular transport and surface chemistry were used. Modeling of the surface chemistry was based on a mean field approach[118] with two different adsorption sites using the activation energies derived in the experimental CO chemisorption studies.[105] The kinetic data for the Langmuir–Hinshelwood surface reaction mechanism used in the modeling calculations was taken from Ref. 119, assuming competitive adsorption of CO and O_2 on both platinum adsorption sites.

The results of the studies presented above clearly demonstrate that SFG surface vibrational spectroscopy, when combined with appropriate calibration measurements, is a promising experimental method for bridging the "pressure gap" as well as the "materials gap" which separate the UHV single-crystal model studies[98] from technical catalytic investigations.[120] Further experimental work is under way in which the method developed in the present work will be applied to investigate the influence of surface structure, mixture composition and pressure on the ignition behavior of the $CO/O_2/Pt$-system and other reactants/surface-systems under technically relevant conditions.

6. Laser Diagnostics of Engine Combustion

Optimization of internal engine combustion in respect of fuel efficiency and pollutant minimization requires detailed insight in the microscopic processes in which complex chemical kinetics is coupled with transport phenomena. Due to the development of various pulsed high power laser sources, experimental possibilities have expanded quite dramatically in recent years. Laser spectroscopic techniques allow nonintrusive measurements with high temporal, spectral and spatial resolution. New *in situ* detection techniques with high sensitivity allow the measurement of multidimensional temperature and species distributions required for the validation of reactive flow modeling calculations.[17] The validated models are then used to find optimal conditions for the various combustion parameters in order to reduce pollutant formation and fuel consumption.

6.1. *Imaging of Fuel and Temperature Distribution Prior to Ignition*

Ignition and flame development in internal combustion (IC) engines depend critically on the starting conditions defined by the fuel and temperature distribution. Modern engine concepts include lean burn engines where the averaged fuel/air mixture does contain enough fuel to ensure reliable ignition. In these engines, the goal is to stratify the fuel inside the cylinder to provide an ignitable mixture around the spark plug. Inhomogeneities, however, do not only apply to the overall air/fuel ratio. When using commercial fuels, components of different volatility classes may be inhomogeneously distributed within the combustion chamber. Further variations in fuel concentration and temperature might be induced by exhaust gas recirculation and residual gases. In all these cases it is desirable to obtain information about the initial fuel and temperature distribution before ignition starts. LIF-imaging using highly defined fluorescent tracers can be used to visualize the mixing process, fuel distribution and temperature distribution during intake and compression.

Commercial fuels contain numerous fluorescing components. Due to their different spectroscopic and thermodynamic properties, however, variations of fluorescence yields with temperature and pressure are hard to quantify. Therefore, the use of nonfluorescent fuels seeded with fluorescence markers of various types have been suggested for mapping fuel vapor distributions, both in fundamental studies[121(a)–(e),122] and practical applications, and especially in IC engines.[123,124] References 121(a)–(d) report on ketones and aldehydes which fluoresce in the spectral region 300-550 nm of the n–π^* transition in the C=O group. Reboux *el al.*[124] investigated toluene fluorescence using excitation at 248 nm, and Itoh *et al.*[122] studied a set of markers with various functional groups using copper vapor laser excitation at 255 nm, and found N,N-dimethylaniline (DMA) to give the strongest fluorescence. With a boiling point of 140°C, DMA was considered to be suitable for the upper end of the gasoline evaporation curve, but it is prone to strong quenching by O_2 which hinders practical application. Fujikawa[125] reports on limitations of the use of 3-pentanone as tracer in the high temperature range which can be corrected for by the methods shown below. In the present work a number of potential tracers with different functional groups were investigated with respect to their fluorescence spectral range and intensity. Most suitable candidates are aromatic markers such as toluene or xylene and ketones like 3-pentanone and acetone.

For simultaneous measurements of different markers, fluorescence signals must be spectrally discriminated upon single laser pulse excitation. Ketones and aromats show emission in significantly different spectral regions after excitation with the same wavelength. Pairs of tracers were selected and investigated in static high pressure experiments[126] which fulfilled the criteria of having (a) fluorescence emission spectra in distinctly different regions and (b) significantly different boiling points which allows for simultaneous measurements of different fuel volatility classes. Tracer pairs investigated in more detail were (a) toluene/acetone (110°C/56°C) (b) *p*-xylene/ acetone (138°C/56°C) and (c) *p*-xylene/3-pentanone (138°C/102°C) given with their boiling points. These tracers cover the typical boiling range of commercial fuels within three different volatility classes. The maximum of fluorescence intensity is found in the wavelength region 280–300 nm and at 350 nm for aromatic and ketonic tracers, respectively.

Simultaneous measurements of the distribution of different fuel volatility classes were carried out in an optically accessible SI engine. The beam of a tunable KrF excimer laser was Raman-shifted to 276 nm (in a H_2-filled cell, 5 bar), formed into a horizontal light sheet and aligned through the engine through parallel entrance and exit windows. Fluorescence was observed via a quartz window in the cylinder head.[126] A dielectric mirror used as beam-splitter directed the signal of two ICCD cameras which were equipped with appropriate filters for selective detection of the signal of the aromatic (band pass filter 295 ± 10 nm) and the ketonic tracer (WG360 and UG5), respectively. When detecting ketone fluorescence, a contribution of the aromatic signal to the overall fluorescence intensity has been observed especially under low pressure conditions (approximately 10% at 140°ca with the mixture of tracers as shown above) and subtracted during data reduction using the simultaneously imaged fluorescence map from the aromatic marker. For all measurements a three component nonfluorescing model fuel consisting of *iso*-pentane, *iso*-octane and *n*-nonane (20%, 70%, 10% volume respectively) was used. The fluorescence markers seeded to the fuel (8% for acetone and *p*-xylene and at 10% for toluene) partially replaced the corresponding aliphatic fuel components of the respective volatility class. The engine was operated at 1200 rpm, with full load under stoichiometric conditions (equivalence ratio $\phi = 1.0$). Ignition was set to 350° crank angle (ca), with 0°ca corresponding to top dead center in the intake stroke. The acquired raw images of markers were corrected for pressure effects[121(a),126] and spatial laser intensity distribution.

Figure 19 shows single-pulse LIF images of simultaneously mapped acetone and *p*-xylene markers as tracers for low and high boiling fuel components at two different crank angle positions in the intake stroke (60°ca and 100°ca). For each timing and marker combination, two pairs of fuel distribution maps are shown. The single pulse images clearly reveal instantaneous fluctuations in spatial fuel distributions of different volatility classes during the intake stroke. Areas showing high concentrations of different markers are not necessarily correlated, indicating substantial local variations in gas-phase composition. Lighter components as represented by acetone show a random distribution yielding an even distribution in the images averaged over many subsequent cycles (Fig. 20).

The distribution of the mid and high boiling components on the other hand reveals systematic variations. The area showing maximum concentrations is moving towards the cylinder wall early in the intake stroke, seems to turn back at around 180°ca and is moving backwards yielding a more homogeneous mixture. This indicates that these components follow the

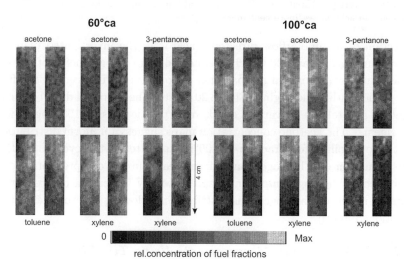

Fig. 19. Distribution of different fuel volatility classes acquired from simultaneous two-tracer measurement in an SI engine fueled with multicomponent fuels. For each combination of tracers two single-shot image pairs are shown for each two different detection timings in the intake stroke (60 and 100°ca).

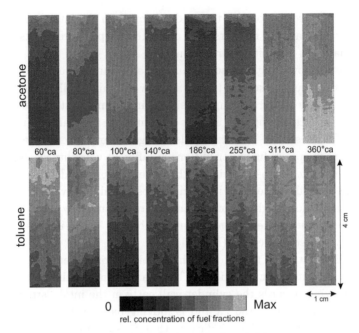

Fig. 20. Averaged distribution of low and mid boiling fuel fractions as represented by toluene and acetone as fluorescent markers.

direction of a tumble flow directed upwards in the investigated plane as shown in Figs. 19 and 20 which is expected from the geometry of the engine. These effects reveal a separation of components induced by the flow properties within the engine. Figure 20 present the averaged distribution of acetone and toluene at various crank angle positions using 100 single measurements. The acetone concentration maps suggest a uniform distribution of the low boiling fuel fraction at 100°ca and later in the cycle indicating the formation of a homogeneous mixture early in the compression stroke. The medium boiling components represented by toluene in contrast indicates a much less homogeneous distribution following the tumble motion mentioned above.

6.2. 2D-Temperature Measurements Using Two-Line Tracer-LIF

UV absorption spectra of 3-pentanone are significantly red-shifted with increasing temperature (Fig. 21). It was found in cell measurements[121(a)] that

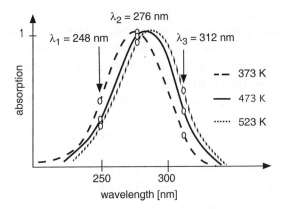

Fig. 21. Absorption spectrum of 3-pentanone at different temperatures at atmospheric pressure.

the maximum of the absorption band, which at $T = 373$ K is located at 275 nm, shifts towards longer wavelengths with increasing temperature while the width of the band remains almost constant (with a full width at half maximum of about 30 nm). Therefore, the fluorescence intensity upon excitation at 248 nm decreases with increasing temperature. Accordingly, if 3-pentanone is excited on the long wavelength side of the absorption maximum, a positive temperature dependence should be observed, as it has been found for excitation at 308 nm.[127] For excitation at 276 nm, which is close to the absorption maximum, a minimal temperature dependence is expected making this wavelength ideal when single-line excitation is applied for fuel vapor concentration mapping like in the work presented above.

The observed temperature dependence of the absorption cross section of 3-pentanone and the corresponding fluorescence intensity offers the possibility for a new type of temperature measurements.[121(a),(e)] This technique gives access to 2D-temperature distributions between 300 and 1000 K relevant for precombustion conditions that could hardly be assessed with other laser spectroscopic techniques developed for combustion thermometry. By calculating temperatures from the ratio of simultaneously acquired intensity distributions, the measurement is independent on local tracer concentrations. Measurements in inhomogeneously mixed environments are therefore feasible.

The first application[128] of this technique was carried out in an optically accessible single-cylinder two-stroke engine (ELO L372)[129] schematically depicted in Fig. 22. *Iso*-octane (p.a) seeded with 10% (v/v) 3-pentanone, was fed into the engine via a carburetor from which the fuel-air mixture piston pumped into the combustion chamber. Measurements were performed both without ignition and in a skip fired mode at 1000 rpm. Two excimer lasers operated on KrF (248 nm) and XeCl (308 nm), respectively, were fired with a fixed delay of 150 ns to prevent cross talk of signals during fluorescence detection. The laser-beams were formed into horizontal light sheets (20 mm × 0.5 mm) and carefully adjusted for spatial overlap. The fluorescence signal was imaged onto two ICCD cameras using a metal coated beam-splitter. In-cylinder pressures and laser pulse energies were stored along with the images. The signal intensities were corrected

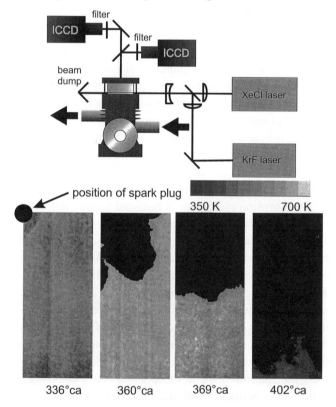

Fig. 22. Experimental setup for 2D-temperature measurement in a optically accessible single-cylinder two-stroke engine. Lower part: Temperature distribution in the unburnt gases for different detection timings obtained from single-shot two-line 3-pentanone LIF measurements. The burned gas area is shown in black.

for pressure influences using the data reported in Refs. 121(a) and (b). With this setup, temperature distributions could be obtained for the unburned gas region throughout the compression stroke even after the onset of combustion.

Resulting 2D-temperature fields are shown in Fig. 22 for different crank angle positions in the compression stroke before and after ignition. In the engine under study, due to the homogeneous load without exhaust gas recirculation, homogeneous temperature distributions were found. The temperature information gained from the two-line measurement in turn can be used to further quantify fuel local concentrations using either of the two single fluorescence maps.[128(b)−(c)]

6.3. *Imaging of NO Concentrations and Temperatures in SI Engines*

The reduction of the emission of NO, as one of the major pollutants in combustion, is of particular interest in the development of internal combustion engines. Further legislative regulations of the NO release from sparkignition (SI) and Diesel engines are expected within the next years. The use of standard three-way catalysts is not possible with future direct injected engines operated under lean burn conditions since the air/fuel ratio deviates significantly from unity. Therefore, NO formation has to be minimized already during the combustion process itself.

For a detailed understanding of the NO formation processes in engines, quantitative NO concentration distributions have to be measured with high spatial and temporal resolution directly in the engine. LIF allows to detect many combustion related species with detection limits in the ppm range.[130] However, quantitative concentration measurements are often complicated by complex pressure, temperature and interference effects. Therefore, the interpretation of LIF measurements performed in technical devices requires detailed knowledge of collisional quenching and spectral line-broadening effects.

Various strategies using different laser excitation wavelengths and fluorescence detection bands were investigated for in-cylinder engine NO diagnostics. NO concentration measurements in SI and Diesel engines using ArF excimer laser radiation tunable around 193 nm to excite the $(D^2\Sigma^+; v' = 0 - X^2\Pi; v'' = 1)$-transition of NO ($\varepsilon$-bands) were reported by several groups.[131] However, the $D\text{--}X$ transition cannot be used for engine measurements during the combustion process since the necessary short wavelength laser radiation is strongly absorbed by transient combustion products. Bräumer *et al.*[132] detected NO via $(A^2\Sigma^+;\ v' = 0 - X^2\Pi;\ v'' = 0)$-excitation around 225 nm (γ-bands) using the Raman-shifted output of a tunable KrF excimer laser. The use of the longer excitation wavelength allowed quantitative NO-LIF measurements in an SI engine fueled with propane/air. However, for engines operated with liquid fuels, attenuation of the 225 nm laser radiation during the combustion phase is still dominant. In a different attempt[133] a tunable KrF excimer laser is used to excite the $\mathrm{NO}(A^2\Sigma^+;v' = 0 - X^2\Pi;\ v'' = 2)$-transition around 248 nm. For this wavelength absorption is further reduced allowing concentration measurement of in-cylinder NO formation in an *iso*-octane driven SI engine throughout the whole combustion chamber.[134,135]

The spectroscopic detection scheme and the calibration methods were investigated using stabilized high pressure flames.[136,137] Fluorescence signals were recorded with an image intensified CCD camera after dispersion with a spectrometer yielding the entire emission spectrum for a given laser wavelength. Tuning the laser and composing emission spectra into an excitation-emission map (Fig. 23) reveals all spectroscopic information necessary to select the best excitation-emission wavelength combination. In a combustion environment, especially at high pressures when the lines are broadened, a laser tuned to a particular NO absorption line can also excite other species. A carefully selected combination of excitation and detection wavelengths is necessary for specific detection of the desired species. In the case of NO, the excitation laser is tuned to the O_{12}-bandhead of the $(A^2\Sigma^+;\ v' = 0 - X^2\Pi;\ v'' = 2)$-transition. This allows efficient excitation because several rovibrational transitions are excited simultaneously. Furthermore, the fluorescence excitation spectrum of molecular oxygen has a local minimum at that wavelength. NO fluorescence emitted in the $(A^2\Sigma^+;\ v' = 0 - X^2\Pi;\ v'' = 0)$ and $(A^2\Sigma^+;\ v' = 0 - X^2\Pi;\ v'' = 1)$ bands at shorter wavelengths, is detected. This scheme, represented by the rectangles in Fig. 23, provides a possibility to avoid interference with O_2, OH, and hydrocarbon fluorescence, and with Rayleigh and Stokes-shifted Raman signals.[135(b)]

High Arrhenius activation energies of the NO forming elementary reactions lead to a strong nonlinearity of the NO formation rate with increasing temperature.[3] Experimental investigations must therefore also

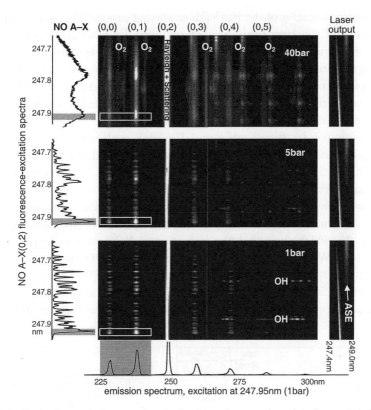

Fig. 23. Excitation-emission spectra obtained in a lean methane/air flame at 1, 5 and 40 bar. The spectra on the left show profiles along the NO A–X (0,1) emission lines for excitation wavelengths between 247.65 and 248.0 nm. The emission spectrum at the bottom was obtained at 1 bar after excitation at 247.95 nm. The NO A–X $(0, v'')$ progression is clearly resolved with the $(0, 2)$ band overlapped with the Rayleigh peak (off-scale). The small frames at the right show enlarged excitation-emission spectra of the laser output obtained via Rayleigh scattering. The intensity of the narrowband part (diagonal line) drops significantly at wavelengths shorter than 247.8 nm.

include *in situ* measurements of the local temperature. Simultaneously recorded temperature and NO concentration data provide a valuable possibility for testing capabilities of the combustion models. The use of a tunable KrF excimer laser for LIF detection of NO allows the simultaneous measurement of Rayleigh signals. With knowledge of local Rayleigh cross sections, Rayleigh signal intensities can be used to determine local total number densities and thus spatially-resolved temperature distributions.[138] Because of the elastic character of the Rayleigh scattering process ($\lambda_{\text{Rayleigh}} = \lambda_{\text{Laser}}$), reflection of the laser-beam at surfaces must be carefully avoided which usually restricts application to specially designed IC engines.

Light from a KrF excimer laser (tuned to the NO bandhead at 247.94 nm with pulse energies of typically 60 mJ) was expanded to form a thin light sheet and directed into the engine (Fig. 24) with optical access to the entire combustion chamber. Perpendicular to the light sheet, two ICCD cameras were mounted for fluorescence and Rayleigh scattering signal detection. NO-LIF signals were corrected for effects of spectral line broadening and quenching[134,139] and a calibration procedure was used for the determination of absolute concentrations.[137]

Image pairs with corresponding NO concentrations and temperature fields, measured simultaneously in the transparent engine, are presented in Fig. 25, which show that the formation of NO occurs in high temperature areas. While the overall spatial distribution of NO and temperature is strongly correlated, the profiles shown below indicate that the temperature distribution is more uniform than the NO concentration distribution. This is an important result for the comparison with mathematical models which

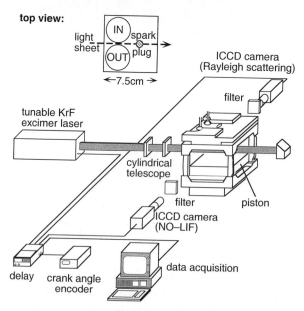

Fig. 24. Experimental setup for simultaneous measurements of temperature and NO concentration fields in a research SI engine.

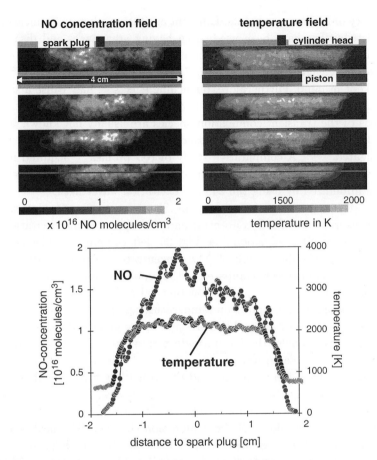

Fig. 25. Simultaneous single-shot measurements of absolute NO concentration and temperature fields in the transparent SI engine fueled with propane/air under stoichiometric conditions.

are developed for engine design. Due to the strong nonlinear temperature dependence of the NO production rate, careful control and homogenization of the combustion conditions should allow significant reduction of primary NO formation.

Detailed analysis of the spatial and temporal distribution of NO during the combustion phase yields information about the formation pathways involved. Nevertheless, as far as processes close to the flame front are concerned, there is a number of effects that cause uncertainties in the reliability of LIF measurements. Different excitation schemes for measuring NO distributions have been used; each having advantages and disadvantages in different technical combustion systems. Besides attenuation of excitation wavelengths,[134] attenuation of emitted fluorescence light[140,141] has to be kept in mind. Furthermore, quenching effects[142] may cause a significant uncertainty around the flame front since collisional quenching cross sections of the transient combustion species are usually not known. Additionally, at high pressures, the interference of LIF from hot O_2 has to be investigated.[136(b)] A comparative study of NO profiles[139] acquired in the same engine under identical operating conditions revealed that these effects can be minimized by using the 248 nm excitation wavelength.

Figure 26 shows NO concentration profiles extracted from typical single-shot measurements of NO concentration fields upon 248 nm excitation for early stages in SI combustion (see Fig. 25, left column). Profiles ranging from the first significant rise of NO concentration to the middle of the combustion chamber were compared to the profiles obtained from a "simple numerical model" (Fig. 26) based on thermal NO formation. This model uses experimental information on temperature and pressure variations with time. As the temperature and the pressure of the unburned gases rise with increasing crank angle, the NO formation rate rises. This means, that in regions where the flame front passes later in the cycle the NO formation rate is higher from the beginning and the formation of NO is enhanced in this area compared to inner parts of the flame. As a result and in agreement with the experiments the simple model predicts a flatter NO concentration profile towards the middle of the combustion chamber. The peak value is calibrated using the experimental data. The contribution of "prompt NO" in the calculation was set to 30 ppm, as predicted by a detailed model with full chemistry[139,143] producing a small offset of the entire NO concentration profile.

Figure 27 shows a series of averaged (over 25–30 single frames) NO concentration images obtained at different times after ignition, taken around top dead center. The engine is fueled with stoichiometric *iso*-octane/air mixtures at 1000 rpm. The region where NO is formed expands rapidly into the combustion chamber. In addition to the spatial growth, the absolute concentration of NO rises rapidly. The total amount of NO within the cylinder increases with time due to the increasing volume of the post flame zone and the slow NO formation in the post flame gases. The highly temperature dependent NO formation rates of the Zeldovich mechanism[3,144] cause a strong dependence of the NO concentration on the residence time of the hot gases. As a result, the highest NO concentrations are found in the center of the combustion chamber where the hot gases have been present for the longest time. Comparison of these experimental data with the results of numerical simulations based on computational fluid dynamics (CFD) models are very promising.[135,145] Further experiments will focus on direct injection gasoline[135(b)] and diesel engines[146] where effects of spatially inhomogeneous mixtures have to be considered very carefully.

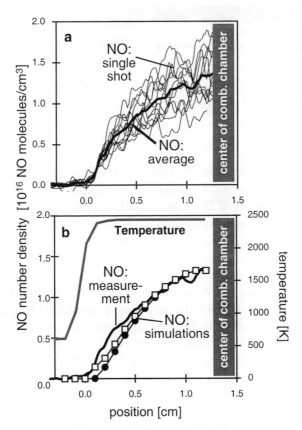

Fig. 26. (a) Single-shot NO concentration profiles along the center of the combustion chamber derived from the 2D-distributions shown in the left part of Fig. 25. The average NO concentration profile is also shown (solid line). (b) Comparison of the average experimental NO concentration profile (solid line) and two profiles obtained from modeling calculations. (•): Thermal NO only, (□): Incl. 30 ppm of prompt NO (for details see text).

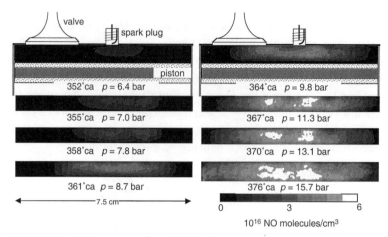

Fig. 27. Absolute NO concentration fields in the transparent SI engine fueled with *iso*-octane/air as function of the crank angle. Ignition at 340°ca, engine speed 1000 rpm.

Acknowledgments

The work was supported by the Deutsche Forschungsgemeinschaft via the Sonderforschungsbereich 359 "Reaktive Strömung, Diffusion und Transport" at the University of Heidelberg. Partial support by the Bundesministerium für Bildung und Forschung, the State of Baden–Württemberg (in the framework of the TECFLAM research association) and the European Union in the framework of the TOPDEC project is also acknowledged. Many thanks to F. Behrendt, R. Kissel–Osterrieder, J. Warnatz (Interdisziplinäres Zentrum für Wissenschaftliches Rechnen, IWR, University of Heidelberg) and P. Lindstedt (Department of Mechanical Engineering, Imperial College of Science, Technology and Medicine, London) for stimulating discussions. Special thanks to U. Mass (ITV, University of Stuttgart) for providing the simulation data shown in Fig. 10(c) and J. Luque (SRI International) for providing the LIFBASE spectral simulation program.

References

1. R. V. Bellomo, *J. Human Evolution* **27**, 173 (1994).
2. B. Dessus, *Energy Strategies for Sustainable Development* (FPH, Lausanne, 1993).
3. W. C. Gardiner, *Combustion Chemistry* (Springer, New York, 1986); J. Warnatz, U. Maas and R. W. Dibble, *Combustion* (Springer, Heidelberg, 1996) and references therein.

4. R. P. Wayne, *The Chemistry of Atmospheres*, 2nd edition (Oxford University Press, Oxford, 1991).
5. S. B. Pope, *23rd Symp. (Int.) Combustion* (The Combustion Institute, Pittsburgh, 1990), p. 591; J. Warnatz, *14th Symp. (Int.) Combustion* (The Combustion Institute, Pittsburgh, 1992), p. 553.
6. R. P. Lindstedt, *27th Symp. (Int.) Combustion* (The Combustion Institute, Pittsburgh, 1998), p. 269 and references therein.
7. M. Bodenstein, *Z. Phys. Chem.* **22**, 23 (1897).
8. F. J. Lipscomb, R. G. W. Norrish and B. A. Thrush, *Proc. Roy. Soc. London Ser.* **A233**, 455 (1956).
9. J. K. Cashion and J. C. Polanyi, *J. Chem. Phys.* **30**, 317 (1959).
10. D. R. Herschbach, *Adv. Chem. Phys.* **10**, 319 (1966); Y. T. Lee and Y. R. Shen, *Phys. Today* **33**, 52 (1980); Y. T. Lee, *Science* **236**, 793 (1987); J. Durup, *Laser Chem.* **7**, 239 (1987).
11. A. Ben-Shaul, Y. Haas, K. L. Kompa and R. D. Levine, *Lasers and Chemical Change*, Springer Series in Chemical Physics, Vol. 10 (Springer, Heidelberg, 1981).
12. J. Wolfrum, H.-R. Volpp, J. Warnatz and R. Rannacher, Ed., *Gas Phase Chemical Reaction Systems: Experiments and Models 100 Years after Max Bodenstein*, Springer Series in Chemical Physics, Vol. 61 (Springer, Heidelberg, 1996).
13. R. D. Levine and R. B. Bernstein, *Molecular Reaction Dynamics and Chemical Reactivity* (Oxford University Press, Oxford, 1987).
14. A. H. Zewail, *Femtochemistry: Ultrafast Dynamics of the Chemical Bond, I and II* (World Scientific, Singapore, 1994) and references therein.
15. H. Eyring, *J. Chem. Phys.* **3**, 107 (1935); M. G. Evans and M. Polanyi, *Trans. Faraday Soc.* **31**, 875 (1935).
16. J. C. Whitehead, Ed., *Selectivity in Chemical Reactions* (Kluwer Academic Publishers, Dordrecht, 1991); M. N. R. Ashfold and J. E. Baggott, Eds., *Advances in Gas-Phase Photochemistry and Kinetics, Bimolecular Collisions* (Royal Society of Chemistry, London, 1989); J. Jortner, R. D. Levine and B. Pulman, Eds., *Mode Selective Chemistry* (Kluwer Academic Publishers, Dordrecht, 1991).
17. J. Wolfrum, V. Sick and K. Kompa, Eds., Laser Diagnostics for Industrial Processes, *Ber. Bunsenges. Phys. Chem.* **97**, 1503 (1993); J. Wolfrum, Hottel Plenary Lecture: Lasers in Combustion: From Basic Theory to Practical Devices, *27th Symp. (Int.) Combustion* (The Combustion Institute, Pittsburgh, 1998), p. 1 and references therein.
18. X. Su, J. Jensen, M. X. Yang, M. Salmeron, Y. R. Shen and G. Somorjai, *Faraday Discuss.* **105**, 263 (1996).
19. G. A. Somorjai, *Appl. Sur. Sci.* **121/122**, 1 (1997).
20. H. Härle, A. Lehnert, U. Metka, H.-R. Volpp, L. Willms and J. Wolfrum, *J. Appl. Phys. B*, Special Issue: Nonlinear Optics at Interfaces **68**, 567 (1999).

21. J. Warnatz, *Modeling of Chemical Reaction System*, Springer Series in Chemical Physics, Vol. 18, Eds. K. H. Ebert, P. Deuflhard and W. Jäger (Springer, Heidelberg, 1980); C. K. Westbrook and F. L. Dryer, *Prog. Ener. Combust. Sci.* **10**, 1 (1984); J. A. Miller, *26th Symp. (Int.) Combustion* (The Combustion Institute, Pittsburgh, 1996), p. 461.

22. F. Haber and H. D. von Schweinitz, *Über die Zündung des Knallgases durch Wasserstoffatome, Sitzungsber. d. Preuss. Akad. d. Wiss., Physik. Math. Kl.* **XXX**, 8 (1928); F. Haber, *Z. Angew. Chem.* **42**, 570 (1929); L. Farkas, F. Haber and P. Harteck, *Z. Elektrochem.* **36**, 711 (1930).

23. F. Haber and K. F. Bonhoeffer, *Bandenspektroskopie und Flammenvorgänge, Sitzungsber. d. Preuss. Akad. d. Wiss.*, Berlin, 1. März 1928; K. F. Bonhoeffer and F. Haber, *Z. Phys. Chem.* **137**, 263 (1926).

24. For recent review see: W. Tsang and A. Lifshitz, *Ann. Rev. Phys. Chem.* **41**, 559 (1990); J. V. Michael and K. P. Lim, *Ann. Rev. Phys. Chem.* **44**, 429 (1993); J. Michael in Ref. 12, p. 177.

25. (a) C. J. Cobos, H. Hippler and J. Troe, *J. Phys. Chem.* **89**, 342 (1985); (b) J. Troe, *J. Phys. Chem.* **90**, 3485 (1986); (c) J. Troe, *22nd Symp. (Int.) Combustion* (The Combustion Institute, Pittsburgh, 1988), p. 843; (d) J. Troe, *Ber. Bunsenges. Phys. Chem.* **94**, 1183 (1990); (e) J. P. Hessler, *J. Phys. Chem.* **102**, 4517 (1998).

26. (a) J. A. Miller, *J. Chem. Phys.* **74**, 5120 (1981); (b) K. Kleinermanns and R. Schinke, *J. Chem. Phys.* **80**, 1440 (1984); (c) A. J. C. Varandas, *J. Chem. Phys.* **99**, 1076 (1993); (d) A. J. C. Varandas, *Chem. Phys. Lett.* **225**, 18 (1994); (e) A. J. C. Varandas, *Chem. Phys. Lett.* **235**, 111 (1995); (f) J. A. Miller and B. C. Garrett, *Int. J. Chem. Kin.* **29**, 275 (1997); (g) A. Lifshitz and H. Teitelbaum, *Chem. Phys.* **219**, 243 (1997).

27. (a) R. T. Pack, E. A. Butcher and G. A. Parker, *J. Chem. Phys.* **99**, 9310 (1993); (b) D. H. Zhang and J. Z. H. Zhang, *J. Chem. Phys.* **101**, 3671 (1994); (c) R. T. Pack, E. A. Butcher and G. A. Parker, *J. Chem. Phys.* **102**, 5998 (1995); (d) B. Kendrik and R. T. Pack, *Chem. Phys. Lett.* **234**, 291 (1995); (e) B. Kendrik and R. T. Pack, *J. Chem. Phys.* **104**, 7475 (1996); *J. Chem. Phys.* **104**, 7502 (1996); (f) J. Q. Dai and J. Z. H. Zhang, *J. Phys. Chem.* **100**, 6898 (1996); (g) B. Kendrik, *Int. J. Quant. Chem.* **64**, 581 (1997); *Int. J. Quant. Chem.* **66**, 111 (1998); (h) G. C. Groenenboom, *J. Chem. Phys.* **108**, 5677 (1998).

28. (a) C. Leforestier and W. H. Miller, *J. Chem. Phys.* **100**, 733 (1994); (b) T. C. Germann and W. H. Miller, *J. Phys. Chem.* **A101**, 6358 (1997); (c) D. E. Skinner, T. C. Germann and W. H. Miller, *J. Phys. Chem.* **A102**, 3828 (1998); (d) A. Viel, C. Leforestier, and W. H. Miller, *J. Chem. Phys.* **108**, 3489 (1998).

29. (a) A. J. H Meijer and E. M. Goldfield, *J. Chem. Phys.* **108**, 5404 (1998); (b) *J. Chem. Phys.* **110**, 870 (1998).

30. R. J. Blint and C. F. Melius, *Chem. Phys. Lett.* **64**, 183 (1979).

31. M. R. Pastrana, L. A. M. Quintales, J. Brañdao and A. J. C. Varandas, *J. Phys. Chem.* **94**, 8073 (1990).

32. B. Kendrik and R. T. Pack, *J. Chem. Phys.* **102**, 1994 (1995).
33. K. Kleinermanns and J. Wolfrum, *J. Chem. Phys.* **80**, 1446 (1984).
34. (a) K. Kleinermanns and E. Linnebach, *J. Chem. Phys.* **82**, 5012 (1985); (b) K. Kleinermanns, E. Linnebach and M. Pohl, *J. Chem. Phys.* **912**, 2181 (1989); (c) M. Bronikowski, R. Zhang, D. J. Rakestraw and R. N. Zare, *Chem. Phys. Lett.* **156**, 7 (1989); (d) A. Jacobs, F. M. Schuler, H.-R. Volpp, M. Wahl and J. Wolfrum, *Ber. Bunsenges. Phys. Chem.* **94**, 1390 (1990).
35. (a) Y. Matsumi, N. Shafer, K. Tonokura and M. Kawasaki, *J. Chem. Phys.* **95**, 4972 (1991); (b) H.-G. Rubahn, W. J. van der Zande, R. Zhang, M. J. Bronikowski and R. N. Zare, *Chem. Phys. Lett.* **186**, 154 (1991).
36. H. L. Kim, M. A. Wickramaaratchi, X. Zheng and G. E. Hall, *J. Chem. Phys.* **101**, 2033 (1994); H.-R. Volpp and J. Wolfrum in Ref. 12, p. 14.
37. R. Fei, X. S. Zheng and G. E. Hall, *J. Phys. Chem.* **A101**, 2541 (1997).
38. (a) S. P. Walch, C. M. Rohlfing, C. F. Melius and C. W. Bauschlinger Jr., *J. Chem. Phys.* **88**, 6273 (1988); (b) S. P. Walch aand C. M. Rohlfing, *J. Chem. Phys.* **91**, 2373 (1989).
39. K. Keßler and K. Kleinermanns, *J. Chem. Phys.* **97**, 374 (1992).
40. (a) A. Jacobs, H.-R. Volpp and J. Wolfrum, *Chem. Phys. Lett.* **177**, 200 (1991); **180**, 613 (1991); (b) S. Seeger, V. Sick, H.-R. Volpp and J. Wolfrum, *Israel J. Chem.* **34**, 5 (1994).
41. R. A. Sutherland and R. A. Anderson, *J. Chem. Phys.* **58**, 1226 (1973).
42. R. D. Levine, *Ann. Rev. Phys. Chem.* **29**, 59 (1978); R. D. Levine and J. L. Kinsey, *Atom–Molecule Collision Theory — A Guide for the Experimentalist*, Ed. R. B. Bernstein (Plenum Press, New York, 1979).
43. R. A. Brownsword, C. Kappel, P. Schmiechen, H. P. Upadhyaya and H.-R. Volpp, *Chem. Phys. Lett.* **289**, 241 (1998); (b) V. Ebert, C. Schulz, H.-R. Volpp, J. Wolfrum and P. Monkhouse, *Israel J. Chem.* **39**, (1999).
44. G. Hilber, A. Lago and R. Wallenstein, *J. Opt. Soc. Am.* **B4**, 1753 (1987).
45. H. Okabe, *The Photochemistry of small Molecules* (Wiley & Sons, New York, 1978); H. Okabe, *Adv. Photochem.* **13**, 17 (1986) and references therein.
46. W. M. Uselman and E. K. C. Lee, *J. Chem. Phys.* **65**, 1948 (1976).
47. J. Miyawaki, T. Tsucizawa, K. Yamanouchi and S. Tsuchiya, *Chem. Phys. Lett.* **165**, 168 (1990).
48. P. Felder, B.-M. Haas and J. R. Huber, *Chem. Phys. Lett.* **204**, 248 (1993) and references therein.
49. M.-H. Hui and S. A. Rice, *Chem. Phys. Lett.* **17**, 474 (1972).
50. (a) Y.-L. Huang and R. J. Gordon, *J. Chem. Phys.* **93**, 868 (1990); (b) M. Abe, Y. Sato, Y. Inagaki, Y. Matsumi and M. Kawasaki, *J. Chem. Phys.* **101**, 5647 (1994).
51. G. N. A. van Veen, K. A. Mohamed, T. Baller and A. E. DeVries, *Chem. Phys.* **74**, 261 (1983).
52. R. Atkinson, D. I. Baulch, R. A. Cox, R. F. Hampson Jr., J. A. Kerr and J. Troe, *J. Phys. Chem. Ref. Data* **21**, 1206 (1992).
53. C. H. Wight and S. R. Leone, *J. Chem. Phys.* **79**, 4823 (1983).

54. M. Abu Baje, M. Cameron, A. Hanf, C. KAppel, H. P. Upadhyaya, H.-R. Volpp and J. Wolfrum, to be presented at the Joint Meeting of the British, German and French Sections of the Combustion Institute (Pittsburgh), Nancy-France, 18–21 May, 1999.

55. (a) F. M. Magnotta, D. J. Nesbitt and S. R. Leone, *Chem. Phys. Lett.* **83**, 21 (1981); (b) Z. Xu, B. Koplitz and C. Wittig, *J. Chem. Phys.* **87**, 1062 (1987).

56. A. J. C. Varandas, *Mol. Phys.* **85**, 1159 (1995).

57. L. R. Thorne, M. C. Branch, D. W. Chandler, R. J. Kee and J. A. Miller, *21st Symp. (Int.) Combustion* (The Combustion Institute, Pittsburgh, 1986), p. 965; J. V. Volponi and M. C. Branch, *24th Symp. (Int.) Combustion* (The Combustion Institute, Pittsburgh, 1992), p. 823; J. Vandooren, M. C. Branch and P. J. Van Tiggelen, *Combust. Flame* **90**, 247 (1992); S. Zabarnick, *Combust. Sci. Tech.* **83**, 115 (1992); B. A. Williams and J. W. Fleming, *Combust. Flame* **98**, 93 (1994); B. A. Williams and J. W. Fleming, *Combust. Flame* **100**, 571 (1995); W. Juchmann, H. Latzel, D. I. Shin, G. Peiter, T. Dreier, H.-R. Volpp, J. Wolfrum, K. M. Leung and P. Lindstedt, *27th Symp. (Int.) Combustion* (The Combustion Institute, Pittsburgh, 1998), p. 469; P. A. Berg, G. P. Smith, J. B. Jeffries and D. R. Crosley, *27th Symp. (Int.) Combustion* (The Combustion Institute, Pittsburgh, 1998), p. 1377.

58. J. A. Miller and C. T. Bowman, *Prog. Ener. Combust. Sci.* **15**, 287 (1989).

59. M. Frenklach, H. Wang, M. Goldenberg, G. P. Smith, D. M. Golden, C. T. Bowman, R. K. Hanson, W. C. Gardiner and V. Lissianski, Gas Research Institute Topical Report, Report No. GRI-95/0058, 1995. http://www.me.berkeley.edu/gri_mech/

60. R. P. Lindstedt, F. C. Lockwood and M. N. Selim, *Combust. Sci. Tech.* **108**, 231 (1995).

61. (a) T. Etzkorn, J. Fitzer, S. Muris and J. Wolfrum, *Chem. Phys. Lett.* **208**, 307 (1993); (b) J. Fitzer, Diploma-Thesis, University of Heidelberg, 1993.

62. R. Farrow, M. N. Bui-Pham and V. Sick, *26th Symp. (Int.) Combustion* (The Combustion Institute, Pittsburgh, 1996), p. 975.

63. (a) J. Luque, G. P. Smith and D. R. Crosley, *26th Symp. (Int.) Combustion* (The Combustion Institute, Pittsburgh, 1996), p. 959; (b) J. Luque and D. R. Crosley, *Appl. Phys.* **B63**, 91 (1996); (c) J. Luque, W. Juchmann and J. B. Jeffries, *Appl. Opt.* **36**, 3261 (1997).

64. J. Luque and D. R. Crosley, LIFBASE: Database and Spectral Simulation Program (Vers. 1.2), SRI International Report MP 96-001 (1996).

65. W. P. Jones and R. P. Lindstedt, *Combust. Flame* **73**, 233 (1988).

66. T. Etzkorn, S. Muris, J. Wolfrum, C. Dembny, H. Bockhorn, P. F. Nelson, A. Attia-Shahin and J. Warnatz, *24th Symp. (Int.) Combustion* (The Combustion Institute, Pittsburgh, 1992), p. 925.

67. V. Sick, F. Hildenbrand and R. P. Lindstedt, *17th Symp. (Int.) Combustion* (The Combustion Institute, Pittsburgh, 1998), p. 1401.

68. R. P. Lindstedt and G. Skevis, *Combust. Sci. Tech.* **125**, 75 (1997) and references therein.
69. J. Warnatz,
    ftp://reaflow.iwr.uni–heidelberg.de/pub/mechanism–for–export/
70. Ch. Dombrowsky and H. Gg. Wagner, *Ber. Bunsenges. Phys. Chem.* **96**, 1048 (1992).
71. P. Frank, K. A. Bashkaran and Th. Just, *J. Phys. Chem.* **90**, 2226 (1986).
72. S. Zabarnick, J. W. Fleming and M. C. Lin, *J. Chem. Phys.* **85**, 4374 (1986).
73. M. Röhrig, E. L. Petersen, D. F. Davidson, R. K. Hanson and C. T. Bowman, *Int. J. Chem. Kin.* **29**, 781 (1997).
74. T. Böhland and F. Temps, *Ber. Bunsenges. Phys. Chem.* **88**, 459 (1984).
75. D. L. Baulch, C. J. Cobos, R. A. Cox, P. Frank, G. Hayman, Th. Just, J. A. Kerr, T. Murrells, M. J. Pilling, J. Troe, R. W. Walker and J. Warnatz, *Combust. Flame* **98**, 59 (1994).
76. M. R. Berman, J. W. Fleming, A. B. Harvey and M. C. Lin, *19th Symp. (Int.) Combustion* (The Combustion Institute, Pittsburgh, 1982), p. 73.
77. M. W. Markus, P. Roth and Th. Just, *Int. J. Chem. Kin.* **28**, 171 (1996).
78. A. J. Dean, R. K. Hanson and C. T. Bowman, *J. Phys. Chem.* **95**, 3180 (1991).
79. A. M. Dean and J. W. Bozzelli, in *Combustion Chemistry II*, Ed. W. C. Gardiner Jr. (Springer, New York, 1999).
80. S. Bauerle, M. Klatt and H. Gg. Wagner, *Ber. Bunsenges. Phys. Chem.* **99**, 97 (1995).
81. P. Glarborg and J. A. Miller, *Combust. Flame* **99**, 475 (1994).
82. G. Kychakoff, R. D. Howe, R. K. Hanson and J. C. McDaniel, *Appl. Opt.* **21**, 3225 (1982).
83. M. J. Dyer and D. R. Crosley, *Opt. Lett.* **7**, 382 (1982).
84. E. W. Rothe and P. Andresen, *Appl. Opt.* **36**, 3971 (1997).
85. T. Heitzmann, J. Wolfrum, U. Maas and J. Warnatz, *Z. Phys. Chem. Neue Folge* **188**, 177 (1995).
86. W. Ketterle, A. Arnold and M. Schäfer, *Appl. Phys.* **B51**, 91 (1990).
87. J. A. Gray and R. L. Farrow, *J. Chem. Phys.* **95**, 7054 (1991).
88. D. E. Heard, D. R. Crosley. J. B. Jeffries, G. P. Smith and A. Hirano, *J. Chem. Phys.* **96**, 4366 (1992).
89. P. M. Doherty and D. R. Crosley, *Appl. Opt.* **23**, 713 (1984).
90. E. W. Rothe, Y. Gu, A. Chryssostomou, P. Andresen and F. Bormann, *Appl. Phys.* **B66**, 251 (1998).
91. A. Arnold, B. Lange, T. Bouché, T. Heitzmann, G. Schiff, W. Ketterle, P. Monkhouse and J. Wolfrum, *Ber. Bunsenges. Phys. Chem.* **96**, 1388 (1992).
92. T. Heitzmannn, J. Wolfrum, U. Maas and J. Warnatz, *Z. Phys. Chem.* **188**, 177 (1995).
93. A. Dreizler, V. Sick and J. Wolfrum, *Ber. Bunsenges. Phys. Chem.* **101**, 771 (1997).

94. B. Bäuerle, F. Hoffmann, F. Behrendt and J. Warnatz, *25th Symp.* (*Int.*) *Combustion* (The Combustion Institute, Pittsburgh, 1994), p. 135.
95. J. Warnatz and J. Wolfrum, *Phys. Blätter* **47**, 193 (1991).
96. W. C. Pfefferle and L. Pfefferle, *Progr. Ener. Comb. Sci.* **12**, 25 (1986).
97. O. Deutschmann, F. Behrendt and J. Warnatz, *Catalysis Today* **21**, 461 (1994); F. Behrendt, O. Deutschmann, U. Maas and J. Warnatz, *J. Vac. Sci. Tech.* **A13**, 1373 (1995); R. Kissel-Osterrieder, F. Behrendt and J. Warnatz, *27th Symp.* (*Int.*) *Combustion* (The combustion Institute, Pittsburgh, 1998), p. 2267.
98. H.-J. Freund, *Ber. Bunsenges. Phys. Chem.* **99**, 1261 (1995); G. Ertl, *Ber. Bunsenges. Phys. Chem.* **99**, 1282 (1995).
99. D. A. King, *Vibrational Spectroscopy of Adsorbates*, Ed. R. F. Willis (Springer, Berlin, 1980); G. Ertl and J. Küppers, *Low Energy Electrons and Surface Chemistry* (VCH-Verlag, Weinheim, 1985); H.-J. Freund and H. Kuhlenbeck, *Applications of Synchrotron Radiation: High Resolution Studies of Molecules and Molecular Adsorbates on Surfaces* (Springer, Berlin, 1995); J. F. McGilp, D. Weaire and C. H. Patterson, *Epioptics: Linear and Nonlinear Optical Spectroscopy of Surfaces and Interfaces* (Springer, Berlin, 1995); H. H. Rotermund, *Surf. Sci. Reports* **29**, 265 (1997) and references therein.
100. Y. R. Shen, *J. Vac. Sci. Tech.* **B3**, 1464 (1984); Z. Rosenzweig, M. Asscher and C. Wittenzeller, *Surf. Sci. Lett.* **240**, 583 (1990); M. Buck, F. Eisert, J. Fischer, M. Grunze and F. Träger, *Appl. Phys.* **A53**, 552 (1991); R. M. Corn and D. A. Higgins, *Chem. Rev.* **94**, 107 (1994); M. Buck, F. Eisert, M. Grunze and F. Träger, *Appl. Phys.* **A60**, 1 (1995).
101. Y. R. Shen, *The Principles of Nonlinear Optics* (Wiley, New York, 1984).
102. J. Hunt, P. Guyot-Sionnest and Y. R. Shen, *Chem. Phys. Lett.* **133**, 189 (1987).
103. X. D. Zhu, H. Suhr and Y. R. Shen, *Phys. Rev.* **B35**, 3047 (1987); O. Du, R. Superfine, E. Freysz and Y. R. Shen, *Phys. Rev. Lett.* **70**, 2315 (1993).
104. C. Klünker, M. Balden, S. Lehwald and W. Daum, *Surf. Sci.* **360**, 104 (1996).
105. H. Härle, A. Lehnert, U. Metka, H.-R. Volpp, L. Willms and J. Wolfrum, *Chem. Phys. Lett.* **293**, 26 (1998).
106. X. Su, P. S. Cremer, Y. R. Shen and G. A. Somorjai, *J. Am. Chem. Soc.* **119**, 3994 (1997).
107. H. Härle, K. Mendel, U. Metka, H.-R. Volpp, L. Willms and J. Wolfrum, *Chem. Phys. Lett.* **279**, 75 (1997).
108. J. Miragliotta, R. S. Polizotti, P. Rabinowitz, S. D. Cameron and R. B. Hall, *Appl. Phys.* **A51**, 221 (1990).
109. J. Hunt, P. Guyot-Sionnest and Y. R. Shen, *Chem. Phys. Lett.* **133**, 189 (1987).
110. X. Su, P. S. Cremer, Y. R. Shen and G. A. Somorjai, *Phys. Rev. Letters* **77**, 3858 (1996).

111. P. Gardner, R. Martin, M. Tüshaus and A. M. Bradshaw, *J. Electron. Spec. Relat. Phenom.* **54/55**, 619 (1990).
112. J. E. Reutt-Robey, D. J. Doren, Y. J. Chabal and S. B. Christman, *Phys. Rev. Letters* **61**, 2778 (1988).
113. A. Szabó, M. A. Henderson and J. T. Yates, Jr., *J. Chem. Phys.* **96**, 6191 (1992).
114. J. Xu and J. T. Yates, Jr., *J. Chem. Phys.* **99**, 725 (1993).
115. A. Bandara, S. Katano, J. K. Onda, A. Wada, K. Domen and C. Hirose, *Chem. Phys. Lett.* **290**, 261 (1998).
116. P. Esser and W. Göpel, *Surf. Sci.* **97**, 309 (1980).
117. F. Behrendt, R. Kissel-Osterrieder, J. Warnatz, H. Härle, A. Lehnert, U. Metka, H.-R. Volpp, L. Willms and J. Wolfrum, *27th Symp. (Int.) Combustion (Book of Abstract: Work-In-Progress Posters)* (The Combustion Institute, Pittsburgh, 1998), p. 269; F. Behrendt, R. Kissel-Osterrieder, J. Warnatz, H. Härle, U. Metka, H.-R. Volpp and J. Wolfrum, *J.* (to be published).
118. J. Warnatz, M. D. Allendorf, R. J. Kee and M. E. Coltrin, *Comb. Flame* **96**, 393 (1994); O. Deutschmann, R. Schmidt, F. Berendt and J. Warnatz, *26th Symp. (Int.) Combustion* (The Combustion Institute, Pittsburgh, 1996), p. 1747.
119. G. Ertl, *Catalysis Sci. Tech. 4*, Eds. J. R. Anderson and M. Boudart (Springer, New York, 1983).
120. K. Christmann, *Surface Physical Chemistry*, Eds. H. Baumgärtel, E. U. Franck and W. Grünbein (Steinkopff/Springer, Darmstadt/New York, 1991); J. M. Thomas and W. J. Thomas, *Principles and Practice of Heterogeneous Catalysis* (VCH-Verlag, Weinheim, 1997) and references therein.
121. (a) F. Großmann, P. Monkhouse, M. Ridder, V. Sick and J. Wolfrum, *Appl. Phys.* **B62**, 249 (1996); (b) F. Ossler and M. Aldén, *Appl. Phys.* **B64**, 493 (1997); (c) A. Lozano, B. Yip and R. K. Hanson, *Exp. Fluids* **13**, 369 (1992); (d) L. S. Yuen, J. E. Peters and R. Lucht, *Appl. Opt.* **36**, 3271 (1997); (e) M. C. Thurber, F. Grisch and R. K. Hanson, *Opt. Lett.* **22**, 251 (1997).
122. T. Itoh, A. Kakuho, H. Hishinuma, T. Urushiahara, Y. Takagi, K. Horie, M. Asano, E. Ogata and T. Yamasita, SAE Paper, No. 952465 (1995).
123. (a) M. Berckmüller, N. P. Tait, R. D. Lockett, D. A. Greenhalgh, K. Iishi, Y. Urata, H. Umiyama and K. Yoshida, *25th Symp. (Int.) Combustion* (The Combustion Institute, Pittsburgh, 1994), p. 151; (b) A. Arnold, A. Buschmann, B. Cousyn, M. Decker, F. Vannobel, V. Sick and J. Wolfrum, SAE Paper, No. 932696 (1993).
124. J. Reboux, D. Puechberty and F. Dionnet, SAE Paper, No. 941988 (1994).
125. T. Fujikawa, Y. Hattori and K. Akihama, SAE Paper, No. 972944 (1997).
126. H. Krämer, S. Einecke, C. Schulz, V. Sick, S. R. Nattrass and J. S. Kitching, SAE Paper, No. 982467 (1998), SAE Transactions, *J. Fuels Lubricants*, **107**, 1048 (1998).
127. R. Tait and D. A. Greenhalgh, *Ber. Bunsenges. Phys. Chem.* **97**, 1619 (1993).

128. (a) S. Einecke, C. Schulz and V. Sick, *Laser Application to Chemical and Environmental Analysis* (Optical Society of America, Washington DC, 1998); (b) S. Einecke, C. Schulz, V. Sick, A. Dreizler, R. Schießl and U. Maas, SAE Paper, No. 982468 (1998), SAE Transactions, Vol. 107, J. of Fuels & Lubricants, p. 1060; S. Einecke, C. Schulz and V. Sick, *Appl. Phys. B*, in press (2000).

129. B. Bäuerle, J. Warnatz and F. Behrendt, *26th Symp. (Int.) Combustion* (The Combustion Institute, Pittsburgh, 1996), p. 2619.

130. (a) K. Kohse-Höinghaus, *Prog. Ener. Combust. Sci.* **20**, 203 (1994); (b) A. Eckbreth, *Laser Diagnostics for Combustion Temperature and Species*, 2nd edition (Gordon & Breach, Amsterdam, 1996); (c) W. Demtröder, *Laser Spectroscopy* (Springer, New York, 1997).

131. (a) P. Andresen, G. Meijer, H. Schlüter, H. Voges, A. Koch, W. Hentschel, W. Oppermann and E. W. Rothe, *Appl. Opt.* **29**, 2392 (1990); (b) A. Arnold, F. Dinkelacker, T. Heitzmann, P. Monkhouse, M. Schäfer, V. Sick, J. Wolfrum, W. Hentschel and K. P. Schindler, *24th Symp. (Int.) Combustion* (The Combustion Institute, Pittsburgh, 1992), p. 1605; (c) T. M. Brugman, R. Klein-Douwel, G. Huigen, E. van Walwijk and J. J. ter Meulen, *Appl. Phys.* **B57**, 405 (1993).

132. A. Bräumer, V. Sick, J. Wolfrum, V. Drewes, R. R. Maly and M. Zahn, SAE Paper, No. 952462 (1995).

133. C. Schulz, B. Yip, V. Sick and J. Wolfrum, *Chem. Phys. Lett.* **242**, 259 (1995).

134. C. Schulz, V. Sick, J. Wolfrum, V. Drewes, M. Zahn and R. R. Maly, *26th Symp. (Int.) Combustion* (The Combustion Institute, Pittsburgh, 1996), p. 2597; F. Hildenbrand, C. Schulz, V. Sick and E. Wagner, *Appl. Opt.* **38**, 1452 (1999).

135. F. Hildenbrand, C. Schulz, V. Sick, G. Josefsson, I. Magnusson, Ö. Andersson and M. Aldén, SAE Paper, No. 980148 (1998), SAE Transactions, **107**, *J. Eng., Sec. 3*, **107**, 205 (1999); F. Hildenbrand, C. Schulz, M. Hartmann, F. Puchner and G. Wawrschin, SAE Paper, No. 1999-01-3545 (1999).

136. (a) C. Schulz, V. Sick, J. Heinze and W. Stricker, *Laser Applications to Chemical and Environmental Analysis*, Vol. 3 (Optical Society of America, Washington DC, 1996); (b) C. Schulz, V. Sick, J. Heinze and W. Stricker, *Appl. Opt.* **36**, 3227 (1997).

137. C. Schulz, V. Sick, U. Meier, J. Heinze and W. Stricker, *Appl. Opt.* **38**, 1434 (1999).

138. A. Orth, V. Sick. J. Wolfrum, R. R. Maly and M. Zahn, *25th Symp. (Int.) Combustion* (The Combustion Institute, Pittsburgh, 1994), p. 143.

139. C. Schulz, J. Wolfrum and V. Sick, *27th Symp. (Int.) Combustion* (The Combustion Institute, Pittsburgh, 1998), p. 2077.

140. M. Knapp, A. Luczak, H. Schlüter, V. Beushausen, W. Hentschel and P. Andresen, *Appl. Opt.* **35**, 4009 (1996).

141. F. Hildenbrand, C. Schulz, V. Sick and E. Wagner, *Appl. Opt.* **38**, 1452 (1999).
142. P. H. Paul, J. A. Gray, J. L. Durant Jr. and J. W. Thoman Jr., *AIAA J.* **32**, 1670 (1994).
143. J. Warnatz, *MixFla: Program Package for Premixed Flames* (Version 8, 1997) IWR, University of Heidelberg (Germany).
144. Y. B. Zeldovich, *Acta Phisicochim. URSS* **21**, 577 (1948).
145. G. Josefsson, I. Magnusson, F. Hildenbrand, C. Schulz and V. Sick, *27th Symp. (Int.) Combustion* (The Combustion Institute, Pittsburgh, 1998), p. 2085.
146. F. Hildenbrand, C. Schulz, J. Wolfrum, F. Keller and E. Wagner, *Proc. Combust. Inst.* **28**, in press (2000).

CHAPTER 6

THE GAS-PHASE CHEMICAL DYNAMICS ASSOCIATED WITH METEORS

Rainer A. Dressler and Edmond Murad

Air Force Research Laboratory, Space Vehicles Directorate
Hansom AFB, MA 01731

Contents

1. Introduction

Meteors are produced by cosmic debris called *meteoroids* that enter Earth's atmosphere at high velocities, typically ranging from 10 to 70 km s^{-1}. It is estimated that the Earth is bombarded by 5 to 200 tons of meteoric material every day. The size of meteoroids can range from micron-sized dust particles that do not reach temperatures exceeding the sublimation point upon atmospheric entry, to larger mm-sized fragments that vaporize completely, to rock-size bodies that do not fully ablate and fall to Earth as *meteorites*. The larger and faster meteoroids give rise to the incandescent trails called *shooting stars* or meteors.

The meteoric vaporization occurs in the Earth's ionosphere, as illustrated schematically in Fig. 1. The vertical nature of the meteor phenomenon makes a physical description highly challenging due to the rapidly changing atmospheric properties with altitude. Meteoroid heating begins below ~ 150 km in the ionospheric F-region. The heating is ensued by ablation and sublimation. The visual meteors are observed between 120 and 60 km, a range comprising the D- and E-regions of the ionosphere. In this interval, the meteoroid traverses the *homopause* at ~ 100 km, which is considered to be the boundary between Earth's atmosphere and space. The pressure at 100 km is about 10^{-3} Torr, which is familiar to most experimental physical chemists as the pressure at which high vacuum pumps, such as diffusion pumps which are based on molecular flow properties, stop functioning. The

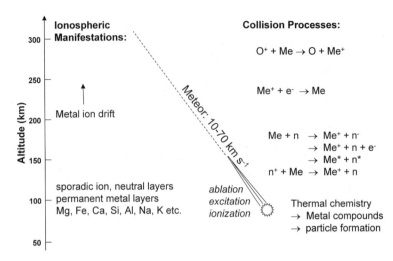

Fig. 1. Schematic representation of the vertical meteor phenomenon and the associated collision processes and ensuing ionospheric manifestations.

homopause is thus the boundary between bulk Eddy diffusion at lower altitudes, where atmospheric mixing is substantially more rapid than molecular diffusion, and molecular transport at higher altitudes, where gravity and the associated *Jeans escape* begin to play a role in the atmospheric composition.

In the vicinity of the homopause is the *mesopause* at about 90 km. The mesopause is the temperature minimum of the atmosphere, with a globally averaged temperature of ~ 195 K. Above the mesopause is the *thermosphere*, where radiation is the most efficient sink of energy, and due to the low abundances of infrared radiators and their declining densities, the temperature rises rapidly with altitude. The increasingly rarefied atmosphere in the thermosphere is no longer in local thermal equilibrium. The degree of nonequilibrium is underscored by the rapidly increasing mixing ratios of atomic oxygen with altitude near 90 km. The altitude regime around 90 to 100 km also represents an important transition for chemical kinetics. Whereas single collision events outweigh the kinetics at altitudes above the homopause, three-body association processes become significant below the mesopause. It is, therefore, not surprising that a number of interesting chemical phenomena related to meteors occur in the region between 90–100 km.

As meteoric material is evaporated, a number of high energy gas-phase collision processes occur. These processes, examples of which are listed in Fig. 1, involve both collisions between neutrals and neutrals and ions. Metal atoms, Me, are abundant in all meteoric bodies, and play an important role for several reasons: (i) Metal atoms and ions frequently have low lying excited states with high oscillator strengths and are, therefore, easily identified and traceable; (ii) Metal atoms have low ionization potentials and can be ionized fairly efficiently in high velocity neutral collisions; (iii) Atomic metal ions have very long lifetimes with respect to ion–molecule reactions and electron–ion recombination compared with molecular ions, which rapidly dissociatively recombine with electrons.

The effect of efficient dissociative recombination is reflected in the daytime ionospheric charged particle densities shown in Fig. 2. In the *E*-region, O_2^+ and NO^+ are the most abundant ions, while in the *F*-region, the higher plasma densities due to increased photoionization rates dramatically reduces the abundance of all molecular ions and O^+ is the principal positively charged species. Figure 2 also depicts a typical metal ion density profile (Me^+) around 100 km. As will be outlined further, this profile can be

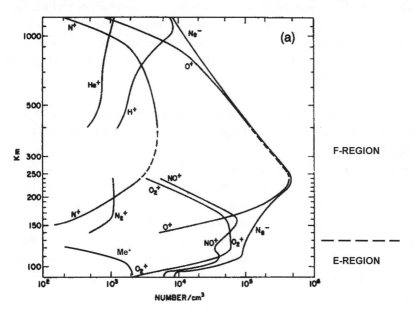

Fig. 2. Daytime ionospheric charged particle densities as a function of altitude.

highly variable. Given the properties of gaseous metals and ions, enhanced loading of metal atoms in the upper atmosphere, as could be the case during a major meteor storm such as the recent Leonids, can have profound effects on the ionospheric plasma properties. The transformation of a molecular ion plasma to an atomic ion plasma due to enhanced metal atom densities can lead to the sudden formation of narrow layers exhibiting order of magnitude increases in plasma densities, so called *sporadic E layers*.[1] As we become increasingly dependent on communications to and from space, there is a greater need for forecasting such ionospheric disturbances. Knowledge of the detailed chemical dynamics of meteor events, such as periodic meteor showers, is necessary to improve model predictions of the ionospheric metal atom loading based on observed ionization trail radar echoes and meteor luminosity.

In summary of the above, there are two main types of chemistry that are induced by meteors: one during the meteor entry that is dominated by *hyperthermal nonequilibrium kinetics* in the ionosphere, and post meteor low energy chemistry centered around 80–90 km where most of the meteoric material is deposited, and where permanent metal atom and ion layers are formed. In the latter, the chemistry is primarily governed by thermally equilibrated kinetics for which the rate coefficients are fairly well established. Meanwhile, the weak electric fields in the ionosphere slowly pull the metal ions to higher altitudes (ionospheric fountain effect) on a time scale long enough to allow electron–ion recombination (see Fig. 1). Metal ions, however, are still observed at altitudes exceeding 300 km.[2-10] The source of ionization is not clear but charge transfer involving ionospheric O^+ may play a role.

This review is organized as follows: in Sec. 2, the large scale phenomenological physics of meteors is discussed, detailing observations as well as the respective current interpretations. Meteorites will not be discussed in this review. Studies of the chemical and physical processes associated with meteorites have been largely concerned with composition, phase transformations and entrapment of noble gases in the solid meteorites. The reader is referred to several reviews and textbooks on the subject.[11-14] The elementary gas-phase molecular dynamics relevant to the meteor environment is discussed in Sec. 3, and will primarily focus on the hyperthermal nonequilibrium processes, in concert with the subject matter of this book.

2. Meteor Phenomenology

2.1. *Origin of Meteors*

Meteors originate mostly from fragmentation of comets as they orbit the Sun, and a smaller fraction can be attributed to asteroid fragmentation. In addition to the meteors that arise from within the solar system, a small amount ($\sim$ 2%) has been identified as arising from interstellar space.[a] A recent review has detailed the characteristics to be expected from cometary meteors.[15] The correct explanation for the composition of comets, namely their being formed of dust particles held together by ice and other low melting point gases (e.g. NH_3, CH_4, CO, CO_2, and more complex organic molecules) was made by Whipple.[16] The physics of comets and their orbital mechanics are discussed in standard texts on comets.[17,18] Figure 3 shows a sketch of a hypothetical comet trajectory as it passes through the inner solar

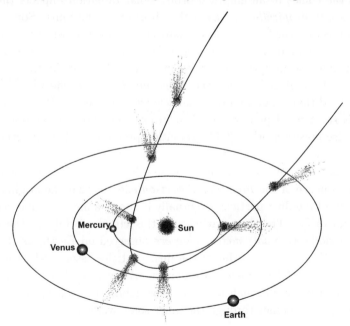

Fig. 3. Sketch showing the orbit of a comet around the Sun and its intersection with the orbits of the inner planets. Note that the comet tail always points away from the Sun due to radiation pressure.

---

[a]Meteorites collected on Earth arise almost entirely from asteroids or asteroid fragments. These are not the subject of this chapter.

system. The comet consists of a head or nucleus surrounded by the coma or the gaseous envelope; the coma is followed by two long tails, one gaseous and the other consisting of dust. The recent apparition of Comet Hale–Bopp has shown the presence of a third tail composed entirely of sodium atoms.[19,20] As the comet nears perihelion (point of closest approach to the Sun), radiation from the Sun heats the comet surface and leads to melting of the ice holding the dust and other chemicals together.

Dust and chemicals are ejected and form the two trails. Radiation pressure from the Sun exerts the largest force on the gaseous and dust particles that have separated from the coma. Hence, as the sketch illustrates, the coma always points towards the Sun, with the gaseous and dust trail always pointing away from the Sun. As the comet usually has an orbit that is elliptical around the Sun (and out of the ecliptic), the dust trail initially follows the comet in an elliptical orbit. Solar radiation impacts the dust particles *anisotropically* (i.e. from the direction of the Sun). Some of the radiation is absorbed and heats the particle; however, most of the incident radiation is reradiated — but *isotropically*. This difference leads to a loss of angular momentum with the consequence that the particle drifts towards the Sun. This effect is called the *Poynting–Robertson* effect.[21–23,24,25] It is estimated that a nominal dust particle will require 10^5–10^7 years to fall towards the Sun, depending on their origin (cometary or asteroidal) and the precise injection orbit.[26] The radiation pressure tends to lengthen this time somewhat.

For the ensemble of particles generated by the dust trail of the comet, the consequence of the *Poynting–Robertson* effect is that the original elliptical orbit of the dust particles eventually becomes a more circularized (less eccentric) orbit. The dust particles in this circular orbit are the source of the periodic (annual) *meteor showers* observed on Earth, as the orbit of the Earth crosses the orbit of the dust particles associated with a particular comet. *Meteor storms* are particularly intense showers that occur when Earth passes through a still narrow debris trail during or following the passage of a comet through the inner solar system.

2.2. *Meteor Physics*

The luminous entry of meteoroids into Earth's atmosphere, known for millennia as the phenomenon of shooting stars, is quite interwoven with myth and folklore. Good historical and physical surveys may be found in the classic texts by Olivier,[27] and Lovell,[28] and in the more recent comprehensive reviews of the physics of meteoroids as they enter the atmosphere.[29,30] A recent pictorial review may also be consulted for beautiful displays of this phenomenon.[31] The term meteor comes from the Greek μετεορον meaning things in the air. The terms *fireball* and *bolide* are also frequently applied. The former refers to brighter events associated with centimeter size meteoroids. The latter term applies to a meteoroid that seems to explode as it gets heated by entry into the Earth's atmosphere.

The physics of meteoric phenomena can be divided into three parts: (1) the *dynamics* of meteoroid motion in the atmosphere; (2) the *small scale physical processes* occurring in the extreme environment of a meteor; and (3) the *ionospheric chemical and plasma kinetics* induced by atoms and molecules deposited by meteors, particularly metals. These will be briefly discussed in the following.

2.2.1. *Meteor Dynamics*

Investigations of the dynamical aspects of the meteor phenomenon dated back to the 19th century. The characteristic properties of a meteor have been visual magnitude, the radiant, and entry velocity, size and mass of the meteoroid. While the first two can be directly observed, the latter three properties must be derived with a model for drag and mass loss from the observed deceleration as a function of momentary velocity and visual magnitude.

Cometary meteoroids enter the Earth's atmosphere at a velocity that varies between 12 and 70 km s^{-1}. This great range arises from the eccentricity of the parent cometary orbit, the point at which the orbits of the meteoroid and Earth cross, and the vector addition between the Earth's Keplerian velocity, ~ 30 km s^{-1}, and that of the meteoroid. In reality, these three factors can add in many combinations leading to the great range of entry velocities. The lowest and highest velocity meteoroids stem from extreme prograde and retrograde encounters, respectively. Table 1 lists some of the important annual meteor showers and their associated parameters. Note the variation in meteoroid velocities for the different showers. Meanwhile, interstellar meteoroids can have velocities approaching 100 km s^{-1}.

Typical visual meteors are associated with particles with representative diameters ranging from 1 mm to 20 cm.[30] The size distribution of interplanetary dust, however, peaks at substantially smaller diameters. Meteoroids

Table 1. List of the important visual meteor showers. The data was obtained from the International Meteor Organization's Web site (www.imo.org). Details in this table correspond to the best information available in April, 1996. v is the meteoroid velocity. r is a measure of intensity: the smaller the number the more intense is the radiant. ZHR is zenithal hourly rate. Some showers have ZHRs that vary from year to year. The most recent reliable figure is given here, except for possibly periodic showers that are noted as "var." = variable. Dates of maximum activity refer to meteor rates from the radiant, not the true maxima.

Shower	v km/s	r	ZHR	Dates	Date of Max. Activity
Quadrantids	41	2.1	120	Jan 01–Jan 05	Jan 03
Lyrids	49	2.9	15	Apr 16–Apr 25	Apr 22
eta-Aquarids	66	2.7	60	Apr 19–May 28	May 05
Southern delta-Aquarids	41	3.2	20	Jul 12–Aug 19	Jul 28
Perseids	59	2.6	200	Jul 17–Aug 24	Aug 12
Draconids (Giacobinids)	20	2.6	var.	Oct 06–Oct 10	Oct 09
Orionids	66	2.9	20	Oct 02–Nov 07	Oct 21
Southern Taurids	27	2.3	5	Oct 01–Nov 25	Nov 05
Leonids	71	2.5	40+	Nov 14–Nov 21	Nov 17
Geminids	35	2.6	110	Dec 07–Dec 17	Dec 13

with sizes down to 50 μm generate sufficiently dense ionization trails to be observed with radar (see Sec. 2.3.2). Meteoroids smaller than 20 μm are efficiently decelerated at higher altitudes and also radiate effectively and, therefore, do not reach a temperature exceeding the vaporization point. These meteoroids become upper atmospheric dust particles that eventually descend to Earth's boundary layer.

The *visual magnitude*, M, is related to the luminous intensity, I, through[32]

$$M = 6.8 - 2.5 \log_{10} I, \tag{1}$$

where I is expressed in watts. Each step increase in M corresponds to an increase of a factor of 2.5 in intensity. As M gets more positive, the object becomes fainter. For comparison, a full Moon has a magnitude of -12.6, while Venus has a magnitude of -4.4. Given the sensitivity of the human eye and atmospheric transmission properties, the limit of observation of visual meteors is $\sim +5$.[33] Very intense meteors have a magnitude of -5. Unusual fireballs, however, can have peak magnitudes exceeding -15.

The *radiant* applies to meteors of a particular meteor stream, and is the point in infinity in the sky from which the meteors appear to come.[27] In the case of the Leonid meteor shower that occurs annually in mid November (see Table 1), the radiant coincides with the constellation, Leo.

Beginning with Olivier,[27] a number of authors have tried to develop meteor theories based on first physical principles.[34-37] The physics of the body is of fundamental importance with respect to the dynamical aspects of the meteor problem discussed above. It concerns the rapidly varying flow-regime with the possible eventual formation of a shock wave, and describes the meteoroid heating that leads to ablation, fragmentation, melting and vaporization. This aerodynamic aspect of the problem will not be discussed in depth. The extreme environmental dynamics of the problem, however, have many similarities with Chaps. 2 of Raz and Levine and 3 of Boyd in this book. Extensive treatises specific to meteor dynamics are found in the reviews by Bronshten[29] and Ceplecha *et al.*[30]

2.2.2. *Small Scale Physical Processes*

The main observables that relate directly to small scale physical processes are *the spectral properties of the meteor emissions* and the *physical properties of ionization trails*. Both luminosity and ionization are derived directly from hypervelocity collisions between vaporized meteoroid species and atmospheric contituents. From Table 1, it is seen that, for example, a Mg atom evaporated from a Draconid meteoroid has a kinetic energy of ~ 50 eV, while the same atom evaporated from a Leonid meteoroid has a kinetic energy of ~ 630 eV when it collides with an atmospheric molecule. In both cases, the translational energy available to reactions is sufficient for inelastic processes such as electronic excitation and collisional ionization. As will be pointed out in Sec. 3, the respective cross sections increase dramatically with collision energy. Both the visual magnitude and the radar echo signatures of equally sized Draconid and Leonid meteoroids will thus differ substantially.

Knowledge of the collision physics is also an important source of information on the dynamical properties of the meteor. In principle, the initial meteoroid mass can be determined from the measured luminous intensity, I:[30]

$$I = -\tau \left(\frac{v^2}{2} \frac{dm}{dt} + mv \frac{dv}{dt} \right), \tag{2}$$

where the first and second term in the parentheses express the energy loss due to ablation and deceleration, respectively. τ is the luminous efficiency that signifies the fraction of kinetic energy converted into observed radiative energy. For high velocities (> 20 km s^{-1}), the first term is found to be dominant, a clear indication that the observed luminosity stems primarily from gas-phase processes as opposed to emissions of the body. The luminous efficiency is the integration over its spectral components, τ_λ, that are given by:

$$\tau_\lambda = \frac{E_\lambda}{E_T} \int_{E_0}^{E_T} \frac{\sigma_\lambda(E)dE}{\sigma_D(E)E} , \tag{3}$$

where E_T is the initial kinetic energy of the vaporized atom or molecule that emits at a wavelength, λ, E_λ is the photon energy, E_0 is the excitation threshold energy, σ_D is the momentum transfer or diffusion cross section, and σ_λ is the emission cross section, a laboratory observable (see Sec. 3.4.2).[38] Given the fact that metal atom lines represent the primary features of the spectral distribution of meteor light, the analysis of meteor spectra based on Eqs. (2) and (3) is rendered complicated because of the significant optical densities associated with lines of high oscillator strengths. Self absorption of the observed lines must, therefore, be accounted for. This requires knowledge of the atomic column densities.

Ionization is treated analogously. It is described in terms of the ionization coefficient, β, which is the ratio of free electrons produced per vaporized atom. The number of free electrons per meter, q, is given by Refs. 29 and 30:

$$q = -\frac{\beta}{\bar{m}v} \frac{dm}{dt} , \tag{4}$$

where $\bar{m}$ is the average mass of ablated meteoroid atoms. β of a particular atom is given by:

$$\beta = 2 \int_{E_{IP}}^{E_T} \frac{\sigma_i(E)dE}{\sigma_D(E)E} , \tag{5}$$

where E_{IP} is the ionization potential of the respective atom, and $\sigma_i(E)$ is the collisional ionization cross section, a laboratory observable (see Sec. 3.4.2).[39] A compilation of observations led to the empirical formula[29]:

$$\beta = 3.02 \times 10^{-17} v^{3.42} , \tag{6}$$

where the units of v are m s^{-1}. Equation (6) predicts that atoms ablated from a 70 km s^{-1} meteoroid ionize with a coefficient larger than 1, signifying that an atom can produce more than one charge pair.

2.2.3. *Ionospheric Chemical and Plasma Kinetics*

The continuous vaporization of cosmic dust in the upper atmosphere leads to permanent metal atom layers at altitudes between 90–100 km. As pointed out in the introduction, the metals play an important role in the lower regions of the ionospheric E-region (see Fig. 2). The first manifestation of upper atmospheric metals was noticed early this century, when coarse spectra of the night sky revealed a line at 589.2 nm. In 1929, Slipher suggested that this emission line was due to sodium.[40] Later studies refined the measurement to 589.3 nm,[41–43] and tne identification of the Na–D emission line was definitely made. Shortly after, the first association of the Na–D line emissions in the nightglow with meteors entering the Earth's atmosphere was made by a number of authors.[44,45] The actual chemical mechanism that can lead to optical emission of the Na–D line was first suggested by Chapman[46]:

$$Na + O_3 \rightarrow NaO + O_2 \tag{7}$$

$$NaO + O \rightarrow Na(^2P_{1/2,3/2}) + O_2 \tag{8}$$

$$Na(^2P_{1/2,3/2}) \rightarrow Na(^2S_{1/2}) + h\upsilon. \tag{9}$$

At that time no data was available for the thermochemical properties of NaO, so the suggestion was based on what turned out to be excellent chemical intuition. The first thermochemical determination of the formation enthalpy of NaO was made in 1970,[47] and it showed that Chapman's mechanism was exquisitely exoergic. Since the early nightglow observations and Chapman's analysis, the number of observations has increased exponentially due to the advent of rocket- and space-borne instrumentation, and the introduction of lidar techniques. These observations will be summarized in the following section.

The metal atoms of the neutral metal layers are subject to charge transfer ionization by the principal molecular ions of the ionospheric E-region, NO^+ and O_2^+. The highly stable atomic metal ions are either transported to higher altitudes, where they can undergo electron–ion recombination, or they can be removed by three-body association reactions with atmospheric molecules at lower altitudes, such as N_2:

$$Me^+ + N_2 + M \rightarrow Me^+(N_2) + M. \tag{10}$$

The product cluster ions can rapidly dissociatively recombine with electrons to produce excited neutrals. Three-body association reactions (10) become increasingly important as one descends from 90 km. Simultaneously, the electron density declines, increasing the lifetimes of the cluster ions. Below 80 km, positive and negative cluster ions become the most abundant charged species.[48]

2.3. *Meteor Field Observations*

Fireballs and shooting star observations have been documented for millennia. According to the review by Kronk,[49] Chladni was the first to conclude that shooting stars came from extraterrestrial sources in 1794. Four years later, two students, Brandes and Benzenberg, determined from triangulation experiments that shooting stars occur at altitudes near 100 km. In the famous Leonid storm of 1833, it was estimated that the zenithal hourly rate was $\sim 150\,000$. Traditional visual observations of meteors are generally limited to visual magnitudes of -5 to $+5$.[33] However, low light TV cameras have extended the observation limit to $\sim +6$ to $+8$.[50] The increased sensitivity has been helpful in observing meteors at higher altitudes. Linblad has reported the sighting of a meteor as high as 138 km.[51]

Conversion of observed intensities into luminous efficiencies Eq. (2) is hampered by the common occurrence of meteoroid fragmentation. Ceplecha *et al.*[30] have compiled a list of analyzed high intensity bolides brighter than magnitude -10. The total luminous efficiencies derived from the analysis of the observations range from 0.2 to 10%. For meteor velocities ≤ 20 km/s, the lumimous efficiency of a meteoritic fireball has been estimated to be in the range 0.04%–0.4% during peak ablation.[52]

Since World War II, the development of sophisticated instrumentation and platforms have dramatically enhanced the field observations of meteors. Spectrographs employing gratings have enabled high resolution spectral analysis of meteoric light, the development of radar techniques have provided for sensitive analysis of meteoric ionization, and the emergence of pulsed laser sources have enabled remote sensing of metal neutral and ion layers. Meanwhile, rocket-launched mass spectrometers have investigated E-region metallic ions, and instruments on orbiting spacecraft provide measurements of ultraviolet emissions of metallic species in the ionosphere that are not observable from the ground. The highlights of these instrumental field observations are briefly discussed below.

2.3.1. *Spectra of Meteors*

Many attempts have been made to obtain spectra of meteor trails. Probably the best spectra ever obtained were by Millman and his colleagues at the Dominion Observatory. Figures 4 and 5 are a comparison of spectra stemming from a slow (probably Giacobinid) and a fast (Perseid) meteor, respectively.[53] The spectrum of the low velocity meteor exhibits neutral emission lines of Mg, Fe, and Ca, the most abundant metals in most meteoroids. The strong Na–D lines as well as a hint at excited N_2 are also visible. By contrast, the spectrum of the high velocity meteor shows very

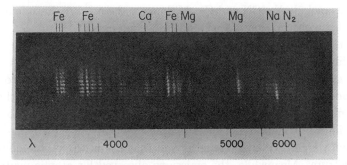

Fig. 4. Photograph taken by Millman showing the spectrum of a slow meteor (average velocity $\sim$ 20 km/s, altitude = 85 km). Note that only the neutral metals are observed. This photograph is shown by permission of the University of Chicago Press and is taken from a paper by Millman.[53]

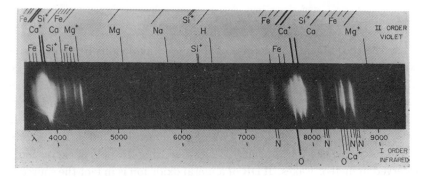

Fig. 5. Photograph taken by Millman showing the spectrum of a fast meteor (Perseid, average velocity of $\sim$ 59 km/s, altitude = 100 km). Note the presence of metallic ions in addition to metallic neutrals, as well as a more complex atmospheric spectrum. This photograph is shown by permission of the University of Chicago Press and is taken from a paper by Millman.[53]

intense emissions of the H and K lines of Ca^+, as well as emissions from Mg^+ in addition to atomic neutrals. Moreover, Fig. 5 shows lines due to emission from atomic N, a species that is not present in the ambient atmosphere at these altitudes. Higher resolution spectra obtained more recently for a Perseid trail clearly show the presence of atmospheric dissociation products.[54] The evidence of ionization and dissociation products is a direct consequence of higher gas-phase collision energies.

Due to the high abundance of Fe in all meteoroids, a large number of Fe lines are observed in all meteor spectra. Molecular metal oxide spectra for FeO, CaO, AlO, and MgO have also been identified.[30] Other observed molecules include N_2, CN and C_2. Millman and coworkers have tentatively reported the observation of O_2^+ and N_2^+ emissions.[55]

Most of the observed luminosity stems from the meteor itself. By using an imager-spectrograph with a shutter, emissions of the much weaker *meteor wake* can be observed. The wake exhibits emissions for up to tenths of seconds after meteor passage.[30] The only ionic lines observed in the wake spectrum are the strong Ca^+ H and K lines. These are the only ionic resonance lines of common meteoritic metal ions that have low lying excited states with high oscillator strengths and that are in the visible domain of the spectrum. Meanwhile, the emissions observed for atomic neutrals do not have excitation energies > 4 eV.[29] This invokes the interesting possibility that the wake spectrum attributable to neutrals is formed in dissociative recombination (DR) reactions of metal oxide ions:

$$MeO^+ + e^- \rightarrow Me^* + O. \tag{11}$$

While the recombination energies of the metal oxide ions lie generally between 7 and 10 eV, the dissociation energies of the neutrals range between 2.7 and 7 eV (see Tables 2 and 3 and associated references). The DR excess energy available for excitation is given by $IP(MeO) - D_0(MeO)$, and ranges from 2.8 to 6.0 eV for relevant metals. Note that 2.8 eV is the value for CaO, and is the only value below 4 eV of those metals listed in Tables 2 and 3. Interestingly, the Ca^+ line at 423 nm, requiring 2.93 eV of excitation, is observed in meteor wakes. If DR of a metal oxide ion is in fact the source for these emissions, it would imply that vibrationally or electronically excited CaO^+ ions lead to the emissions.

In addition to the spectral observations of meteor trails, there have been a number of episodic reports of long lived (10^1 to 10^3 s) persistent trains

Table 2. Thermochemistry of meteoric metal oxides and dioxides.[195]

Metal	IP(eV)	$\Delta_f H^0$(kcal/mol)				
		Me	MeO	MeO$_2$	MeO$^+$	MeO$_2^+$
Mg	7.65	35.3	13.4	–	236	–
Fe	7.87	99	60	18 ± 5^b	265	273^c
Si	8.15	108	−24	–	240	–
Al	5.99	78.2	16	−31	234	200
Ni	7.64	102.8	71	–	290	–
Ca	6.11	42.6	6	–	184	–
Na	5.14	25.6	20	-13 ± 3^d	191	139.5 ± 2^d
K	4.34	21.3	15	–	178	–
Ti	6.82	112	13	−71	164.3	149
Co	7.86	102	72	–	277	–
Zn	9.39	31.2	72.6^a	–	275	–
Mn	7.44	67	41	–	240	–
Cr	6.77	94.8	52	−15	233	223

[a] Reference[135]
[b] References[250]
[c] From bond dissociation energy calculated by Schroeder *et al.*[251]
[d] *Ab initio* calculations by Lee *et al.*[142]

Table 3. Bond dissociation energies determined by Armentrout and coworkers[135] (unless otherwise noted) for meteoric metal oxides and the respective ions.

Metal	D_0^0(MO$^+$) (eV)	D_0^0(MO) (eV)
Mg^a	2.5 ± 0.10	3.72 ± 0.13^b
Fe	3.53 ± 0.06	4.21 ± 0.09
Al^c	1.50 ± 0.12	5.26 ± 0.04^d
Ni	2.78 ± 0.07	3.91 ± 0.17
Ca	3.57 ± 0.05	4.12 ± 0.17
Ti	6.93 ± 0.10	6.92 ± 0.10
Co	3.29 ± 0.06	3.94 ± 0.14
Zn	1.65 ± 0.12	< 2.77
Mn	2.95 ± 0.13	3.82 ± 0.08
Cr	3.72 ± 0.12	4.41 ± 0.30

[a] Reference[192]
[b] Reference[252]
[c] Reference[194]
[d] Reference[253]

associated with a meteor. Baggaley has discussed in detail the persistent red afterglow reported by many observers.[56] Likewise, green emission at 557.7 nm from the $O(^1S \rightarrow {}^1D)$ transition has been analyzed.[56,57] In both of these emissions, the proposed mechanisms involve ion chemistry. Baggaley suggested that fast atomic oxygen released by a meteoroid is ionized in collisions with atmospheric constituents. O^+ then undergoes charge transfer with O_2 which is followed by DR:

$$O_2{}^+ + e^- \rightarrow O({}^1S, {}^1D) + O. \tag{12}$$

Reaction (12) is also the primary source of atomic oxygen emissions in Earth's day and nightglow spectra.

2.3.2. *Radar Observations of Ionization Trails*

In contrast with optical meteor observations, radar measurements can be conducted both in day and night skies. Radar measurements are based on the principles of radiowave scattering by electrons. The scatter cross section is directly related to the electron density. One of the earlier uses of radar technology following the end of World War II was the study of meteor *ionization trails*. In so called underdense trails, corresponding to line densities below 10^{13} electrons per meter, the radiowave is not significantly attenuated by the meteor, and the observed radar echo can be considered to be the sum of contributions from individual electrons. At line densities above $\sim 10^{15}$ electrons per meter, the radar no longer penetrates the trails, and the reflection mimics that of a solid metal surface. The observed echo intensities of underdense and overdense trails depend on different physics (see review of Ceplecha *et al.*[30]) from which various meteoric parameters are derived. Of particular interest is the monitoring of the depletion processes of meteoric ionization. These processes can change dramatically around 90 km, i.e from ambipolar diffusion at higher altitudes (> 90 km) to metal atom ion chemistry at lower altitudes leading to the formation of metal oxide ions that undergo rapid DR.[58] This will be discussed further in Sec. 3.2.1.

Radar measurements enable the precise determination of meteor velocities as well as their radiants. An excellent discussion of the early uses of the technique as well as a summary of the earlier studies can be found in the complilation by Lovell.[28] As summarized in that compendium, the early work established beyond doubt that meteor showers lead to an enhancement of electron densities at night. Recent improvements in the technique have greatly enhanced the sensitivity; it is now estimated that the EISCAT (European Incoherent Scatter) radar facility can track meteors corresponding to a visual magnitude of $+10$.[59] Even more impressive are the results from the special incoherent scatter radar (ISR) at Arecibo, where meteors with visual magnitudes of $+16$ have been reported.[60] This sensitivity to smaller meteoroids will substantially improve estimates of the total extraterrestrial mass flux to Earth.

From an analysis of 6000 meteors by radio echo techniques, Verniani[61] concluded that the mean velocity was 34 km/s, and that their mean mass was $\sim 10^{-4}$ g. Verniani also found that the velocity distribution of the meteors shifted with meteor mass distribution. The physics of radar scatter from meteor trails extracts the ionization coefficient, Eq. (5), from the observed echo intensities and associated electron densities. More recent analysis of the ionization trails of meteors[62] has indicated a slightly different expression than Eq. (6) for the ionization coefficient associated with faint, low velocity (< 35 km s^{-1}), radio meteors, namely $\beta = 9.4 \times 10^{-6}$ $(v-10)^2 v^{2.8}$, where v is in km s^{-1}. The magnitude and velocity dependence of β remains a major puzzle. It must be noted that the calculations assume no chemical reactions in the trail; this assumption is needed to make the calculations tractable and because many of the reaction cross sections are not known. Collision processes leading to ionization are discussed in Secs. 3.2.2 and 3.4.

2.3.3. *Airglow and Lidar Observations of Steady State Metal Layers*

Optical studies of steady state metal layers generally pertain to observations of the meteor metals in the altitude regime 80–100 km, where the densities are highest. Satellite and Space Shuttle measurements, however, have yielded airglow data at altitudes as high as 300 km. While the *dayglow*, best observed from the ground in twilight, consists of solar-induced fluorescence of atomic emission lines, the *nightglow* results from chemiluminescent processes such as the one producing the Na–D-line emissions in the Chapman mechanism discussed in Sec. 2.2.3. With the advent of increasingly sensitive instrumentation, it became possible to study optical spectra of airglow metals.[63-66] The Earth's atmosphere limits transmission of radiation to the near-ultraviolet, visible range and to the infrared, the latter in specific bands. For meteor metals, this limitation means that only those atoms with lines in the spectral region 325–850 nm are readily observed from the ground. Thus ground based airglow measurements have been dominated by the study of the sodium dayglow and nightglow, as described in the aforementioned reviews. A worthy complement to the study of sodium has been the dayglow measurement of emissions from Ca^+ at $\lambda = 393.4$ nm[67] and from Fe at $\lambda = 386$ nm.[68]

The development of rockets following World War II made it possible to deploy spectrographs and mass spectrometers beyond the lower atmosphere (i.e. above ~ 70 km), thereby circumventing the atmospheric transmission limitations. Using a rocket deployed spectrograph, Anderson and Barth[69] obtained the first dayglow data for the intense resonance lines of Mg ($\lambda = 285.2$ nm) and Mg^+ (doublet at $\lambda = 280$). Rocket borne measurements are limited by shortness of flight times, specificity of location, season, and altitude, and rocket induced disturbances of the local environment. Spacecraft in low earth orbit (altitudes between 200 and 1000 km) remove some of these restrictions. An example can be seen in Mg^+ and Fe^+ probes aboard the TD-1 satellite[7,70] that showed Mg^+ extending above 540 km altitude. This was surprising because Mg^+ is expected to recombine with an electron during the time it takes it to migrate to the respective higher altitudes from the ionospheric E-region. Later, measurements aboard the OGO-4[8] and Atmospheric Explorer-E[71] satellites confirmed the earlier observations and showed an asymmetry in latitudinal mapping. The data from the AE-E observations indicated that at 150–200 km, the Mg^+ density is approximately 100 ions cm^{-3} in the afternoon.[71] More recent measurements aboard the Nimbus satellite have provided more detailed information on the magnesium ion distributions, in particular information about the seasonal and geomagnetic dependence of the Mg^+ abundance.[72] These observations eventually led to a good understanding of the interplay between geophysical parameters (electric and magnetic fields, neutral winds) affecting the transport of meteor metal ions to high latitudes and into structured regions.[10,73,74]

Space Shuttle flights also brought new dimensions to the observations of the metals, especially to the study of Mg^+.[75] A typical spectrum of the thermospheric dayglow observed from the Shuttle is shown in Fig. 6 and shows the prominence of the Mg^+ resonance line as well as the lines of other metals. The spectrograph was aimed at a tangent height of 136 km. Particularly exciting and important for the development of adequate models has been the recent simultaneous observations of Mg and Mg^+.[76–78] These

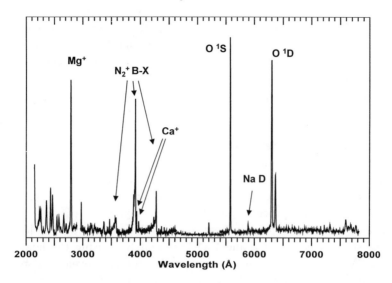

Fig. 6. Dayglow spectrum observed for a tangent height of 136 km on STS-53 using the GLO1 experiment.[76]

latter measurements have shown the presence of columns of Mg^+ that extend from ~ 100 km to the Space Shuttle altitude, ~ 350 km. The columns seem to have dimensions of ~ 200 km diameter. There is no adequate explanation for the mechanism for generating these columns at this time.

In lidar experiments, specific atmospheric species are probed actively by monitoring the time-resolved fluorescence return signal induced by a monochromatic pulsed laser source tuned to the resonance line of the atom of interest. The intensity of the light echo carries number density information, while the echo time is converted into the altitude of the emitter. The high sensitivity of the lidar technique (< 1 atom cm^{-3}) permits monitoring of the steady state metal layers with high time resolution. Early lidar measurements of the neutral sodium layer revealed variations of the altitude density profiles attributable to gravity waves,[79] zonal and meridional winds, sporadic-E events,[80] and horizontal winds temperature.[81] These effects have been studied by several investigators and many reviews have been published.[82-86] Perhaps the most stunning feature in the lidar observations of the sodium layer is the sudden appearance of thin layers (FWHM ~ 1 km) with densities that are up to an order of magnitude higher than the normal quiescent densities.[87,88] The layers can appear within minutes, last for several hours, before vanishing in a matter of minutes. The features of the layers are similar to the sporadic E layers that form when the plasma dynamics and/or neutral wind shear around the homopause compress metal ions into thin layers of enhanced density. The features of sporadic E, observed with radar, and sporadic metal atom layers are highly latitude and season dependent. The broad horizontal distribution of the layers rules out direct deposition by meteor events. In several instances, the neutral and ionic sporadic layers have been linked.[76,89] This has prompted the postulation of an ion–molecule mechanism that links a descending sporadic metal ion layer via association reactions (10), which become faster at lower altitudes, and rapid DR to a sporadic neutral layer.[90] There are still many unanswered questions with respect to sporadic layers.

The advent of new intense laser sources allowed probing of additional metals, such as Fe,[91,92] Ca and Ca$^+$,[92,93] and K,[94] using the lidar techniques. All of the investigated layers exhibit sporadic events. Von Zahn and coworkers have improved the time resolution of their experiments to 1 s, allowing them to observe individual meteor trails and the associated local metal atom density increases.[95]

2.3.4. *Rocket- and Space-Borne Mass Spectrometer Measurements*

Following World War II, the limitations of the atmosphere became surmountable in that rocket-borne spectrometers (optical- and mass-) could be launched into the upper atmosphere. The first mass spectrometric observations of the ion composition of the D-region of the ionosphere were carried out by Johnson *et al.*[96,97] who flew a Bennett mass spectrometer. Following these studies, Istomin and Pokhunkov,[98] also using a Bennett-type mass spectrometer, obtained definite evidence for the presence of Fe$^+$ and Ca$^+$ metal ions at an altitude of 105 km. Following these measurements, Narcisi and his colleagues[99,100] obtained very detailed information about the altitude profiles of the various metal ions. These latter studies were facilitated by the use of a miniature quadrupole mass filter with increased sensitivity. Subsequently, a number of mass spectrometric identifications of metal ions in the upper atmosphere have been achieved.[101–104] A correlation of the metal ion layer with the Perseid meteor shower was also demonstrated.[105,106]

A very important contribution to understanding the dynamics of metals in the D- and E-regions of the ionosphere has been the near-simultaneous measurement of Fe$^+$ and Fe.[89] The neutral layer was monitored using lidar and a rocket with mass spectrometer was launched at the onset of a sporadic Fe layer. While the steady state Fe layer peaks at $\sim$ 85 km, the sporadic neutral and ion layer peaked around 98 km. The ratio of densities, [Fe$^+$]/[Fe], at the peak of the sporadic layer was determined to be 1.75. This is consistent with a model in which the neutral layer is caused by the ionic layer, possibly through an ion–molecule mechanism such as that proposed by Cox and Plane.[90] Interestingly, in the peak of the ion layer, the density ratio [Mg$^+$]/[Fe$^+$] was 4.2, despite the comparable abundances of the two metals in typical chondritic meteorites. This could be the direct consequence of substantially higher ion–metal charge transfer cross sections for Mg. Indeed, the measurements of Rutherford and coworkers, to be discussed in Sec. 3.3.3, show that the combined charge transfer reactions with O$_2{}^+$ and NO$^+$ are approximately 5 times larger for Mg than for Fe.[107,108]

Hanson and coworkers used a retarding potential analyzer aboard the OGO-6 satellite to look at ion distributions in the ionospheric F-region. They found that Fe$^+$ is a minor ionic constituent at 492 km (a few one-hundredths of a percent of the major ambient ion, O$^+$, measured at a density of $\sim$ 105 cm^{-3}).[2] In a later study using the same technique, Fe$^+$ was observed at altitudes as high as 1000 km, with densities near 100 cm^{-3}. An extremely high density of 2000 cm^{-3} was also reported.[3,4] A Bennett ion mass spectrometer aboard the *Atmospheric Explorer-C* satellite confirmed the presence of Fe$^+$ at altitudes between 220 and 320 km as well as at high latitudes (near 60°).[5] These ions were found to be present in patches at altitudes between 250 and 600 km, with densities as high as 10^4 cm^{-3}.[109] This work not only clearly demonstrated the drift of E-region ions towards higher altitudes, but to higher latitudes as well. These observations were explained by the fountain effect, where $\mathbf{E} \times \mathbf{B}$ drift pulls the ions upwards in the mid latitudes followed by neutral wind transport towards high latitudes. This dynamo effect will be further addressed in the following section.[b]

2.4. *Ionospheric Models*

A key goal of ionospheric modeling efforts is to determine from meteor observations the influx of extraterrestrial matter, and how the matter that is deposited between altitudes of 80 and 120 km couples to ionospheric disturbances such as sporadic E layers that seriously affect communications

---

[b]Meanwhile, patch structure has been found at altitudes as low as 100 km using ground-based radar observations.[110]

to and from space. The total mass flux and the mass distribution of incoming extraterrestrial particles is still not fully established. It is also unsure whether larger meteor storms, while they dramatically enhance the visual rate of meteors, have a serious impact on the daily global mass influx, particularly in the range of particle sizes that vaporize in the lower ionosphere. If the impact of meteor storms is severe, it is important to be able to predict the time evolution and intensity of the resulting ionospheric disturbances.

The development of models that can quantitatively explain the sodium nightglow, in particular, and the other meteor metal species, in general, proceeded along several fronts beginning with the pioneering work of Öpik.[35] Starting from first principles, Öpik treated the structure of a meteoroid, its heating by the Earth's atmosphere upon entry, and its subsequent degree of vaporization. He introduced the concept of dustballs, namely that meteoroids consist of solid oxides and silicates held together by ice. Upon entry into the Earth's atmosphere, the binder (water) holding the dust balls together vaporizes first, followed by the vaporization of the individual dust kernels. This resulted in a coherent picture of most of the observations associated with entry of meteoroids into the Earth's atmosphere. More quantitative analysis confirmed the physical picture provided by Öpik for the vaporization of meteoroids.[111,112]

The end result of the Chapman mechanism, and of the modified Chapman processes as elucidated by later investigators[113–115] is that sodium and other metals (if Chapman-like mechanisms hold for the other metals) are abundant in atomic form. The kinetic studies show that molecular compounds are not present in large quantities above ~ 85 km. Further work[36,116] showed that following the formation of dense atomic trails during the vaporization process, molecular recombination in the wake occurs to form smoke or dust that can then act as a delayed source of sodium (and other) atoms. It was then suggested that NaO reacts with atmospheric H_2O to form gaseous NaOH, with the latter reacting with atmospheric CO_2 in a three-body reaction to form $NaHCO_3$[117]:

$$NaO + H_2O \rightarrow NaOH + OH \tag{13}$$

$$NaOH + CO_2 + M \rightarrow NaHCO_3 + M. \tag{14}$$

Later model calculations[118–121] and laboratory measurements[122,123] confirmed the $NaHCO_3$ hypothesis. It can thus be assumed that $NaHCO_3$ is the lower altitude end product.

Modeling of the morphology of the other meteor metals is not as well developed as that of sodium. The kinetic measurements made by Plane and coworkers[120,121,124] have shown that there are viable chemical mechanisms that can lead to the eventual removal of the other metal atoms once they are deposited in the atmosphere. The modeling of the other metals is hampered by reduced availability of atmospheric data in comparison with sodium. Nevertheless, early attempts to understand the nature of high altitude metal–ion layers (described above), principally Fe^+ and Ca^+, led to the development of a model in which ions formed by charge exchange at ~ 100 km are pumped upwards by the equatorial fountain effect and are then released into higher altitudes from which they can be deposited into the mid latitude regions.[2-4] A three-dimensional code that included $\mathbf{E} \times \mathbf{B}$ drift and winds was found to yield agreement with the observations provided that the initial conditions consisted of ~ 100 ions cm^{-3} at ~ 100 km.[10,71,74] A less empirical one-dimensional model that did not include winds also provided agreement with measured densities.[125]

More recently, two important studies have been published that, in time, should provide good models for the metals in the upper atmosphere.[126,127] In the first case, a 1-dimensional model was developed, where a deposition profile and a meteoroid size distribution is assumed (based on various astronomical observations) followed by attempts to directly derive deposition profiles for the various metals based on: (1) the frictional heating and slowing down of the meteoroids as they enter the atmosphere; and (2) on the thermochemical properties of the silicates.[126] This treatment included a comprehensive chemical reaction scheme and averaged electric fields, however, it neglected the effect of winds (both zonal and meridional) which are generally of horizontal nature. The profiles obtained for Ca, Mg, and Na agree reasonably well with the observed abundances. This study highlighted the fractionation that takes place when the meteoroids are heated upon entry into the Earth's atmosphere. A direct result of fractionation is that the low boiling element Na is deposited at a higher altitude than the more refractory metals Ca and Mg. The model can then include specifically the chemical reactions for both neutral and ionized Mg to obtain altitude profiles. An earlier version gave very good agreement with the observations.[125]

A current limitation of the model is the omission of hyperthermal collision effects, particularly with respect to the formation of ions. Another serious disadvantage is the exclusion of winds, with the result that the global transport and distribution of the metals is not understood within the framework of this model. However, global transport is not an issue in the analysis of meteor trail data due to the much shorter associated time scales. The model was, therefore, successfully applied to the analysis of potassium lidar data obtained during the 1996 Leonid shower.[95] Through the use of parametric calculations, a very good estimate of the meteoroid sizes that can give rise to the observed trails was obtained.

The second study[127] by Carter and Forbes is more detailed in that it considers the modeling within the framework of a global circulation model; thus the role of winds is explicitly investigated. Both zonal (longitudinal direction) and meridional (latitudinal direction) winds[128] are found to have important effects on the global transport of the metals. In the model of Carter and Forbes, the method of finite elements is used to solve the coupled continuity equations for Fe and Fe^+. The source function for Fe neutrals and ions is a crucial element in the analysis. It is based on an influx of 200 tons of extraterrestrial dust per day over the Earth's surface. 11.5% of this influx is attributed to Fe, and of this, 5% is assumed to be released in atomic vapor state. The chemical processes included in this model are those that affect the ionic component. It is assumed that below 85 km, the neutral chemistry to form metal compounds is a perfect sink for Fe. Thus, the neutral reactions are not needed in the model because it emphasizes the high altitude composition, which is dominated by ion chemistry. This model equally does not include hyperthermal chemical reactions that take place when evaporation occurs at the initial stages of entry into the atmosphere. Nevertheless, even within these limitations, the profiles obtained for altitudes between 90 and 250 km agree very well with observations.

3. Chemical Dynamics

3.1. *Overview*

In this section, the chemical dynamics of meteor events are examined in greater detail. The present discussion addresses those processes that play an important role in the observations treated earlier. As iterated above, due to their high reactivity, high luminosity and low ionization potentials, the collision dynamics of metal atoms and ions play a prominent role in the ionospheric meteor phenomenology, and are therefore the focus of this section. There is a considerable body of literature on gas-phase metal chemistry in thermal equilibrium. During the past 15 years, these thermally equilibrated studies have resulted in significant gas-kinetic models capable of explaining the observed properties of permanent metal layers on a gas-kinetic level. The large amount of experimental kinetic data has also permitted the generation of predictive empirical models for activation barriers based on known metal and neutral physical constants,[129] or the testing of more elaborate statistical approaches such as that provided by the Troe-formalism.[130,131] Given these inputs, the modeling of the steady state, quiescent upper atmospheric metal chemistry has enjoyed considerable success. A noteworthy achievement is the previously discussed sporadic sodium layer model recently put forward by Cox and Plane[90] that is based on experimental ion–molecule rate coefficient data.

Due to the present focus on extreme environments, however, the extensive research on thermally equilibrated metal chemistry will only be mentioned in cases where this work provides a baseline to understanding the hyperthermal collision dynamics that occur during and immediately after a meteor event. For a more comprehensive treatment of upper atmospheric metal chemistry under thermalized conditions, the reader is referred to numerous reviews in the book of Fontijn,[132] as well as to review articles by Plane,[118,119] Brown,[133] and Ferguson.[134] Efforts to model meteor luminosity and ionization efficiencies, as well as metal ion densities and emissions at altitudes exceeding 100 km, are far more challenging. The calculation of these quantitative processes relies on knowledge of the dynamics occurring over a wide range of collision energies at conditions far from equilibrium. A lack of integral and differential cross section data for metal atom and ion collisions with atmospheric species at relative velocities as high as meteoroid atmospheric entry velocities handicaps many modeling efforts. In contrast to the thermally equilibrated studies, concerted experimental and theoretical efforts on the collision dynamics of metals and metal ions covering a collision energy range of 1 to 1000 eV (corresponding to relative velocities of $\sim$ 3–100 km s^{-1}) are notably sparse. When multiple studies can be compared, there are frequently order of magnitude differences in results. Probably the most noteworthy body of work stems from the extended metal–ion studies of Armentrout and coworkers. Their ion beam work has been mostly motivated by the need for gas-phase thermochemical information on neutral and ionic metal compounds. This information can be obtained from the reaction thresholds observed in the translational energy dependence of cross sections for endothermic processes. While these studies have normally been carried out at collision energies below 20 eV, respective reactions like $M^+ + O_2 \rightarrow MO^+ + O$ (M = Ca through Zn)[135] can represent important sinks for atomic metal ions. This is particularly true at lower energies (and altitudes), where charge transfer between metal ions and more abundant atmospheric species is improbable. Furthermore, the information gained by Armentrout's group on the chemical bonding and orbital correlations of molecular and ionic metal species is invaluable for the modeling of higher energy metal dynamics and for the validation of high level *ab initio* calculations of the low lying potential energy surfaces.

As seen in most endothermic chemical reactions leading to molecular products (see for example the metal ion studies of Fisher *et al.*[135]), as the collision energy is raised far above threshold, the cross sections eventually decline. This can be attributed to the increasing internal excitation of the molecular reaction products, eventually leading to dissociative decay. If applied to the previously noted metal ion–oxygen reactions, this amounts to inelastic metal ion scattering, and, therefore, no change in the identity of the charged particle. Thus, as the collision energy approaches 100 eV, chemical reactions, consisting of breaking and establishing new chemical bonds to form new molecular species, are no longer important processes. Instead, electronically inelastic collisions involving nonadiabatic transitions to excited potential energy surfaces of intermediates become probable, leading to electronically excited atomic products that can possibly luminesce, or are able to charge transfer, both processes that do not necessarily require penetration to the repulsive potential. Predictions of cross sections require precise knowledge of the electronic structure of the collision complex at high levels of excitation and over broad ranges of the nuclear coordinate. Even for diatomic systems, there are few examples where this information is available.

Experimental hyperthermal ion–neutral collision studies are carried out in ion beam experiments. While the generation of well-characterized ion beams can be considered standard technology, this cannot be said of the generation of high energy neutral beams. There are few reliable measurements of hyperthermal metal + neutral collision cross sections. This is particularly problematic with respect to modeling meteor ionization and luminosity, where hypervelocity collisions between ablated metal atoms and atmospheric neutrals are the excitation process. Most of the work on fast neutral beams dates back to the 1960s and early 1970s, during a period where the Defense Nuclear Agency generously supported high energy gas-kinetics in an attempt to understand the effects of nuclear explosions at higher altitudes. Following the change in NATO defense strategy and the associated cuts in support of this type of research, the physics community has diverted its attention away from hyperthermal neutral beam dynamics. The more recent discovery of the Shuttle-glow phenomenon[136] (see Chaps. 9 and 7 by Minton and Garton, and Jacobs in this volume, respectively) has somewhat revived the interest in fast neutral beam experiments, in particular 5 eV atomic oxygen beam experiments.

In Sec. 3.2, we will discuss the dynamics of hyperthermal collisions involving metal species. Section 3.3 is a review of ion–neutral collision dynamics involving metallic species, and Sec. 3.4 examines what is known of hyperthermal metal atom collisions with atmospheric species.

3.2. *Hyperthermal Collision Processes*

3.2.1. *Reactive Collisions*

At collision energies below 20 eV, reactive collisions leading to metal compounds or metal compound ions may play an important role. Such processes represent potential sources and sinks for metal atoms or ions. By far the most important reactive processes to be considered in rarefied (only two-body processes) upper atmospheric conditions at hyperthermal energies are collisions involving molecular and atomic oxygen:

$$Me + O_2 \quad \rightarrow MeO + O \tag{15}$$

$$\rightarrow MeO_2^+ + e^- \tag{16}$$

$$\rightarrow MeO^+ + (O + e) \tag{17}$$

$$Me + O \quad \rightarrow MeO^+ + e^- \tag{18}$$

$$Me^+ + O_2 \quad \rightarrow MeO^+ + O. \tag{19}$$

Table 2 lists the thermochemistry of those metal oxides that are relevant to meteor studies. Reaction (15) is the only process above that does not involve ionization. Bands of MgO, FeO, CaO, and AlO have been identified in the spectra of bolides (the larger bodies that enter the atmosphere and glow bright white, reaching temperatures as high as 4000 K).[30] It is not clear whether hyperthermal processes, Eq. (15), or direct ablation of the oxides lead to the emissions. Another possibility is the reaction of weakly bound metal compounds, such as Al_2, with atomic oxygen to form metal monoxides with substantial excess energy. Such processes have been suggested to be the source of intense AlO chemiluminescence observed upon release of aluminum vapor or volatile aluminum compounds such as trimethylaluminum (TMA) in rocket experiments.[137]

It is seen in Table 2, that only for Al and Si does the stability of the oxide with respect to the gaseous metal exceed the formation enthalpy of an oxygen atom (59.6 kcal/mol). Only the $Al + O_2$ and $Si + O_2$ systems are, therefore, exothermic, and exhibit sizeable rate coefficients at thermal energies.[138,139] Except for rate coefficient measurements at high temperatures,[138] there are essentially no dynamics studies of reaction (15) at hyperthermal energies. This can be attributed to sensitivity problems when detecting quantitatively neutral molecular products in hyperthermal neutral beam experiments (see Sec. 3.4). Metal atom reactions with O_3, as seen in the Chapman mechanism, Eqs. (7)–(9), are generally exothermic, however, and represent a viable source of the observed oxides, as well, particularly at lower altitudes.

The chemiionization processes, Eqs. (16)–(18), are endothermic for meteoric metals. It is worth noting, however, that some heavy metals, such as U, Ta, and Th, form exceptionally stable MeO_2^+ ions, resulting in exothermic associative ionization reactions (16).[140] Reactive ionization (17) can potentially include an ion-pair formation channel, $MeO^+ + O^-$. This process exhibits a thermodynamic threshold that lies 1.46 eV (corresponding to the electron affinity of O) below that for forming a free electron. Interestingly, the hyperthermal studies that have so far detected MeO^+ channels observe a threshold corresponding to free electron formation, $MeO^+ + O + e^-$. This implies that reactive ionization occurs preferentially via autoionization to a MeO_2^+ surface followed by dissociation.[141] Of the primary meteoric metals, only Al, Fe, Ti, and Cr have been found experimentally to form stable dioxide ions. Recently, Lee *et al.* have calculated NaO_2^+ to be bound by 0.2 eV.[142] $Al + O_2$ scattering studies by Cohen and coworkers,[143] however, have failed to observe ionized products aside from Al^+ while minor TiO_2^+ signals were observed in corresponding $Ti + O_2$ studies.

The alkaline earth metals and transition metals form stable oxide ions, while oxide ions of alkali metals are very weakly bound, and their observation has thus been somewhat controversial. Hildebrand and Murad[47] derived an upper limit of 0.5 eV for $D_0^0(Na^+-O)$, while Rol *et al.*

estimated this bond energy to be 0.8 ± 0.3 eV[144]; by comparison O'Hare and Wahl,[145] and very recently Soldan *et al.*[146] calculated this bond energy to be 0.05 eV and 0.29 ± 0.02 eV, respectively. For most metal oxide ions of relevance to meteor studies, the bond energy is less than the O_2 bond energy, $D_0(O_2) = 5.16$ eV, and reactions (19) are, therefore, endothermic reactions. Consequently, these reactions occur only at hyperthermal collision energies above thresholds of a few electron volts corresponding to $D_0(O_2)-D_0(MeO^+)$. The cross section maxima are generally less than 3 Å^2.[135] It is worth noting that Ti and the fourth-period metals Sc and V form oxide ions with bond energies exceeding $D_0(O_2)$.[135] For these metals, reaction (19) is exothermic and close to collisional cross sections are found at near-thermal collision energies.[135]

Given the relatively low cross sections of reactions (19), exothermic reactions with trace species such as ozone:

$$Me^+ + O_3 \rightarrow MeO^+ + O_2 \tag{20}$$

become more important at lower altitudes. The respective Mg^+, Ca^+ and Fe^+ reactions have been observed to proceed at near collisional rates at thermal energies.[147] It is believed that metal–ion ozone reactions, and the short lifetimes of the product metal oxide ions with respect to dissociative recombination, are responsible for reducing the duration of overdense ionization trail radar-echoes below 90 km.[58]

Ion–metal atom reactions involving the E-region ion, O_2^+, can also lead to oxide ions:

$$O_2^+ + Me \rightarrow MeO^+ + O. \tag{21}$$

Since the dissociation energy of O_2^+, 6.16 eV, is higher than that of O_2, this reaction is equally or less probable than reaction (19). To date, it has only been observed by Rol and Entemann[148] in the reactions of $O_2^+ + Na$. The former merged beam study, however, is questionable given the failure to observe NaO^+ in a number of sodium studies based on reactions (19) and (20)[147,149,150] as well as in high temperature mass spectrometry measurements.[47] Lo *et al.*[149] have also reported CaO^+ formation in thermal $NO^+ + Ca$ collisions. Given the known thermochemistry of CaO^+ (see Table 3), however, it must be concluded that this observation is an experimental artifact.

At altitudes above 90 km, metal oxide ions are subject to fast dissociative electron–ion recombination reactions. At altitudes below 90 km, the reaction

$$MeO^+ + O \rightarrow Me^+ + O_2 , \qquad (22)$$

can effectively compete with dissociative recombination rates provided they proceed with near collisional rate coefficients. No rate coefficients or cross sections have been measured for reactions of type (22).

3.2.2. *Electronically Inelastic Processes*

3.2.2.1. *Overview*

At low collision energies, reactive processes apart from chemiionization reactions can usually be regarded as proceeding on a single potential energy surface. As the collision energy is increased above 1 eV, however, models involving a single potential energy surface eventually break down. This is especially the case in metal atom collisions where many low lying electronic states exist, and for which the ionization continuum becomes accessible at low collision energies. The elevated radial velocities at hyperthermal collision energies can induce nonadiabatic transitions between potential energy surfaces, and autoionizing regions of the potential energy surface can be accessed in low impact parameter collisions. Following electronically inelastic processes, sorted by collision pairs involving metals, Me, and target species, n, can play an important part in the collision dynamics at sub-keV energies:

$$Me + n \quad \rightarrow Me^* + n^* \qquad (23)$$

$$\rightarrow Me^+ + (n + e^-) \qquad (24)$$

$$\rightarrow n^+ + (Me + e^-) \qquad (25)$$

$$Me^+ + n \quad \rightarrow Me^{+*} + n^* \qquad (26)$$

$$\rightarrow Me + n^+ \qquad (27)$$

$$\rightarrow Me^+ + n^+ + e^- \qquad (28)$$

$$n^+ + Me \quad \rightarrow n^{+*} + Me^* \qquad (29)$$

$$\rightarrow n + Me^+ \qquad (30)$$

$$\rightarrow n^+ + Me^+ + e^- . \qquad (31)$$

An asterisk indicates potential electronic excitation, which may include dissociation in the case of molecules. Asterisks could have also been applied to the charge transfer and ionizing collisions, but are left out for clarity. The processes can be summarized as excitation processes (23, 26, 29), charge transfer processes (27, 30, and negative ion formation processes, 24 and 25), and collisional ionization (28, 31, and free electron formation processes 24 and 25).

3.2.2.2. *Atomic processes and semiquantitative predictions thereof*

To illustrate the mechanisms leading to fast neutral collision channels 23–25, representative potentials of a collision between Na and O atoms are shown in Fig. 7. Even for this simple diatomic system, the dynamics leading to ionization or excitation involve multiple asymptotic limits, all correlating to a number of molecular states for which accurate potential curves are necessary. For this particular collision system, only the X, A and C doublet states have been calculated at a high level. The respective curves in Fig. 7 are the CASSCF/MRCI calculations by Langhoff *et al.*[151]

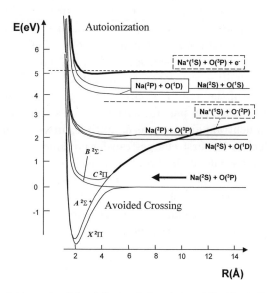

Fig. 7. Representative potentials of the low lying states of the Na + O collision system. The *X*, *A*, *B*, and *C* states are curves calculated by Langhoff and coworkers.[151] The others are estimated curves.

The other curves are speculations based on a more thorough study of the K + O system.[151] The curves representing neutral excitation channels are in reality split into multiple components, including spin-orbit states.

The $Na(^2S_g) + O(^3P_g)$ reactants give rise to 4 states, $^2\Pi$, $^2\Sigma^-$, $^4\Pi$, and $^4\Sigma^-$. The latter two are presumably repulsive and not shown in Fig. 7 for clarity. The ion pair $Na^+(^1S_g) + O^-(^2P_u)$ asymptote lies only 3.67. eV above the neutral reactants. Consequently, the respective Coulomb potential crosses the reactant potential at the relatively large interatomic distance of ~ 4 Å. The ion pair gives rise to $^2\Pi$ and $^2\Sigma^+$ states. The covalent and ionic $^2\Pi$ states undergo an exchange interaction resulting in the avoided crossing seen in Fig. 7. The $X^2\Pi$ state, therefore, has an ion-pair configuration at the equilibrium distance of 2.06 Å, but correlates adiabatically to the covalent $Na(^2S_g) + O(^3P_g)$ asymptote.

The Coulombic potential also crosses the $Na(^2P_u) + O(^3P_g)$ and $Na(^2S_g) + O(^1D_g)$ excitation channels [(process (23)] at long range. The charge transfer state, thus, represents a conduit to these excitation channels. Note that the two remaining excitation channels below the $Na^+ + O + e^-$ continuum are not accessed via the $Na^+ + O^-$ curve. These states must be accessed via crossings in the repulsive region of the interaction potential or through a Demkov-type mechanism discussed later. As will be further discussed in Sec. 3.4, experiments on alkali + O_2 collisions provide evidence for alkali excitation via ion pair states. This is not the case in the extensively studied alkali + halogen systems, where the substantially higher electron affinities of halogen atoms causes the ion pair asymptotes to lie below those of the first ^2P excited states of the alkalis.[152,153]

We can thus regard ion pair formation [process (24)] and excitation to the first excited neutral pair states as a multiple curve-crossing process, the gateway to which is the first charge-transfer crossing at ~ 4 Å. The accurate adiabatic potentials of the avoided crossing enable the application of semiclassical Landau–Zener[154,155] (LZ) type two-state curve-crossing models to estimate transition probabilities from the $X^2\Pi$ to the $C^2\Pi$ state. Approximate cross sections can then be calculated in an impact parameter approach. The LZ transition probability is given by:

$$P_{LZ} = \exp\left[-2\frac{\pi H_{12}^2}{\hbar \dot{R} \Delta F}\right], \tag{32}$$

where H_{12} is the electronic coupling between diabatic states, ΔF is the difference of the slopes of the diabatic curves at the crossing point, R_c, and $\dot{R}$

is the relative velocity along the interatomic coordinate R and R_c. The parameters in Eq. (32) can be obtained from the adiabat potentials, provided a reasonable description of the diabatic potentials exists. Since the chemical interaction between the reactants is negligible at $R_c \approx 4$ Å, a first approximation of the diabatic potentials near the crossing is given by a very weak long range covalent interaction (i.e. $\partial V(R)/\partial R \approx 0$) and the Coulombic ion pair potential. The energy gap between adiabatic states at R_c then represents $2H_{12}$. This approach is validated by the fact that the calculated adiabatic potentials lie equally above and below the asymptotic reactant potential at $R_c = 3.95$ eV by a value of 0.39 eV corresponding to H_{12}.

In case of the crossings leading to excitation channels, no accurate calculations of the couplings are available. The calculation of excitation cross sections, therefore, depends on empirical estimates of the respective couplings. Olson, Smith and Bauer[156] have derived the following semiempirical expression (in atomic units) for couplings based on a large number of measured and calculated values:

$$H_{12}^* = R_c^* \exp(-0.86 R_c^*) \,, \tag{33}$$

where H_{12}^* and R_c^* are the reduced couplings and crossing radii, respectively, and are given by:

$$H_{12}^* = \frac{H_{12}}{\sqrt{I_1 I_2}} \tag{34}$$

$$R_c^* = (\alpha + \gamma) R_c / 2 \,, \tag{35}$$

where $I_1 = \alpha^2/2$ and $I_2 = \gamma^2/2$ are the effective ionization energies of those reactant and product states from which an electron is removed. Although very helpful in assessing cross sections of excitation processes, an assessment of the accuracy of Eq. (33) is provided by comparing the Na + O couplings at 3.95 Å determined with the semiempirical expression (33), 0.47 eV, to the high level *ab initio* value of 0.39 eV. Equation (33) provides couplings of 0.017 and 0.034 eV for the $O(^1D) + Na(^2S)$ and $Na(^2P) + O(^3P)$ crossings with the ion pair potential.

Figure 8 plots the relative velocity dependence of Na + O excitation cross sections of the 3 lowest channels shown in Fig. 7. The cross sections are calculated using simple classical straight line trajectories in which

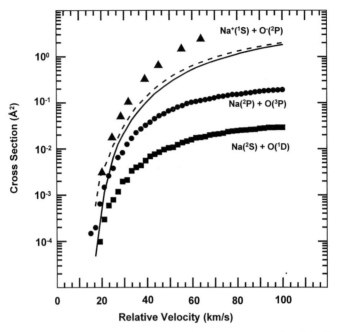

Fig. 8. Relative velocity dependence of integral cross sections calculated for Na + O collisions for the indicated exit channels. The solid curve is the charge transfer cross section calculated using a multichannel Landau–Zener formalism (see text). The dashed curve is the two-state Landau–Zener cross section. Charge transfer calculations by van den Bos[157] are indicated by triangles. Full circles and squares are the respective excitation channels as determined using the multichannel Landau–Zener model.

impact parameters are randomly chosen between 0 and $b_{\max} = 3.95$ Å, and at each successive crossing, surface hopping probabilities according to Eq. (32) are computed for the effective relative velocity:

$$\dot{R} = v\sqrt{1 - V_c/E_T - b^2/R_c^2}\,, \tag{36}$$

where v is the relative velocity at infinite R, b is the impact parameter, V_c is the potential energy at the crossing point, and E_T is the collision energy in the center-of-mass frame. A Monte Carlo approach is used to determine the event of a transition. Cross sections, $\sigma_i(E_T)$, are then computed from

$$\sigma_i(E_T) = \pi b_{\max}^2 N_i/N\,, \tag{37}$$

where N_i is the number of trajectories ending in product asymptote i, and N is the total number of trajectories. The charge transfer cross section approaches 1 Å^2 near 80 km/s, while the excitation cross sections exhibit a similar energy dependence but are at least an order of magnitude smaller at high relative velocities. Note, however, that at low relative velocities, the cross sections to form Na(^2P), the source of Na–D-line emissions, exceed the ion pair formation cross section. At the respective collision energies, the line-of-centers velocity, $\dot{R}$, can be very low at large impact parameters, leading to a prefered exit trajectory involving an adiabatic passage of the outer crossing on the ion pair potential.

Figure 8 also includes a Landau–Zener calculation assuming a simple two-state charge transfer system. In this case, the cross section is calculated from the simple double passage probability, $P = 2p(1 - p)$:

$$\sigma_{LZ}(E_T) = 4\pi \int_0^{R_c} p_{LZ}(E_T)(1 - p_{LZ}(E_T))b\,db. \qquad (38)$$

The figure also includes calculations by van den Bos[157] conducted in 1969 based on the Landau–Zener–Stueckelberg[154,155,158] approximation. The significantly higher cross section can be primarily attributed to the smaller coupling of 0.356 eV calculated by van den Bos.

A more accurate calculation would involve the wave properties of the trajectories. The present calculations do not include trajectory phases, as introduced by Stueckelberg.[158] The wave properties of the trajectories are not expected to affect the results of the two-state system due to averaging over a large range of impact parameters. They can, however, introduce oscillatory behavior in the energy dependence of multistate system cross sections at low collision energies. Zhu and Nakamura[159] have derived analytical formulae for two-state curve crossing problems from which exact transition probabilities can be calculated directly from adiabatic curves. The formulae account for the quantum mechanical behavior of the colliding particles, and overcome the many restrictions of the LZ model, such as the limitation to linear curves, and the constant electronic coupling with respect to the interfragment coordinate. These advances are particularly valuable for crossings at shorter range, where the choice of diabatic potentials is not as obvious as in the Na + O example above. In the case of oxygen atom collisions with alkaline earth and transition metals that have higher ionization potentials than the alkali metals, the ion pair crossings shift to substantially shorter range. Accurate quantum chemical calculations of excited state potentials are required in order to determine excitation and ionization cross sections. Metal oxide potentials of low lying excited states could be validated by photoelectron spectroscopic studies of metal oxide negative ions such as the extensive work by Wang and coworkers.[160-165]

As mentioned earlier, quartet and sigma collision pairs do not encounter the long range $^2\Pi$ avoided crossing that lead to ion pair formation and excitation. It follows that only 2 out of 9 collisions can contribute to ion-pair formation or excitation, while all collisions below a certain impact parameter have access to the repulsive wall of the potential where autoionization as well as curve crossings to excited states may occur. The latter are neglected in the present treatment, but could be of importance, particularly near threshold.[166] Given the small fraction of reactants experiencing the long range charge transfer interaction, repulsive wall processes including autoionization can, therefore, potentially compete with the longer range processes in this collision system.

Collisional autoionization is not a process governed by isolated crossings as those discussed above. Few theories exist on free-electron formation, and detailed experiments are scarce as well. Early studies involve the statistical model of Firsov[167] that provides order of magnitude predictions for a number of higher energy measurements.[168] Demkov and Komarov have considered the problem of two-atom free-electron formation from the point of view of a crossing between the neutral and ionic diabatic states, as depicted in Fig. 9(a).[169] The problem is reduced to the survival probability of the collision system as it crosses an infinite number of Rydberg levels converging to the ionization limit prior to interacting with the continuum. Once in the continuum, ionization is assumed to occur with a probability of 1. Thus, the ionization probability consists of a product of Landau–Zener probabilities involving ground-state–Rydberg-state transitions. In its simplest form, the Demkov–Komarov model can then be written[170]:

$$p_{DK} \approx \exp(-v_c/v_r) \,, \tag{39}$$

where v_c is an average coupling term and v_r is an average relative velocity in the crossing region. So far, experiments on H^+ rare gas atom collisions could only confirm that the energy dependence of free-electron formation cross sections corresponds to those obtained when integrating Eq. (39) over impact parameters:

$$\sigma_{e^-} = 2\pi \int_0^{r_c} p_{DK} b\, db \,, \tag{40}$$

and that the parameters v_c and v_r that fit the experimental results take on reasonable values.[170] It is obvious that Eq. (39) is highly approximate, and that several aspects of autoionization are not included in the model. For example, contrary to the Demkov–Komarov model, the interaction with the continuum results in characteristic autoionization lifetimes, τ_a, and the autoionization probability is given by:

$$p_{AI} = 1 - \exp(-t_a/\tau_a),\tag{41}$$

where t_a is the collision complex residence time in the autoionization region. Autoionization lifetimes[171] at short interaction distances are typically 10^{-15} s. Incorporation of autoionization lifetimes is critical for determining electron energy distributions which are of importance for the modeling of radar cross sections and their time evolution.

More recently, Solov'ev has reexamined the theory of collisional ionization using the example of $H^+ + H$ collisions.[172] A model was established involving an infinite series of quasi-intersections in an adiabatic framework, in analogy to the Demkov–Komarov model.[169] A simple analytical expression for the ionization cross section was derived for this two-Coulomb center problem based on an infinite series of Landau–Zener curve crossings. The calculations, however, are in poor agreement with experiment, which is explained by interference from charge transfer terms. Meanwhile, the two-Coulomb center problem provides a further ionization mechanism involving so called saddle-point electrons.[173] In this mechanism, the departing electron takes the position of the center-point on the line of centers of the two recoiling protons.

The Demkov–Komarov model employs a diabatic description of the potential energy surfaces, as shown in Fig. 9(a). If the coupling takes on an exponential radial form, as is frequently assumed, very large splittings would be expected at the repulsive part of the potential. Given the low nuclear velocities, $\dot{R}$, in this region of the potential, it can, therefore, be argued that an adiabatic approach may be preferable in treating the autoionization problem. In an adiabatic framework, the neutral and ionic curves do not cross in the repulsive regions of the potential due to the interaction with the infinite series of Rydberg levels converging to the ionic states. The adiabatic potentials are, therefore, nearly parallel at short range and approach each other gradually with increasing energy. This is illustrated in Fig. 9(b). One can, therefore, imagine a mechanism in which a nonclassical position hop to the ionization continuum becomes probable at and above energies,

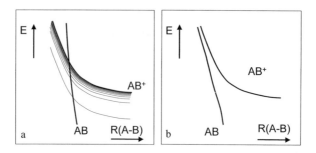

Fig. 9. (a) Representative diabatic curves for the collisional ionization of an A + B collision system. (b) Representative adiabatic curves for the collisional ionization of an A + B collision system.

E_c, at which the uncertainly in R, ΔR, is comparable to the separation of the ionic and neutral potential. The uncertainty of R can be regarded to be analogous to the range parameter, L, of the Demkov–Komarov model that defines the width of the crossing region. In the work of Aberle and coworkers[170] on rare gas + H collision systems, a range parameter, L, of ≤ 0.1 Å is determined.

In order to explore the plausibility of a position hop model, we have conducted density functional calculations of the ArH and ArH$^+$ ground state potentials at short range.[c] The calculations at the B3LYP/6-31G(3d) level[174] are shown in Fig. 10 and indicate that the adiabatic potentials are already very close below the atomic ionization limit of 13.6 eV, and that above 13 eV, ΔR is less than 0.1 Å. Upon inspection of the corresponding density functional potentials for the Na + O example in Fig. 10, the situation is rather different compared with Ar + H. Despite the fact that the ionization limit lies significantly lower than in the ArH case, the neutral and ionic curves of NaO exhibit significant separation in the repulsive region and approach to within 0.1 Å only above ~ 48 eV, corresponding to $R_c = 0.8$ eV. Taking further into account that the uncertainty in position can be expected to be higher for the Ar + H collision system given its smaller reduced mass, we can conclude that free-electron formation is not of importance in this collision system at relative velocities pertinent to meteor problems. Consequently, the calculated ionization cross sections

---

[c]The H + Ar collission system exhibits measurable collisional ionization cross sections at sub-keV energies.[170]

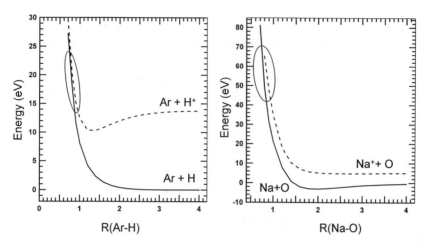

Fig. 10. Short range adiabatic potentials calculated for ArH/ArH$^+$ and NaO/NaO$^+$ using density functional theory at the B3LYP/6-31G(3d) level.

are almost entirely governed by the long range crossing leading to Na$^+$ + O$^-$. Free-electron formation from this atomic system would, therefore, be the result of collisional or associative detachment of the negative ion.

The present low level density functional calculations demonstrate that higher level quantum chemical calculations at very short range may provide valuable information regarding the dynamics in energetic regions of the repulsive interaction. More sophisticated models are needed to treat large atomic and molecular autoionization systems. Although simple in its physical picture, the Demkov–Komarov model also invokes the fact that Rydberg atom formation can be efficient at small impact parameters. The metastable character of Rydberg atoms may result in efficient ionizing secondary collisions. The true challenge of the future obviously lies in determining cross sections of excited species, a necessary aspect of most extreme environmental problems.

3.2.2.3. *Near resonant processes*

Most neutral charge transfer systems, as indicated by the cases above, involve crossings between curves with substantially different slopes, resulting in highly localized transitions that warrant the application of a Landau–Zener formalism. In ion–neutral charge transfer systems involving singly-charged species, however, reactant and product potentials frequently have similar asymptotic energies (near resonant charge transfer systems), and given the weak ion–induced dipole interaction compared with the Coulombic ion pair interaction, the potential inclinations are frequently small in areas of significant exchange interaction. Even if such near parallel diabatic states do not cross, there can be a resonant exchange interaction at an interfragment distance, R_c, at which $\Delta = 2H_{12}(R_c)$ holds, where Δ is the difference between the diabatic state energies. Nonadiabatic transition probabilities for near parallel states are given by the Rosen–Zener[175] or Demkov[176] model:

$$p_D = \left[1 + \exp\left(\frac{\pi\Delta}{\hbar a \dot{R}}\right)\right]^{-1}, \tag{42}$$

where a is the parameter in an exponential expression for the electronic coupling:

$$H_{12}(R) = A\exp(-aR). \tag{43}$$

Note that unlike the Landau–Zener case, where p_{LZ} ranges from 0 to 1, signifying adiabatic and diabatic passage of the crossing point, respectively, the Demkov probability, p_D, ranges from 0 to 0.5, the maximum corresponding to a complete mixing of states. The accurate determination of $H_{12}(R)$ is considerably more critical in Demkov systems compared with those of curve crossings because the transition probability, p_D, as well as the crossing radius, R_c, depends on it. Fortunately, integral cross section measurements for ion–neutral systems at hyperthermal energies are substantially more tractable than for their neutral counterparts.

3.2.2.4. *Molecular processes*

The theoretical treatment of electronically inelastic scattering cross sections for *molecular systems* are considerably more complicated than those of atomic systems. The additional internal degrees of freedom of the molecular fragments dramatically increases the density of energy levels accessible to the collision system, and the motion along vibrational coordinates can play an important role with respect to electronic transition probabilities. Unlike many atomic charge transfer studies, successful theoretical analysis of molecular charge transfer systems is still quite rare and limited to small triatomic systems. This is the case both for ionic processes, where absolute integral cross section measurements are abundant, and for neutral systems, where reliable cross sections are hard to find, as further iterated in Sec. 3.4. There are no semiclassical formulae, such as the Landau–Zener and Demkov formulae, that reasonably reproduce absolute integral and vibrationally state resolved cross sections over a large range of hyperthermal energies.

The most frequently cited model, the Bauer–Fisher–Gilmore (BFG) model,[177] arose from a study of Na* + N_2 quenching collisions, which have a direct bearing on meteor luminosity measurements. In this model, the quenching process was regarded as equivalent to charge transfer transitions to $Na^+ + N_2^-$ states that occur at a multitude of crossings between vibronic curves correlating to individual N_2 and N_2^- vibrational states. The coupling between vibronic states $|\psi_i \chi_i\rangle$ is given by the product of the electronic coupling and the Franck–Condon overlap between reactant and product vibrational wavefunctions:

$$H_{12v'v''}(R) = H_{12}(R)\langle \chi_{1v'} | \chi_{2v''} \rangle. \tag{44}$$

Based on Eq. (44), Spalburg, Los and Gislason[178] have derived the following analytical expression for vibrationally state-resolved change transfer cross sections in the weak-coupling limit:

$$\sigma_{v' \to v''} = \sigma_0 \frac{\langle \chi_{1v'} | \chi_{2v''} \rangle^2}{[1 + (\Delta E / a\hbar v)^2]^3}, \tag{45}$$

where σ_0 is the total cross section, ΔE is the energy gap between reactant and product asymptotic levels, v is the relative velocity at infinite interfragment distances, and a is the electronic coupling parameter defined previously in Eq. (43). Equation (45) indicates that low energy molecular charge transfer systems favor transitions between near resonant vibronic states with significant Franck–Condon overlap. At high collision energies, Eq. (45) reduces to:

$$\sigma_{v' \to v''} = \sigma_0 \langle \chi_{1v'} | \chi_{2v''} \rangle^2, \tag{46}$$

which amounts to the Franck–Condon approximation. This is a direct consequence of the uncertainty principle according to which the dynamic range of accessible states in near resonant transitions, corresponding to the uncertainty of the resonant energy levels, increases with decreasing charge transfer interaction time. At the very high energy limit, the dynamic range covers the entire Franck–Condon envelope, leading to state-to-state cross sections described by Eq. (46).

At low collision energies, Eq. (45) provides a reasonable prediction of the vibrational product distributions, but fails to correctly reproduce the energy dependence of state-to-state cross sections. This can in part be explained by the fact that the motion along nuclear coordinates orthogonal to the interfragment coordinate, i.e., vibrational coordinate, cannot be considered frozen during the collision. Another reason for discrepancies is the large density of states in molecular systems, due to which the molecular charge transfer system represents a quasiresonant charge transfer system provided the Franck–Condon overlap between reactant and product vibrational wavefunctions is significant.[179]

In short range processes, such as neutral excitation collisions, approaches such as Eq. (45) become inaccurate, since asymptotic Franck–Condon factors are no longer meaningful. In this case, predictions of molecular excitation cross sections must rely on detailed calculations of the multidimensional potential energy surfaces and the associated crossing seams. Given the fact that crossings are no longer localized, transition probabilities are likely to "optimize" at a particular interfragment distance. Molecular charge transfer transitions, therefore, tend to be substantially more efficient than their atomic counterparts. This is confirmed in neutral excitation systems, as well as in ionic systems, and will be further discussed in Secs. 3.3 and 3.4. The same applies to the type of autoionization processes described above, where the various nuclear degrees of freedom increase the probability of resonance between ionic and neutral potentials. Meanwhile, if an ion pair formation channel with a bound molecular negative ion exists, autoionizing vibrational states can be accessed in charge transfer collisions, greatly enhancing free-electron formation.

3.3. *Ion Beam Studies Involving Metallic Species*

3.3.1. *Overview of Ion Beam Techniques*

Measurements of ion–neutral cross sections in the 1 to 500 eV collision energy range require the use of ion beam experiments. Ion beams are ideally suited for studying collision dynamics as a function of translational energy. In a generic integral cross section measurement, a mass and energy selected ion beam propagates through a cell containing the neutral target gas, and intensity events such as the production of an ion with a mass different from the primary ion mass, or the emission of luminescence photons, are monitored. Provided the effective solid angle of signal collection is known, integral cross sections, σ, can then be determined from the Lambert–Beer expression:

$$I_0 - I_s = I_0 \exp(-\sigma n l), \qquad (47)$$

where I_0 is the total ion current, I_s is the secondary current (in case of luminescence experiments corresponding to that of the light-emitting species), n is the number density of the target gas and l is the effective target gas path length. At single-collision conditions, Eq. (47) reduces to:

$$\sigma = \frac{I_s}{I_0 n l}. \qquad (48)$$

Another approach is the crossed-beam method[180] in which angular scattering distributions are determined that are highly sensitive to details in the interaction potential. Integration over all angles leads in principle to integral cross sections. The difficulty of determining the target column density, nl, and the limited range of available angles in all experiments, reduce the utility of this approach for determining absolute integral cross sections. The theoretical reproduction of observed distributions, however, results in accurate potential energy surfaces, from which integral and differential cross sections can be derived. There are few cases, however, where this has systematically been exploited.

Some of the problems associated with determining integral cross sections using crossed beam techniques have been overcome using merged beam techniques.[181,182] Hyperthermal energy experiments, however, have traditionally been limited to collision energies below ~ 20 eV, and, due to the generation of the neutral target through ion beam neutralization, suffer from low secondary ion signals.

Most integral ion–neutral cross section measurements have been obtained in ion-cell arrangements. The key difficulty in measuring secondary ion formation cross sections is ensuring a 4π solid angle collection of secondary ions, or, if this is not possible, determining the effective solid angle of collection. The latter is only useful if the laboratory-frame anisotropy of the angular distribution is known. The field of low energy ion dynamics has made remarkable progress since the introduction of the *guided-ion beam (GIB) technique* developed by Gerlich.[183,184] The principle of a GIB experiment is illustrated in Fig. 11. A mass and energy selected ion beam

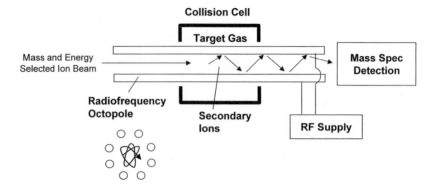

Fig. 11. Schematic representation of a guided-ion beam experiment.

is injected into a radio-frequency octopole ion guide consisting of 8 parallel rods symmetrically spaced on a circle. The rf frequency is chosen to be substantially higher than the frequency associated with the oscillatory motion of the ions in the guide. In these conditions, ions exposed to the inhomogeneous electric fields of a multipole can be described as moving in an effective potential that is proportional to $(r/r_0)^{n-2}$, where r is the radial position of the ion, r_0 is the inner radius of the octopole, and n is the number of rods. An ideal multipole consists of hyperbolically shaped rods. With the choice of an optimal ratio between rod diameter and multipole inner radius of approximately 0.67, however, cylindrical rods approach the hyperbolic ideal closely. Note that an octopole generates an $(r/r_0)^6$ potential, and that the ions mostly move in a near field-free area. This is contrary to the motion of ions in a quadrupole, which generates a quadratic effective potential.

As depicted in the figure, the rf octopole guides the ion beam through a collision cell, and secondary ions are collected irrespective of scattering angle. Laboratory back-scattered ions are reflected in the forward direction by applying a high potential on the injection electrode, thus ensuring a 4π solid collection angle. Primary and secondary ions are then extracted and mass analyzed prior to ion detection. In order to minimize gas losses through the collision cell orifices, the cross sectional dimensions of the octopole must be kept as small as possible. Typically, rod diameters of 2 mm and inner radii of 3 mm are chosen.

Given its sensitivity that is almost independent of ion beam energy, the GIB technique has bridged the gap between thermal collision energy ion–molecule experiments, traditionally investigated using flowing-afterglow[185] and ICR[186] techniques, and the hyperthermal studies using electrostatic ion beam experiments. The latter are generally insensitive at collision energies below 1 eV due to beam divergence problems. The use of this technique has allowed the investigation of ion–molecule reactions in the collision energy range corresponding to chemical bond energies with unprecedented sensitivity and accuracy. However, it has been used less extensively at collision energies above 20 eV, despite its many advantages over common ion beam techniques.

The virtues of the GIB experiment are not compatible with the requirements of luminescence measurements, where the small signal levels mandate high ion beam intensities that would result in space-charge problems in a GIB approach. Luminescence measurements can either be conducted in cell experiments, where an intense ion beam passes through a target cell, and luminescence photons are collected either through a slit or through fiber optics, or in crossed beam arrangements, where the collision region is more readily viewed. The latter is better suited for measurements of neutral metal atom targets because effusive metal vapors inevitably coat optical surfaces. The emission cross section, σ_λ, is determined from:

$$I_\lambda = I_0[1 - \exp(-\sigma_\lambda nl)], \qquad (49)$$

where I_λ is the photon intensity emitted into a solid angle of 4π at a particular wavelength, λ.

Recent advances in hyperthermal metal ion chemistry, as applicable to meteoroid problems, will be summarized in Sec. 3.3.2. Ion collision studies involving metal vapors are discussed in Sec. 3.3.3.

3.3.2. *Metal Ion Collisions with Atmospheric Atoms and Molecules*

Given the low ionization potentials of metals, the endothermic charge transfer processes with atmospheric species such as O_2 and N_2 only occur at high collision energies, while chemical reaction channels are usually closed at collision energies below 1 eV. This, in conjunction with the long lifetime with respect to electron–ion recombination, is responsible for the stability of thermalized metal ions in the ionosphere. Very few studies exist on charge transfer and ionizing collisions of metal ions with atmospheric

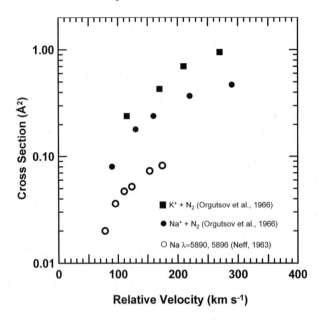

Fig. 12. Relative velocity dependence of metal ion charge transfer[187] and excitation[188] collisions with N_2.

species at sub-keV energies. An indication of the magnitude of alkali ion + N_2 charge transfer cross sections at meteoric energies can be obtained from the higher energy work by Ogurtsov *et al.*[187] These authors used a conceptually simple beam cell arrangement in which all charged particles are deflected after passage of the beam through the cell, and the neutral beam intensity is measured. The relative velocity dependence of the Na^+ and K^+ + N_2 charge transfer cross sections determined by Ogurtsov *et al.* is shown in Fig. 12. Interestingly, the K^+ + N_2 cross section is larger despite the higher endothermicity. This can possibly be explained by a substantially stronger electronic coupling in a long range Demkov-type mechanism in the K^+ system,[176] or by the larger radius of the potassium ion, implying a mechanism involving the repulsive part of the potential.

Neff measured excitation cross sections of resonance lines observed in meteors in collisions of Na^+, K^+, Mg^+, and Ca^+ with N_2 in a beam cell arrangement equipped with optical detection.[188] The observed sodium D-line excitation cross sections are also shown in Fig. 12, and are approximately a factor of 3 smaller than the total charge transfer cross sections.

Neff estimated that up to $\sim 10\%$ of charge transfer sodium atoms are produced in the excited 3^2P state, which is consistent with the measurements of Ogurtsov *et al.* The experiment was not sensitive enough to measuring the corresponding emission cross sections for $K^+ + N_2$, where the optical transitions are at 770 and 766 nm, respectively. The cross section for exciting the K $5p(^2P_j) - 4p(^2S)$ transitions at ~ 405 nm was found to be only $\sim 10^{-3}$ Å^2 at 70 km s^{-1}. The largest emission cross sections were found for excitation of the Ca$^+$ K-line at 394.0 nm, where a cross section of ~ 0.2 Å^2 was observed between 50 and 100 km s^{-1}. As related to meteor studies, an important element of the work by Neff was the detection of atomic nitrogen ion emission lines at 567.5, 500.5 and 399.5 nm in Na$^+$ + N$_2$ collisions. The reported excitation cross sections lie between 10^{-4} and 10^{-3} Å^2 at Na$^+$ energies ranging from 400 to 800 eV. N$_2$$^+$ first negative band emissions were also observed. Interestingly, atomic emissions from N species were not detected in N$_2$ collisions with other meteoric metal ions.

Rutherford and Vroom[189] studied Al$^+$ collisions with O$_2$ and N$_2$ at ion energies ranging from 1 to 5000 eV in a crossed beam apparatus involving a modulated neutral beam.[107] The aluminum ions were produced by surface ionization of AlCl$_3$ vapor on a hot tungsten filament and product ions were detected mass spectrometrically. In the Al$^+$ + O$_2$ collision system, the O$_2$$^+$ formation cross section was found to be at the detection limit of 0.01 Å^2 at 1 keV ion energy (~ 115 km/s), after which it rose to values above 0.1 Å^2 at 5 keV. N$_2$$^+$ formation in Al$^+$ + N$_2$ collisions was only measurable above 1.5 keV (~ 150 km/s). The dissociative charge transfer processes:

$$Al^+ + O_2 \rightarrow Al + O^+ + O \qquad (50)$$

$$Al^+ + N_2 \rightarrow Al + N^+ + N, \qquad (51)$$

were not observed. That the Al$^+$ cross sections are substantially lower than those of the alkali metals is somewhat surprising. While Al$^+$ has several excited states at excitation energies below the charge transfer endothermicity, there are no such states for the alkali ions. Al$^+$ excitation channels may, therefore, play a role in suppressing charge transfer.

At low collision energies (< 100 eV), metal ions can undergo chemical reactions with oxygen and other oxygen containing molecules to form metal oxide ions. In the Al$^+$ + O$_2$ work by Rutherford and Vroom,[189] AlO$^+$ formation was observed at the thermodynamic threshold, reaching a peak cross section near 0.5 Å² at 10 eV, above which the cross section rapidly declines reaching the detection limit at 50 eV. Armentrout, Halle and Beauchamp[190] investigated oxide formation for a number of transition metal ions, including Fe^+, using a tandem mass spectrometer with a collision cell interaction region. A similar instrument was used by Murad[191] to investigate Ca^+ abstraction reactions with O_2, CO_2, and H_2O. These studies had the primary goal of elucidating the metal oxide thermochemistry from the observed thresholds. Assuming that there are no barriers in excess of the threshold, E_0, the oxide ion dissociation energy is obtained from:

$$D_0^0(MO^+) = D_0^0(O_2) - E_0 \,. \tag{52}$$

The dissociation energy of the neutral oxide is then obtained from the thermochemical cycle:

$$D_0^0(MO) = D_0^0(MO^+) + IP(MO) - IP(M) \,. \tag{53}$$

The determination of E_0 from visible onsets introduces considerable error in the derived bond dissociation energies. This problem is primarily attributed to experimental kinematic broadening introduced through the ion beam energy distributions and the thermal motion of the target gas in the case of beam cell experiments. Armentrout *et al.*[190] have fit various functional forms of the cross section collision energy dependence, $\sigma(E_T)$, convoluted with the experimental broadening conditions, to the observed threshold curves. The best results were obtained with a classical line-of-centers model:

$$\sigma(E_T) = \sigma_0 \frac{(E_T - E_0)}{E_T}, \quad E_T > E_0$$
$$\sigma(E_T) = 0, \quad E_T \leq E_0, \tag{54}$$

where σ_0 represents a hard-sphere cross section, πd^2. The line-of-centers model assumes that all of the translational energy with respect to the line-of-centers of the reactants is converted into internal energy at the collision turning point.

More recently, Armentrout's group at the University of Utah has applied the guided ion beam (GIB) technique[183,184] to determine thermochemical information for a vast number of metal compounds.[135,192–194] The high sensitivity at low collision energies, as well as the high collection efficiency of their experiment, coupled with a number of innovative techniques to produce thermalized as well as excited metal ions, has permitted a substantially more rigorous analysis of observed reaction thresholds. The analysis involves a nonlinear least-square fitting procedure using a modified line-of-centers functional form:

$$\sigma(E_T) = A \sum_i g_i \frac{(E_T + E_i - E_0)^n}{E_T^m} \,, \tag{55}$$

where A is a scaling parameter, n and m are curvature parameters, and the sum is over internal states with energies, E_i, and relative populations, g_i. Normally, m is set to 1 and a fit of the remaining parameters is conducted. Implicit in Eq. (55) is the assumption that n, m and A do not vary for all internal states of the reactants involved. Accurate application of expression (55) depends on high signal-to-noise in the threshold region of the respective product ion yields. It is at these energies where the translational energy dependence of the cross section is most sensitive to the curvature parameters. At higher energies, the cross correlation of the three fitting parameters, A, E_0 and n is very high, and free fits of the three parameters can result in substantial errors. Triple-quadrupole experiments have also been used to determine thresholds in ion–molecule collisions. These experiments have been hampered by the fact that the ion energy distribution is poorly preserved in quadrupoles.[184] Rf-heating produces high energy tails in the ion energy distribution that result in signals below threshold, thereby encumbering the analysis.

Dissociation energies of meteoric metal oxide ions determined by the group of Armentrout,[135,192–194] along with those of the neutrals, are listed in Table 3. Except for the Al^+ measurements, where a model with $m = 3$ provided the best fit of the energy dependence near threshold, all thresholds were modeled using $m = 1$, and the curvatures range between $n = 1$ and $n = 2$. Note that the dissociation energies of Table 3 imply some differences in the neutral thermochemistry as reported by Lias *et al.*[195] and listed in Table 2. It can be stated that the GIB-derived dissociation energies are usually approximately 0.2 eV higher than those determined at the highest level of theory. (See for example Dalleska and Armentrout[192] and Bauschlicher *et al.*[196].)

An example of a threshold curve is shown in Fig. 13 for the reaction[192]:

$$Mg^+ + O_2 \rightarrow MgO^+ + O \,. \tag{56}$$

The ground state magnesium ions were produced by argon ion sputtering of a magnesium cathode, and through thermalization of the ions in a 1 m long flow tube operated at $\sim$ 0.5 Torr He. Dalleska and Armentrout[192] interpreted the observations with two dynamic reaction pathways. In the first, labeled T in the figure, an intermediate complex dissociates into products at threshold. In a second *impulsive* or stripping mechanism labeled I, the collision is described to occur between Mg^+ and a single O atom. The effective "pairwise energy", E_p, available to the collision pair is then given in relation to the center-of-mass collision energy, E_T, by:

$$E_p = \frac{E_T M}{2(m_{Mg} + m_O)}, \tag{57}$$

where M is the total mass, and m_{Mg} and m_O are the masses of the magnesium ion and oxygen atom, respectively. As a result, the MgO^+ threshold produced in an impulsive mechanism is delayed by a factor of 1.43, as shown in Fig. 13. The MgO^+ cross section has a maximum of 1 Å^2 at $E_T \approx 5.0$ eV, after which a prompt decline is observed with energy. The

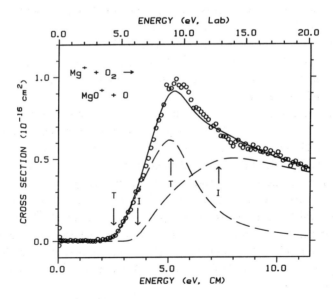

Fig. 13. The collision energy dependence of the $Mg^+ + O_2 \rightarrow MgO^+ + O$ reaction cross section. Figure printed with permission from Elsevier Science Publishers.[192]

onset of the decline is consistent with the 5.12 eV threshold for:

$$Mg^+ + O_2 \rightarrow Mg^+ + O + O. \tag{58}$$

The dynamics are substantially more complicated for transition metal ions, where a number of low lying excited states with high multiplicities, each split into a series of spin-orbit states, can be involved that exhibit different reactivities. For example, the atom transfer reaction of ground state $Fe^+(^6D)$ with H_2 is approximately 15 times slower than that of the first excited state, $Fe^+(^4F)$, that lies only 0.3 eV above the ground state.[193] Meanwhile, the presence of the ^4D and ^4P states of Fe^+ with excitation energies of 1.03 and 1.69 eV, respectively, may enhance the $Fe^+(^4F)$ reactivity at hyperthermal energies.

3.3.3. *Ion–Metal Vapor Collision Studies*

From the point of meteoroid ablation to the subsequent thermalization of the metal atoms, charge transfer reactions (30) can occur and contribute to the dynamics of ionization trails and in the formation of the equilibrium background ionization. From a simple examination of the time constants for these reactions, it can be concluded that metal atom formation through hyperthermal charge transfer reactions with E-region ions, NO^+ and O_2^+, is not important on the time scale of a meteor. Assuming a very fast reaction rate coefficient ($k \sim 1 \times 10^{-9}$ cm^3 s^{-1}), the time constant, $\tau = 1/(k^*[NO^+ + O_2^+])$, for the charge exchange reaction with NO^+ or O_2^+ is ~ 1000 s for a total ion density $\sim 10^6$ cm^{-3}. The primary role of reactions (30) is, therefore, formation of the steady state metal ion layers. Since charge transfer between atmospheric ions and metal atoms is highly exothermic, large long range exchange probabilities are possible over the entire range of hyperthermal energies, leading to large cross sections. This is particularly the case for molecular ions. Usually, metal atom charge transfer cross sections involving atomic ions, such as $O^+(^4S)$ at altitudes above 250 km (ionospheric F region), are small, unless there are accidental resonances between charge transfer states with equal spin. For the molecular E-region ions, energy resonance and Franck–Condon overlap primarily dictate the magnitude of molecular charge transfer cross sections, as discussed in Sec. 3.2.2.4. If the Franck–Condon factors are favorable between near-resonant states, charge transfer cross sections can exhibit collisional efficiencies at near-thermal energies.

Measurements of ion–metal atom charge transfer cross sections at hyperthermal energies are rare due to the experimental difficulties associated with metal vapors. Even in the case of thermal measurements, where the flowing afterglow technique[185] has been highly prolific, only one ion + metal atom experiment has been reported, that of Farragher and coworkers on reactions with Na.[197] The first beam measurements involving metal vapors were conducted by Fite and coworkers in a study of Na charge transfer collisions with N_2^+ and O_2^+.[198] Cross sections were measured at ion energies between 20 and 200 eV in a crossed beam arrangement. The interaction region consisted of a cylindrical slow ion collector with axis along the ion beam. A collimated and chopped metal beam passed through apertures in the cylinder. The density of the metal beam in the interaction region was derived from the metal furnace temperature and the surface ionization current produced when the transmitted neutral beam impinged on a hot wire. The observed cross sections between 1 and 3 $Å^2$ are generally perceived to be too low.

Merged beam experiments have been applied to atmospheric ion–metal atom collision studies, as well.[148,199,200] In these experiments, an ion beam is merged with a neutral metal beam using a magnetic sector analyzer. The neutral metal atom beam is produced by passing a metal ion beam through a symmetric charge transfer oven cell. The beams are merged following the charge transfer region and enter a well-defined interaction zone in which the collision velocity is given by the difference in reactant velocities. Product ions are then energy analyzed. Even though the beams are operated at keV energies, this methodology permits experiments at very low collision energies (~ 0.05 eV) with remarkable resolution. The main difficulty is accurately determining the volume over which the reactant beams interact.

Given the complexity of these experiments, the results from the cited merged-beam experiments are somewhat disappointing. It is difficult to extract absolute cross sections from experiments, and in the particular case of atmospheric ion + metal atom reactions, charge transfer cross sections could not be recorded due to high metal vapor backgrounds in the interaction region.[148,199,200] The experiments were, therefore, limited to examining reactive collisions and the associated energy partitioning. Rol and Entemann[148] investigated the reaction

$$O_2^+ + Na \rightarrow NaO^+ + O, \tag{59}$$

and reported a cross section of 10 Å^2 at 0.05 eV. This implies a NaO$^+$ bond energy exceeding 5 eV, which is in stark contradiction to all other measurements which lie below 1 eV. It is probable that interference from potassium ions, which have the same mass as NaO$^+$, was responsible for the observed signal.

The most comprehensive body of work on ion–metal vapor collisions at hyperthermal energies was carried out by Rutherford and coworkers, who applied a crossed beam double mass spectrometer[107,108,150,189,201] with a collimated, chopped neutral beam produced in a Knudsen cell furnace. Rutherford and coworkers used two methods to determine the neutral beam density, a key problem in metal vapor experiments. In the first approach, the density was calculated from the furnace temperature, the experimental geometry, and the known vapor pressure temperature dependence. In the second, the transmitted metal was collected on a cold surface, and the amount deposited was determined using activation analysis. Another difficulty of the double-mass spectrometer crossed beam experiment is the assessment of the collection efficiency, which can be low in the case of slow charge transfer ions. The collection efficiency was estimated based on other charge transfer cross section measurements. Given the considerable sources of error, an error of $\sim 60\%$ is reported for the lowest collision energies in the range 1–500 eV.

Nonetheless, using this technique, Rutherford and coworkers investigated sodium,[150] magnesium,[107] calcium,[201] and iron[108] ion formation by a number of primary ions including N$_2$$^+$, NO$^+$ and O$_2$$^+$. The cross sections for sodium charge transfer with N$_2$$^+$ and O$_2$$^+$ were found to be approximately an order of magnitude higher than the first measurements by Fite and coworkers.[198] For all metallic systems, molecular charge transfer cross sections are found to be large, in particular if resonances with product states exist. Interestingly, the cross sections measured for O$_2$$^+$ and NO$^+$ charge transfer reactions with Fe are considerably smaller than those found for Mg, Ca and Na. This could partially explain the low Fe$^+$ abundance observed in a sporadic E layer[89] despite the high Fe abundance in meteoroids.

In all of these studies, the cross sections for reaction with O$^+$ were also measured. Mg and Na were found to yield no measurable cross section while cross sections exceeding 100 Å^2 and 10 Å^2 were measured below ion energies of 10 eV for Ca and Fe, respectively. While accidental resonances are not surprising in the case of Fe, given the high density of ionic and neutral Fe states, the very large $O^+ + Ca$ charge transfer cross section is likely due to the highly resonant process:

$$O^+(^4S) + Ca(^1S_0) \rightarrow O(^3P) + Ca^+(5p,^2 P_{1/2}) + 0.005 \text{ eV}. \qquad (60)$$

It is worth noting that the small asymptotic energy gap can grow dramatically as the reactants approach, if one assumes that the long range interaction of the reactants, involving the high polarizability of Ca atom $(\sim 25\text{Å}^2),$[202] is much stronger compared to that of the products. It is, however, difficult to predict how the polarizability of the excited $Ca^+(5p,^2 P_{1/2})$ ion, which unlike the ground state, cannot be considered a point charge, affects the long range interaction. The decreasing cross section with translational energy between ion energies of 2 and 500 eV, typical of resonant charge transfer systems, indicates that the product interaction is similar to that of the reactants at radii of strong exchange interaction.

Attempts at bridging thermal charge transfer cross section measurements and ion beam measurements through extrapolation of the beam data result in considerable discrepancies between the two techniques. This, coupled with the considerable discrepancies between the different beam experiments, prompted Levandier and coworkers to develop a new technique for metal vapor studies based on the venerable GIB technique.[203] Hitherto, GIB experiments had been primarily dedicated to volatile samples. Anderson and coworkers[204–206] have conducted ion–C_{60} studies using a GIB experiment equipped with an oven cell, but no attempts were made to determine the absolute cross sections. Sunderlin and Armentrout[207] have constructed a temperature variable octopole where the rod support structure and the collision cell have thermal contact to tubing through which a temperature controlled liquid is passed. The experiment was used for cooling the ion guide down to ~ 100 K using liquid nitrogen, but no measurements are reported in which the system was heated.

The high temperature octopole developed by Levandier *et al.*[203] is depicted schematically in Fig. 14. A metal sample is placed in a tantalum oven cell that is heated with thermocoax heaters. In order to generate a well-characterized temperature from which the metal vapor density can be determined, the octopole rods that pass through the oven cell need to be heated to temperatures equal or higher than that of the oven cell. Heating of the rods is also important to minimize the deposit of metal vapor on the rods. This is accomplished through the use of thermocoax heater tubes as

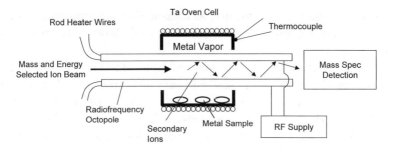

Fig. 14. Schematic representation of the high temperature octopole ion guide developed by Levandier *et al.*[203]

octopole rods, thereby allowing heating that is independent of the rf circuitry. Measurement of the temperature at the coolest point of the collision cell yields the vapor pressure of the metal. The ion source and ion detector of the apparatus of Levandier and coworkers are situated off the octopole axis. Consequently there is a free line-of-sight through the octopole that can be used for metal vapor absorption measurements if a strong resonance line exists. This provides an additional means of determining the metal vapor density. Technically, the system can be operated to temperatures as high as 1200 K.

Levandier *et al.* used the high temperature GIB technique to study sodium ion formation in charge transfer collisions with O_2^+, NO^+ and N_2^+ at center-of-mass collision energies ranging between 0.2 and 10 eV.[208] The sodium density was determined by measuring the oven cell temperature and by measuring the white light transmission bands of the Na–D lines. The curve-of-growth of absorption equivalent widths was calculated through numerical integration of the absorption Voigt profile. The two methods provided consistent measurements of the sodium density. The high temperature GIB technique is nevertheless not problem-free. The trapping properties of the octopole create an extreme sensitivity to surface ionization of sodium on the hot rods. Sodium ion backgrounds similar to the charge transfer signal are typical. Meanwhile, despite the elevated rod temperatures, metal deposits are not avoidable, and oxide formation on the rods creates substantial potential barriers after prolonged operation, rendering low energy measurements inaccurate. This problem is particularly acute when O_2 is the precursor gas, as in the O_2^+ + Na studies, in which O_2^+ was produced by electron impact ionization of O_2. Even though the chamber background pressure, as measured by an ionization gauge, is $\sim 10^{-8}$ Torr, this is sufficient to generate an oxide layer on the rods within hours of operation.

The determined GIB cross sections are depicted in Fig. 15 and compared with flowing afterglow[197] and crossed-beam measurements.[150] The energy range of the GIB measurements lies nicely between the thermal flowing-afterglow and the crossed beam experiments. The near $E_T^{-0.5}$ dependence below 1 eV is characteristic of a near-resonant, long range molecular charge transfer mechanism. The GIB cross sections are generally larger than the earlier experiments. In comparison with the crossed-beam experiments, it may be argued that no assumptions of the collection efficiency are necessary in the GIB experiments. Also, Rutherford and coworkers reported problems with sodium oxide coatings preventing them from determining the target density using Knudsen cell temperature measurements.

The fact that the O_2^+ measurements exceed the Langevin–Giomousis–Stevens (LGS) capture cross section[209] raise a suspicion that these GIB cross sections may be too high. On the other hand, the simple LGS capture model may not be applicable in a case where rapidly dissociating states are produced, as is the case in the $O_2^+ + Na$ system. Inspection of the O_2/O_2^+ vibrational coordinates on a charge transfer energy scale

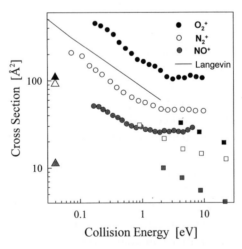

Fig. 15. Sodium charge transfer cross sections with atmospheric molecular ions as measured using the high temperature guided-ion beam experiment (circles).[208] Triangles are the flowing-afterglow measurements by Farragher *et al.*[197] and squares are the crossed-beam measurements by Rutherford *et al.*[150]

demonstrates that charge transfer can lead to $O_2 A^3\Sigma_u^+$, $C^3\Delta_u$ and $c^1\Sigma_u^-$ formation at their repulsive walls in the respective dissociation continua. Levandier *et al.* also argue that the NO^+ + Na collision system is likely to be affected by substantial vibrational effects. The fact that electron-impact ionization produces NO^+ in excited vibrational levels implies that the cross section for ground state NO^+ could be lower than the GIB measurements in Fig. 15. The discrepancy between the GIB and previous experiments calls for further experiments to measure these vital cross sections.

McNeil, Lai and Murad[210] have used the cross sections of Levandier *et al.* in models for the calculation of permanent sodium and sodium ion metal layers. Figure 16 compares the nighttime altitude density profiles of sodium and sodium ion calculated using the thermal flowing afterglow measurements of Farragher *et al.*[197] and the cross sections of Levandier *et al.*[208] including thermal extrapolations of the respective data. The sodium profiles are compared to a typical sodium layer as observed in a lidar experiment. It is seen that the charge transfer cross sections have the biggest effect on the topside of the neutral layer, and that the GIB measurements provide an accurate representation of the neutral density profile. The steady-state ionic layer has been measured by a number of authors in mass spectrometric rocket experiments.[99,101,211]

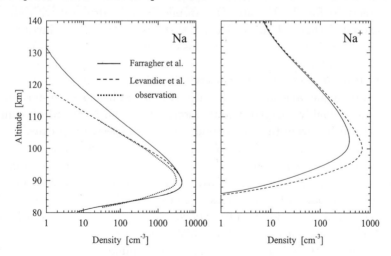

Fig. 16. Na and Na^+ density profiles calculated using the one-dimensional model developed by McNeil *et al.*[254] The calculations are conducted for different rate coefficients for sodium charge transfer with atmospheric ions, NO^+ and O_2^+. The dashed line is a typical neutral profile as observed in a lidar experiment.

In a series of publications, Boitnott and Savage have reported emission cross sections for numerous ion–metal atom collision systems.[38,212,213] The experiments were specifically targeted at meteor problems and consisted of crossed-beam experiments at ion energies ranging from 200 to 2000 eV covering many atomic lines observable from the ground. The ions were produced in an rf ion source. Although tests were made with respect to the cross section energy dependence on rf amplitude, it cannot be excluded that metastable ions as well as doubly charged ions were present in their beams, even at the lowest operatable rf levels. Among the largest ion–neutral emission cross sections observed were ion–Na induced Na–D-line emissions.[212] For N_2^+, N^+, O_2^+, and O^+, a cross section that increases with ion energy is observed in the vicinity of 10 Å^2. The cross sections can be associated with inelastic ion scattering.

In Mg collisions,[214] the highest emission cross sections stemmed from ionic emissions at 448 nm. These emissions are associated with the Mg^+ $4f(^2F)$–$3d(^2D)$ transitions. The $4f(^2F)$ state lies 11.6 eV above the ionic ground state. The respective charge transfer reactions with N_2^+, N^+, O_2^+, and O^+ are consequently endothermic. The observed cross sections ranged between 0.1 and 1 Å^2. Interestingly, while the N_2^+ and O_2^+ emission cross sections increased with energy, as expected in an endothermic process, the cross sections for the N^+ and O^+ atomic systems dropped with energy and exceeded those of the molecular systems at the lowest investigated energies. This is rather surprising, considering the arguments of Sec. 3.2.2 and the fact that for O^+, the endothermicity for producing $Mg^+4f(^2F)$ is 5.7 eV. It must, therefore, be suspected that these measurements are either affected by metastable ions, or that the atomic ion beams have substantial fractions of doubly charged diatomic ions, O_2^{2+} and N_2^{2+}, which cannot be distinguished from the monoatomic ions mass spectrometrically.

The intense $Ca^+4p(^2P)$–$4s(^2S)$ resonance lines at 393.4 and 396.8 nm were observed in exothermic charge transfer reactions involving all of the atmospheric ions investigated by Savage and Boitnott.[214] Whereas no cross sections are reported for O^+, cross sections around 10 Å^2 are reported for N_2^+, O_2^+ and N^+ at ion energies above 200 eV. The same ions including O^+ produce emissions at 370.6 and 373.7 nm, corresponding to the Ca^+ $5s(^2S)$–$4p(^2P)$ transition, with cross sections around 1 Å^2. The results for O^+ are not fully consistent with the very large near-resonant charge transfer observed for $O^+ + Ca$ by Rutherford and coworkers.[201] Resonant charge transfer leads to the Ca^+ $5p(^2P)$ level, which can, however, produce the

~ 370 nm and ~ 395 nm emissions via cascading initiated by an infrared transition at 1184 and 1195 nm to the Ca^+ $5s(^2S)$ level. The measurements of Savage and Boitnott assume that the radiative lifetimes are fast enough that all photons are emitted while the ions are in the viewing region of their apparatus. If this assumption is correct for cascading transitions as well, higher emission cross sections for O^+ would be expected if the measurements of Rutherford *et al.* are correct.

Although very important in reconciling radar observations of ionization wakes and optical observations of luminous trails of meteors, there are still significant discrepancies in existing charge transfer and luminescence measurements. Most of the experiments, however, date back to the 1960s and early 1970s. There is, therefore, hope that the recent advancements in experimental capabilities will eventually be applied to these problems in order to improve the predictive capability of existing meteor models.

3.4. *Fast Neutral Collisions Involving Metallic Species*

3.4.1. *Overview of Fast Neutral Beam Experiments*

The body of literature on fast neutral beam studies is far smaller than that for hyperthermal ionic studies given the associated experimental difficulties. Reviews of earlier fast beam techniques have been given by Kempter[153] and Wexler.[140] Four methods have successfully been used to produce fast neutrals at sufficient intensities for scattering studies: (i) neutralization of ion beams; (ii) ion beam sputtering; (iii) laser ablation; and (iv) seeded supersonic beams. The most straightforward way of generating fast neutrals is directing a collimated ion beam of known mass and energy through a resonant charge transfer cell following which the residual ion beam is deflected off the main axis of the experiment.[215] Since resonant charge transfer processes are accompanied by minimal momentum transfer, a neutral beam is generated with energy distribution close to that of the original ion beam. Usually, symmetric charge transfer systems are chosen. Neutrals produced in nonresonant processes typically scatter into large solid angles, and, therefore, do not affect the internal state purity of the beam provided the interaction region is at sufficient distances from the charge transfer cell. Meanwhile, absolute intensities can be determined directly by measuring the current of slow charge transfer ions produced in the charge transfer cell.[216]

In more recent years, fast neutral beams have been generated through photodetachment of negative ion beams. Photodetachment experiments have the advantage of producing pure ground state species in the case of atomic beams, and they have essentially no momentum transfer losses. Van Zyl *et al.*[217] generated intense hydrogen atom beams by passing an H^- beam through a cw YAG laser cavity. This technique, however, is experimentally much more demanding than that using charge transfer cells, and depends on the ability to produce intense precursor negative ion beams, which is not possible for species with positive electron affinities, like nitrogen.

The main advantage of ion beam neutralization is the ease of energy selection, and the high energy range. Below 10 eV, however, the neutral beam intensities become too weak for meaningful scattering results. More recently, the interest in low energy atomic oxygen beams for space research purposes (see Chap. 9 by Minton and Garton in this book) has called for higher intensities below 10 eV. Orient, Chutjan and Murad[218] have developed an experiment that magnetically confines an O^- beam using a strong axial magnetic field generated by a superconducting magnet. The negative ion beam is then photodetached in a cw-multipass arrangement.

In ion beam sputtering experiments, fast metal atom beams are generated by directing a keV ion beam with mA cm^{-2} current densities at a solid metal target.[219] The energy distribution of the sputtered neutrals is broad, and energy selecting devices consisting of high speed rotating disks are necessary. This dramatically reduces the intensity. Ion beam sputtering, however, provides higher intensities below 10 eV than beam neutralization. At energies below 1 eV, however, clustering becomes a serious problem, while at energies above 10 eV, metastable species associated with the ion beam are present in the neutral beam. Metal atom beams produced by laser ablation of metals[220,221] or metal compounds[222] can also produce high intensities of pulsed atomic beams. The results of dynamics studies obtained from such experiments, however, indicate substantial contributions from metastables.

Beams of heavier species with energies around 1 eV can be produced by seeding supersonic jets of He or H_2.[223] By seeding $\sim 1\%$ of the species of interest in the light gas, the heavier species adopts a velocity close to that of the carrier gas. Heating the jet nozzle can control the energy of the seeded neutral beam. Energies as high as 15 eV have been achieved for Xe in H_2.[224] Seeded beams have the highest hyperthermal neutral intensities.

A common drawback of all fast neutral beam experiments is the difficulty in determining absolute cross sections. This can be primarily attributed to problems in deriving the neutral beam intensity in the interaction region and product detection efficiencies. As will be discussed in the following section, estimated cross sections for examined processes have been reported, but they can differ by orders of magnitude from one fast neutral beam experiment to another.

The above beam techniques can be categorized as single-collision experiments. In direct application to meteor ionization, Friichtenicht and coworkers[225,226] have measured the degree of ionization, β, of fast particles injected into a thick gas target. "Thick" implies a gaseous column density sufficient to thermalize most of the vaporized atoms. These experiments consist of producing fast submicron sized particles using a charged particle injector coupled with an electrostatic particle accelerator. The particle charge, mass and velocity are determined through capacitively induced voltage pulses on cylinders surrounding the beam. The beam then passes a differentially pumped gas chamber, where the particle is vaporized. Ion pair formation is measured on opposite parallel plates following the collision chamber. Because the initial particle velocity is much larger than the thermal velocity of the target gas, and given the approximate $\sim v^4$ dependence of the ionization cross sections (see Sec. 2.2.2) implying that most ionization occurs at large relative velocities, it is assumed that all ions are produced with sufficient velocities to emerge from the collision chamber. The degree of ionization is then given by the ratio of ion pairs formed to the total number of atoms in the particle.

3.4.2. *Neutral Beam Studies Involving Metallic Species*

Neutral beam experiments covering a large range of kinetic energies are necessary to study processes (15)–(18) and (23)–(25). A wealth of *fast* neutral beam studies stem from the early molecular beam studies in the 1960s and 1970s. Extensive reviews of this body of research have been published by Baede[152] and Kempter.[153] A large fraction of this work focused on the dynamics of alkali metals, given the relative ease in producing high intensity beams, as well as the ability to efficiently measure the beam intensities using surface ionization detectors.[227]

Bydin and Bukhteev[228,229] and Bukhteev *et al.*[230] reported the earliest studies on collisional ionization of fast alkali metal beams with various gases including N_2 and O_2 at beam energies between 150 and 2000 eV. The neutral beams were produced by ion beam neutralization. Ten years later, Cuderman[231] used improved instrumentation to determine the ionization cross sections of K atoms in collisions with gases including N_2 and O_2 from 20 to 1000 eV. The two sets of data are compared in Fig. 17. The K measurements of of Cuderman are significantly higher than those of Bukhteev and coworkers at velocities exceeding 20 km s^{-1}. In both the Na and K measurements, the O_2 cross sections are more than an order of magnitude higher than those of N_2. This can partly be related to the longer range ion-pair formation crossing in the O_2 system (see Sec. 3.2.2.2), as well as to the lower respective thresholds. The higher K ionization cross sections compared with Na can be equally argued on the basis of the lower K ionization potential, thereby shifting the charge transfer crossing to longer range where diabatic passage of crossings is more likely.

From the point of view of meteor modeling, the comparison of the cross sections in Fig. 17 to the charge transfer cross sections in Fig. 12 is important. While the Na + N_2 collisional ionization and Na$^+$ + N_2 charge transfer cross sections are comparable at ~ 70 km s^{-1}, collisional ionization

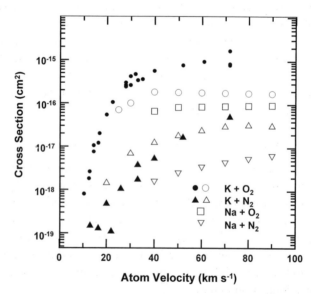

Fig. 17. Velocity dependence of ionization cross sections for Na and K collisions with O_2 and N_2 as measured using neutralized alkali ion beams. ▲● Cuderman[231] ○□△▽ Bydin and Bukhteev.[229]

is significantly more efficient in the corresponding $(K + N_2)$ collision pairs. Bukhteev *et al.*[230] also used instrumentation with a mass spectrometric detection system that allowed them to differentiate between free electron formation and ion pair formation. In the $K + O_2$ system, ion-pair formation was the most important process until 50 km s^{-1}, after which electron loss becomes the dominant channel. The ion pair formation cross section peaks at about 35 km s^{-1} before it drops quite dramatically. The drop coincides with the plateau of the total ionization cross section in Fig. 17. The maximum could be explained by the traversal of a Landau–Zener maximum, or by increasing degrees of O_2^- vibrational excitation, eventually leading to autoionization.

The first detailed molecular dynamics studies on the effects of translational energy on chemical reactivity were conducted with alkali metal atomic beams. Of particular mention are the studies by the group of Herschbach[232] at Harvard University, scientists at the FOM institute in Amsterdam, and scientists at the University of Freiburg, Germany. These three groups made a concerted effort in conducting fast neutral studies in the translational energy range symptomatic of chemical bonding, with particular emphasis on observing thresholds of processes. Extensive studies were conducted on alkali + halogen collisions, where thresholds provide the first measurements of a number of halogen electron affinities.[152]

In the first hyperthermal K atom studies involving beams produced by ion beam neutralization, Herschbach and coworkers[233] compared collisional excitation functions for exciting the K (4^2P) lines in collisions with the rare gases and N_2, O_2, H_2, and Cl_2. While the observed excitation thresholds for the atomic target gases were far higher than the endoergicity, this was not the case for N_2 and O_2, where substantial relative cross sections were measured at the then lowest available beam velocity of 20 km s^{-1}. The results were interpreted with respect to ion pair states, and the high thresholds of the H_2 and Cl_2 collision systems in contrast to those of N_2 and O_2 were rationalized on the basis of poor electron attachment Franck–Condon factors of the former. The latter argument is questionable in the case of H_2 and N_2, where the negative ion ground states are shape resonances.

In an ensuing study, Lacman and Herschbach[234] studied collisional excitation and ionization of K with improved neutral beam intensities, permitting studies at beam energies as low as 2 eV. The study involved a number of diatomic molecules. Both ionization and excitation thresholds close to the thermochemical limits were observed for NO, CO, N_2 and O_2, while for HCl and Cl_2, only the ionization thresholds were observed. In parallel, the Freiburg group conducted excitation studies with a sputtering source that produces higher intensities at low energies and clearly identified the thresholds as corresponding to thermochemical onsets.[235] These authors estimated the excitation cross section in K + O_2 collisions to be on the order of 2 Å^2 at $\sim$ 7 km s^{-1}.

Los and coworkers[236] at the FOM institute also used a sputtering source to study ionization cross sections of alkali metal atom collisions with O_2 at relative velocities below 13 km s^{-1}. These authors used magnetic fields in the charged particle detection part of their experiment, allowing them to differentiate between negative ions and e$^-$. The K + O_2 measurements of Moutinho *et al.* are shown in Fig. 18. They are compared with the K($4^2P \rightarrow 4^2S$) resonance line excitation cross sections of Lacmann and Herschbach[234] after scaling them to the estimated cross section of Kempter *et al.*[235] A dotted line indicates the measurements of Kempter *et al.* indicating that discrimination, possibly related to determining the surface ionization detection efficiency of the neutral beam, may have affected one or the other experiment. It is important to note that the excitation and ionization cross sections exhibit distinct thresholds and have comparable

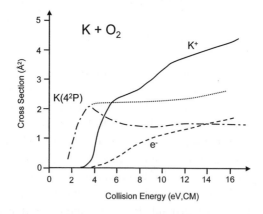

Fig. 18. Collision energy dependence of the near threshold ionization and excitation cross sections in K + O_2 collisions. The K$^+$ and e$^-$ cross sections are those of Moutinho *et al.*[236] The excitation cross sections are those of Kempter *et al.* (higher dotted curve) and Lacman and Herschbach (dot-dashed curve).[234] The latter were reported as relative cross sections, while the former provided an estimated absolute cross section.

cross sections. Near threshold, the negative ions are produced in the ion pair formation channel:

$$K + O_2 \rightarrow K^+ + O_2^- , \tag{61}$$

and the observed threshold is consistent with the respective thermochemical limit of 3.9 eV. Given the relatively small electron affinity of O_2 of 0.44 eV, the observed onset for free electron formation:

$$K + O_2 \rightarrow K^+ + O_2 + e^- , \tag{62}$$

lies very close to this thermochemical limit. The resonant line excitation cross section exhibits a clear drop (leveling off, in the case of the Kempter data) at the onset where negative ions are formed.

The above observations are fully consistent with the type of curve-crossing model discussed for the Na + O atomic system in Sec. 3.2.2.2 and illustrated in Fig. 7. The excitation channel is accessed via the ion pair state. The distinct appearance at threshold illustrates nicely the substantially higher charge transfer efficiency for molecular systems vs. that of atomic systems. This is not unreasonable considering the fact that the inefficiency of the Na + O system at energies near threshold can be attributed to the high charge transfer coupling, while the vibronic coupling in the molecular system can be substantially reduced through unfavorable Franck–Condon factors [see Eq. (45)] or substantial increases in the equilibrium distance of the molecular negative ion (corresponding to an increase in electronegativity with interatomic distance), leading to an increase in the crossing radius, R_c, as the diatomic negative ion stretches during the collision. Meanwhile, considering the autoionization arguments put forth in Sec. 3.2.2.2, the appearance of free electrons at threshold implies a mechanism involving autoionization of a vibrationally excited negative ion. This is in stark contrast to atomic systems, where single-collision free-electron formation implies a mechanism involving the repulsive part of the potential that is substantially less efficient.

Maybe the most astonishing feature of the cross sections in Fig. 18 are the high absolute magnitude of the cross sections compared with the measurements of Buktheev and coworkers[229,230] and Cuderman[231] for total ionization shown in Fig. 17 (16 eV c.m. corresponds to ~ 13 km s^{-1}). The cross sections of the Amsterdam group are nearly two orders of magnitude larger than those obtained using ion beam neutralization methods at corresponding low energies. The competition between excitation and ion pair formation shown in Fig. 18 indicates that the magnitudes of the cross sections reported by Moutinho *et al.*[236] and Kempter *et al.*,[235] both involving sputtering sources, are consistent. Meanwhile, the agreement between observed and thermochemical thresholds suggests that excited states in the sputtering experiments are not of importance.

Moutinho *et al.*[236] measured analogous behavior for Na + O_2. The cross sections are approximately a factor of 2 smaller than those of K + O_2. Note that the cross sections of Bukhteev and coworkers[229,230] for Na + O_2 are similar to those predicted in Fig. 8 for the atomic Na + O system. Because molecular charge transfer systems usually exhibit higher cross sections, the comparison with the calculated atomic cross sections may indicate that the measured values for the molecular systems are too low. Given the fact that the Russian experiments had a narrow acceptance angle, it can be assumed that they suffered serious discrimination at low energies, where large angle scattering is more probable. The Na + O_2 collision system was also investigated by Neynaber and coworkers using a merged beam experiment.[237] Although their instrument was more designed to investigate differential scattering at lower energies, these authors report an estimated cross section of ~ 0.05 Å^2 at 10 eV which is about 40 times lower than the total ionization cross section reported by Moutinho and coworkers.

In several ensuing studies with alkali beams from charge transfer cells, Los and coworkers[238–240] determined cross sections of alkali–O_2 systems at higher energies than the work by Moutinho *et al.*[236] The higher energy range enabled the observation of Landau–Zener maxima ($p_{LZ} = 0.5$). Consequently, an excellent estimate of the electronic coupling parameters Eq. (33) was possible. Simple trajectory models,[241] including the vibrational motion of the diatomic molecule, provided good agreement with the structure observed in the energy dependence of the cross sections.[238] The structure was attributed to the effect of vibrational motion of O_2^- on R_c. The calculated higher energy cross sections of the K + O_2 system were found to be in good agreement with those of Cuderman.[231] The absolute cross sections calculated at low energies agreed with the low energy cross sections measured by Moutinho *et al.*[236] Meanwhile, a recent analysis of meteor observations[242] and their relationship to laboratory ionization coefficient measurements,[226] further discussed in Sec. 3.4.3, also appear consistent with large ionization cross sections. The marked difference between the molecular and atomic free-electron cross sections, as predicted in Sec. 3.2.2.2, implies that atomic oxygen + metal ionization systems contribute substantially less to the observed ionization than the molecular systems.

Few experiments exist on nonalkali metal collisions at sub-keV collision energies. This can be attributed to the difficulty in generating intense beams of these metals. Bukhteev and Bydin reported measurements of fast Ca, Mg, Si and Fe ionization cross sections in N_2 and O_2 in ion beam neutralization experiments.[243] The measured cross sections are shown in Fig. 19. Only for Ca was the sensitivity sufficient for cross section measurements in the velocity range of interest to meteor research. The cross sections are also considered to be of order of magnitude accuracy. The authors do not rule out a reduced collection efficiency due to large angle scattering. As seen in Fig. 19, except for Fe collisional ionization, the cross sections are found to be generally lower than those observed for the alkali metals in Fig. 17. This can in part be attributed to the higher ionization potentials of the metals. In case of the Fe work, it must be assumed that a number of excited states are produced in both the ion precursor beam and the charge transfer neutral beam.

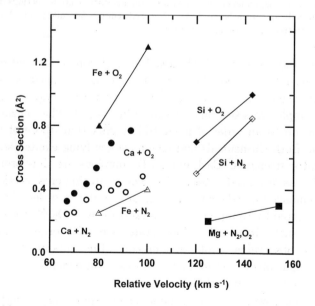

Fig. 19. Velocity dependence of collisional ionization cross sections of nonalkali metal atoms in collisions with O_2 and N_2, measured using a fast alkali metal atom beam apparatus of Bukhteev and Bydin.[243]

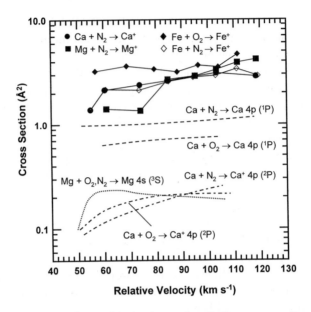

Fig. 20. Velocity dependence of ionization and excitation cross sections for several fast nonalkali collisions with N_2 and O_2, measured using a fast alkali metal atom beam apparatus of Boitnott and Savage.[38,213,214]

Boitnott and Savage[38,213,214] looked at both ionization and excitation cross sections of Mg, Ca, and Fe + N_2, O_2 collision systems using a crossed-beam arrangement and neutralized ion beams with energies from 200 to 2400 eV. Sample cross sections are shown in Fig. 20. The ionization cross sections are up to an order of magnitude higher than those of Bukhteev and Bydin. Both Ca neutrals and ions have low lying excited states with high oscillator strengths. Ca I and Ca II emissions are thus prominent in meteors,[29,30] as discussed in Sec. 2.3.1. The excitation cross sections of the neutral resonance lines shown in Fig. 20 are similar in order of magnitude to those reported by Kempter and coworkers for K + O_2.[235] Cross sections for charge transfer into excited ionic states are a full order of magnitude smaller. The observable visible Mg emissions of meteors involve highly excited states that involve excited lower states. Consequently, the respective cross sections are significantly lower.

Wexler and coworkers have conducted ion sputtering experiments on a large number of nonalkalis, including Al, Fe, Mg and Ca.[140,141] These authors did not report absolute cross sections and were more interested in chemiionization processes (Reactions 16–18). For those metallic species that exhibit chemiionization channels, they are of importance below collision energies of ~ 20 eV. At higher energies, electron transfer ionization rapidly becomes the dominant ionic excitation channel. For all relevant meteoric metals, chemiionization is found to be negligible compared with electron transfer ionization. Chemiionization, however, can be very important for all metallic species when electronically excited species are involved, such as $O_2(a^1\Delta_g)$.[244,245]

Low energy chemical reactions and quenching of excited metal atoms has been investigated in crossed-beams. Examples are the studies of the sodium-nightglow precursor reaction:

$$Na + O_3 \rightarrow NaO + O_2, \tag{63}$$

by Herschbach, Kolb and coworkers,[114,115,246] Dyke and coworkers[113,247] and Lee and coworkers.[248] The combined low pressure flow tube and molecular beam experiments of Herschbach and coworkers, and photoionization studies of the products of reaction (7) led to the conclusion that the product monoxide is formed with a branching ratio exceeding 0.8 in the first excited $A^2\Sigma^+$ state. This finding explained the inconsistency between the observed sodium nightglow yields and the Chapman models based on the NaO + O $\rightarrow Na(^2P) + O_2$ rate coefficient (Reaction 8) measured for $NaO(X^2\Pi)$.[249] The latter was measured in an atmospheric heat-pipe experiment in which NaO(A) is efficiently quenched.

3.4.3. *Fast Particle Experiments: The Work of Früchtenicht, Slattery, and Hansen*

Früchtenicht and coworkers[225,226] have investigated the degree of ionization of Fe particles passing through air and various pure gaseous targets at initial velocities ranging between 15 and 45 km s^{-1}. The observed ionization coefficient in O_2 is ~ 0.1 at 20 km s^{-1} and rises almost exponentially with energy to 1.0 at ~ 33 km s^{-1}. This translates into a $\sim v^{3.12}$ ionization efficiency dependence. The observed ionization coefficient in N_2 is ~ 0.45 at 20 km s^{-1} and passes 1.0 at 44 km s^{-1}. In an attempt to compare the determined ionization coefficients with the cross section measurements of Bukhteev and Bydin,[243] Früchtenicht derived the ionization cross section from the measured ionization coefficients assuming that only the first collision produced ionization, in which case the ionization coefficient is given by:

$$\beta \approx \frac{\sigma_I}{\sigma_D}, \tag{64}$$

where σ_I is the ionization cross section and σ_D is the momentum transfer cross section.

Assuming that σ_D is approximately equal to a hard sphere cross section, these assumptions lead to an ionization cross section of approximately 15 Å^2 for Fe + N$_2$ collisions at 40 km s^{-1}. Interestingly, this large cross section, which must be considered an upper limit povided the ionization coefficients are accurate, is in the range of magnitudes reported for fast alkali + O$_2$/N$_2$ experiments such as those depicted in Fig. 18. Friichtenicht and coworkers concluded that this cross section was not dramatically inconsistent with the measurements of Bukhteev and Bydin if the experiments were subject to discrimination given the narrow acceptance angle of 1°48' of the respective instrument. The conclusion was based on a large angle scattering correction that assumed isotropic scattering in the center-of-mass frame of reference. This assumption may be valid at low energies near threshold, but is questionable if ionizing transitions occur primarily in the attractive part of the potential, as is clearly the case in the discussed alkali + O$_2$[236] and the Na + O collision systems.[157] In this case, straight-line trajectories are a more accurate approximation and the collection correction to the high energy measurements of Bukhteev and Bydin is not as large as determined by Friichtenicht *et al.* In a recent analysis, Jones has found that the measurements of Friichtenicht *et al.* are consistent with meteor observations.[242]

4. Conclusions

This chapter demonstrates the important role chemical dynamics play in the phenomenology of meteors and the associated ionospheric consequences. Although considerable work has been done to derive the important parameters of the relevant molecular and atomic collisions under extreme conditions, there are still many poorly known cross sections that are required to properly model the macroscopic observables based on microscopic processes. The current state of knowledge is best for low energy metal ion collisions, where the extensive work of the Armentrout group at the University of Utah (see Sec. 3.3.2) has made the most valuable and extensive contribution. A missing component are cross sections for Me$^+$ + O$_3$ reactions which could be an efficient source of metal oxide ions that dissociatively recombine to form excited metal atoms, thus providing a source for observed atomic emissions. Decisive fast neutral beam experiments are still needed to fully establish the relevant collisional ionization cross sections. The recent progress in calculating excited state potentials should make accurate theoretical treatments of the respective high energy nonadiabatic dynamics feasible. Maybe the most serious shortcoming is the lack of data on excited state collisions. Given the extreme difficulty in state-selected high energy experiments, future progress will depend strongly on theory. Models for excitation and ionization are already sufficiently developed when long range curve crossings govern the dynamics, although there are few rigorous calculations for hyperthermal molecular systems. Free-electron formation, important with respect to radar observations, appears to be most efficient through metal atom collisions with atmospheric molecules, where the efficiency can be rationalized through vibrational autodetachment of an intermediate negative ion formed in a curve-crossing process. The theory to predict free-electron formation in atomic collisions is still underdeveloped. Improved insight is necessary on the dynamics occurring high on the repulsive wall of the interfragment potential where transitions from the ground state to an infinite series of Rydberg states and the ionization continuum are possible.

Acknowledgments

This work was supported by AFOSR. We thank the University of Chicago Press for permission to use Figs. 5 and 6 and Elsevier Science Publishers for permission to use Fig. 13. We thank Prof. Peter Armentrout for helpful suggestions, a thorough review of the manuscript, and for providing Fig. 13. We are deeply indebted to Dr. Charles Kolb for his scrutiny of the manuscript and for valuable additions to this chapter. Many thanks to Dr. Detlef Schröder for his input on iron oxide thermochemistry, and Prof. Joachim Grosser for his helpful comments regarding ionizing collisions. We benefited immensely from spirited discussions with Dale Levandier, Shu Lai and William J. McNeil, Dr. Zdeněk Ceplecha, Dr. Ivo Čermák, Prof. Iwan Williams, Dr. Asta Pellinen-Wannberg, Prof. William J. Baggaley, and Prof. J. A. M. McDonnel.

References

1. M. C. Kelley and R. A. Heelis, *The Earth's Ionosphere — Plasma Physics and Electrodynamics* (Academic, San Diego, CA, 1989).

2. W. B. Hanson and S. Sanatani, *J. Geophys. Res.* **75**, 5503 (1970).
3. W. B. Hanson and S. Sanatani, *J. Geophys. Res.* **76**, 7761 (1971).
4. W. B. Hanson, D. L. Sterling and R. F. Woodman, *J. Geophys. Res.* **77**, 5530 (1972).
5. J. M. Grebowsky and H. C. Brinton, *Geophys. Res. Lett.* **5**, 791 (1978).
6. J. M. Grebowsky and M. W. Pharo III, *Planet. Space Sci.* **33**, 807 (1985).
7. J.-C. Gérard and A. Monfils, *J. Geophys. Res.* **A79**, 2544 (1974).
8. J.-C. Gérard, *J. Geophys. Res.* **81**, 83 (1976).
9. J.-C. Gérard and A. Monfils, *J. Geophys. Res.* **A83**, 4389 (1978).
10. C. L. Fesen and P. B. Hays, *J. Geophys. Res.* **87**, 9217 (1982).
11. E. Anders, in *The Moon, Meteorites, and Comets*, Vol. 3, Eds B. M. Middlehurst and G. P. Kuiper (University of Chicago Press, Chicago, 1963), p. 402.
12. J. A. Wood, in *The Moon, Meteorites, and Comets*, Vol. 3. Eds. B. M. Middlehurst and G. P. Kuiper (University of Chicago Press, Chicago, 1963), p. 337.
13. J. T. Wasson, *Meteorites — Their Record of Early Solar System History* (W. H. Freeman, New York, 1985).
14. H. Y. McSween, Jr., *Meteorites and Their Parent Planets* (Cambridge University Press, New York, 1987).
15. H. Campins and T. D. Swindle, *Meteoritics Planet. Sci.* **33**, 1201 (1998).
16. F. L. Whipple, *Astrophys. J.* **111**, 375 (1950).
17. K. S. K. Swamy, *Physics of Comets*, 2nd edition (World Scientific, Singapore, 1997).
18. M. E. Bailey, S. V. M. Clube and W. M. Napier, *The Origin of Comets* (Pergamon Press, Oxford, United Kingdom, 1990).
19. S. Battersby, *Nature* **387**, 23 (1997).
20. J. K. Wilson, J. Baumgardner and M. Mendillo, *Geophys. Res. Lett.* **25**, 225 (1998).
21. J. H. Poynting, *Phil. Trans. Roy. Soc.* **A202**, 525 (1904).
22. H. P. Robertson, *Mon. Not. Roy. Astron. Soc.* **97**, 423 (1937).
23. D. W. Hughes, in *Cosmic Dust*, Ed. J. A. M. McDonnel (Wiley, New York, 1978).
24. A. Evans, *The Dusty Universe*, 1st edition (Wiley, West Sussex, United Kingdom, 1994).
25. J. S. Lewis, *Physics and Chemistry of the Solar System* (Academic Press, New York, 1995).
26. B. J. Gladman, F. Migliorini, A. Morbidelli, V. Zappalà, P. Michel, A. Cellino, C. Froeschlé, H. F. Levison, M. Bailey and M. Duncan, *Science* **277**, 197 (1997).
27. C. P. Olivier, *Meteors* (Williams and Wilkins, London, 1925).
28. A. C. B. Lovell, *Meteor Astronomy* (Oxford University Press, London, 1954).
29. V. A. Bronshten, *Physics of Meteoric Phenomena* (Kluwer, Dordrecht, Holland, 1983).

30. Z. Ceplecha, J. Borovicka, W. G. Elford, D. O. Revelle, R. L. Hawkes, V. Porubcan and M. Simek, *Space Sci. Rev.* **84**, 327 (1998).
31. R. J. M. Olson and J. M. Pasachoff, *Fire in the Sky — Comets and Meteors, the Decisive Centuries, in British Art and Science* (Cambridge Univeristy Press, Cambridge, United Kingdom, 1998).
32. D. W. R. McKinley, *Meteor Science and Engineering* (McGraw-Hill, New York, 1961).
33. D. W. Hughes, *Space Res.* **XV**, 531 (1975).
34. F. L. Whipple, *Rev. Mod. Phys.* **15**, 246 (1943).
35. E. J. Öpik, *Physics of Meteor Flight in the Atmosphere* (Interscience Publishers, New York, 1958).
36. J. Rosinski and R. H. Snow, *J. Meteorology* **18**, 736 (1961).
37. D. M. Hunten, *Geophys. Res. Lett.* **8**, 369 (1981).
38. C. A. Boitnott and H. F. Savage, *Astrophys. J.* **167**, 349 (1971).
39. H. S. W. Massey and D. W. Sida, *Phil. Mag.* **46**, 190 (1955).
40. V. M. Slipher, *Publ. Astro. Soc. Pacific* **41**, 262 (1929).
41. L. Vegard and E. Tønsberg, *Z. Physik* **94**, 413 (1935).
42. J. Cabannes and J. Dufay, *Compte Rendus* **206**, 221 (1938).
43. R. Bernard, *Z. Physik* **110**, 291 (1938).
44. R. Bernard, *Astrophys. J.* **89**, 133 (1939).
45. J. Cabannes, J. Dufay, and J. Gauzit, *Astrophys. J.* **88**, 164 (1938).
46. S. Chapman, *Astrophys. J.* **90**, 309 (1939).
47. D. L. Hildenbrand and E. Murad, *J. Chem. Phys.* **53**, 3403 (1970).
48. A. A. Viggiano and F. Arnold, in *Handbook of Atmospheric Electrodynamics*, Vol. 1, Ed. H. Volland (CRC Press, 1995), p. 1.
49. G. W. Kronk, *Meteor Showers — A Descriptive Catalog* (Enslow Puablishers, Hillside, New Jersey, 1988).
50. Y. Fujiwara, M. Ueda, Y. Shiba, M. Sugimoto, M. Kinoshita and C. Shimoda, *Geophys. Res. Lett.* **25**, 285 (1998).
51. B. A. Linblad, *Earth, Moon, and Planets* **68**, 405 (1995).
52. D. O. ReVelle and R. S. Rajan, *J. Geophys. Res.* **84**, 6255 (1979).
53. P. M. Millman and D. W. R. McKinley, in *The Moon, Meteorites, Comets*, Vol. IV, Eds. B. M. Middlehurst and G. P. Kuiper (University of Chicago, Chicago, IL, 1963), p. 674.
54. O. G. Ovezgeldyev, S. Mukhamednazarov, R. I. Shafiev and N. V. Maltsev, in *Handbook for Middle Atmosphere Program*, Vol. 25, Ed. R. G. Roper (SCOSTEP Secretariat, University of Illinois, Urbana, IL, 1987), p. 422.
55. P. M. Millman, A. F. Cook and C. L. Hemenway, *Can. J. Phys.* **49**, 1365 (1971).
56. W. J. Baggaley, *Bull. Astron. Inst. Czechoslovakia* **28**, 356 (1977).
57. W. J. Baggaley, *Bull. Astron. Inst. Czechoslovakia* **29**, 59 (1978).
58. J. Jones, B. A. McIntosh and M. Simek, *J. Atmos. Terr. Phys.* **52**, 253 (1990).

59. A. Pellinen-Wannberg and G. Wannberg, *J. Geophys. Res.* **A99**, 11379 (1994).
60. Q. H. Zhou and M. C. Kelley, *J. Atm. Solar Terr. Phys.* **59**, 739 (1997).
61. F. Verniani, *J. Geophys. Res.* **78**, 8429 (1973).
62. W. Jones, *Mon. Not. Roy. Astron. Soc.* **288**, 995 (1997).
63. D. M. Hunten, *Science* **145**, 26 (1964).
64. D. M. Hunten and L. Wallace, *J. Geophys. Res.* **77**, 69 (1967).
65. J. E. Blamont, *J. Atm. Terr. Phys.* **37**, 927 (1975).
66. D. R. Bates, in *Applied Atomic Collision Physics, — Atmospheric Physics and Chemistry*, Vol. 1, Eds. H. S. W. Massey and D. R. Bates (Academic Press, New York, 1982), p. 482.
67. A. L. Broadfoot, *Planet. Space Sci.* **15**, 503 (1967).
68. A. L. Broadfoot and A. E. Johanson, *J. Geophys. Res.* **81**, 1331 (1976).
69. J. G. Anderson and C. A. Barth, *J. Geophys. Res.* **76**, 3723 (1971).
70. A. Boksenberg and J.-C. Gérard, *J. Geophys. Res.* **78**, 4641 (1973).
71. J.-C. Gérard, D. W. W. Rusch, P. B. Hays and C. L. C. L. Fesen, *J. Geophys. Res.* **A84**, 5249 (1979).
72. J. Joiner and A. C. Aiken, *J. Geophys. Res.* **101**, 5239 (1996).
73. S. Kumar and W. B. Hanson, *J. Geophys. Res.* **A85**, 6783 (1980).
74. C. Fesen, P. B. Hays and D. N. Anderson, *J. Geophys. Res.* **88**, 3211 (1983).
75. S. B. Mende, G. R. Swenson and K. L. Miller, *J. Geophys. Res.* **A90**, 6667 (1985).
76. J. A. Gardner, R. A. Viereck, E. Murad, D. J. Knecht, C. P. Pike, A. L. Broadfoot and E. Anderson, *Geophys. Res. Lett.* **22**, 2119 (1995).
77. J. A. Gardner, E. Murad, D. J. Knecht, R. A. Viereck, C. P. Pike and A. L. Broadfoot, *SPIE Proc.* **2830**, 64 (1996).
78. J. A. Gardner, E. Murad, R. A. Viereck, D. J. Knecht, C. P. Pike and A. L. Broadfoot, *Adv. Space Res.* **21**, 867 (1998).
79. D. C. Senft, C. A. Hostetler and C. S. Gardner, *J. Atm. Terr. Phys.* **55**, 425 (1993).
80. J. D. Mathews, T. J. Kane, C. S. Gardner and Q. Zhou, *J. Atm. Terr. Phys.* **55**, 363 (1993).
81. R. Neuber, P. von der Gathen and U. von Zahn, *J. Geophys. Res.* **93**, 11 (1988).
82. U. von Zahn, R. A. Goldberg, J. Stegman and G. Witt, *Planet. Space Sci.* **37**, 657 (1989).
83. U. von Zahn and C. Tilgner, 8th ESA Symposium on Rocket and Balloon Programmes and Related Research, ESA SP-270 (ESA, Sunne, Sweden, 1987), pp. 107–112.
84. U. von Zahn, G. Hansen and H. Kurazawa, *Nature* **331**, 594 (1988).
85. C. S. Gardner, *Proc. IEEE* **77**, 408 (1989).
86. G. Hansen and U. von Zahn, *J. Atm. Terr. Phys.* **52**, 585 (1990).
87. U. von Zahn and T. L. Hansen, *Planet. Space Sci.* **50**, 93 (1988).
88. B. R. Clemesha, *J. Atm. Terr. Phys.* **57**, 725 (1995).

89. M. Alpers, T. Blix, S. Kirkwood, D. Krankowsky, F.-J. Lübken, S. Lutz and U. von Zahn, *J. Geophys. Res.* **A98**, 275 (1993).
90. R. M. Cox and J. M. C. Plane, *J. Geophys. Res.* **103**, 6349 (1998).
91. C. Granier, J. P. Jegou and G. Megie, *Geophys. Res. Lett.* **16**, 243 (1989).
92. C. S. Gardner, T. J. Kane, D. C. Senft, J. Qian and G. C. Papen, *J. Geophys. Res.* **98**, 16 (1993).
93. C. Granier, J. P. Jégou and G. Mégie, *Geophys. Res. Lett.* **12**, 655 (1985).
94. V. Eska, J. Höffner and U. von Zahn, *J. Geophys. Res.* **103**, 29 (1998).
95. J. Höffner, U. von Zahn, W. J. McNeil and E. Murad, *J. Geophys. Res.* **104**, 2633 (1999).
96. C. Y. Johnson and E. B. Meadoows, *J. Geophys. Res.* **60**, 193 (1955).
97. C. Y. Johnson, J. P. Heppner, J. C. Holmes and E. B. Meadows, *Ann. Geophys.* **14**, 475 (1958).
98. V. G. Istomin and A. Z. Pokhunkov, *Space Res.* **3**, 117 (1963).
99. R. S. Narcisi, *Ann. Geophys.* **22**, 224 (1966).
100. R. S. Narcisi, *Space Res.* **8**, 360 (1968).
101. D. Krankowsky, F. Arnold, H. Wieder and J. Kissel, *Int. J. Mass Spectrom. Ion Phys.* **8**, 379 (1972).
102. A. C. Aikin and R. A. Goldberg, *J. Geophys. Res.* **78**, 734 (1973).
103. A. D. Zhlood'ko, V. N. Lebedinets and V. B. Shushkova, *Space Res.* **9**, 277 (1974).
104. P. A. Zbinden, M. A. Hidalgo, P. Eberhardt and J. Geiss, *Planet. Space Sci.* **23**, 1621 (1975).
105. U. Herrmann, P. Eberhardt, M. A. Hidalgo, E. Kopp and L. G. Smith, *Cospar Space Res.* **18**, 249 (1978).
106. E. Kopp, *J. Geophys. Res.* **102**, 9667 (1997).
107. J. A. Rutherford, R. F. Mathis, B. R. Turner and D. A. Vroom, *J. Chem. Phys.* **55**, 3785 (1971).
108. J. A. Rutherford and D. A. Vroom, *J. Chem. Phys.* **57**, 3091 (1972).
109. J. M. Grebowsky and N. Reese, *J. Geophys. Res.* **94**, 5427 (1989).
110. D. F. Bedey and B. J. Watkins, *J. Geophys. Res.* **102** (A5), 9675 (1997).
111. S. G. Love and D. E. Brownlee, *Icarus* **89**, 26 (1991).
112. J. Fegley, B. and A. G. W. Cameron, *Earth Planet. Sci. Lett.* **82**, 207 (1987).
113. J. M. Dyke, A. M. Shaw and T. G. Wright, in *Gas-Phase Metal Reactions*, Ed. A. Fontijn (Elsevier, The Netherlands, 1992), p. 467.
114. D. R. Herschbach, C. E. Kolb, D. R. Worsnop and X. Shi, *Nature* **356**, 414 (1992).
115. C. E. Kolb, D. R. Worsnop, M. S. Zahnhiser, G. N. Robinson, X. Shi and D. R. Herschbach, in *Gas-Phase Metal Reactions*, Ed. A. Fontijn (Elsevier, Amsterdam, 1992), p. 15.
116. D. M. Hunten, R. P. Turco and O. B. Toon, *J. Atm. Sci.* **37**, 1342 (1980).
117. E. Murad and W. Swider, *Geophys. Res. Lett.* **6**, 929 (1979).
118. J. M. C. Plane, *Int. Rev. Phys. Chem.* **10**, 55 (1991).

119. J. M. C. Plane and M. Helmer, in *Research in Chemical Kinetics*, Vol. 2, Eds. R. G. Compton and G. Hancock (Elsevier, Amsterdam, 1994), p. 313.
120. J. M. C. Plane and M. Helmer, *Faraday Discuss.* **100**, 411 (1995).
121. J. M. C. Plane and R. J. Rollaston, *J. Chem. Soc. Faraday Trans.* **92**, 4371 (1996).
122. I. Ager, J. W. and C. J. Howard, *J. Geophys. Res.* **92**, 6675 (1987).
123. C. E. Kolb and J. B. Elgin, *Nature* **263**, 488 (1976).
124. R. J. Rollaston and J. M. C. Plane, *J. Chem. Soc. Faraday Trans.* **94**, 3067 (1998).
125. W. J. McNeil, S. T. Lai and E. Murad, *J. Geophys. Res.* **A101**, 5 (1996).
126. W. J. McNeil, S. T. Lai and E. Murad, *J. Geophys. Res.* **D103**, 10899 (1998).
127. L. N. Carter and J. M. Forbes, *Ann. Geophysicae* **17**, 190 (1999).
128. G. Brasseur and S. Solomon, *Aeronomy of the Middle Atmosphere —* *Chemistry and Physics of the Stratosphere and Mesosphere* (D. Reidel, Dordrecht, The Netherlands, 1984).
129. A. Fontijn and P. M. Futerko, in *Gas-Phase Metal Reactions*, Ed. A. Fontijn (Elsevier Science Publishers, Amsterdam, 1992).
130. C. Vinckier, P. Christiaens and M. Hendrickx, in *Gas-Phase Metal Reactions*, Ed. A. Fontijn (Elsevier Science Publishers, Amsterdam, 1992).
131. J. Troe, *J. Phys. Chem.* **83**, 114 (1979).
132. A. Fontijn, *Gas-Phase Metal Reactions* (Elsevier Science Publishers, Amsterdam, 1992).
133. T. L. Brown, *Chem. Rev.* **73**, 645 (1973).
134. E. E. Ferguson, *Radio Sci.* **7**, 397 (1972).
135. E. R. Fisher, J. L. Elkind, D. E. Clemmer, R. Georgiadis, S. K. Loh, N. Aristov, L. S. Sunderlin and P. B. Armentrout, *J. Chem. Phys.* **93**, 2676 (1990).
136. E. Murad, *Annu. Rev. Phys. Chem.* **49**, 73 (1998).
137. J. L. Gole and C. E. Kolb, *J. Geophys. Res.* **86**, 9125 (1981).
138. A. Fontijn, W. Felder and J. J. Houghton, *Symp. Int. Combust. Proc.* **16**, 871 (1977).
139. D. Husain and P. E. Norris, *J. Chem. Soc. Faraday Trans. 2* **74**, 93 (1978).
140. S. Wexler, *Ber. Berliner Bunsenges. Physik. Chem.* **77**, 606 (1973).
141. C. E. Young, R. B. Cohen, P. M. Dehmer, L. G. Pobo and S. Wexler, *J. Chem. Phys.* **65**, 2562 (1976).
142. E. P. F. Lee, P. Soldan and T. G. Wright, *Chem. Phys. Lett.* **301**, 317 (1999).
143. R. B. Cohen, C. E. Young and S. Wexler, *Chem. Phys. Lett.* **19**, 99 (1973).
144. P. K. Rol, E. A. Entemann and K. L. Wendell, *J. Chem. Phys.* **61**, 2050 (1964).
145. P. A. G. O'Hare and A. C. Wahl, *J. Chem. Phys.* **56**, 4516 (1972).
146. P. Soldan, E. P. F. Lee and T. G. Wright, *J. Phys. Chem. A* in press (1999).
147. E. E. Ferguson and F. C. Fehsenfeld, *J. Geophys. Res.* **73**, 6215 (1968).
148. P. K. Rol and E. A. Entemann, *J. Chem. Phys.* **49**, 1430 (1968).

149. H. H. Lo, L. M. Clendenning and W. L. Fite, *J. Chem. Phys.* **66**, 947 (1977).
150. J. A. Rutherford, R. F. Mathis, B. R. Turner and D. A. Vroom, *J. Chem. Phys.* **56**, 4654 (1972).
151. S. R. Langhoff, H. Partridge and C. W. Bauschlicher, *Chem. Phys.* **153**, 1 (1991).
152. A. P. M. Baede, *Adv. Chem. Phys.* **30**, 463 (1975).
153. V. Kempter, *Adv. Chem. Phys.* **30**, 417 (1975).
154. L. D. Landau, *Phys. Z. Sowjetunion* **2**, 46 (1932).
155. C. Zener, *Proc. Roy. Soc. London Ser.* **AB7**, 696 (1932).
156. R. E. Olson, F. T. Smith and E. Bauer, *App. Optics* **10**, 1848 (1971).
157. J. v. d. Bos, *J. Chem. Phys.* **52**, 3254 (1969).
158. E. C. G. Stueckelberg, *Helv. Phys. Acta* **5**, 369 (1932).
159. C. Zhu and H. Nakamura, *J. Chem. Phys.* **102**, 7448 (1995).
160. H. Wu, S. R. Desai and L.-S. Wang, *J. Am. Chem. Soc.* **118**, 5296 (1996).
161. S. R. Desai, H. Wu and L.-S. Wang, *Int. J. Mass Spectrom. Ion Proc.* **159**, 75 (1996).
162. H. Wu and L.-S. Wang, *J. Chem. Phys.* **107**, 16 (1997).
163. H. Wu and L.-S. Wang, *J. Chem. Phys.* **107**, 8221 (1997).
164. S. R. Desai, H. Wu, C. M. Rohlfing and L.-S. Wang, *J. Chem. Phys.* **106**, 1309 (1997).
165. H. Wu and L.-S. Wang, *J. Chem. Phys.* **108**, 5310 (1998).
166. V. Kempter, B. Kuebler and W. Mecklenbrauck, *J. Phys.* **B7**, 2375 (1974).
167. O. B. Firsov, *Sov. Phys. JETP* **36**, 1076 (1959).
168. H. S. W. Massey and H. B. Gilbody, *Electronic and Ionic Impact Phenomena*, 2nd Edition. (Oxford University Press, London, 1974).
169. Y. N. Demkov and I. V. Komarov, *Sov. Phys. JETP* **23**, 189 (1966).
170. W. Aberle, J. Grosser and W. Krueger, *Chem. Phys.* **41**, 245 (1979).
171. P. E. Siska, *Rev. Mod. Phys.* **65**, 337 (1993).
172. E. A. Solov'ev, *Sov. Phys. JETP* **63**, 893 (1981).
173. E. A. Solov'ev, *Phys. Rev.* **A42**, 1331 (1990).
174. M. J. Frisch, G. W. Trucks, H. B. Schlegel, P. M. W. Gill, B. G. Johnson, M. A. Robb, T. K. J. R. Cheeseman, G. A. Petersson, J. A. Montgomery, K. Raghavachari, V. G. Z. M. A. Al-Laham, J. V. Ortiz, J. B. Foresman, J. Cioslowski, A. N. B. B. Stefanov, M. Challacombe, C. Y. Peng, P. Y. Ayala, W. Chen, J. L. A. M. W. Wong, E. S. Replogle, R. Gomperts, R. L. Martin, D. J. Fox, D. J. D. J. S. Binkley, J. Baker, J. P. Stewart, M. Head-Gordon, C. Gonzalez, and J. A. Pople, Revision D. 1st edition (Gaussian, Inc., Pittsburgh PA, 1995).
175. N. Rosen and C. Zener, *Phys. Rev.* **40**, 502 (1932).
176. Y. U. Demkov, *Sov. Phys. JETP* **18**, 138 (1964).
177. E. Bauer, E. R. Fisher and F. R. Gilmore, *J. Chem. Phys.* **51**, 4173 (1969).
178. M. R. Spalburg, J. Los and E. A. Gislason, *Chem. Phys.* **94**, 327 (1985).
179. P. Tosi, C. Delvai, D. Bassi, O. Dmitriev, D. Cappelletti and F. Vecchiocattivi, *Chem. Phys.* **209**, 227 (1996).

180. J. M. Farrar, in *Techniques for the Study of Ion–Molecule Reactions*, Vol. 20, Eds. J. M. Farrar and J. W. H. Saunders (Wiley, New York, 1988), p. 325.
181. W. R. Gentry, D. J. McClure and C. H. Douglass, *Rev. Sci. Instr.* **46**, 367 (1975).
182. S. M. Trujillo, R. H. Neynaber and E. W. Rothe, *Rev. Sci. Instr.* **37**, 1655 (1966).
183. E. Teloy and D. Gerlich, *Chem. Phys.* **4**, 417 (1974).
184. D. Gerlich, in *State-Selected and State-to-State Ion-Molecule Reaction Dynamics: Experiment*, Part I, Vol. 82, Eds. C.-Y. Ng and M. Baer (Wiley, New York, 1992), p. 1.
185. E. E. Ferguson, F. C. Fehsenfeld and A. L. Schmeltekopf, *Adv. Atm. Mol. Phys.* **5**, 1 (1969).
186. P. R. Kemper and M. T. Bowers, in *Techniques for the Study of Ion–Molecule Reactions*, Vol. 20, Eds. J. M. Farrar and J. W. H. Saunders (Wiley, New York, 1988).
187. G. N. Ogurtsov, B. I. Kikiani and I. P. Flaks, *Soviet Physics Technical Physics* **11**, 362 (1966).
188. S. H. Neff, *Astrophys. J.* **140**, 348 (1963).
189. J. A. Rutherford and D. A. Vroom, *J. Chem. Phys.* **65**, 4445 (1976).
190. P. B. Armentrout, L. F. Halle and J. L. Beauchamp, *J. Chem. Phys.* **76**, 2449 (1982).
191. E. Murad, *J. Chem. Phys.* **78**, 6611 (1983).
192. N. F. Dalleska and P. B. Armentrout, *Int. J. Mass Spectrom. Ion Process.* **134**, 203 (1994).
193. S. K. Loh, E. R. Fisher, L. Lian, R. H. Schultz and P. B. Armentrout, *J. Phys. Chem.* **93**, 3159 (1989).
194. M. E. Weber, J. L. Elkind and P. B. Armentrout, *J. Chem. Phys.* **84**, 1521 (1986).
195. S. G. Lias, J. E. Bartmess, J. F. Liebman, J. L. Holmes, R. D. Levin and W. G. Mallard, *J. Chem. Ref. Data* **17**, suppl. 1, 1 (1989).
196. J. C. W. Bauschlicher, S. R. Langhoff and H. Partridge, *J. Chem. Phys.* **101**, 2644 (1994).
197. A. L. Farragher, J. A. Peden and W. L. Fite, *J. Chem. Phys.* **50**, 287 (1969).
198. W. R. Henderson, J. E. Mentall and W. L. Fite, *J. Chem. Phys.* **46**, 3447 (1967).
199. P. K. Rol, *Adv. Mass Spectrom* **5**, 189 (1971).
200. R. H. Neynaber, G. D. Magnuson and S. M. Trujillo, *Phys. Rev.* **A5**, 285 (1972).
201. J. A. Rutherford, R. F. Mathis, B. R. Turner and D. A. Vroom, *J. Chem. Phys.* **57**, 3087 (1972).
202. A. A. Radzig and B. M. Smirnov, *Reference Data on Atoms, Molecules, and Ions* (Springer-Verlag, Berlin, 1985).
203. D. J. Levandier, R. A. Dressler and E. Murad, *Rev. Sci. Instr.* **68**, 64 (1997).
204. Z. Wan, J. F. Christian and S. L. Anderson, *J. Chem. Phys.* **96**, 3344 (1992).

205. Z. Wan, J. F. Christian and S. L. Anderson, *Phys. Rev. Lett.* **69**, 1352 (1992).
206. J. F. Christian, Z. Wan and S. L. Anderson, *J. Chem. Phys.* **99**, 3468 (1993).
207. L. S. Sunderlin and P. B. Armentrout, *Chem. Phys. Lett.* **167**, 188 (1990).
208. D. J. Levandier, R. A. Dressler, S. Williams and E. Murad, *J. Chem. Soc. Faraday Trans.* **93**, 2611 (1997).
209. G. Gioumousis and D. P. Stevenson, *J. Chem. Phys.* **29**, 294 (1958).
210. W. McNeil, S. T. Lai and E. Murad, private communication (1998).
211. E. Kopp and U. Herrmann, *Ann. Geophys.* **2**, 83 (1984).
212. C. A. Boitnott and H. F. Savage, *Astrophys. J.* **161**, 351 (1970).
213. C. A. Boitnott and H. F. Savage, *Astrophys. J.* **174**, 201 (1972).
214. H. F. Savage and C. A. Boitnott, *Astrophys. J.* **167**, 341 (1971).
215. I. Amdur and J. E. Jordan, *Adv. Chem. Phys.* **10**, 29 (1966).
216. N. G. Utterback and G. H. Miller, *Rev. Sci. Instr.* **32**, 1101 (1961).
217. B. V. Zyl, N. G. Utterback and R. C. Amme, *Rev. Sci. Instr.* **47**, 814 (1976).
218. O. J. Orient, A. Chutjian and E. Murad, *Phys. Rev.* **A41**, 4106 (1990).
219. J. Politiek, P. K. Rol, J. Los and P. G. Ikelaar, *Rev. Sci. Instr.* **39**, 1147 (1968).
220. J. F. Friichtenicht, *Rev. Sci. Instr.* **45**, 51 (1974).
221. G. B. Wicke, *J. Chem. Phys.* **78**, 6036 (1983).
222. T. L. Thiem, L. R. Watson, R. A. Dressler, R. H. Salter and E. Murad, *J. Phys. Chem.* **98**, 11931 (1994).
223. N. Abuaf, J. B. Anderson, R. P. Andres, J. B. Fenn and D. G. H. Marsden, *Science* **155**, 997 (1967).
224. E. K. Parks and S. Wexler, *Chem. Phys. Lett.* **10**, 245 (1971).
225. J. C. Slattery and J. F. Friichtenicht, *Astrophys. J.* **147**, 235 (1967).
226. J. F. Friichtenicht, J. C. Slattery and D. O. Hanson, *Phys. Rev.* **163**, 75 (1967).
227. J. Politiek and J. Los, *Rev. Sci. Instr.* **40**, 1576 (1969).
228. Y. F. Bydin and A. M. Bukhteev, *Dokl. Akad. Nauk, SSSR* **119**, 1131 (1958).
229. Y. F. Bydin and A. M. Bukhteev, *Sov. Phys. Tech. Phys.* **5**, 512 (1960).
230. A. M. Bukhteev, Y. F. Bydin and V. M. Dukelski, *Sov. Phys. Tech. Phys.* **5**, 496 (1961).
231. J. F. Cuderman, *Phys. Rev.* **A5**, 1687 (1972).
232. D. R. Herschbach, *Adv. Chem. Phys.* **10**, 319 (1966).
233. R. W. Anderson, V. Aquilanti and D. R. Herschbach, *Chem. Phys. Lett.* **4**, 5 (1969).
234. K. Lacmann and D. R. Herschbach, *Chem. Phys. Lett.* **6**, 106 (1970).
235. V. Kempter, W. Mecklenbrauck, M. Menzinger, G. Schuller, D. R. Herschbach and C. Schlier, *Chem. Phys. Lett.* **6**, 97 (1970).
236. A. M. C. Moutinho, A. P. M. Baede and J. Los, *Physica* **51**, 432 (1970).
237. R. H. Neynaber, B. F. Meyers and S. M. Trujillo, *Phys. Rev.* **180**, 139 (1969).
238. A. W. Kleyn, M. M. Hubers and J. Los, *Chem. Phys.* **34**, 55 (1978).

239. U. C. Klomp and J. Los, *Chem. Phys.* **71**, 443 (1982).
240. U. C. Klomp and J. Los, *Chem. Phys.* **83**, 19 (1983).
241. J. A. Aten, M. M. Hubers, A. W. Kleyn and J. Los, *Chem. Phys.* **18**, 311 (1976).
242. W. Jones, AIAA Paper **99–0501** (1999).
243. A. M. Bukhteev and Y. F. Bydin, *Bull. Acad. Sci. USSR. Phys. Ser.* **27**, 985 (1963).
244. A. Fontijn, *Progr. React. Kin.* **6**, 75 (1972).
245. A. Fontijn, *Pure Appl. Chem.* **39**, 287 (1974).
246. X. Shi, D. R. Herschbach, D. R. Worsnop and C. E. Kolb, *J. Phys. Chem.* **97**, 2113 (1993).
247. T. G. Wright, A. M. Ellis and J. M. Dyke, *J. Chem. Phys.* **98**, 2891 (1993).
248. M. H. Covinsky, A. G. Suits, H. F. Davis and Y. T. Lee, *J. Chem. Phys.* **97**, 2515 (1992).
249. J. M. C. Plane and D. Husain, *J. Chem. Soc. Faraday Trans.* **82**, 2047 (1986).
250. D. L. Hildenbrand, *Chem. Phys. Lett.* **34**, 252 (1975).
251. D. Schroeder, A. Fiedler, J. Schwarz and H. Schwarz, *Inorg. Chem.* **33**, 5094 (1993).
252. J. B. Pedley and E. M. Marshall, *J. Phys. Chem. Ref. Data* **12**, 967 (1964).
253. J. Drowart, *Faraday Symp. Chem. Soc.* **8**, 165 (1974).
254. W. J. McNeil, E. Murad and S. Lai, *J. Geophys. Res.* **100**, 16847 (1995).

CHAPTER 7

DYNAMICS OF HYPERVELOCITY
GAS/SURFACE COLLISIONS

Dennis C. Jacobs

Department of Chemistry and Biochemistry
University of Notre Dame, 251 Nieuwland Science Hall
Notre Dame, IN 46556-5670
E-mail: Jacobs.2@nd.edu

Contents

1. Introduction

This chapter reviews experimental and theoretical studies pertaining to the interaction of hypervelocity atoms/molecules/ions with surfaces. Examining the scattering dynamics of hypervelocity gas/surface collisions reveals a variety of relevant fundamental mechanisms including energy transfer, charge transfer, and reactions under nonthermal conditions. Special attention is given here to the deleterious effects these processes have on spacecraft, primarily, aerodynamic drag and vehicle charging; materials degradation resulting from prolonged O-atom exposure in low Earth orbit is discussed at length in Chap. 9. Although direct connections are made primarily with spacecraft applications, many of the conclusions from this chapter are relevant to other technologies, e.g., plasma processing in the fabrication of microelectronics devices.

1.1. *Orbital Environment*

An orbiting space vehicle is continuously exposed to a flux of energetic neutral atoms/molecules, ions, electrons, charged particle radiation, electromagnetic radiation, meteoroids, and orbital debris.[1] This hostile environment can adversely impact the longevity and success of a spacecraft's mission through material degradation, radiation damage, electrostatics charging/arcing, aerodynamic drag, etc. A complete understanding of the gas/surface processes relevant to low Earth orbit (LEO) and geosynchronous Earth orbit (GEO) environments can aid in the design of spacecraft materials resistant to these deleterious effects.

In LEO (ca. 300 km altitude), the ambient gas density and temperature are typically 10^9 cm^{-3} and 1200 K, respectively; although solar activity and diurnal phase can strongly perturb these values. With an orbital velocity of 7.5–8 km/s, the flux of neutral particles on ram surfaces is $\sim 10^{20}$ m^{-2}s^{-1}. This corresponds to approximately ten collisions per atomic surface site per second. Atmospheric collisions lead to satellite drag and torques which hinder attitude control.[2] Degradation of materials on ram-facing surfaces can compromise a spacecraft's flight duration as well.[3]

In addition to the neutral atmospheric flux of particles, a cold, weakly ionized plasma exists in LEO.[4] The ionospheric plasma density varies dramatically with altitude and latitude. Near the equator at an altitude of 300 km, peak plasma densities and temperatures are 10^6 cm^{-3} and 2000 K, respectively. The motion of charged particles is strongly influenced by the Earth's magnetic field. An orbiting spacecraft will develop an electrostatic potential as it interacts with the neutral plasma. In GEO (ca. 35 000 km altitude), geomagnetic substorms during the premidnight to dawn hours produce a plasma density and temperature of ~ 1 cm^{-3} and $> 10^4$ eV, respectively. Significant electrostatic charging within this environment can lead to kilovolt-level difference potentials across the spacecraft. Plasma/ surface interactions play an important role when reusable spacecraft reenter the earth's atmosphere. The severe aerothermal load during reentry induces thermionic emission and associative ionization to form NO^+, $N_2{}^+$, and $O_2{}^+$ within the shock layer. This results in a dense plasma sheath around the reentry vehicle causing radio blackout.[5]

Studying the dynamics of hypervelocity gas/surface collisions under well-characterized conditions can provide insight into how a spacecraft's external surfaces will interact with the ionosphere/magnetosphere in orbit.

1.2. *Historical Background to Hyperthermal Energy Gas/Surface Dynamics*

Over the last three decades, tremendous advances have been made in understanding the interaction of gases with solid surfaces.[6] Historically, these investigations were motivated by the desire to optimize heterogeneous catalysis in the petrochemical industry. As such, the majority of work had originally focused on elucidating the structures, bonding energetics, and rates of reaction for adsorbates bound to single-crystal metal surfaces. In recent years, the surface science community has turned its attention to metal oxide and semiconductor surfaces; the former presents a more realistic prototype of supported catalyst systems; the latter makes connections with the microelectronics industry.

Gas/surface reactions are commonly studied under thermal conditions, where neutral reagents are deposited on a surface sample at a temperature, T_s. Thermal conditions imply that the distribution of available reactant energies is given by Boltzmann statistics. Incident energies are limited (ca. < 1 eV) and stochastically distributed across all degrees of freedom (translation, vibration, rotation and electronic). Consequently, measuring a reaction rate as a function of temperature does not allow one to definitively assign a reaction mechanism, because the specific nuclear motion associated with reactants transforming into products remains unresolved.

An exciting alternative to driving surface reactions under thermal conditions involves the use of hyperthermal energy reagents; one can span a wider range of initial energies and selectively introduce energy into specific modes of the system. If energy is placed into a mode coupled strongly to the "reaction coordinate," then a strong enhancement in the reaction rate is often observed. Incident hypervelocity particles may accept electrons from the surface, fragment, react with an adsorbate, or induce a surface reaction. In addition, hypervelocity reagents will transfer a significant amount of their translational energy to the surface.

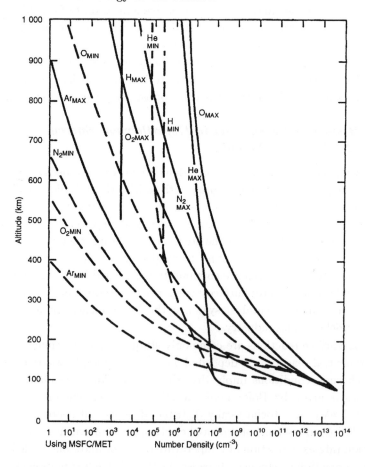

Fig. 1. Number density of neutral atmospheric constituents versus altitude (from Ref. 1).

2. Orbital Decay from Aerodynamic Drag

Premature reentry of spacecraft can be very costly. Yet the best available models for predicting satellite drag presently have only a 15% accuracy.[7] In low earth orbit (ca. 300 km), a typical spacecraft velocity is $\sim$ 7.7 km/s. The relative velocity between ambient gaseous species and the external surfaces of the spacecraft will be slightly less ($\sim$ 7.3 km/s) because of corotation of the atmosphere.[8] The ambient temperature of gases at low orbital altitudes ranges from 700 K at nighttime to 2000 K during the day. The neutral and cationic compositions of the terrestrial atmosphere are shown in Figs. 1 and 2, respectively.[9] At altitudes between 100–350 km, O, N_2, O_2, He and Ar are the major neutral species in the ambient atmosphere. At altitudes between 100 and 500 km, O^+, H^+, and NO^+ are the predominant ions incident on the spacecraft's exterior.

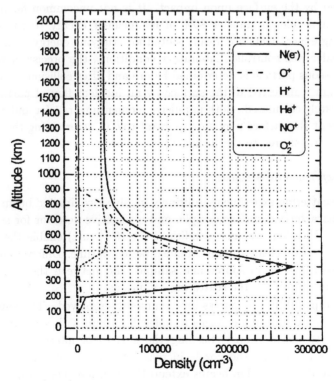

Fig. 2. Number density of plasma constituents versus altitude (from Ref. 4).

Predicting vehicle drag for a hypersonic spacecraft is difficult for a variety of reasons. Variations in solar activity and geomagnetic storms introduce large fluctuations in the number density, temperature, and wind currents of the ionosphere. Moreover, gas/surface collisions significantly perturb the concentrations of both neutral and ionic species proximate to the spacecraft.[10] The specific amount of momentum transferred by a hypervelocity particle impinging the spacecraft exterior depends on the detailed microscopic trajectory of the particle. The following sections highlight experiments and models addressing momentum transfer in hypervelocity gas/surface collisions.

2.1. *Ion/Surface Scattering Experiments*

A universal process associated with hypervelocity gas/surface collisions is energy transfer. A sizable fraction of a particle's incident energy can be dissipated by the surface upon impact. It is not uncommon for a hypervelocity molecule to transfer 80% of its incident energy to the surface. Scattering experiments are ideally suited for studying energy transfer, because under well-controlled laboratory conditions, the precise momentum of the gaseous particle is determined before and after collision. Hence, the momentum transferred to the surface by an incident flux of particles can be accurately calculated. Both impulsive momentum transfer and chemical reactions at the surface play an important role in determining the amount of energy transferred to the surface upon collision.

2.1.1. *Experimental Methods*

Surface scattering experiments routinely involve a collimated beam source, a well-characterized surface sample, and a sensitive detector for scattered particles. Ultrahigh vacuum conditions ($< 10^{-9}$ torr) enable the surface sample to remain clean over an extended period of time. Differential pumping between the beam source and sample, and between the sample and detector further define the scattering conditions.

Two types of beam sources are commonly employed. The supersonic expansion of a neutral gas forms a molecular beam with a relatively narrow velocity distribution.[11] Seeding the species of interest in a lighter carrier gas or heating the nozzle increases the range of accessible beam energies to 0.1–10 eV. Ion beams extend the collision energy range to keV or higher. Ions are routinely formed by electron impact, plasma discharge, or laser ionization methods. Although ions can easily be accelerated to arbitrarily high velocities using electrostatic optics, it becomes difficult to handle ion beams below 10 eV because of space charge broadening, i.e. the electrostatic repulsion between ions causes significant divergence within an intense ion beam. Mass selection of the mononergetic ions can be performed with a quadrupole filter, a magnetic sector, or a Wien filter. An example apparatus is shown in Fig. 3.

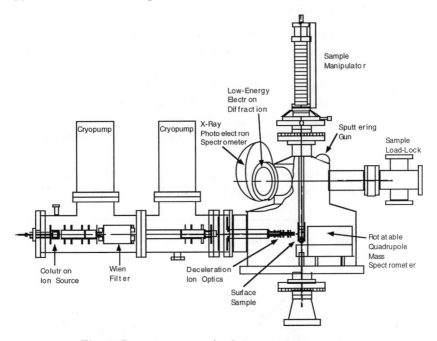

Fig. 3. Representative ion/surface scattering apparatus.

Ideally, the vacuum chamber is designed to independently vary the angle between the incident beam and the surface normal as well as the angle between the surface normal and the detector axis. Whereas the former is easily achieved with a standard surface manipulator, most detectors are mounted at a fixed angle relative to the incident beam axis. Independent rotation of the detector allows one to map out a complete angular distribution for each angle of incidence. Detectors achieve velocity resolution by either time-of-flight analysis or through the use of an electrostatic energy analyzer. Time-of-flight methods require that the incident beam be pulsed:

neutral molecular beams are mechanically chopped; whereas ion beams are electrostatically chopped with pulsed deflector plates. In a time-of-flight analysis, the arrival time at the detector is directly related to the path length and the initial and final velocities. Electrostatic energy analyzers disperse ions according to their kinetic energy by passing them through a curved electrostatic field.

2.1.2. *Atom/Surface Scattering*

Hurst *et al.* extensively studied the momentum transferred in collisions of Ar on Pt(111) at impact energies below 1.7 eV.[12] Rettner and coworkers reexamined this system and extended the range of collision energies to 7.3 eV.[13] In both cases, a chopped, supersonic molecular beam of Ar was directed at the surface, and the time-of-flight spectra of scattered Ar were recorded for a series of scattering angles. Here, the scattering angle, χ, is defined as the deflection of the projectile from its original velocity direction (see Fig. 4).

For an incident angle, $\theta_i = 45°$, Fig. 5(a) shows the angular distribution of scattered Ar at collidion energies of 0.13 eV and 7.3 eV.[12,13] The two

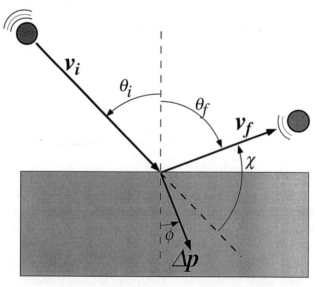

Fig. 4. Scattering geometry with initial and final particle velocity vectors and a vector representing the momentum transferred to the solid upon collision.

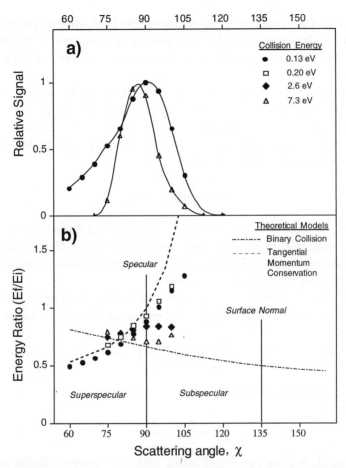

Fig. 5. Scattering data for Ar incident at 45° on Pt(111). (a) Angular distributions are plotted for two incident energies. (b) The average energy of scattered Ar divided by the incident energy is plotted as a function of the scattering angle. (Adapted from Refs. 12 and 13). Two theoretical models (dashed curves) predict the fraction of incident energy retained after scattering.

in-plane scattering distributions are scaled to one another. The angular distributions are peaked near the specular angle: $\theta_i = \theta_f = 45°$, or $\chi = 90°$. As the collision energy is lowered, the angular distribution shifts toward the surface normal and broadens.

Figure 5(b) displays the ratio of the particle's final energy over collision energy as a function of the scattering angle. The data at 0.13 eV

corresponded to a surface temperature of 500 K,[12] whereas the 0.20, 2.6
and 7.3 eV data were recorded at a surface temperature of 800 K.[13] In
general, the kinetic energy of scattered Ar increases as the scattering angle
increases.

The lowest energy, Ar/Pt(111) scattering data can be explained with a
simple model which assumes that the surface is perfectly flat — a reasonable assumption for the close packed Pt(111) surface. Translational energy
can be divided into components associated with motion parallel to and
perpendicular to the surface plane:

$$E = \frac{1}{2}m(v_\parallel^2 + v_\perp^2) \tag{1}$$

where $v_\perp = v\cos\theta$ and $v_\parallel = v\sin\theta$. For a flat surface without lateral
friction, tangential momentum, $mv_\parallel$, will be conserved. Only momentum,
$mv_\perp$, directed along the surface normal can be transferred to and from the
surface by coupling with surface phonons.

Under this constraint of "tangential momentum conservation," a simple
relationship between final energy, initial energy, and scattering angles arises:

$$\frac{E_f}{E_i} = \left(\frac{\sin\theta_i}{\sin\theta_f}\right)^2. \tag{2}$$

For the present case where $\theta_i = 45°$, Eq. (2) can be recast in terms scattering angle, χ:

$$\frac{E_f}{E_i} = \frac{1}{2}(1 + \cot^2\chi). \tag{3}$$

Equation (3) is drawn on Fig. 5 along with the "binary collision model."
These two models will be discussed at length in Secs. 2.2.1 and 2.2.2. Under the constraint of "tangential momentum conservation," superspecular
angles arise when normal momentum is lost to surface phonon excitation;
subspecular angles result from a transfer of thermal surface motion into
normal momentum of the scattered particle. As the incident energy is lowered, a more significant fraction of the incident Ar atoms undergo multiple
bounces on the surface. Thermal accommodation leads to angular distributions centered around the surface normal and to Boltzmann energy distributions at the surface temperature, T_s. Trapping occurs more frequently
in systems where the projectile has a strong binding energy with the surface. For example, in the scattering of Xe on Pt(111), a bimodal angular

distribution is seen.[14] The specular peak is assigned to inelastic scattering, while the broad peak centered around the surface normal is associated with trapping-desorption.

At higher collision energies, the data in Fig. 5 is better fit by the binary collision model. This model does not require the conservation of tangential momentum. Instead it assumes that both kinetic energy and total momentum are conserved as the impinging atom undergoes a single impulsive collision with one surface atom. Figure 6 shows data from the scattering of Ne^+ on Si(100),[15] a more corrugated surface than either Ag(111) or Pt(111). Over the range of collision energies explored, the energy ratio (E_f/E_i) varies systematically with the scattering angle (χ). The data all fall below the binary collision model (solid curve) for two major reasons. First, the scattered ions decelerate along the exit trajectory as they overcome the attraction to their image charge. The image

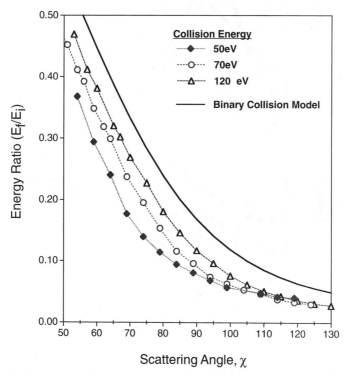

Fig. 6. The relative final energy versus scattering angle of Ne^+ incident on Si(100) at 45°. The solid line represents the binary collision model. (Data from Ref. 15).

charge arises from the dielectric response of the silicon to screen the ion's electric field. Second, the collision may incur inelastic losses (e.g. electron–hole pair formation, phonon dissipation during the collision) which are not included in the binary collision model but can be treated separately.[15]

2.1.3. *Molecule/Surface Scattering*

Further complications arise when molecules, rather than atoms, scatter on surfaces, because the incident translational energy can couple with

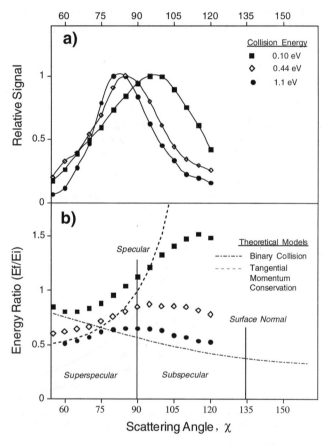

Fig. 7. Scattering data for NO incident at 45° on Ag(111). (a) Relative signal and (b) relative energies are plotted as a function of the scattering angle for three incident energies. (Adapted from Ref. 19) Two theoretical models (dashed curves) predict the fraction of incident energy retained after scattering.

molecular vibration and rotation as well as with surface phonons. Many research groups have studied rotational and vibrational excitation in molecule/surface collisions.[16–18] In the scattering of NO on Ag(111), Kleyn and coworkers measured the time-of-flight distribution of scattered NO as a function of scattering angle.[19] Figure 7 shows the NO/Ag(111) scattering data for a surface temperature of 670 K and incident angle of 45°. Similar to the Ar/Pt(111) data shown in Fig. 5, the NO/Ag(111) angular distributions are peaked near the specular angle and shift towards the surface normal as the collision energy is reduced. The angular distributions for NO/Ag(111) scattering are broader than those seen for the Ar/Pt(111) system, because molecules approaching the surface with different molecular orientations will transfer varying amounts of tangential and normal momentum. Remarkably, the binary collision model and the "tangential momentum conservation" model are in qualitative agreement with the data. As in the case of Ar/Pt(111), tangential momentum conservation provides a reasonable fit at low collision energies; the binary collision model becomes more appropriate at higher collision energies. More sophisticated models have also been applied to this system.[20,21]

Hyperthermal molecule/surface collisions introduce the possibility of dissociative scattering. The two most common mechanisms for dissociative scattering are impulsive dissociation and dissociative electron transfer. Impulsive dissociation occurs when collisional impact transfers incident translational energy into rovibrational energy that, in turn, ruptures the molecular bond. This mechanism is often termed collision induced dissociation (CID) or surface induced dissociation (SID). SID has been implicated in the scattering of O_2^+ on Ag(111),[22,23] CO^+ and N_2^+ on Pt(100),[24,25] BF_2^+ on Au(111),[26] H_2^+ on Ag(111),[27] and NO^+ on GaAs(110) and Ag(111).[28] Unlike binary gas-phase collisions, the center-of-mass velocity in a gas/surface collision is ill-defined, because the surface target may not simply be represented by a single surface atom. Consequently, a threshold in surface scattering is not reported as an energy within the center-of-mass frame; instead, it represents the minimum translational energy, in the laboratory frame, for which a reaction can still be observed. In determining a dissociation threshold, authors will sometimes linearly extrapolate their data to zero signal or fit their data to a hyperbolic tangent function[29] or an erf function.[30] The measured values for SID thresholds are often 4–5 times greater than the molecule's bond dissociation energy. This disparity arises because a majority of the collision energy is dissipated by the surface through excitation of lattice vibrations (phonons) and/or electron–hole pairs. The remaining energy retained by the molecule after collision is redistributed into translational, vibrational, and rotational energy. Only the latter two forms of energy participate in activating dissociation.[31] Classical trajectory calculations are able to accurately predict CID thresholds using simple pairwise-additive potentials.[32,33]

The most complete study of CID at hyperthermal energies involved NO^+ scattering on either GaAs(110) or Ag(111) at normal incidence conditions for $T_s = 298$ K.[28,34] A laser-based technique, resonance enhanced multiphoton ionization (REMPI), was employed to form the incident NO^+ ions in virtually a single rovibrational quantum state.[35] The relative yield and kinetic energy of scattered NO^- parents and O^- fragments were measured

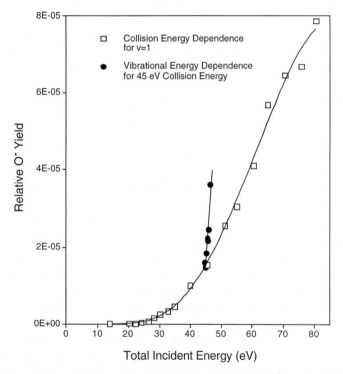

Fig. 8. The relative yield of scattered O^- from collisions of NO^+ on GaAs(110). The open squares represent the $v = 1$ level of incident NO^+ at various collision energies. The closed circles show the effect of initial vibrational quanta ($v = 0 - 6$) at a fixed 45 eV collision energy. (Adapted from Ref. 28).

as a function of collision energy, the initial number of vibrational quanta, and the alignment of the incident molecule's internuclear axis.[36] Figure 8 shows the relative yield of O^- as a function of the total energy initially within NO^+. The translational energy threshold to dissociation is 25 eV, four times the bond dissociation energy of NO. At 45 eV collision energy, the solid circles in Fig. 8 represent the effect of initial vibrational energy. Figure 8 suggests that vibrational energy is an order of magnitude more effective than translational energy in promoting O^- formation. Although it is intuitively reasonable to expect a large enhancement in dissociation when the molecule is initially excited along the reaction coordinate, our classical trajectory calculations failed to reproduce the experimental findings. The classical mechanical calculations predicted that even if the equivalent of fifteen quanta of vibrational energy were introduced into the incident molecule, the NO dissociation probability would increase only modestly.[32] However, the classical trajectory calculations did not allow for electron transfer in the system. To explore the role of charge transfer in the dissociation dynamics, we developed a time-dependent quantum mechanical model that explicitly treated coupling between the three different molecular charge states. The quantum trajectories accurately reproduced the experimentally observed vibrational effect and demonstrated that electron transfer, immediately prior to surface impact, creates a vibrational coherence in the molecule.[37] In summary, impulsive energy exchange and electron transfer are inextricably linked in the dynamics of O^- formation.

In collisions of NO^+ on GaAs(110), the mean kinetic energies of scattered O^- and NO^- are 13% and 6% of the NO^+ collision energy, respectively.[28] Although the nitrogen fragment could not be detected, kinematic arguments suggest that it should scatter with a similar kinetic energy as O^-, because it is comparable in mass to oxygen. Thus, a dissociative NO^+/GaAs(110) collision imparts, on average, 25% of the incident energy into the kinetic energy of the fragments. A nondissociative NO^+/GaAs(110) collision retains only 6% of the collision energy in the form of translational energy within the scattered parent molecule. In both cases, the majority of the collision energy is transferred into surface phonons or the creation of electron–hole pairs within the solid. This general behavior should be observed for most normal incidence collisions. Furthermore, the most efficient coupling of translational energy into surface phonon excitations occurs when the incident particle has a mass similar to that of an individual surface atom.

Dissociation of NO^+ on Ag(111) was studied by Greeley *et al.*[34] as a function of collision energy and the incident diatom's internuclear-axis direction. Incident NO^+ ions were photoselected through REMPI to have a preferred internuclear axis direction relative to the laser polarization vector. In this way, Greeley *et al.* compared the relative reactivity of molecules striking the surface with their internuclear axes pointing parallel versus perpendicular to the surface normal. The emergence of scattered O^- products was strongly enhanced when NO^+ approached the surface with an "end-on" rather than a "side-on" orientation. This finding is consistent with an impulsive dissociation mechanism.

In contrast to the aforementioned systems, relatively low translational energy thresholds for fragmentation are observed in the scattering of H_2^+, N_2^+, and O_2^+ on Ni(111),[38,39] O_2^+ on Ag(111),[40] H_2^+ on Cu(111),[41,42] and OCS^+ on Ag(111).[43] In each of these systems, a dissociative neutralization mechanism has been assigned. Dissociative neutralization occurs when a surface electron transfers to an incident cation, whereupon a repulsive

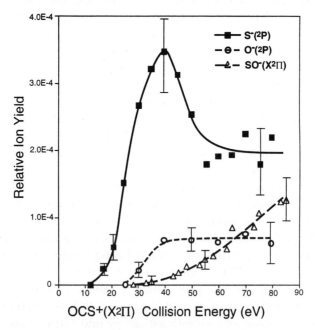

Fig. 9. Relative product yields of S^-, O^-, and SO^- as a function of collision energy for OCS^+ incident on Ag(111) (from Ref. 43).

electronic state of the neutral molecule is populated. In the scattering of OCS^+ on Ag(111) at normal incidence, three fragmentation pathways have been identified (see Fig. 9).[43] The formation of S^- from OCS^+ collisions on Ag(111) has a threshold of 11 eV and is assigned to a dissociative neutralization mechanism. The assignment could be made only after comparing detailed scattering experiments involving state selected OCS^+ and S^+ ions. The O^- and SO^- products arise from impulsive dissociation and rearrangement, respectively. This study demonstrates the benefit of using state selective ionization when investigating the fragmentation dynamics for a polyatomic projectile.[43] The S^-, O^-, and SO^- fragments were all found to scatter with a mean velocity that is $\sim 40\%$ of the incident OCS^+ velocity. This implies that the sum of the fragment kinetic energies equals $\sim 17\%$ of the collision energy.

Despite the variation in scattering results for the aforementioned prototypical systems, theoretical models have been successful in predicting the degree of energy and momentum transfer in gas/surface collisions. This will be discussed in the following section.

2.2. *Models of Collisional Energy Transfer*

The primary energy loss mechanism in a hypervelocity gas/surface collision is an impulsive transfer of momentum from the gas projectile to one or more surface atoms. At thermal energies, an incident particle couples with surface phonons (delocalized lattice vibrations analogous to normal vibrational modes in gas-phase molecules). On a microscopic scale, this coupling begins as the approaching particle is attracted to a cluster of surface atoms proximate to the point of impact. Upon impact, short range repulsion between the gas atom and the closest surface atom dominates. As the surface atom recoils from the impulsive collision, it transfers its momentum to neighboring surface atoms. This localized "hot spot" propagates as a shock wave through the solid and dissipates its energy on a picosecond timescale.[44] Classical trajectory calculations have proven accurate in predicting the amount of mechanical energy transferred during a collision; however, molecular dynamics simulations are computationally intensive and require some knowledge of the molecule/surface interaction potential.[45,46] Simple analytical models are useful for capturing qualitative, if not quantitative, trends observed in the data. The most common analytical models are described below.

2.2.1. *Hard Cube Model*

In scattering on an uncorrugated surface, a gas atom will transfer momentum in the direction along the surface normal but will conserve tangential momentum.[47] The portion of the surface that interacts with the impinging atom is referred to as a cube with an associated mass, M. In an impulsive collision, the gas surface potential is simply approximated as a hard wall. The vibrating surface cube is given an initial velocity consistent with a one-dimensional Maxwellian distribution at the surface temperature, T_s. The hard cube model predicts that the mean energy of the scattered gas atom will be:[48]

$$\langle E_f \rangle = \left(1 - \frac{4\mu}{(\mu+1)^2}\cos^2\theta_i\right) E_i + \frac{2\mu k_B T_s(2-\mu)}{(\mu+1)^2} \qquad (4)$$

where k_B is Boltzmann's constant and μ is the mass of the gas atom, m, divided by the effective surface mass, M. The first term in Eq. (4) evaluates the fraction of incident energy, in the laboratory frame, that is retained after a purely elastic collision with the hard cube. The second term includes the amount of transferred energy originating from thermal motion of the hard cube, i.e. surface phonons. Increasing M has the effect of lowering the amount of energy lost to the surface upon collision. Typically, M takes on a value between one and three times the mass of a single surface atom. Improvements can be made to the hard cube model by adding a uniform attractive potential,[48] or by replacing the hard wall potential with a more realistic Morse or Lennard–Jones potential (soft cube model).[49] However, these enhancements add more parameters to the model, and a reasonable fit to the data is often achieved by simply adjusting M within the hard cube model.

2.2.2. *Binary Collision Model*

At higher collision energies, energy transfer is dominated by the short range repulsive interaction between the gas atom and the closest surface atom. In the impulsive limit where a light gas atom collides with a heavier surface atom, the gas atom scatters from the surface before the target surface atom has had time to recoil and transfer its nascent momentum to neighboring surface atoms. Thus the hypervelocity gas/surface collision resembles that of an isolated collision between two gas phase atoms. An elastic collision between two gas-phase particles conserves both total energy and momentum. This constraint gives rise to the binary collision model:

$$\frac{E_f}{E_i} = \left(\frac{m}{m_s + m}\right)^2 \left(\cos\chi + \sqrt{\frac{m_s^2}{m^2} - \sin^2\chi}\right)^2. \qquad (5)$$

Equation (5) is an exact analytical expression for the fraction of initial energy, (E_f/E_i), remaining in a gas atom of mass, m, after elastically scattering at an angle, χ, from a surface atom of mass, m_s, initially at rest. Figures 5–7 illustrate the functional form of Eq. (5) for various gas/surface systems. The binary collision model does not explicitly depend on the incident angle, θ_i. However, the angle of incidence and the surface morphology will affect the number of scattering events in the overall trajectory. Equation (5) treats only single scattering events. If a gas atom sequentially scatters from two or more surface atoms, its energy may be more or less than for a single scattering event that leads to the same final scattering angle. For example, a gas atom which undergoes two successive collisions, each with a 20° scattering angle, will lose less energy in total than a gas atom which undergoes a single collision with a 40° scattering angle. In contrast, a trajectory which describes a gas atom chattering within a surface cavity may scatter to a final angle, χ, with less energy than Eq. (5) predicts.

The binary collision model has been successfully applied to a variety of systems: from the scattering of noble gases off liquid surfaces to the scattering of polyatomic molecules off thin polymer films.[50,51]

2.2.3. *Multiple Collision Simulations*

Conceptually, the simplest way to simulate gas particles scattering on a realistic solid is to perform a molecular dynamics (MD) calculation.[52] Here, one integrates the equations of motion for an ensemble containing the projectile and a large number of solid atoms. At a series of infinitesimal time steps, the positions of the nuclei are updated until the trajectory is finished. MD simulations typically involves averaging over 10^4–10^6 such trajectories. Molecular dynamics is computationally demanding, because the potential energy is computed from pairwise additive interactions summed over all neighboring atoms, and because thousands of coupled differential equations must be solved at every time step.

An alternative simulation method, based on the binary collision approximation, is conceptually more complicated but computationally more efficient. In this approach, the trajectory is viewed as a cascade of collisions, each collision involving just two atoms. Instead of having to calculate the complete path of all the nuclei, as in a MD simulation, the trajectory is approximated as a set of straight line paths. Each change of direction in an atom's path is due to a binary collision, for which the scattering angle and energy can be computed analytically. The widely distributed MARLOWE code simulates cascades in a crystalline lattice; binary collisions between atoms are governed by a Moliére central potential.[53]

When simulating a projectile impacting an amorphous solid, Monte Carlo (MC) algorithms are the method of choice. Here, the projectile collides with a sequence of randomly selected collision partners in the solid. A popular MC code, TRIM, requires the user to simply specify the incident energy, angle, and mass of the projectile, along with the mass and binding energies of the solid.[54,55] As in the case of MD, $\sim 10^6$ trajectories need to be performed to gather reliable statistics. MC methods can also be used to predict the depth profile of projectiles that implant beneath the surface.

2.3. *Implications for Satellite Drag*

An orbiting satellite experiences a drag force caused by collisions with atmospheric particles. In the reference frame of the satellite, incident particles are impinging on the ram surfaces of the satellite with energies ranging from 10^0–10^6 eV, depending on the altitude. From the experiments and models described in Secs. 2.1 and 2.2, we can recast the scattering problem in terms of the momentum, rather than the energy, that the particle transfers to the surface. Momentum conservation necessitates that any change in particle momentum during the collision corresponds with a change of momentum in the solid. By vectorially subtracting the average momentum of a scattered particle from its incident momentum, one calculates a vector corresponding to the momentum transferred to the solid during the collision. In the case where an incident particle becomes trapped at the surface, the momentum transferred upon collision equals the incident momentum of the particle. If a particle is incident at 45° to the surface and scatters at the specular angle without any energy loss, then the momentum transferred to the surface would be $\sqrt{2}$ or 141% of the incident momentum. For a variety of scattering systems, Table 1 provides both the percentage of incident momentum transferred to the solid and the angle, ϕ, that the transferred momentum vector makes with the surface normal (see Fig. 3). At low collision energies, the momentum transfer is directed close to the surface normal. The hard cube model does not permit momentum to be transferred parallel

Table 1. Linear momentum transfer in gas/surface collisions.

Incident Conditions			Experimental Results		Theoretical Predictions	
System	Collision Energy (eV)	θ_i	Percent of Particle Momentum Transferred	Momentum Transfer Angle, ϕ	Hard Cube Model (Mass $= 1.5 \times$ surface atom)	Binary Collision Model
NO/Ag(111)	0.10	45°	156%	0.8°	142%	
NO/Ag(111)	0.44	45°	132%	2.7°	125%	
NO/Ag(111)	1.16	45°	124%	5.5°	121%	
NO/Ag(111)	0.10	35°	174%	0.6°	169%	
NO/Ag(111)	0.44	35°	147%	1.6°	143%	
NO/Ag(111)	1.08	35°	142%	3.4°	140%	
NO$^+$/GaAs(110) $\to$ NO$^-$	10–80	0°	124%	0°		141%
NO$^+$/GaAs(110) $\to$ N + O$^-$	10–80	0°	150%	0°		141%
OCS$^+$/Ag(111) $\to \Sigma$ fragments	20–80	0°	142%	0°		128%
Ne$^+$/Si(100)	10	45°	104%	15°		107%
Ne$^+$/Si(100)	40	45°	105%	26°		107%
Ne$^+$/Si(100)	40	55°	94%	34°		94%
Ne$^+$/Si(100)	40	25°	113%	15°		111%

to the surface; thus the hard cube model always predicts that $\phi = 0$. The general trend is that for decreasing incident angles and decreasing collision energies, the relative fraction of incident momentum transferred to the surface increases.

Assuming a symmetric shape to the satellite, the drag force need only be calculated in the direction of the incident particle flux. Forces orthogonal to this direction will be canceled by equal and opposite forces on the other side of the satellite. For each particle incident on the surface, the satellite drag is proportional to the projection of the particle's change in momentum along the incident velocity direction. The ratio of momentum, Δp, transferred along the satellite's orbital velocity direction divided by the momentum, p_i, of the incident gaseous particle is give by:

$$\frac{\Delta p}{p_i} = 1 - \sqrt{\frac{E_f}{E_i}} \cos\chi. \tag{6}$$

In the limit where scattering is centered around the specular angle, the relative momentum transferred is related to the incident angle.

$$\frac{\Delta p}{p_i} \approx 1 + \sqrt{\frac{E_f}{E_i}} \cos(2\theta_i). \tag{7}$$

A qualitative, if not quantitative, measure of E_f/E_i at hyperthermal energies is given by the binary collision model. Inserting Eq. (5) into Eq. (7), yields:

$$\frac{\Delta p}{p_i} \approx 1 + \left(\frac{m}{M+m}\right)\left(\sqrt{\frac{M^2}{m^2} - \sin^2(2\theta_i)} - \cos(2\theta_i)\right)\cos(2\theta_i). \tag{8}$$

Figure 10 is a plot of Eq. (8) for four different values of the surface/particle mass ratio. This analysis may have important implications for the selection of external satellite materials that minimize aerodynamic drag. Note that for obtuse scattering angles, i.e $\chi > 90°$ or $0° < \theta_i < 45°$, the amount of momentum transferred increases with increasing surface mass. Although a larger surface mass leads to less energy transfer at all scattering angles, backscattered particles will undergo a larger change in momentum when striking a high mass surface atom than they will with a low mass surface atom. Consequently, under conditions where backscattering is dominant, aerodynamic drag is reduced when the exterior material of the spacecraft is composed of low mass elements and when the surface is highly corrugated. In contrast, for more glancing incidence angles, $45° < \theta_i < 90°$, forward scattering dominates. Here, larger surface masses result in less momentum transferred. Therefore, external surfaces of the spacecraft that make oblique angles with the vehicle's velocity vector are best composed of highly polished materials containing high mass elements. Surface defects can certainly alter the scattering dynamics; however, scattering experiments performed on surfaces for which atomic steps were intentionally introduced showed that atoms will scatter specularly from the local surface plane, even when this differs from the macroscopic surface plane.[56] Although many approximations are intrinsic to Fig. 10, the general conclusions are likely to be robust for hypervelocity satellites.

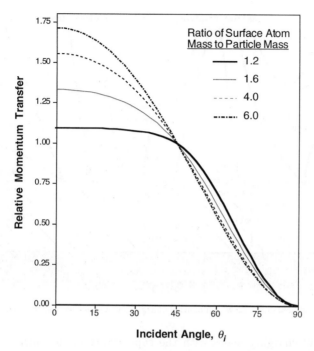

Fig. 10. Predicted relative momentum transferred to a satellite surface by an incident particle as a function of the incident angle. Each curve corresponds to a different surface-to-particle atom mass ratio.

3. Vehicular Charging in Earth Orbit

The surfaces of an orbiting spacecraft will develop an electrostatic charge arising from their interaction with the ionosphere and magnetosphere. Figure 11 illustrates the primary charging mechanisms associated with ions, electrons, and photons bombarding the exterior surface of the space-craft. Each type of charging process can be represented by an electrical current between the plasma and the vehicle's surface. Modelers calculate the net electrical current as a summation over the currents generated by primary electrons, secondary electrons, backscattered electrons, primary and secondary ions, photoemission, and charge transport between different spacecraft surfaces:

$$j_{net} = j_{electrons} + j_{secondary} + j_{backscattering} + j_{ions}$$

$$+ j_{photoemission} + j_{conduction} \cdot \tag{9}$$

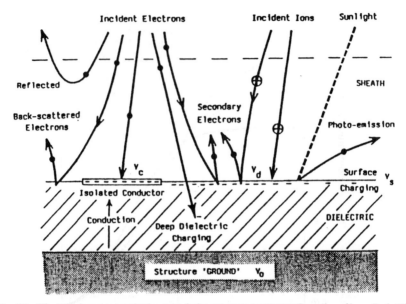

Fig. 11. Electrical currents which control charging at a spacecraft surface (from Ref. 57).

The rate of charging scales with the capacitance of the surface. There is a feedback in the circuit, because as the spacecraft's electrostatic potential increases, the flux and energy of the incident and outgoing charged particles will be altered.[57] The surface of the spacecraft will reach a "steady state" electrostatic potential, when the net current equals zero.[58]

The ionospheric plasma is generally considered to be overall neutral, i.e the density of cations equals the density of electrons. The characteristic temperatures of both the ions and electrons are $\sim 10^{-1}$ eV in LEO.[59] In GEO, the electron temperature is $\sim 10^4$ eV, but the ion temperature is centered around 10^6 eV. The GEO plasma environment is composed primarily of H^+ and alpha particles, He^{2+}, with trace amounts of C, O, and Fe ions.[1] Because velocity is inversely proportional to the square root of the particle mass, the flux of electrons incident on the spacecraft is two orders of magnitude greater than the flux of ions. Consequently, as a general rule, the spacecraft will adopt a negative surface potential close to the electron temperature. Thus, it is not uncommon for a geosynchronous spacecraft to become charged to $-10\,000$ Volts relative to plasma neutral.[58] Spacecraft in LEO develop charges of only a few volts, except in polar orbits where occasional auroral events may charge vehicles to keV levels.[61]

Absolute charging of a spaceraft is not as deleterious as differential charging between surface regions on the same spacecraft. Solar radiation plays a major role in the charging dynamics through photoemission. Differential charging occurs between surfaces that are illuminated by the sun and surfaces that are in shadow, or between neighboring conducting and insulating surfaces. Differential potentials can reach hundreds to thousands of Volts.[57] Discharges between neighboring surfaces present a serious risk to sensitive on-board instrumentation.

Accurate modeling of charging effects can assist in establishing construction practices designed to minimize spaceraft charging problems.[60] Charging models require a complete description of the fundamental pathways by which charge is transferred at the plasma/surface interface. Ultraviolet photons can eject electrons, neutrals, and ions from the surface. Energetic electrons will inelastically scatter at the surface and produce secondary electrons. For > keV electrons incident on a variety of materials, it is common for more than one secondary electron to be emitted.[61] Emitted electrons may not always escape the magnetic fields around the spacecraft; nevertheless, hot electrons can create local excitations of surface species, leading to the desorption of neutral and ionic particles. The latter process, known as Electron Stimulated Desorption (ESD), is discussed extensively elsewhere.[62] The following sections focus on recent experimental and theoretical advances in ion/surface scattering and refine some of the terms appearing in Eq. (9) that involve ions in the charging of space vehicles.

3.1. *Energetic Ion/Surface Collisions*

In GEO, cations will be accelerated into a negatively charged spacecraft and impact the surface with collision energies of 10^3–10^6 eV. The impinging ions may neuteralize, undergo charge inversion, implant, sputter, initiate a shower of secondary electrons, and/or generate defects in the solid.

After an energetic ion strikes a surface, it will decelerate as its kinetic energy is dissipated via elastic (nuclear collision) and inelastic processes. A small fraction of the ions may reflect from the surface; the majority are implanted beneath the surface.[63] The average distance an ion travels before it is stopped within a solid increases with collision energy. Figure 12 shows the mean depth of penetration achieved by He^+ incident on a silicon surface.[64,65] Molecular dynamics simulations reveal that within the first picosecond after a keV ion impacts a solid, a local hot spot develops at the

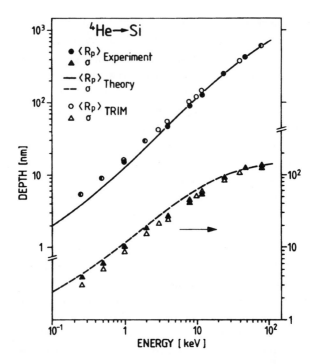

Fig. 12. Mean penetration depth (R_p) and fluctuations (σ) in the energy loss associated with He[+] implanting into Si as a function of the collision energy of He[+]. (Ref. 64). Theoretical data are from Ref. 65.

surface. In the near surface region, hundreds of atoms are excited in the collision cascade following impact, and the local instantaneous temperature can exceed the solid's melting point.[66]

The collision energy of an impinging ion is not completely transferred to the surface, because the primary ion may reflect from the surface or sputter secondary particles which escape with kinetic energy. An energy deposition coefficient represents the fraction of collision energy which is deposited into the solid. The energy deposition coefficient depends on the mass of the primary ion and the surface target atoms, the collision energy, the angle of incidence, and the surface morphology. Figure 13 shows the energy deposition coefficient for three primary ions impacting a polycrystalline gold film at normal incidence.[67,68] Note that energy deposition increases with increasing mass of the incident particle and with increasing collision energy.

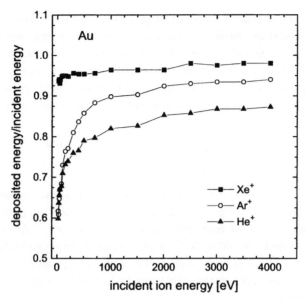

Fig. 13. Fraction of the incident energy of Xe^+, Ar^+, and He^+ deposited in a gold film for normal incidence (from Refs. 67 and 68).

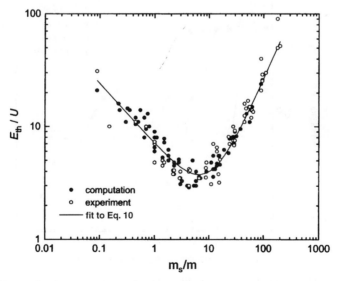

Fig. 14. The relative threshold energy for sputtering versus the ratio of the surface to ion masses (from Ref. 70).

The energy threshold for sputtering depends on the masses of both the incident ion and the surface atoms as well as the binding energy, U, of a surface atom to the substrate. For pure materials, the value for U is often taken as the cohesive energy of the solid.[69] Figure 14 shows the relative threshold energy, E_{th}, for sputtering as a function of the "surface atom" to "incident ion" mass ratio.[70] An empirical fit to the data has the following functional form:[71]

$$\frac{E_{th}}{U} = 7.0\left(\frac{m_s}{m}\right)^{-0.54} + 0.15\left(\frac{m_s}{m}\right)^{1.12} \tag{10}$$

The number of atoms sputtered by a single incident ion varies with an ion's individual trajectory. Figure 15 shows the distribution of sputtered particles per incident ion at four different primary energies.[72] As the collision energy increases from 100 to 1000 eV, the most probable yield shifts from 0 to 5 sputtered atoms per ion impact. The computer program, TRIM, mentioned earlier, provides quantitative simulations of both sputtering thresholds and yields.[54,55]

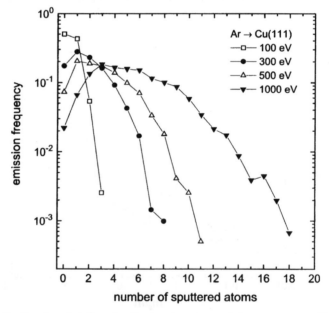

Fig. 15. Predicted probability distribution for sputtered Cu atoms per collision of an incident Ar$^+$ ion. Molecular dynamics calculations are from Ref. 72.

3.2. *Charge Transfer at Surfaces*

Sputtered particles may emerge from the surface as neutrals or as positive or negative ions. Whereas emitted positive ions will not escape the electrostatic field of the predominantly negatively charged spacecraft, the ejection of negative ions will alter the vehicle's surface potential. Consequently, the following section will focus on neutralization and charge inversion of reflected incident cations and on electron attachment to sputtered particles in the near surface region.

As a cation approaches a metal or semiconducting surface, an electron from the solid may resonantly tunnel through the vacuum to fill a vacant level on the ion. This process of resonant neutralization becomes facile, when (1) the affinity level of the ion is resonant with the occupied levels of the solid, and (2) the ion is sufficiently proximate to the surface

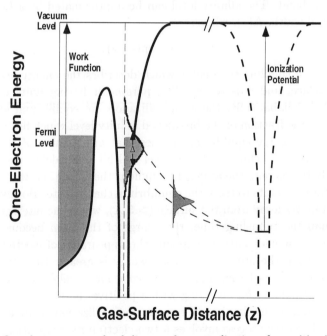

Fig. 16. One-electron energy level diagram for neutralization of a positive ion near a metal surface. The metal conduction band is shown on the left. The dashed and solid curves represent the potential energy for an electron when the ion is at infinite and close distances to the surface, respectively. The affinity level for the ion shifts upward and broadens as the ion approaches the surface.

that there is an appreciable overlap of the wavefunctions associated with the occupied band of the solid and the affinity level of the ion. It is useful to draw a one-electron energy level diagram for modeling charge transfer processes. On the left side of Fig. 16 are the occupied states of a metal conduction band. The ion is shown at a distance, z, above the surface. The attraction of an electron to the ion's nucleus is depicted by a generic potential well on the right. The affinity level represents the lowest lying bound state available to an electron if it is captured by the ion. At large z, the ion's affinity level is located below the vacuum level by an amount equal to the atom's ionization potential. As the ion approaches the metal surface, the affinity level increases in energy as a result of the ion's attractive interaction with it's image charge in the surface. For $z > 3\,\text{Å}$, the shift is approximately 3.6 eVÅ/z. Additionally, the affinity level broadens near the surface as it couples more strongly with nearly degenerate states in the conduction band. The affinity level can be approximated by a Lorentzian with a level width, $\Delta(z)$, where:

$$\Delta(z) = \Delta_o \exp(-\gamma z). \qquad (11)$$

Here, Δ_o is a coupling parameter which describes the energy level width at the surface, and γ is a tunneling parameter having units of inverse length.[73,74] Adiabatically, the atom will adopt an equilibrium state represented by the fraction of the broadened affinity level which lies above the Fermi level. Charge equilibrium implies that any portion of the affinity level that extends below the Fermi level will fill as a valence electron can resonantly tunnel onto the atom. Throughout the trajectory, the fractional charge state attempts to track the equilibrium charge state. However, when the atom moves away from the surface ($> 5\,\text{Å}$), where the interaction time is less than the tunneling time, the charge of the atom becomes frozen. The resulting nonadiabaticity explains the appearance of ejected positive ions even when the atom's ionization energy is greater than the surface work function. Particles leaving the surface at larger velocities will have an increased probability for emerging with a positive charge.

An alternative mechanism to resonant neutralization is Auger neutralization. The latter process involves a two-electron reorganization whereby one electron from the solid tunnels across to the ion while a second electron is excited out of the conduction band of the metal. If the second electron has sufficient energy, it may be ejected from the surface and is referred to as an Auger electron.

Many empirical models have been proposed to treat both resonant and Auger electron transfer processes at surfaces. Hagstrum assumed that the rate of neutralization has an exponential dependence on ion/surface distance.[75] He then showed that the ion survival probability, P_+, can be approximated by:

$$P_+ = \exp\left(\frac{-v_c}{v_\perp}\right) \tag{12}$$

where $v_\perp$ is the velocity component along the surface normal and v_c is an empirical parameter known as the characteristic velocity. Equation (12) can be expressed in terms of $\Delta(z_c)$, the level width, Eq. (11), evaluated at the point where the affinity level crosses the Fermi level[76]

$$P_+ = \exp\left(\frac{-\Delta(z_c)}{\gamma v_\perp}\right) . \tag{13}$$

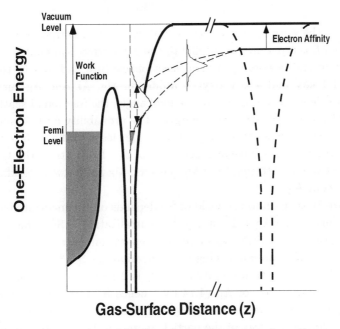

Fig. 17. One-electron energy level diagram for negative ion formation near a metal surface. The dashed and solid curves represent the potential energy for an electron when the atom is at infinite and close distances to the surface, respectively. The affinity level for the neutral atom shifts downward and broadens as the atom approaches the surface.

Negative ion formation is analogous to resonant neutralization and exhibits similar behavior as described in Eq. (13). Figure 17 shows the affinity level of a neutral atom as a function of z. Here, tunneling of an electron into the affinity level forms a negative ion near the surface. In contrast to Fig. 16, the affinity level of the neutral atom decreases as the atom approaches the surface, because the anion is stabilized by its image charge in the metal. For $z > 3\,\text{Å}$, the shift is approximately -3.6 eVÅ/z. It is not uncommon for the equilibrium charge state of an atom near a surface to be predominantly negative, even when the gas-phase electron affinity of the atom is much less than the work function of the metal. If the particle leaves the surface slowly, it will reneutralize before emerging into the vacuum. For relatively fast exit velocities, and modest work function changes ($|\Delta\phi| < 2$ eV), the negative ion yield scales as:

$$Y_- \propto \exp\left(\frac{-(\phi - \text{EA})}{kv_\perp}\right) \qquad (14)$$

where ϕ is the work function of the metal, EA is the electron affinity of the atom, and k is a parameter specific to the gas/surface system.[77]

Esaulov and coworkers studied electron capture and loss in the scattering of 1 keV and 4 keV oxygen atoms on Mg, Al, and Ag surfaces.[78] Figure 18 shows, as a function of the final angle, the fraction of scattered O atoms which emerge negatively charged. The probability for O^- emergence is greater when the atom scatters at an angle closer to the surface normal, because these species have a larger value of $v_\perp$. In addition, the O^- yield is greatest when scattering from Mg, because Mg has a lower workfunction, ϕ, than Al or Ag.

Figure 19 shows how the yield of S^- depends on the incident velocity of S^+ colliding with a Ag(111) surface.[43] Neutralization of S^+ is facile on the incoming trajectory. State selective ion preparation restricts the electronic state of the sulfur atom following neutralization, i.e. $S^+(^4S) + e^- \to S(^3P)$ exclusively. The electron affinity of $S(^3P)$ is 2.1 eV, a value much smaller than the workfunction of Ag (4.74 eV); hence S^--formation is not adiabatically favored. Nevertheless, the probability for S^- emergence increases when the collision energy of the particle increases, because the sulfur atom penetrates deeper into the surface before scattering away. As the atom is brought closer to the surface, the negative ion is stabilized by its image charge in the metal, and electron attachment occurs. In comparison, Fig. 19

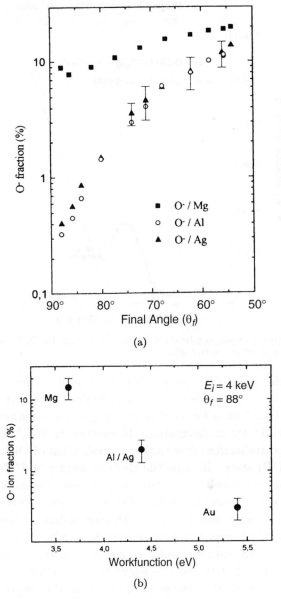

Fig. 18. (a) Negative ion fraction of scattered O versus exit angle for 1 keV oxygen atoms incident on Mg, Al, and Ag. (b) Ion fractions for different metals as a function of the metal's workfunction (from Ref. 78).

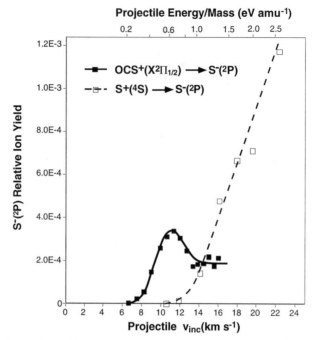

Fig. 19. Relative S^- yield as a function of collision velocity for OCS^+ and S^+ incident on Ag(111) at $\theta_i = 0°$ (from Ref. 43).

also shows the yield of S^- arising from dissociative scattering of OCS^+. The onset for S^- emergence occurs at a collision energy that is 8 eV lower for incident OCS^+ than for incident S^+, despite the additional energy required by OCS^+ for fragmentation. In contrast to the ^3P state formed by $S^+(^4S)$ neutralization, dissociative neutralization of $OCS^+(X^1\Sigma)$ produces the $S(^1D)$ state. Because the electron affinity of $S(^1D)$ is 3.25 eV, the atom does not have to penetrate as deeply into the metal before electron attachment becomes favorable. The $S(^1D)$ intermediate captures an electron more readily than does the $S(^3P)$ intermediate. Consequently, the threshold for S^- emergence is lower for incident OCS^+ than S^+.[43]

Negative ion formation is sensitive not only to the electronic state of the incident atom but to the site where the impinging atom strikes the surface. For example, the emergence of O^- from NO^+ dissociatively scattering from GaAs(110) occurs only when the oxygen atom scatters from a dangling bond site on the semiconductor surface.[79] These sites correspond with surface states that are energetically situated in the band gap and that have wavefunctions extending far into the vacuum. Both of these qualities make surface states ideal conduits for electron attachment. A local effect in electron transfer was also observed in the neutralization of Li^+ on an alkali-covered Al(100) surface.[80]

Insulating surfaces are characterized as having a broad band gap and a deep valence band maximum. Although these properties seem unfavorable to resonant electron transfer, alkali–halide surfaces have demonstrated a remarkable ability to produce negative ions under grazing angle scattering conditions. Winter *et al.* report a O^- yield of up to 70% when 20 keV oxygen atoms were incident on a LiF(100) surface at glancing angles $(87° - 89°)$.[81] Similarly, they found that 10 keV fluorine atoms incident at 89° on LiF(100) and KI(100) surfaces demonstrated 80% and 98.5% F^- yields, respectively.[82] In these systems, the mechanism for electron capture involves a localized collision between the incident atom and a surface halide atom. The "Demkov model" for gas-phase charge transfer (see Chap. 6) can be adapted to predict the probability of forming a negative ion on an insulating surface for grazing angles of incidence:[83,84]

$$P_- = \frac{1}{2}\text{sech}^2 \left(\frac{\pi}{v\left(\sqrt{2\text{EA}_s} + \sqrt{2\text{EA}_i}\right)} \left[\Delta E + \frac{v^2}{2}\right] \right) \qquad (15)$$

where v is the incident velocity, EA_i and EA_s are the electron affinities of the incident and surface atoms, respectively, and ΔE is the energy change between the initial and final states. The unusually high negative ion yields arise in part because once an electron is captured by a scattered ion, there are no resonant states at the surface to which the electron can be transferred back.

3.3. *Secondary Ion Production*

As described in Sec. 3.1, an energetic ion will produce a collision cascade upon surface impact. Recoiled surface atoms within the nascent hot spot may escape into the vacuum. As the sputtered atoms emerge from the surface, they can capture or lose an electron to the surface. On metal surfaces where conduction band electrons are delocalized, secondary ions are formed when sputtered atoms undergo resonant electron transfer along their outgoing trajectory. The relations expressed by Eqs. (12)–(14) in Sec. 3.2 approximate the charge transfer probability; more sophisticated models have also been developed.[77,85]

Although clean metal surfaces are often inefficient at producing secondary electrons and negative ions, the introduction of surface adsorbates can dramatically enhance the yield of negatively charged secondary particles. A submonolayer coverage of alkali atoms (e.g. Li, K, Na, or Cs) will reduce the work function of a metal surface by up to 3 eV. As predicted from Eq. (14), lowering the work function enhances negative ion production by as much as four orders of magnitude.[86]

The addition of oxygen, an electronegative element, to transition metal surfaces also increases negative ion production.[87,88] Figure 20 shows the

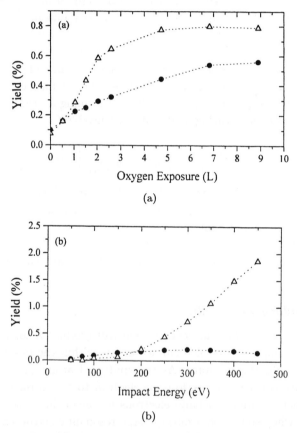

(a)

(b)

Fig. 20. (a) Secondary electron ($\triangle$) and O^- ($\bullet$) yields as a function of oxygen exposure on Mo(100) at a fixed collision energy of 250 eV. (b) The yields as a function of collision energy at a 3 L oxygen exposure on Mo(100) (from Ref. 88).

yield of secondary electrons and negative ions as a function of both oxygen exposure and primary energy for Na^+ incident on $Mo(100)$.[88] In this, and other systems where electrons are bound within covalent or ionic bonds, a localized electronic excitation picture is appropriate. The Landau–Zener model has been evoked to describe the probability of moving between covalent and ionic diabatic potential energy surfaces near the region where the two surfaces cross. The Landau–Zener survival probability can be expressed as

$$P_- \cong \frac{g_-}{g_0} \exp\left(\frac{2\pi z_c^2 H_{12}^2}{e^2 v_\perp}\right) \tag{16}$$

where z_c is the crossing distance, H_{12} is transition matrix element, and g_- and g_0 are the degeneracies of the negative ion and neutral atomic states, respectively. Note that Eqs. (12), (13), (14), and (16) show a similar dependence on the perpendicular velocity of the sputtered atom. The mechanism for secondary ion production in the ion bombardment of $O/Mo(100)$ was inferred from a careful study of the kinetic energy distributions for the sputtered O^- ions.[88] Collisional excitation of an ionic (MoO^-) bond into a repulsive (MoO^{-*}) electronic state is responsible for both oxygen ejection and the final O^- charge state. Secondary electrons arise from the decay of the ionic (MoO^{-*}) electronic state to a covalent (MoO) electronic state, accompanied by ejection of the detached electron into the vacuum. This is in sharp contrast to the mechanism for secondary ion production on clean metals: on metals, a collisional cascade (momentum transfer) provides the sputtered atom with the kinetic energy to leave; as the sputtered atom departs, it resonantly captures an electron.

4. Summary

The external surfaces of orbiting spacecraft are constantly bombarded by neutral and charged particles. Atmospheric drag and vehicular charging are two phenomena that are moderated by hypervelocity gas/spacecraft collisions. Understanding the fundamental dynamics of how atoms and molecules interact with surfaces is of principal importance to modeling and minimizing the detrimental effects that atmospheric drag and vehicular charging can have on a spacecraft's mission.

The results of detailed gas/surface scattering experiments reveal the precise momentum transferred to a surface by an incident gas particle. Relatively simple models, such as the binary collision approximation, provide valuable predictions of how the compositional and morphological characteristics of spacecraft materials affect atmospheric drag.

Plasma/surface interactions are paramount to the electrostatic charging of spacecraft in orbit. Incident ions undergo facile charge exchange as they collide with the surface. They may reflect from the surface or become implanted beneath the surface. Hypervelocity gas/surface collisions frequently produce sputtered particles that become charged as they depart from the surface. Understanding and controlling these charge transfer phenomena will help to reduce the risk of damaging electrical discharges on the surfaces of satellites.

Acknowledgments

The author gratefully acknowledges the Air Force Office of Scientific Research for their primary support of this work studying the interactions of ions with hypervelocity vehicles. The National Science Foundation and the Alfred P. Sloan Foundation are also acknowledged for their support of the author's general research in ion/surface chemistry.

References

1. B. J. Anderson, *Natural Orbital Environment Guidelines for Use in Aerospace Vehicle Development* (NASA Technical Memorandum 4527, 1994).
2. R. F. Fischell, *Torques and Attitude Sensing in Earth Satellites*, 1964, p. 13.
3. T. Ngo, E. J. Snyder, W. M. Tong, R. S. Williams and M. S. Anderson, *Surf. Sci.* **L314**, L817 (1994).
4. E. Murad, *Ann. Rev. Phys. Chem.* **49**, 73 (1998).
5. C. Park, *J. Thermophys. Heat Transfer* **7**, 385 (1993).
6. C. T. Rettner and M. N. R. Ashfold, *Dynamics of Gas–Surface Interactions* (Royal Society of Chemistry, Cambridge, 1991).
7. T. L. Killeen, A. G. Burns, R. M. Johnson and F. A. Marcos, *Modeling and Prefiction of Density Changes and Winds Affecting Satellite Trajectories*, AGU Monograph No. 13, *Environmental Effects on Spacecraft Positioning and Trajectories*, Vol. 13, 83–108, 1993.
8. E. Murad, *J. Spacecraft Rockets* **33**, 131 (1996).
9. D. Smith and P. Spanel, *Mass Spectrom. Rev.* **14**, 255 (1995).
10. C. Park, *J. Thermophys.* **3**, 233 (1989).
11. J. B. Anerson, R. P. Andres and J. B. Fenn, *Adv. Chem. Phys.* **10**, 275 (1996).
12. J. E. Hurst, L. Wharton, K. C. Janda and D. J. Auerbach, *J. Chem. Phys.* **78**, 1559 (1983).
13. D. Kulginov, M. Persson, C. T. Rettner and D. S. Bethune, *J. Phys. Chem.* **100**, 7919 (1996).

14. K. C. Janda, J. E. Hurst, C. A. Becker, J. P. Cowin, D. J. Auerbach and L. Wharton, *Surf. Sci.* **93**, 270 (1980).
15. C. Quinteros, T. Tzvetkov, S. Tzanev and D. C. Jacobs, in preparation.
16. G. D. Kubiak, J. E. Hurst, H. G. Rennagel, G. M. McClelland and R. N. Zare, *J. Chem. Phys.* **79**, 5163 (1983).
17. D. J. Auerbach, *Physica Scripta*, **T6**, 122 (1983).
18. C. T. Rettner, F. Fabre, J. Kimman and D. J. Auerbach, *Phys. Rev. Lett.* **55**, 1904 (1985).
19. E. W. Kuipers, M. G. Tenner, M. E. M. Spruit and A. W. Kleyn, *Surf. Sci.* **189/190**, 669 (1987).
20. B. M. Rice, B. C. Gareett, P. K. Swaminathan and M. H. Alexander, *J. Chem. Phys.* **90**, 575 (1989).
21. J. B. C. Pettersson, *J. Chem. Phys.* **100**, 2359 (1994).
22. P. H. F. Reijnen, P. J. van den Hoek, A. W. Kleyn, U. Imke and K. J. Snowdon, *Surf. Sci.* **221**, 427 (1989).
23. A. W. Kleyn, *J. Phys.* **C4**, 8375 (1992).
24. H. Akazawa and Y. Murata, *J. Chem. Phys.* **92**, 5560 (1990).
25. Y. Murata, *Unimolecular and Bimolecular Reaction Dynamics*, Eds. C. Y. Ng, T. Baer and I. Powis (John Wiley and Sons Ltd, New York, 1994).
26. Y. G. Shen, I. Bello and W. M. Lau, *Nucl. Instrum. Meth.* **B73**, 35 (1993).
27. U. V. Slooten, D. R. Andersson, A. W. Kleyn and E. A. Gislason, *Surf. Sci.* **274**, 1 (1992).
28. J. S. Martin, J. N. Greeley, J. R. Morris, B. T. Feranchak and D. C. Jacobs, *J. Chem. Phys.* **100**, 6791 (1994).
29. C. T. Rettner, D. J. Auerbach and H. A. Michelsen, *J. Vac. Sci. Techn.* **A10**, 2282 (1992).
30. H. A. Michelsen, C. T. Rettner, D. J. Auerbach and R. N. Zare, *J. Chem. Phys.* **98**, 8294 (1993).
31. A. W. Kleyn, *J. Phys. Condens. Matter* **4**, 8375 (1992).
32. J. S. Martin, B. T. Feranchak, J. R. Morris, J. N. Greeley and D. C. Jacobs, *J. Phys. Chem.* **100**, 1689 (1996).
33. J. R. Morris, J. S. Martin, J. N. Greeley and D. C. Jacobs, *Surf. Sci.* **330**, 323 (1995).
34. J. N. Greeley, J. S. Martin, J. R. Morris and D. C. Jacobs, *J. Chem Phys.* **102**, 4996 (1995).
35. S. L. Anderson, *State Selected and State to State Ion Molecule Reaction Dynamics*, Eds. C. Y. Ng and M. Baer (Wiley and Sons, New York, 1992).
36. D. C. Jacobs, *J. Phys. Condens. Matter* **7**, 1023 (1995).
37. J. Qian, D. C. Jacobs and D. J. Tannor, *J. Chem. Phys.* **103**, 10764 (1995).
38. B. Willerding, W. Heiland and K. J. Snowdon, *Phys. Rev. Lett.* **53**, 2031 (1984).
39. B. Willerding, K. J. Snowdon, U. Imke and W. Heiland, *Nucl. Instrum. Meth.* **B13**, 614 (1986).
40. P. H. F. Reijnen and A. W. Kleyn, *Chem. Phys.* **139**, 489 (1989).

41. J. H. Rechtien, R. Harder, G. Herrmann and K. J. Snowdon, *Surf. Sci.* **272**, 240 (1992).
42. R. H. Rechtien, R. Harder, G. Herrman, C. Rothing and K. J. Snowdon, *Surf. Sci.* **269/270**, 213 (1992).
43. J. R. Morris, G. Kim, T. L. O. Barstis, R. Mitra and D. C. Jacobs, *J. Chem. Phys.* **107**, 6448 (1997).
44. Y. Zeiri and R. R. Lucchese, *J. Chem. Phys.* **94**, 4055 (1991).
45. B. H. Cooper, C. A. DiRubio, G. A. Kimmel and R. L. McEachern, *Nucl. Instr. Meth. Phys. Res.* **B64**, 49 (1992).
46. J. R. Morris, J. S. Martin, J. N. Greeley and D. C. Jacobs, *Surf. Sci.* **330**, 323 (1995).
47. R. M. Logan and R. E. Stickney, *J. Chem. Phys.* **44**, 195 (1996).
48. E. K. Grimmelmann, J. C. Tully and M. J. Cardillo, *J. Chem. Phys.* **72**, 1039 (1980).
49. R. M. Logan and J. C. Keck, *J. Chem. Phys.* **49**, 860 (1968).
50. M. E. Saecker and G. M. Nathanson, *J. Chem. Phys.* **99**, 7056 (1993).
51. W. R. Koppers, J. H. M. Beijersbergen, T. L. Weeding, P. G. Kistemaker and A. W. Kleyn, *J. Chem. Phys.* **107**, 10736 (1997).
52. R. Smith, *Atomic and Ion Collisions in Solids and at Surfaces* (Cambridge University Press, Cambridge, United Kingdom, 1997).
53. M. Hou, W. Eckstein and M. T. Rovinson, *Nucl. Insrum. Meth.* **B82**, 234 (1993).
54. J. F. Ziegler, J. P. Biersack and U. Littmark, *The Stopping and Range of Ions in Matter*, Vol. 1 (Pergamon, New York, 1985).
55. W. Eckstein, *Computer Simulation of Ion Solid Interactions* (Springer, 1991).
56. T. Hecht, C. Auth and H. Winter, *Nucl. Instrum. Meth.* **B129**, 194 (1997).
57. G. L. Wrenn and A. J. Sims, *The Behavior of Systems in the Space Environment*, Eds. R. N. DeWitt, D. Duston and A. K. Hyder (Kluwer Academic Publishers, 1991).
58. S. E. Deforest, *J. Geophys. Res.* **77**, 651 (1972).
59. I. Katz, *The Behavior of System in the Space Environment*, Eds. R. N. DeWitt, D. Duston and A. K. Hyder (Kluwer Academic Publishers, 1991).
60. C. K. Purvis, H. B. Garrett and A. G. Whittlesey, *Design guidelines for assessing and controlling spacecraft charging effects* (NASA Technical Paper 2361, 1984).
61. E. J. Daly and D. J. Rodgers, *The Behavior of Systems in the Space Environment*, Ed. R. N. DeWitt (Kluwer Academic Publishers, 1993).
62. A. R. Burns, E. B. Stechel and D. R. Jennison, Eds., *Desorption Induced by Electronic Transition* (Springer, New York, 1993).
63. H. Gnaser, *Low Energy Ion Irradiation of Solid Surface* (Springer, Berlin, 1999).
64. H. Gnaser, H. L. Bay and W. O. Hofer, *Nucl. Instrum. Meth.* **B15**, 49 (1986).
65. J. F. Ziegler, J. Pl. Biersack and U. Littmark, *The Stopping and Range of Ions in Matter*, Vol. 1 (Pergamon, New York, 1985).

66. T. J. Colla and H. M. Urbassek, *Rad. Eff. Def. Solids* **142**, 439 (1997).
67. H. F. Winters, H. Coufal, C. T. Rettner and D. S. Bethune, *Phys. Rev.* **B41**, 6240 (1990).
68. H. Coufal, H. F. Winters, H. L. Bay and W. Eckstein, *Phys. Rev.* **B44**, 4747 (1991).
69. H. H. Andersen, *Nucl. Instrum. Meth.* **B18**, 321 (1987).
70. C. Garcia-Rosales, W. Eckstein and J. Roth, *J. Nucl. Mat.* **218**, 8 (1994).
71. W. Eckstein, C. Garcia-Rosales, J. Roth and J. Laszlo, *Nucl. Instrum. Meth.* **B83**, 95 (1993).
72. G. Betz, R. Kirchner, W. Husinsky, R. Rüdenauer and H. M. Urbassek, *Rad. Eff. Def. Solids* **130/131**, 251 (1994).
73. P. Nordlander and J. C. Tully, *Surf. Sci.* **211/212**, 207 (1989).
74. J. P. Gauyacq and A. G. Borisov, *J. Phys.: Condens. Matter* **10**, 6585 (1998).
75. H. D. Hagstrum, *Inelastic Ion–Surface Collisions*, Eds. J. N. H. Tolk, J. C. Yully, W. Heiland and C. W. White (Academic, New York, 1977).
76. J. Los and J. J. C. Geerlings, *Phys. Rep.* **190**, 133 (1990).
77. N. D. Lang, *Phys. Rev.* **B27**, 2019 (1983).
78. M. Maazouz, L. Guillemot, T. Schatholter, S. Ustaze and V. A. Esaulov, *Nucl. Instrum. Meth.* **B125**, 283 (1997).
79. J. R. Morris, J. S. Martin, J. N. Greeley and D. C. Jacobs, *Surf. Sci.* **330**, 321 (1995).
80. J. A. Yarmoff, *Surf. Sci.* **348**, 359 (1996).
81. C. Auth, A. G. Borisov and H. Winter, *Phys. Rev. Lett.* **75**, 2292 (1995).
82. H. Winter, A. Mertens, C. Auth and A. G. Borisov, *Phys. Rev.* **A54**, 2486 (1996).
83. V. N. Ostrowski, Proc. 17th Int. Conf. Phys. Electron. Atom. Collisions, Brisbane, 1991, edited by W. R. MacGillivray *et al.* (Adam Hilger, Bristol, 1992).
84. S. J. Pfeifer and J. D. Garcia, *Phys. Rev.* **A23**, 2267 (1981).
85. I. F. Urazgil'din and A. G. Borisov, *Surf. Sci.* **227**, L112 (1990).
86. M. Bernheim and F. Le Bourse, *Nucl. Instrum. Meth.* **B27**, 94 (1987).
87. M. L. Y, *Phys. Rev. Lett.* **47**, 1325 (1981).
88. J. C. Tucek, S. G. Walton and R. L. Champion, *Surf. Sci.* **410**, 258 (1998).

CHAPTER 8

SURFACE CHEMISTRY IN THE JOVIAN MAGNETOSPHERE RADIATION ENVIRONMENT

Robert E. Johnson

Engineering Physics, Thornton Hall B103
University of Virginia, Charlottesville, VA 22903
Tel.k: (804) 924-3244. Fax: (804) 924-1353
E-mail: rej@virginia.edu

Contents

1. Introduction

One of the most exciting areas of research in Planetary Science is the study of the chemistry induced in the surfaces of the icy moons of Jupiter by the Jovian magnetospheric particle radiation. Observations by Galileo of Io, Europa, Ganymede and Callisto (Fig. 1) using the newly discovered telescope initiated enormous controversy and changed our image of the solar system over three centuries ago. Now observations of these moons using the

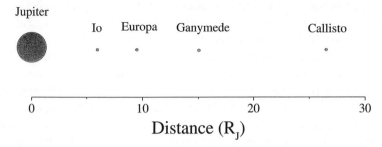

Fig. 1. A schematic diagram of the orbital positions of the moons of Jupiter discovered by Galileo. Their positions are scaled to Jupiter's radius (R_J) with properties given in Table 1. Note: these satellites, like our moon, are all phase locked to the parent planet, hence the same side faces Jupiter throughout each moon's orbit. Since the magnetic field is attached to Jupiter, it rotates faster, therefore, in addition to ions and electrons moving up and down the field lines and impacting the satellites, there is a net preferential flow onto the hemisphere trailing the satellite's motion.

Table 1. Surface properties of Galilean satellites.

	$R_{JM}(R_J)$[a]	R(km)[b]	$T(K)$[c]	Constituents[d]	Atmosphere[e]
Io	5.90	1820	80–150	SO_2, S_x, Na K, Cl	SO_2, Na K, Cl
Europa[f]	9.40	1570	80–130	H_2O, SO_2, Na K, H_2O_2, S_x	O_2, Na, K
Ganymede[g]	15.0	2630	80–150	H_2O, O_2 O_3, CO_2	O_2, Na, O
Callisto[f]	26.4	2400	80–150	H_2O, CO_2	CO_2, H_2O

[a]Distance from Jupiter in Jupiter radii, $R_J = 7.14 \times 10^4$ km.
[b]Radius of moon in km.
[c]Average temperature range. Note: surfaces segregated into bright, cold volatile regions and dark refractory regions with different temperatures.
[d]Surface constituents are at present being identified using NIMS and the UVS instruments on Galileo. Hydrated mineral bands are seen on Europa, Ganymede and Callisto. On Europa suggestive of frozen dilute sulfuric acid,[7] but other models have been proposed.[14] CH, CN, and OH bands seen. Some species inferred from atmosphere and plasma.
[e]Species identified to date in atmosphere. Plasma is primarily H^+, O^{+z}, S^{z+}.
[f]Suggested as having an underground "ocean."
[g]Has its own magnetic field.

instruments on the Galileo spacecraft and Hubble Space Telescope (HST) are again leading to radically new insights on the origins of bodies in our solar system and, possibly, to new insights into the origins of biologically active molecules. This understanding is expected to grow dramatically again when the CASSINI spacecraft examines Saturn's moons beginning in 2004.

A critical issue for interpreting the new observations is to obtain an understanding of the chemistry induced in the surfaces of the Galilean satellites by the energetic ions and electrons trapped in the giant magnetosphere of Jupiter. These energetic particles produce new molecular species in the surface and cause desorption of atoms and molecules. The desorbed species form an ambient gas around each object, which contributes to the tenuous atmospheres and ionospheres on these moons.[1] Therefore, if the interaction of the radiation with the candidate surface materials is understood, direct collection of the ionized component by mass spectrometry from an orbiting probe can be used to determine the surface composition.[2,3]

Whereas the dominant surface material on the outer three Galilean moons is ice, the inner moon, Io, is the most active volcanic object in the solar system. Therefore, it has been totally dehydrated and has a surface coated with sulfur dioxide.[4] This surface is exposed to significant levels of ion, electron and UV photon radiation. At the time of this writing, one of the moons, Ganymede, is the first nonplanetary body in the solar system on which an intrinsic magnetic field has been discovered.[5] This field provides a useful tool. Since it partially deflects the incident radiation, observers can compare the surface chemistry in the highly irradiated regions with that in lightly irradiated regions.

Also remarkable is the fact that the other two icy moons, Europa and Callisto, have unusual conducting properties as indicated by the magnetometer measurements on the Galileo spacecraft. This effect has been tentatively attributed to the presence of underground, tidally heated oceans,[6] although we recently suggested that an irradiation-produced material[7] can act as a conducting layer. However, the ocean hypothesis is attractive as it is based on models for their formation, differentiation and tidal heating by Jupiter.[8] Also, the magnitude of the conductivity of the object deduced from the time variability of the local fields appears to be appropriate for a "salty" ocean. As a further support of this hypothesis, the surface of Europa appears to be young and chaotic suggestive that fresh material has "recently" reached the surface.[9-13] One model for such activity is that ice warmed by tidal heating under a few kilometers-thick frozen layer, can be extruded onto the surface in some regions with the surface subducted causing burial of material in other regions.

Recent infrared spectra of the shifted suppressed water bands indicate that the surface of Europa contains hydrated minerals. These have been suggested to be hydrated salts and organics,[14] which would also be consistent with material from an underground ocean. Because of this strong, but indirect, evidence for an ocean, Europa is now considered to be an object on which biological materials could have evolved. Such an evolution, if it occurred, could have been driven by the heat created by the tidal interaction with Jupiter, although recently this has been suggested to be too small.[15] It has also been suggested that the energy of the Jovian magnetospheric particle radiation incident on to the surface could drive chemistry needed for initiating biological evolution.[16,17]

A recent laboratory comparison of the shape of the water of hydration bands with the spectra obtained by the Galileo spacecraft indicates that the hydrated material mentioned above might in fact be frozen, hydrated H_2SO_4.[7] This material would be produced by the charged particle irradiation of sulfate salts, sulfur or SO_2 in an ice matrix.[2] Therefore, a sulfur chemical cycle is maintained by the incident radiation, a cycle similar to that occurring at higher temperatures in the atmospheres of the Earth and Venus.[7] On Europa, this involves oxidants such as SO_4^{2-}, H_2O_2[18] and O_2[19] which have potential importance for the proposed prebiotic chemistry.

In this chapter, I will briefly review the observations relevant to the Jovian radiation environment and the suggested surface materials on the large moons. I will then summarize what is known about the radiation-induced surface chemistry and, finally, I will suggest laboratory studies and molecular dynamics simulations that might be carried out to contribute to an understanding of the surface chemistry relevant to this exciting planetary environment. More extensive reviews of various aspects exist and will be referred to below rather than repeating those summaries.[1,19,20] In addition, a review of the physics of the sputtering process was presented recently[21] and an overview was provided in the Reviews of Modern Physics.[22]

2. Observations

When the Pioneer and Voyager spacecraft passed through the Jovian system in the 1970's a much more intense radiation environment was found than was expected.[23] The particle radiation consisted of energetic ions trapped in

Jupiter's giant magnetic field. These ions had as their source the icy satellites that orbit within the Jovian magnetosphere and, hence, the plasma composition was determined by the surface composition.[3] Using the data from the Energetic Particle Detector on the Galileo spacecraft, the ion and electron fluxes[16,24] are given in Fig. 2. In addition, a low energy plasma component exists.[25,26] Since the dominant volatile on Io is SO_2 (with S_x) and with H_2O the dominant volatile on the other three moons, the composition of the plasma is predominantly H^+, O^{z+}, S^{z+}. A smaller component of undissociated molecular ions (SO_2^+, SO^+, NaX^+, where X is O or S) is found, primarily, close to Io, the principal source of the plasma. In addition to ionized Na, Cl^{+z} was recently discovered, indicating Cl is present in Io's surface.[27] Although the flux of ions and electrons in this plasma is not large by laboratory standards, the plasma energy flux incident on

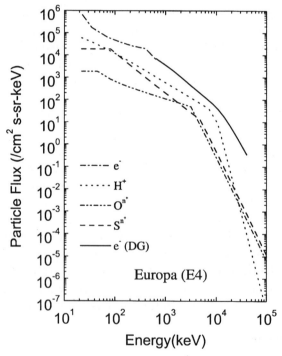

Fig. 2. The ambient, energetic plasma flux at Europa, assumed isotropic, of energetic ions (H^+, O^+, S^+) and electrons: taken from Cooper *et al.*[16] Measured by the Energetic Particle Detector (EPD) on Galileo. Curve labeled DG are the higher electron energies based on Voyager data.[28]

the surfaces or atmospheres of Europa and Io is much larger than the solar UV flux.[16]

An important issue for irradiation chemistry is the dose versus depth into the surface. This was computed by J. Cooper *et al.*[16] and the results for Europa are shown in Fig. 3. Here the vertical axis is the time to achieve a dose of 100 eV per molecule in H_2O. This is a useful form since we note below that radiation chemists typically give radiation effects as *G*-values, the number of a particular molecular species created or destroyed for each 100 eV of energy deposited. Geologist have suggested that the youngest surface ages are $\sim 10^{6-7}$ years on Europa,[13] hence, the significant dose, as

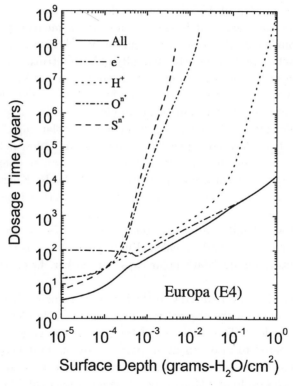

Fig. 3. Time versus depth for the ions and electrons in Fig. 2: taken from Cooper *et al.*[16] The time on the vertical axis is that time it takes for each molecule to receive, on the average, a dose of 100 eV. This is a significant dose for which each molecule has likely been dissociated and it allows easy use of a *G*-value (probability of a molecular change per 100 eV deposited). The horizontal axis gives the depth scaled by density.

expressed in Fig. 3, is achieved at all depths above ~ 1 cm at unit density in geologically relevant times. Since over most of the surface the solar photon penetration depths are much smaller than this, the optical layer is totally altered by the particle irradiation in relatively short times. Proposed plans for a mission to find chemical products of prebiologic activity or, possibly, biological activity in Europa's surface will probably require sampling at depths > 10 cm based on a porosity ~ 0.8. That is, a critical issue for planning future missions is determining the radiation dose (hence, depth) that biologically important molecules could receive and still be identified by their radiation-induced decomposition products.

The most penetrating of the detected incident particles are electrons with energies of 10's of MeV[28] and the dominant carrier of energy are the keV–MeV electrons.[16] It is seen in an electron microscope that energetic electrons cause the growth of voids in ice.[29] These are formed by defect production and by inducing mobility of both intrinsic and radiation-induced defects.[30] The formation of such voids can produce an efficient light-scattering and, hence, optically bright surface.[31] Therefore, I have suggested that this effect produces the bright surface in the polar regions on Ganymede, often referred to as "Ganymede's polar caps." Since the energetic electrons have long penetration depths and follow field lines, the cap boundary is found to coincide with the regions where open field lines intersect Ganymede's surface as modeled recently by K. Khurana and R. Pappalardo (private communication). At Io, on the other hand, radiation darkens the sulfur containing surface materials by producing molecular chains. At low latitudes, any radiation damage of Io's surface material is rapidly covered by SO_2 and annealed. But at the poles, this process is very slow, accounting for its "dark polar caps" or, rather, brighter equatorial region.[31]

Matson *et al.*[32] first realized that the incident magnetospheric ion radiation could account for the observation of the desorbed Na seen as a "cloud" near the moon Io.[33] The discovery of the intense energetic particle environment appeared to also explain in part two of the best known features of these moons: hemispherical differences in their reflectance and the polar spectral features on Io and Ganymede.[4] The hemispherical differences have been attributed to preferential plasma energy deposited onto the hemisphere that trails the orbital motion on these moons which are tidally locked to Jupiter.[1,4,34,35] The Ganymede polar spectral feature appear to be due to particles flowing along the magnetic field lines to the poles, as discussed above, and the efficient sublimation and annealing of radiation damage in the equatorial regions.[31]

The discovery of the energetic Jovian plasma by Voyager and Pioneer spacecraft led Brown, Lanzerotti and coworkers[36,37] to study the sputtering of the principal surface constituents of the Galilean satellites, H_2O and SO_2 ices at $T < 150$ K. The sputtering yield for ice is given in Fig. 4(a). They found, surprisingly, that whole molecules dominated the ejecta (Fig. 4(b)) and that the yields (number of molecules ejected per ion incident) were much larger than expected. They also found that sputter ejection of molecules was initiated not only by momentum transfer to the atoms in the solid but also by the electronic ionization and excitation produced by the fast ions incident on these ices. They named the latter process electronic sputtering. This is a process closely related to electronically

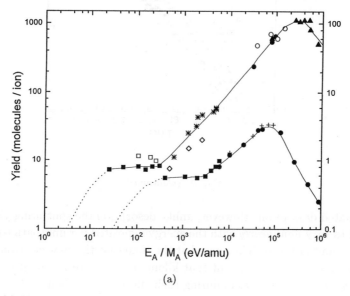

(a)

Fig. 4. (a) The ice sputtering yield (number of equivalent H_2O molecules ejected per ion incident) versus the ion energy: taken from Shi *et al.*[103] The top curve and left hand axis is for incident O^+ and the bottom curve and right hand axis is for incident H^+. The structure at low velocity is sputtering due to momentum transfer and that at higher velocities is due to electronic excitation and ionization, which is the dominant sputtering process in ice and closely related to electronically-stimulated desorption.[1,19] (b) The sputtering yield of D_2O versus surface temperature for 1.5 MeV He^+ ions (Dots: mass 20; Square: mass 4; Triangle: mass 32). There is a T independent component but the decomposition products, D_2 and O_2 are T dependent: taken from Brown *et al.*[38]

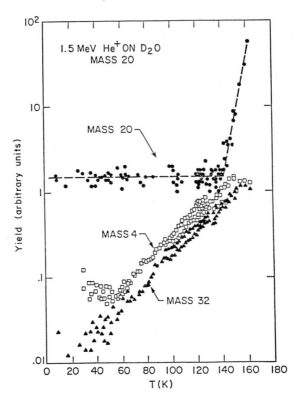

Fig. 4. (*Continued*).

stimulated desorption. However, unlike desorption the sputtering yield was found to vary *nonlinearly* with the energy deposited per unit path length by the incident fast ions. They noted at the time that this new electronic sputtering process was so efficient that kilometers of material could have been removed from these objects during their lifetime.[37] Although this was certainly an overestimate due to the extrapolations made and the incomplete knowledge of Europa's surface, sputtering is a dominant process at Europa. In addition, the electronic sputtering process is in fact closely related to the radiation chemistry produced in ice by the fast ions and electrons.[20]

Subsequent work has shown that new more volatile molecules are also produced and desorbed by energetic ions [Fig. 4(b)], and the desorption efficiencies depended on the material temperature.[38] This dependence is due to the radicals or defects becoming mobile. Since the principal products ejected from ice, in addition to H_2O, are H_2 and O_2, we predicted that the icy satellites could have tenuous O_2 atmospheres.[1,39] That is, the desorbed H_2 is light and escapes readily from a moon's gravitational field but the heavier O_2 produced on the Galilean satellites would not. However, unlike H_2O, once formed O_2 would not condense at the surface temperatures on these moons, $>\sim 80$ K. Remarkably an O_2 atmosphere was recently observed by HST both on Europa and Ganymede.[40] An even more tenuous CO_2 atmosphere was observed on Callisto by the Galileo spacecraft.[41] Because CO_2 does not condense efficiently at the surface temperatures on Callisto, it must be trapped in the ice or be created by the radiation prior to desorption. Finally, the gravity filter that allows H_2 to escape much more efficiently than O_2 also means that the D/H ratio in the surface layer will be much larger than the solar values.[1,2,19]

In a parallel set of discoveries, a reflectance band in the visible, similar to that for solid O_2,[42] was seen at low latitudes on Ganymede. In addition, a UV feature associated with O_3 was seen on Ganymede[43,44] and on the icy satellites of Saturn.[43] Coupled with these observations was the much earlier discovery of a band indicative of SO_2 in ice at Europa[45] and Callisto[46] and the recent discovery of CO_2 trapped in the icy surfaces.[41] The SO_2 was initially assumed to be due to sulfur ions originating at Io implanted into the ice at Europa,[45,47] but the SO_2 is also a radiation decomposition product like the O_2, as discussed below.[2,7] The CO_2 source is probably internal as carbon ions have not yet been seen in the plasma.

Although the presence of these species *suggests* that the ions actually bombard the surface, the clearest signature that radiation-induced chemistry is occurring in these icy surfaces was recently reported. That is, the radiation-induced product, H_2O_2, has been seen in both the IR and the UV in the surface ice of Europa.[18] Therefore, it is now established that the radiation-induced changes in the surface occur at a rate that significantly exceeds the geologic resurfacing rate. However, accurate radiation-induced yields (or G-values, yield/100 eV deposited) are typically *not available* to successfully determine surface dose. This is unfortunate as such information can be used to determine the age of the optical surface.

Recent data from the near IR mapping spectrometer (NIMS) of the Galileo spacecraft has also identified areas on the moons in which the water bands are shifted and suppressed, consistent with large surface areas containing hydrated minerals.[14] The presence of SO_2 in a water ice surface containing H_2O_2 and subject to irradiation suggested to us that hydrated H_2SO_4 should also be present, as in cloud particles at the Earth and Venus. During the writing of this manuscript we showed that bands associated with water of hydration, seen in the IR by NIMS on Galileo, were reasonably well fit by the spectra of frozen $H_2SO_4 \cdot XH_2O$.[7] This data had been used to suggest the presence on the outer three moons of organics in ice and certain hydrated salts and carbonates: $Na_2CO_3 \cdot XH_2O$ (62%), $Na_2Mg(SO_4)^2 \cdot XH_2O$ (30%), $Na_2SO_4 \cdot XH_2O$ (8%).[48] The latter identifications were based on models of satellite formation and evolution that predicted the presence of such materials.[8] In addition, Na_2SO_4 is a material suggested to be on the dehydrated moon Io[49] and was noted to be radiation modified by Nash and Fanale.[35] If sulfate salts are present, radiation-induced decomposition is the likely source of the SO_2 and Na seen at Europa[2,50,51] and the observed CO_2[2,50] may be the product of an irradiated organic, like the carbonate suggested above.

Due to tidal heating and radiation-induced desorption, Io has lost most of its water and other light species. Therefore, in addition to SO_2 and its radiation products (e.g. S_2O or SO_3), Na_2SO_4,[52] Na_2S_x,[53] S_x and NaCl[27] are suggested surface constituents. Decomposition of these materials is the likely source for the well known Na "cloud" at Io, extensively studied from earth. Observations and models also suggest that there should be silicate intrusions. Recently, a silicate flow from an Io volcano was identified.[54] However, over most of the satellites' surfaces the rocky material appears to be covered by the dominant volatile: SO_2 at Io and H_2O at Europa and Ganymede as ice or a hydrated species. Callisto was seen to have clean water ice bands but the surface is dominated by an unidentified dark contaminant, possibly micrometeorite debris collected on its surface, and its physical surface features appear to be modified by outgassing of a volatile like CO_2.[41,55] To interpret the new spectral data, a clear understanding of the surface chemistry of the absorption bands and of the desorbed species is now required. Such an analysis can, in principle, be used to date geologically young features on these moons.[16]

3. Radiation Chemistry

The study of the irradiation of ice has a long history due, in part, to the considerable interest in radiation biology in the 1950's and 1960's. Freeze-dried samples of biomolecules were often irradiated and compared, spectrally and by using ESR, to irradiated, degassed ice samples.[56] In spite of this history, useful quantitative information for radiolysis relevant to the planetary problems described above is scarce. Therefore, at the time of Voyager, Brown, Lanzerotti and coworkers[36,37] initiated studies of the ices and Nash and Fanale[35] initiated studies of the spectral changes due to irradiation of those refractory materials thought to exist at Io. We recently reviewed the database on the radiation chemistry of ice as it relates to the icy satellites.[20] I also recently reviewed the database for the sputtering of ices by energetic ions.[19] These papers and the exciting spacecraft and HST data have stimulated a large amount of recent work which is summarized below.

The temperature range of interest is between $\sim$ 70–90 K near the poles and varies from $\sim$ 90 K nightside near the equators to $\sim$ 120–150 K dayside depending on the albedo in the equatorial region. This is a range of temperature variation over which radicals can be made mobile and react. There are a number of agreed upon radiation products which are both relevant to the planetary problem and reasonably well studied. Whereas it was thought that protons are mobile even in an amorphous ice, recently it has been suggested that they trap efficiently.[57] Following recombination, the principal products in photolysis or radiolysis of ice, H and OH, come from the primary dissociation channel: $H_2O \rightarrow H + OH$. Whereas H is thought to diffuse effectively at most relevant temperatures, at low temperatures the OH traps in ice, as indicated by the ESR spectra and by the blue shifted OH absorption band at $\sim$ 0.28 μm.[56,58] On the other hand the more mobile H atoms diffuse and can react forming H_2. When the sample is in a good vacuum, as it is on the surfaces of these moons, the H_2 diffuses to the vacuum interface and escapes leaving behind a chemically altered ice. A second channel, that has $\sim$ 10% branching ratio in the gasphase under solar UV is $H_2O \rightarrow H_2 + O$. Single event production of H_2 from ice, seen by Reimann,[59] was recently confirmed to occur even with low energy electrons, at least at the ice–vacuum surface.[60] The loss of H_2 leaves behind a reactive O. Recent experiments suggest the irradiation-induced formation of a species such as H_2O–O.[61]

Because any H_2 produced diffuses efficiently and is lost to the vacuum, the ice is permanently altered. Therefore, there is a depletion of H and formation of "trapped" O and OH, so the surface layers of the grains are oxidizing. They also should have an enhanced D/H ratio.[1] The O and OH radicals can react directly: $OH + OH \rightarrow H_2O_2$ and $OH + O \rightarrow HO_2$ or

$O_2 + H$. In addition, lattice relaxation around a species like H_2O-O might lead to H_2O_2 and HO_2. The latter products have been identified by ESR and by absorption features in the UV.[56,58,62]

The source of the observed O_2 is less certain. $H_2 + \frac{1}{2}O_2$ are the decomposition products of irradiated ice.[56,59] O_2 is formed in the gas-phase by reactions involving the products above. However, following exposure to ionizing radiation, O_2 is found to evolve from an initially degassed ice sample on warming.[56] G-values (or yields) larger than those for the gas-phase were found for ice although *simple* chemical rate equations predict otherwise. In studying the radiation effects on the icy satellites, Brown, Lanzerotti and coworkers showed O_2 is directly produced and ejected into the vacuum by energetic ions.[38,63] More recently, UV photons[64] and low energy electrons[65] were shown to produce O_2 with a threshold energy ~ 10 eV.

Relatively high G-values ($\sim 0.15/100$ eV) are obtained if ice is irradiated at low temperatures in a closed system and then warmed.[56] However, recent laboratory data suggest lower G-values, so that the early experiments may have been affected by small amounts of contaminants such as CO_2. For instance, Baragiola and coworkers suggest very small G-value ($\sim 0.003 O_2/100$ eV) for a thin ice layer irradiated by keV protons.[66] Assuming that the low energy electron desorption data of Seiger *et al.*[65] are applicable to the secondary electrons produced by a fast ion, and using a W-value (average energy per ion–electron pair produced) for light ions of ~ 26 eV, a somewhat larger G-value ($\sim 0.01 O_2/100$ eV) is obtained. But this is still smaller than in the early experiments. Since production and loss of hydrogen occurs with higher G-values, implying O must eventually be produced, there are some inconsistencies to be resolved. Therefore, additional measurements are needed for the amount of O_2 produced per incident fast ion, both that directly desorbed (sputtered) and that produced and trapped in the solid. This is needed as a function of temperature in order to determine the amount of ambient gas and the potential for chemical synthesis. Since there is a dE/dx dependence in the production of O_2, experiments are needed to determine the efficiency of the production of O_2 for a broad range of ion types and velocities or a clear model for O_2 formation is needed. In addition, since radiation followed by annealing occurs in the equatorial regions of these moons, the earlier experiments, on gas evolving from warmed ice or from a hydrated salt irradiated at low temperature, need to be repeated.

Reimann *et al.*[59] measured the fluence and temperature dependence for the production of O_2 from a thin ice sample in a vacuum exposed to energetic Ne^+ [Fig. 5(a)]. These ions were used to represent the energetic O^+ ions in the Jovian plasma. They found a correlation between the loss of

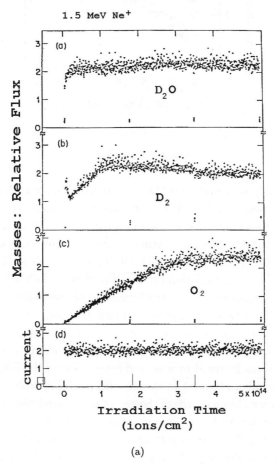

(a)

Fig. 5. (a) The production of D_2, O_2 and D_2O from a D_2O ice sample by energetic Ne^+ (used to model the effect of the O^+ flux in Fig. 2) as a function of irradiation time given as number of impinging ions per square centimeter. If beam is turned off at any point, each signal immediately restores when beam is turned on, indicating the ice is altered: taken from Reimann *et al.*[59] (b) The production of O_2 by low energy electron irradiation of ice: taken from Seiger *et al.*[65] Curves are for 100 eV and 50 eV electrons, as indicated, and the symbols are for different temperatures showing the scaling of the yields.

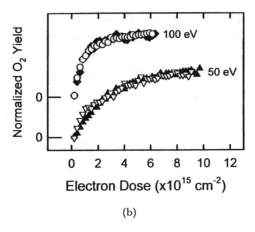

(b)

Fig. 5. (*Continued*).

H_2 and the production and loss of O_2. It is seen in Fig. 5(a) that H_2 exhibits a prompt component followed by a gradually increasing component. The former is due to a single event near the surface and the latter depends on the film thickness for ions penetrating to the substrate. The O_2 signal was initially zero but grew linearly with the irradiation dose. This indicates that O_2 is not a direct radiation product (as discussed above), but requires that the ice is first altered: e.g. the formation of trapped radicals or other precursors. This was confirmed by shutting the beam off and letting the sample sit.[59] On turning the beam back on, the O_2 signal immediately restored indicating that a "permanent" change occurred. The initial linear increase in O_2 was followed by a second region of growth. Finally, both the H_2 and O_2 signals decreased as the film thinned. For low energy electron impact on a very thin ice surface, Seiger *et al.*[65] found results surprisingly similar to those for MeV Ne$^+$ on ice as seen in Fig. 5(b).

Reimann *et al.*[59] noted an activation barrier of 0.05–0.07 eV and considered diffusion of the radicals produced. Seiger *et al.*[65] concluded that a nondiffusing precursor is first formed (cross section $\sim 10^{-18}$ cm^2 in the surface layer) in a manner that is temperature dependent, accounting for the scaling of the results at different T in Fig. 5(b), and a subsequent impacting electron directly produces O_2 (cross section $\sim 10^{-16}$ cm^2). [A candidate precursor may be the H_2O–O mentioned earlier[61]]. They pointed out that the low fluence dependence seen in Figs. 5(a) and 5(b) is inconsistent with the diffusion model. This is the case unless "diffusion" is fast and temperature independent over the region studied. Therefore, a "hot O" atom reacting with trapped OH, $O(hot) + OH \rightarrow H + O_2$, is a candidate.

In these studies the role of the surface, the sample temperature and the sample structure (defects, voids and crystallinity) appear to be important but require further study. As mentioned earlier, void formation has been imaged when ice is placed in an electron microscope.[29] However, the relationship between the formation of interior surfaces and O_2 production has not been studied, although it appears that O_2 may be formed more efficiently at a surface.

The radiation product H_2O_2 is seen in the photolysis and radiolysis of ice,[56,67] and the role of electron scavengers has been studied.[68] H_2O_2 has now been identified in UV and IR spectra of Europa's surface where it appears to be ubiquitous.[18] As stated earlier, this confirms that energetic ions are reaching the surface and that the radiolysis of ice is occurring. It is now generally accepted that the very thin O_2 atmosphere at Europa[40] is produced by the radiolysis of ice, as suggested earlier.[1,39] At Ganymede, sublimation may be more vigorous due to higher average surface temperature, although the temperatures in the icy patches may be similar to those on Europa. In any case, the production of O_2 at Ganymede could involve gas-phase chemistry[69] at low temperatures, which would also require new rate coefficients. In addition, a pair of weak bands in the visible have been associated with the presence of O_2 inclusions in the surface of Ganymede.[30,42] Because of the surface temperatures (> 80 K), it has been suggested that the codeposited O_2 may be trapped by water molecules, or it may be formed and trapped in voids in an ice[30] or in a hydrated mineral surface.[70] In Fig. 6, the data for O_2 codeposited with ice and deposited as solid O_2 are compared to the reflectance spectrum. The former appear to this author to show reasonable agreement with the space observations. However, in a series of papers,[66] Baragiola and coworkers suggested that using the peak positions was most important in the comparison and concluded that solid O_2 exists on the surface in cracks or crevices which are only temporarily exposed to sun light. If correct, this would require very low (< 40 K) surface temperatures.[70] If trapping O_2 in ice is problematic, as they suggest, it is possible that the O_2 may be formed and trapped in a hydrated mineral.[2,70] In any case, the physical-chemical explanation of this band is not available. It may be associated with a specific terrain type.

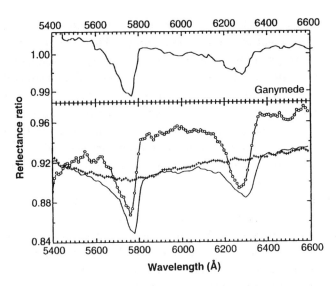

Fig. 6. Upper panel: absorption bands of Ganymede which are attributed to "solid O_2"[42] (obtained as a ratio of Ganymede's trailing hemisphere reflectance to that of Callisto's). Lower panel: pure O_2 at 26 K (line with circles), codeposited $O_2 + H_2O$ ice film at 26 K (solid curve). Same after warming to 100 K, (pluses): taken from Vidal et al.[66]

HST observations in the UV of a broad band roughly consistent with the Hartley band, suggest the presence of O_3 in the surface of Ganymede.[43] The possible presence of O_3 is important, since O_3 is not a direct radiation product of ice,[56,71] but forms readily in O_2.[71-73] Therefore, its presence reenforces the concept that a condensed form of O_2 exists: e.g. trapped in voids in the ice[30] or other irradiated surface material.[70] Initially, a broadened O_3 band was used to roughly fit the data, but evidence for another band at longer wavelengths is clear.[43] In Fig. 7, the solid O_3 band, which differs little from O_3 trapped in O_2, is shown. The second band may be associated with trapped OH but here I show a–CH_2O band in an organic compound. The presence of such a band would not be surprising as CO_2 and H_2O both exist in the irradiated surface.

A number of studies have been carried out on the sputtering of CO_2, SO_2, and S_8, all proposed or observed volatiles on these moons.[19,74] The ejecta, as in the sputtering of water ice, are the parent molecules plus those decomposition products having high volatility[19] allowing them to be easily driven into the gas-phase (e.g. CO and O_2 from CO_2; and SO and O_2 from

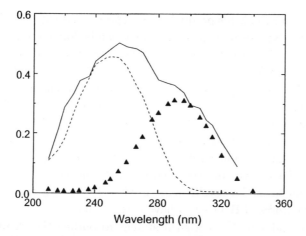

Fig. 7. Hubble Space Telescope UV spectrum (solid line), ratio of trailing hemisphere reflectance converted to an absorbance.[43] Broken line: condensed form of O_3[72]; triangles: $-CH_2O$ in isobutyraldehyde.[105]

SO_2; S_2 from S_8). Equally interesting are the potential refractory products such as polymerized carbon and sulfur, and the sulfur and carbon sub-oxides.[75] However, there is very little data on the physical, chemical or spectral properties of the refractory materials formed in an ice that contains CO_2, SO_2 or sulfur.[74] Complex irradiation produced, carbon based materials, named Tholins (muddy) by C. Sagan, have been suggested as being present on the surfaces by the Galileo NIMS spectra.[14] Earlier Nash and Fanale[35] studied irradiation alteration of reflectance of potential satellite minerals to fit Io's spectrum. The general reddening of the visible-near-UV reflectance at Europa has been attributed to irradiation-induced changes in grain size[47] and to radiation-induced chemistry forming a polymerized sulfur.[34,76,77] Agreement with the optical reflectance data requires < 1% of the refractory component of a photolyzed, frozen $H_2S + H_2O$ mixture.[7,78] However, definitive identifications and quantitative evaluations under known radiation environments have not been made and competing evaporative processes have also been suggested to account for the reddening of the optical spectra.[79]

There are fewer studies giving absolute yields for mixed ices subjected to charged particle radiation. Two relevant mixtures for Europa, Ganymede and Callisto are $H_2O + SO_2$ and $H_2O + CO_2$. The latter may also be important for the Uranian moons.[80] Chemical pathways have been studied but not the absolute yields for producing new species. Clearly, complex hydro-carbons of the type seen in cometary comas can be formed in Europa's surface and then subducted into the putative underground ocean. Moore and coworkers[81–83] and Strazzulla and coworkers[84–86] have studied mixtures of H_2O with CO and CO_2 as well as other carbon containing molecules and complex mixtures relevant to comets at low temperatures (< 20 K). They have shown that among other species, carbonic acid, H_2CO_3, is formed. This is a species not yet identified on the icy satellites but should also be a product of radiation-induced desorption of Na from $Na_2CO_3 \cdot XH_2O$ (see Appendix). They have obtained relative yields but data at those temperatures relevant to the icy satellites are needed. In a recent paper[83] absolute yields for mixtures of H_2O + CO have been obtained.

Delitsky and Lane[87] used such studies to outline the chemical pathways that might be relevant on these moons. However, detailed quantitative information for the separate pathways is needed. Earlier, Haring and coworkers,[88] among others, measured ejected molecule mass spectra and ejecta energy spectra for a number of mixtures. These studies indicate that a variety of products are formed and ejected into the gas-phase. However, absolute yields for desorbed species from mixed ices were not given but are needed to understand the ambient gas at each of the icy satellites.[1,2,19]

The plasma ion bombardment also leads to implantation of reactive atoms into the surface. Therefore, quantitative data is also needed on the chemistry induced by reactive atoms Na and S. These atoms are introduced as energetic ions from Io and are implanted into Europa's low temperature water/ice or into its hydrated mineral surface. On implantation into and annealing of an ice depleted in H, S is likely to form SO^{47} and then annealing or further bombardment produces SO_2 giving the 0.28 μm band,[46] although recent Galileo data[89] suggests the SO_2 is associated with the non-ice regions as in the model in Ref. 7. Incident Na is likely to form NaOH. This latter molecule has not yet been seen by NIMS on Europa and, therefore, would have to be present at less than a few percent if it exists in the surface material.

4. Hydrated Minerals

The hydrated minerals, suggested as being present on Europa by the Galileo NIMS data and by the possible presence of an underground ocean, were described earlier. One model has them being hydrated versions of the minerals expected on Io. The principal chemical issue learned from the above is that plasma bombardment causes changes in the spectra[35] and the preferential loss of H (as H_2) which leads to a surface layer which is oxidizing.[20] Therefore, the favored chemical pathways are fairly clear[19] and are considered in the Appendix. The presence of both SO_2 and H_2O_2 led us to look at the sulfur chemistry in Europa's radiation environment.[7] A principal product is a sulfate, which is a good oxidant, probably in the form of hydrated H_2SO_4 as mentioned earlier. The comparison of this species to the water of hydration bands in the NIMS spectra is shown in Fig. 8. Although the agreement is far from perfect, small shifts are expected due to ion radiation[35] and agreement with other radiation products reenforces the presence of $H_2SO_4 \cdot XH_2O$[7] as discussed below. Radiation yields for producing hydrated H_2SO_4 at the temperatures and for the relevant incident particle energies are not available. Moore[90] irradiated an H_2O/SO_2 mixture and observed a refractory product that was consistent with the presence of H_2SO_4, but did not identify the product. Also, in pure SO_2, SO_3 is produced efficiently $(G \sim 5/100 \text{ eV})$.[90] In the presence of H_2O this exothermically yields H_2SO_4. Similarly, sulfur particles in water produce H_2SO_4 under irradiation with quite high G-values.[91]

Therefore, sulfur may be introduced into Europa's system as implanted sulfur, as a sulfate or sulfide from an underground ocean or as SO_2 and S_x gas as at Io, as shown schematically in Fig. 9. The sulfate, either delivered to the surface or formed by irradiation as discussed above is very stable

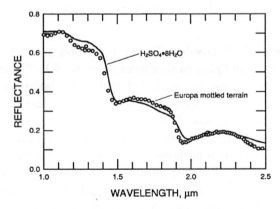

Fig. 8. Diffuse reflectance versus wavelength in microns. A comparison of a dilute (1/8) frozen sulfuric acid spectrum (line) with the spectrum from the NIMS instrument on Galileo (dots): taken from Carlson *et al.*[7]

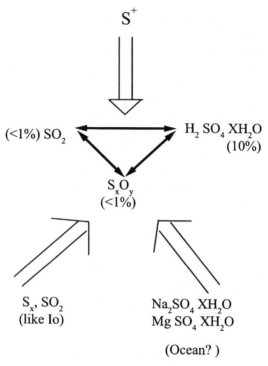

Fig. 9. The sulfur cycle as deduced from approximate radiolysis yields. Sulfur can enter the system by implantation from Io, possibly gases from beneath the surface as at Io, or as an ocean salt exposed to the surface. Radiation chemistry is fast, so the species are cycled through the various sulfur forms shown here. The relative amounts agree with the fits in Ref. 7.

under irradiation.[92-95] G-values for decomposition are $\sim 10^{-3}$ to $10^{-4}/$ 100 eV for X-rays and gamma-rays incident on Li_2SO_4.[96] Ions would produce decomposition more efficiently but still with relatively low G values. Therefore, we suggest that hydrated H_2SO_4 is the dominant species in the dark/chaos areas on Europa. This decomposes into SO_2 ($< 1\%$) and polymerized sulfur ($< 1\%$). Spectral comparisons are given in Ref. 7 and rough quantitative agreement is achieved for all three species. Since Europa appears to have a single principal darkening agent, this scenario would suggest it is polymerized sulfur. If this analysis is correct, the dark and young areas would appear to be regions in which a sulfur containing material is brought to the surface[7] either by venting or by break through of a melt.[11] Therefore, as illustrated in Fig. 9, the sulfur species in the chaos areas[9,10] mentioned above could be brought to the surface as hydrated salts, as SO_2 trapped in ice, or as polymerized sulfur all leading to the *same mix* of SO_2, frozen dilute sulfuric acid and polymerized sulfur. Because the radiation doses in the optical layer (Fig. 3) are large for the suggested geologic ages,[16] radiation-induced equilibrium is established among these sulfur forms.

Thermal processing has also been proposed[79] and a recent model suggests organic species are dominant in the ocean. Since similarly reddened spectra can be obtained for carbon containing species,[97] radiation cycling of a carbonate producing frozen dilute carbonic acid, CO_2, and carbon suboxides is also possible (Appendix).

Hydrated sulfates and carbonates in the form of salts had been suggested as the dominant materials in the geologically young regions of Europa as mentioned earlier.[14] On an oxidizing surface, irradiation of hydrated Na_2CO_3 should produce Na_2O, Na_2O_2 and CO_2 in ice (Appendix) in addition to free Na. Although the hydrated sulfates are much more stable, as discussed above for frozen dilute (hydrated) H_2SO_4, a hydrated Na_2SO_4 will also readily lose Na.[51,52,98] In fact, an atomic Na component is seen in the very tenuous ambient gas over Europa's surface.[99] Having lost Na, the hydrated sulfate can form hydrated $HNaSO_4$ first and then hydrated H_2SO_4, although this needs to be confirmed in the laboratory. In competition with this is implantation of Na from the neighboring satellite Io as shown in Fig. 10. Recent analysis indicates Europa is a net source of Na[51,99] (i.e. the loss rate is larger than the implantation rate) suggesting an internal source of Na.

Models of the icy satellites also suggest that hydrated Mg SO_4 should also be present.[8,14,35] Decomposition[100] of this species could again lead to hydrated H_2SO_4, in which case MgO or $Mg(OH)_2$ should be present in the surface (Appendix) since Mg is desorbed much less efficiently. However, Mg has not yet been seen spectrally in the surface, ambient gas or plasma.

Finally, interesting "aging" effects have been noted by a number of authors.[97] That is, the geologically youngest material appears to have more of the "dark/red" contaminant. With geological aging these terrains appear to have brightened in time.[11,12] This can occur by a "volcanic" emplacement of dark particulates that are gradually mixed with the underlying icy material by micrometeorites or buried by vapor deposition. Transport of sputtered and thermally desorbed H_2O can slowly coat a surface[13,31,101] or the radiation bombardment of a clear transparent (hence, dark) ice formed

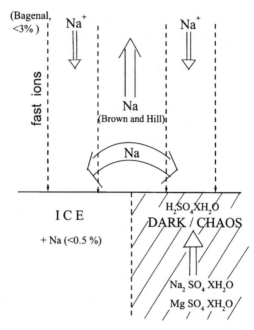

Fig. 10. The redistribution of Na on Europa's surface. Na is implanted from Io either into ice or into the dark terrain, or it is brought to the surface as a hydrated mineral. Sputtering and photodesorption[106] release the Na into the gas phase so it is redistributed[51] and seen in the gas phase as escaping Na.[99] If Na comes from a sulfate, the processing of the sulfate is described in Fig. 9.

from a melt could lead to brightening, as discussed above.[31] In addition, on changing the amount of hydration, minerals can change color. For instance, P. Williams and coworkers (private communication) showed that hydrated copper sulfate lost its blue color under irradiation and became clear. Then on exposure to H_2O, its color was restored. Therefore, in radiation equilibrium, the level of hydration in a refractory region depends on transport of water from the icy regions. Finally, if the red material in the fresh surface is a sulfur contaminant, the radiation-induced conversion of sulfur imbedded in water ice into hydrated sulfuric acid and SO_2 would appear to brighten the surface.[7] These are all potential radiation-induced aging processes.

5. Summary

There is recent spectral data on the icy moons of Jupiter that is not yet understood and the spectral data base will continue to increase with new data from Galileo, HST, CASSINI and the proposed Europa probe. Therefore, understanding the formation and evolution of the surface composition, the ambient gases and the local plasmas for the Jovian satellites is now becoming possible.[2,87] Such an understanding is critical for determining the origins and evolutions of these moons. Of particular interest to NASA and the public is the possibility of prebiological or biologic activity in the solar system.[16,17] In this respect, the moon Europa is a prime target and there are intriguing spectral signatures as discussed.

Since the surfaces of these objects are bombarded by UV photons and energetic ions and electrons, interesting low temperature, surface chemistry occurs which can both obscure the understanding of the intrinsic processes and can play an important role in the chemical evolution of the satellite. Compared to geologic time scales, the surface of Europa is rapidly processed by radiolysis and photolysis, so that even geologically young surfaces are in radiation-induced chemical equilibrium. Io's surface is a possible exception as it is continuously replenished by volcanism. At Ganymede, the fact that poles and equators experience very different radiation dose rates will be useful as the new spectral reflectance data is analyzed. Recently, Denk and coworkers identified a color change across the magnetic boundary in a *single* geologic unit.[102]

At Europa, the intense radiation bombardment also produces an ambient gas of sputtered molecules[1,19,103] and volatile decomposition products.[2] Only a couple of species in this atmosphere have been identified: O_2 from water/ice and Na and K from decomposition of hydrated minerals and, likely, H_2O, SO_2 and CO_2. However, this atmosphere must contain many other molecular species, either intrinsic or formed locally, and these will be observable either telescopically or by *in situ* sampling on a future probe.[2] This presents the exciting possibility of identifying surface constituents by detecting the gas-phase products. Because the sputtering yields by heavy energetic ions (O^+, S^+) are very large (Fig. 4), even massive organic molecules imbedded in the ice can be ejected into the gas-phase[104] and be detected.[2]

What is surprising is that the database for absolute product yields from energetic charged-particle-induced solid state chemistry is very poor in spite of the years of work on radiolysis and photolysis. Even the yields of H_2O_2 and O_2 from ice by radiolysis are not agreed upon. This lack may simply be a matter of focusing the attention of the chemistry community on the relevant materials and relevant incident radiation. Since definitive space data is now becoming available, and since the questions, such as the origins of these moons, the possibility of underground oceans, and the potential of Europa for prebiotic chemistry, are fundamental, this is an opportune time for new work on radiation-induced solid state chemistry.

Acknowledgments

I would like to acknowledge the support of the NSF Chemistry and Astronomy Division and the Planetary Geology Division of NASA.

Appendix

Low temperature (< 130 K) hydrated salts[8,48] form surface brines from material that may have been extruded onto Europa's surface by volcanism or due to the huge tidal flexing (10's of meters/day). The exposure of these hydrated minerals to radiation in an environment in which H_2 is lost to space produces preferred radiation chemical pathways. Three types of materials were recently suggested[14]: $Na_2SO_4 \cdot XH_2O$, $MgSO_4 \cdot XH_2O$ and $Na_2CO_3 \cdot XH_2O$ with $X \sim 6$–8. Ion bombardment can remove the water of hydration by sputtering and can dissociate molecules producing chemistry in an oxidizing surface. Some pathways, ordered according to increasing endothermicity, are listed below.[2]

$$MgSO_4 + 2H_2O \rightarrow MgO + H_2SO_4 + H_2O$$
$$\rightarrow MgO + SO_3 + 2H_2O$$
$$\rightarrow MgO + O_2 + SO_2 + H_2 + H_2O^*$$
$$\rightarrow MgS + 2O_2 + 2H_2O$$
$$\rightarrow MgO + 2O_2 + H_2S + H_2O$$
$$\rightarrow Mg(OH)_2 + SO_2 + O_2 + H_2^* .$$

Because H_2 is lost permanently during irradiation, the processes with the asterisks are favored, although often they do not have the smallest endothermicity. More simply, an oxidizing surface can give

$$MgSO_4 + O \rightarrow MgO + SO_2 + O_2$$
$$MgSO_4 + 2OH \rightarrow Mg(OH)_2 + SO_2 + O_2 .$$

Therefore, if the magnesium sulfate reaches the surface and is exposed to irradiation, $Mg(OH)_2$, MgO and, possibly, MgS should be present on the surface of Europa. For the sodium containing salt

$$Na_2SO_4 + 2H_2O \rightarrow NaHSO_4 + NaOH + H_2O$$

$$\rightarrow 2NaOH + H_2SO_4$$

$$\rightarrow 2NaOH + SO_3 + H_2O$$

$$\rightarrow Na_2O + H_2SO_4 + H_2O$$

$$\rightarrow Na_2O_2 + SO_2 + 2H_2O$$

$$\rightarrow Na_2O_2 + H_2SO_4 + H_2^*$$

$$\rightarrow 2NaOH + SO_2 + O_2 + H_2^* .$$

Na plasma ions implanted from the Io torus into an ice rich region can also give $NaOH$ or NaO_2.[2] Again, the asterisks indicate favored chemical pathways due to permanent loss of H_2. Also SO_3 in the presence of H_2O immediately forms H_2SO_4.[7] Therefore, the sulfates are driven to dilute frozen H_2SO_4, SO_2, and polymerized sulfur suboxides. Finally for the proposed carbonate $Na_2CO_3 \cdot XH_2O$[14] $X > 6$ exposed to the radiation in the presence of water and with the loss of hydrogen gives

$$Na_2CO_3 \rightarrow Na_2O + CO_2$$

$$Na_2CO_3 + H_2O \rightarrow Na_2O + H_2CO_3$$

$$Na_2CO_3 + O \rightarrow Na_2O_2 + CO_2 .$$

Removal of Na from the Na containing species by electronic excitations appears to occur readily on surfaces.[106] Therefore, Na is desorbed principally as Na atoms[52] by either charged particles or photons[106] contributing to the Na "clouds" seen at Io[33] and Europa.[99] Similarly, K is also seen.[99] This can be assisted by replacement by O or H from dissociated H_2O: e.g. $Na_2O + O \rightarrow NaO_2 + Na$ and $Na_2O + H \rightarrow NaOH + Na$.

References

1. R. E. Johnson, *Energetic Charge-Particle Interaction with Atmosphere Surface* (Springer-Verlag, Berlin, 1990).
2. R. E. Johnson, R. M. Killen, J. H. Waite and W. S. Lewis, *Geophys. Res. Lett.* **25**, 3257 (1998).

3. R. E. Johnson and E. C. Sittler, *Geophy. Res. Lett.* **17**, 1629 (1990).
4. J. A. Burns, *Satellites* (University of Arizona Press, Tucson, 1986), pp. 1–39.
5. M. G. Kivelson, K. K. Khurana, F. V. Coroniti, S. Joy, C. T. Russell, R. J. Walker, J. Warnecke, L. Bennett and C. Polanskey, *Geophys. Res. Lett.* **24**, 2155 (1997).
6. M. G. Kivelson, K. K. Khurana, D. J. Stevenson, L. Bennett, S. Joy, C. T. Russell, R. J. Walker and C. Polanskey, *Goephys. Res. Lett.* **104** (A3), 4609 (1999).
7. R. W. Carlson *et al.*, *Science* **286**, 97 (1999).
8. J. S. Kargel, *Icarus* **94**, 368 (1991); F. P. Fanale *et al.*; D. Morrison Ed., Satellites of Jupiter (University of Arizona Press, Tucson, 1982), pp. 756–781.
9. P. E. Geissler *et al.*, *J. Geophys. Res.* **135**, 107 (1998).
10. R. T. Pappalardo *et al.*, *Nature* **391**, 365 (1998).
11. P. Helfenstein *et al.*, *Icarus* **135**, 41 (1998).
12. B. E. Clark *et al.* and the Galileo SSI Team, *Icarus* **135**, 95 (1998).
13. R. T. Pappalardo *et al.*, *J. Geophys. Res.* **104** (E4), 10524 (1999).
14. T. B. McCord *et al.*, *J. Geophys. Res.* **104**, 11827 (1999).
15. E. J. Gaidos, K. H. Nealson and J. L. Kirschvink, *Science* **284**, 1631 (1999).
16. J. H. Cooper, R. E. Johnson, B. H. Mauk and N. Gehrels, *Icarus*, in press (2000).
17. C. Chyba, *Nature* **403**, 381 (2000).
18. R. W. Carlson *et al.*, *Science* **283**, 2062 (1999).
19. R. E. Johnson, *Solar System Ices*, Eds. B. Schmitt, C. beBergh and M. Festou (Kluwer Acad. Pub., Netherlands, 1998), pp. 303–334.
20. R. E. Johnson and T. I. Quickenden, *J. Geophys. Res.* **102**, 10985 (1997).
21. R. E. Johnson and J. Schou, Sputtering of Inorganic Insulators, in *Fundamental Processes in the Sputtering of Atoms and Molecules* (SPUT 92). Ed. P. Sigmund (Royal Danish of Academic Sciences, Copenhagen) [Mat. -fys. Medd 43, (1993)] pp. 403–494.
22. R. E. Johnson, *Rev. Mod. Phys.* **68**, 305 (1996).
23. A. J. Desseler, *Physics of The Jovian Magnetosphere* (Cambridge University Press, Cambridge, 1983).
24. W.-H. Ip, D. J. Williams, R. W. McEntire and B. H. Mauk, *Geophys. Res. Lett.* **25**, 829 (1998).
25. F. Bagenal, *J. Geophys. Res.* **99**, 11043 (1994).
26. C. Paranicas, W. R. Paterson, A. F. Cheng, B. H. Mauk, R. W. McEntire, L. A. Frank and D. J. Williams, *J. Geophys. Res.* **104**, 17459 (1999).
27. M. Kueppers and N. M. Schneider, AGU Abstract, Eos Supp. (1999).
28. N. Devine and H. B. Garrett, *J. Geophys. Res.* **88** (A9), 6889 (1983).
29. H.G. Heide and E. Zeitler, *Ultramicroscopy* **16**, 151 (1985).
30. R. E. Johnson and W. A. Jesser, *Astrophys. J. Lett.* **480**, L79 (1997).
31. R. E. Johnson, *Icarus* **62**, 344 (1985); R. E. Johnson, *Icarus* **128**, 469 (1997); M. Wong and R. E. Johnson, *J. Geophys. Res.* **102**, 23, 25523 (1997).

32. D. L. Matson, T. V. Johnson and F. P. Fanale, *Astrophys J.* **192**, L43 (1974).
33. R. A. Brown and F. H. Chaffee Jr., *Astrophys J.* **187**, L125 (1974).
34. R. E. Johnson, M. Nelson, T. B. McCord and J. Gradie, *Icarus* **75**, 423 (1988).
35. D. Nash and F. P. Fanale, *Icarus* **31**, 763 (1977).
36. W. L. Brown, L. J. Lanzerotti, J. M. Poate and W. M. Augustyniak, *Phys. Rev. Lett.* **49**, 1027 (1978).
37. L. J. Lanzerotti, W. L. Brown, J. M. Poate and W. M. Augustyniak, *Geophys. Res. Lett.* **5**, 155 (1978).
38. W. L. Brown *et al.*, *Nucl. Instrum. Meth.* **B1**, 307 (1982).
39. R. E. Johnson, L. J. Lanzerotti and W. L. Brown, *Nucl. Instrum. Meth.* **198**, 147 (1982).
40. D. T. Hall *et al.*, *Science* **273**, 677 (1995).
41. R. W. Carlson, *Science* **283**, 820 (1999).
42. W. M. Calvin, R. E. Johnson and J. A. Spencer, *Geophys. Res. Lett.* **23**, 673 (1996); J. Spencer, W. Calvin and J. Person, *J. Geophys. Res.* **100**, 19049 (1995).
43. K. S. Noll, R. E. Johnson, A. L. Lane, D. L. Dominigue and H. A. Weaver, *Science* **273**, 341 (1996); K. S. Noll, T. L. Rousch, D. P. Cruikshank, R. E. Johnson and Y. J. Pendleton, *Nature* **388**, 45 (1997).
44. A. R. Hendrix, C. A. Barth and C. W. Hord, *Geophys. Res.* **104** (E6), 14169 (1999).
45. A. L. Lane, R. M. Nelson and D. L. Matson, *Nature* **292**, 38 (1981).
46. K. S. Noll, H. A. Weaver and A. M. Gonella, *J. Geophs. Res.* **100**, 19057 (1995).
47. N. J. Sack, R. E. Johnson, J. W. Boring and R. A. Baragiola, *Icarus* **100**, 534 (1992).
48. T. B. McCord *et al.*, *Science* **280**, 1242 (1998).
49. F. P. Fanale, T. V. Johnson and D. L. Matson, *Science* **186**, 922 (1974).
50. R. E. Johnson, *Brazilian J. Phys.* **29**, 444 (1999).
51. R. E. Johnson, *Icarus*, in press (2000).
52. R. C. Weins, D. S. Burnett, W. F. Calaway, C. S. Hansen, K. R. Lykke and M. L. Pellin, *Icarus* **128**, 386 (1997).
53. D. B. Chrisey, R. E. Johnson, J. W. Boring and J. H. Phipps, *Icarus* **75**, 233 (1988).
54. A. S. McEwen *et al.*, *Geophys. Res. Letts.* **24**, 2443 (1997).
55. J. M. Moore *et al.*, *Icarus* **140**, 294 (1999).
56. E. J. Hart and R. L. Platzman, *Physical Mechanisms in Radiation Biology*, Eds. M. Errera and A. Forssberg (Academic, San Diego, California, 1961), pp. 93–120; B. G. Ershov and A. K. Pikaev, *Radiation Research Review* (Elsevier, Amsterdam, 1969).
57. J. P. Cowin, A. A. Tsekouras, M. J. Iedema, K. Wu and G. B. Ellison, *Nature* **398**, 405 (1999).
58. I. A. Taub and K. Eiben, *J. Chem. Phys.* **49**, 2499 (1968).

59. C. T. Reimann, J. W. Boring, R. E. Johnson, J. W. Garrett and K. R. Farmer, *Surf. Sci.* **147**, 227 (1984).

60. G. A. Kimmel and T. A. Orlando, *Phys. Rev. Lett.* **75**, 2606 (1995).

61. L. Khriachtchev, personal communication (1999); T. Orlando, personal communication (1999).

62. J. A. Ghormley and C. J. Hochanadel, *J. Phys. Chem.* **75**, 40 (1971).

63. W. L. Brown, W. M. Augustyniak, E. Simmons, K. J. Marcantonio, L. J. Lanzerotti, R. E. Johnson, J. W. Boring, C. T. Reimann, G. Foti and V. Pirronello, *Nucl. Instrum. Meth.* **198**, 1 (1982).

64. M. S. Westley, R. A. Baragiola, R. E. Johnson and G. A. Baratta, *Planet. Space Sci.* **43**, 1311 (1995).

65. M. T. Sieger, W. C. Simpson and T. M. Orlando, *Nature* **394**, 554 (1998).

66. R. A. Baragiola, C. L. Atteberry, D. A. Bahr and M. Peters, *J. Geophys. Res.* **104** (E6), 14183 (1999); R. A. Baragiola and D. A. Bahr, *J. Geophys. Res.* **103**, 25865 (1998); R. A. Vidal, D. B. Bahr, R. A. Baragiola and M. Peters, *Science* **276**, 1839 (1997).

67. A. J. Matich, M. G. Bakker, D. Lennon, T. I. Quickenden and C. G. Freeman, *J. Phys. Chem.* **97**, 10539 (1993).

68. M. H. Moore and R. L. Hudson, *Icarus* **145**, 282 (2000).

69. M. B. McElroy and Y. L. Yung, *J. Astrophys.* **196**, 227 (1975).

70. R. E. Johnson, *J. Geophys. Res.* **104** (E6), 14179 (1999).

71. P. A. Gerakines, W. A. Schutte and P. Ehrenfreund, *Astron. Astrophys.* **312**, 289 (1996).

72. V. Vaida, D. J. Donaldson, S. J. Strickland, S. L. Stephens and J. W. Birks, *J. Phys. Chem.* **93**, 506 (1989).

73. R. A. Bargiola, C. L. Atteberry, D. A. Bahr and M. Jakas, *Nucl. Instrum. Meth.* **B157**, 233 (1999).

74. G. Strazzulla, *Solar System Ices*, Eds. B. Schmitt *et al.* (Kluwer, Netherlands, 1998), pp. 281–302.

75. D. B. Chrisey, W. L. Brown, J. W. Boring, *Surf. Sci.* **225**, 130 (1990).

76. D. J. O'Shaugnessy, J. W. Boring and R. E. Johnson, *Nature* **333**, 240 (1988).

77. W. M. Calvin, R. N. Clark, R. H. Brown and J. A. Spencer, *J. Geophys. Res.* **100**, 19041 (1995).

78. L. A. Lebofsky and N. B. J. Fegley, *Icarus* **28**, 379 (1976).

79. F. P. Fanale *et al.*, *Lunar Planet. Sci. Conf.* [CD-ROM] XXIX abstract. p. 1248 (1998).

80. L. J. Lanzerotti, W. L. Brown, C. G. Macclennan, A. F. Cheng, S. M. Krimigis and R. E. Johnson, *J. Geophys. Res.* **92**, 14949 (1987).

81. M. H. Moore, R. K. Khanna and B. Donn *J. Geophys. Res.* **96**, 17541 (1991); M. H. Moore and R. K. Khanna, *Spectrochimica Acta* **A47**, 255 (1991).

82. M. H. Moore and R. L. Hudson, *Icarus* **135**, 518 (1998).

83. R. L. Hudson and M. H. Moore, *Icarus* **140**, 451 (1999).

84. J. R. Brucato, A. C. Castorina, M. E. Palumbo, M. A. Satorre and G. Strazzulla, *Planet Space Sci.* **45**, 835 (1997).

85. G. Strazzulla and M. E. Palumbo, *Planet Space Sci.* **46**, 1339 (1998).

86. J. R. Brucato, M. E. Palumbo, G. Strazzulla, *Icarus* **125**, 135 (1997).

87. M. L. Delitsky and A. L. Lane, *J. Geophys. Res.* **103**, 31391 (1998); M. L. Delitsky and A. L. Lane, *J. Geophys. Res.* **102**, 16385 (1997).

88. R. A. Haring, R. Pedrys, D. J. Oostra, A. Haring and A. E. deVries, *Nucl. Instrum. Meth. Res. Phys. Sect.* **B5**, 476 (1984).

89. A. R. Hendrix, *EOS Trans. AGU 79* (fall meeting supp.) F335 (1998).

90. M. H. Moore, *Icarus* **59**, 114 (1984).

91. G. W. Donaldson and F. J. Johnson, *J. Phys. Chem.* **72**, 3552 (1968).

92. J. Ganjei, R. Colton and J. Murray, *SIMS*, (1978), pp. 221–229.

93. C. J. Hochancel, J. A. Ghormley and T. J. Sworski, *J. Am. Chem. Soc.* **77**, 3215 (1955).

94. E. R. Johnson and A. O. Allen, *J. Am. Chem. Soc.* **74**, 4147 (1952).

95. R. A. Haring, A. Haring, F. S. Klein, A. C. Kummel and A. E. deVries, *Nucl. Instrum. Meth.* **211**, 529 (1983).

96. T. Sasaki, R. S. Williams, J. S. Wing and D. A. Shirley, *J. Chem. Phys.* **68**, 2178 (1978).

97. D. A. Geissler, W. A. Schutte and P. Ehrenfreund, *Astron. Astrophys.* **312**, 289 (1996).

98. A. Z. Benninghoven, *Z. für Naturforsch* **A21**, 859 (1969).

99. M. E. Brown and R. E. Hill, *Nature* **380**, 229 (1996); M. E. Brown, *Icarus*, in press (2000).

100. A. Z. Benninghoven, F. G. Rudenauer and H. W. Werner, *Secondary Ion Mas. Spectrom.* (John Wiley, New York, 1987), p. 721.

101. E. M. Sieveka and R. E. Johnson, *J. Geophys. Res.* **90**, 5327 (1985).

102. T. Denk, G. Neaken, J. W. Head and R. Pappalardo, *Bull. AAS* **31** (4), 1182 (1999).

103. M. Shi, R. A. Baragiola, D. E. Grosjean, R. E. Johnson, S. Jurac and J. Schou, *J. Geophys. Res.* **100**, 26387 (1995).

104. R. E. Johnson, B. U. R. Sundqvist, A. Hedin and D. Fenyo, *Phys. Rev.* **B40**, 49 (1989); R. E. Johnson and B. U. R. Sundqvist, *Physics Today*, March 1992, p. 28.

105. J. G. Calvert and J. N. Pitts, Photochemistry (John Wiley and Sons, New York, 1996), pp. 368–380.

106. T. E. Madey, B. V. Yakshinsky, V. N. Ageev and R. E. Johnson, *J. Geophys. Res.* **103**, 5873 (1998); B. V. Yakshinsky and T. E. Madey, *Nature* **400**, 642 (1999).

CHAPTER 9

DYNAMICS OF ATOMIC-OXYGEN-INDUCED POLYMER DEGRADATION IN LOW EARTH ORBIT

Timothy K. Minton and Donna J. Garton

Department of Chemistry and Biochemistry
Montana State University, Bozeman, MT 59717 USA
E-mail: *tminton@montana.edu*

Contents

1. Introduction

1.1. *The Atomic Oxygen Problem*

Spacecraft in low Earth orbit (LEO) must function in a harsh oxidizing environment. At LEO altitudes, ranging from ~200 to ~700 km, the residual atmosphere is predominantly neutral, with the dominant component being atomic oxygen. A typical O-atom number density at space shuttle altitudes (~300 km) is on the order of 10^9 cm^{-3}.[1-4] At orbital altitudes of ~300–400 km, the LEO environment subjects materials on the ram side of a spacecraft to collisions with ambient oxygen atoms that have an average impact velocity of ~7.4 km s^{-1},[5-7] corresponding to a mean collision energy of ~450 kJ mol^{-1}. The combination of impact velocity and O-atom density yields a flux of approximately 10^{15} O atoms cm^{-2} s^{-1} on ram surfaces. The roughly 1000 K kinetic temperature[1] of the ambient atmosphere gives an energy spread (full width at half maximum) of 200 kJ mol^{-1} to the collisions.[5,7] The atomic oxygen concentration is dependent on many factors besides altitude, including solar activity, season, local time, latitude, and variations in the Earth's magnetic field.[8-10] Model calculations, such as MSIS-83 or MSIS-86,[11,12] are usually used to calculate atomic oxygen number densities encountered for a particular mission.[10,13] Figure 1 shows typical number densities as a function of altitude for the predominant species in LEO.[1]

Atomic-oxygen is considered to be one of the most important hazards to spacecraft in LEO.[14,15] Energetic collisions of O atoms with spacecraft surfaces can degrade materials through oxidation and erosion (etching). The detrimental effects of atomic oxygen in LEO were first recognized after postflight analyses of polymer and paint surfaces that were exposed during the earliest space shuttle flights.[10,16] Polymers showed a loss of surface gloss and concomitant weight loss, while paint surfaces exhibited premature aging. Concern over the degradation of materials by atomic oxygen sparked a huge effort, involving space- and ground-based studies, that has been aimed at the identification, understanding, and mitigation of problems caused by atomic oxygen in LEO.[17-19] These problems can be classified and summarized as follows (see Fig. 2):

Erosion. Reactions involving hyperthermal collisions of O atoms with various polymeric surfaces (e.g. optical and thermal control coatings, carbon-based composites, thermal blankets, solar panels, optical components) lead to the production of volatile reaction products, which carry mass

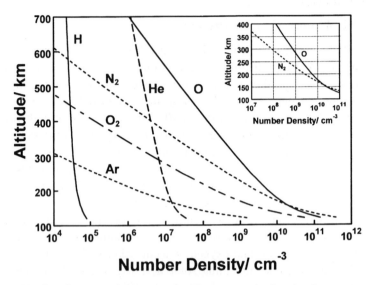

Fig. 1. Number densities of the most abundant atomic and molecular species at low Earth orbital altitudes as a function of altitude.

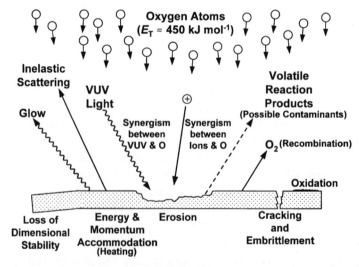

Fig. 2. A pictorial representation of the possible interactions on spacecraft surfaces in low Earth orbit that involve atomic-oxygen.

Fig. 3. Scanning electron micrograph of a Kapton polyimide surface that was exposed to the low Earth orbital ram environment for 40 hours on the STS-46 space shuttle mission. The smooth portion of the Kapton surface was covered during exposure and therefore shows no erosion. The rougher portion of the surface shows the topography and extent of erosion (6 μm deep) resulting from the exposure.

away from the surface. These gaseous products can subsequently be a source of contamination to critical components on a spacecraft (e.g. optics). The remaining material will not only be structurally weakened, but the surface will be roughened (Fig. 3), thus reducing specular reflectance. Erosion may also release macroscopic particles that are hazardous to other aspects of spacecraft function. For example, erosion of the matrix in a carbon fiber reinforced composite material can release carbon fibers, which may cause short circuiting in nearby electronic components.

Spacecraft flying in LEO are exposed not only to neutral atomic and molecular species, but also high fluxes of vacuum ultraviolet (VUV) radiation and relatively low fluxes of ions, both of which might contribute to the degradation of materials present on spacecraft surfaces.[20-24] It has been reported that fluorinated ethylene-propylene copolymer (FEP Teflon), which is used as a thermal blanketing material, erodes under the combined exposure of VUV light and atomic oxygen, while the erosion by atomic oxygen alone is negligible.[20,25,26] However, the reported synergistic enhancement in the erosion rate of perfluorinated hydrocarbons may not be relevant at the VUV fluxes typical of missions in LEO.[22] Ions may also aid in the erosion of some polymer surfaces.[23,27] The most abundant ion in the natural LEO environment is O^+.[3,13,27,28] Although relative ion concentrations in LEO are low (much less than one percent of the flux onto a surface), a study that examined the dependence of FEP Teflon erosion rate on O^+ mole fraction suggested that the combined exposure of this material to O^+ and O can increase the erosion rate by one or more orders of magnitude over the erosion yield with no ions present.[27] Nevertheless, most polymers, especially hydrocarbons, are eroded by atomic oxygen in the absence of VUV light or ions, and the erosion rates of these polymers are relatively insensitive to synergistic effects at the fluxes of VUV light and ions that are present in LEO.

Oxidation. Eroded materials generally show an increase in surface oxide content, thus altering their surface properties. In addition, many materials become oxidized without eroding. Such materials may take on undesirable optical and mechanical properties during a mission. Oxidation may lead to a change in the absorptivity or emissivity of a material, which will consequently alter its thermal characteristics. A lubricant, such as molybdenum disulfide (MoS_2), can be oxidized to an abrasive oxide.[29] Dimensional changes can also occur; for example, silicone will form a surface oxide layer that can contract and crack, and oxidation of silver will cause expansion and spalling.

Inelastic Scattering. Incident oxygen atoms may scatter from a surface without reaction (see Sec. 3.3), and subsequent collisions of these scattered O atoms can cause oxidation and, possibly, erosion of materials that are shadowed from direct attack. The surface may experience some heating as the scattered O atoms transfer some of their incident translational energy to the surface. Scattering of oxygen atoms may also affect the evolution of surface roughness on a material that erodes. Directed impingement tends to produce very rough eroded surfaces, as seen in Fig. 3, while isotropic impingement tends to yield relatively smooth surfaces.

Glow. Chemiluminescent products formed in surface-assisted reactions involving atomic oxygen are believed to be a source of space vehicle glow, which may seriously interfere with the function of sensitive optical instruments designed to detect low light levels.[6,28,30] The most widely accepted explanation for space vehicle glow is emission from NO_2^* formed from the reaction of O atoms with NO adsorbed on a spacecraft surface.[30,31] It has also been speculated that atomic oxygen could react with a hydrocarbon surface, or adsorbed water or hydrocarbons on a surface, to produce vibrationally excited OH, which subsequently emits light and contributes to spacecraft glow.[30,32]

Oxygen Atom Recombination. Formation of molecular oxygen from the recombination of oxygen atoms on a spacecraft surface has been proposed to occur.[6] The extent to which this recombination will affect the erosion and/or oxidation of a surface is only speculative at this time.

1.2. *Mitigation Approaches*

Mitigation of polymer degradation in space has emphasized function over basic understanding. Controlled experiments are difficult to perform in space, and sources of hyperthermal O atoms for ground-based studies usually contain byproducts whose effect on a material under atomic oxygen attack is not well known. Most data therefore tend to be phenomenological, and the approach of the user community has generally been to expose candidate materials in space and/or in ground-based test environments and carry out functional analyses to assess any changes from the pre-exposure properties. If the functional property of a candidate material is found to degrade as a result of the exposure, then the material must either be replaced or protected.

Polymers that are nominally atomic-oxygen resistant have been engineered and include polymers with silicon[33-35] or phosphorous[36-38] incorporated into the backbone. Upon atomic oxygen exposure, both the silicon and phosphorous react with incident oxygen atoms to form an oxide layer on the polymer surface. Once the oxide layer is formed, it acts as a protective coating, and the polymer is effectively shielded from further atomic oxygen degradation. Atomic-oxygen-resistant layers have also been created by implanting various types of ions, including carbon, boron, silicon, and aluminum, into a polymer and exposing the polymer surface to atomic oxygen.[39,40] As in the polymers that contain silicon and phosphorous, an oxide layer is formed that protects the polymer from further erosion. With the formation of atomic-oxygen-resistant layers in both mitigation methods mentioned above, regeneration of the protective surface is possible if a portion of the layer is removed by a crack or by impact with an orbiting debris particle. However, the incorporation of chemistries into a polymer that will lead to formation of a protective oxide layer can result in a material whose functional properties are less desirable than those of the native polymer.

The most common method of mitigating polymer degradation is to deposit a protective coating, such as a metal[41-43] or an inorganic oxide,[41,44-46] on the surface of the polymer. Coatings are widely used but have the distinct disadvantage of being permanently lost if removed by debris impacts, cracking, or delamination in the LEO environment.

1.3. *Gas-Phase Atomic-Oxygen Reactions with Hydrocarbons*

Hydrocarbon polymers are particularly susceptible to attack by atomic oxygen in LEO. The reactions of atomic oxygen with hydrocarbon molecules in the gas phase serve as models for the relatively unstudied reactions of atomic oxygen with a hydrocarbon surface. A wealth of knowledge of gas-phase reactions is available, largely because these reactions are important in combustion and in atmospheric chemistry. Studies of both reaction kinetics and dynamics have revealed many of the mechanisms by which atomic oxygen reacts with gaseous alkanes and alkenes.[47-49] A summary of probable reaction pathways is presented in Fig. 4.

Dynamical studies of O-atom reactions with various alkanes have shown that a ground state $O(^3P)$ atom typically reacts by abstracting a hydrogen atom to form an OH radical product.[50-52] The barrier to this reaction is significant: 28.9, 18.8, and 13.8 kJ mol^{-1}, depending on whether the abstracted hydrogen atom is primary, secondary, or tertiary.[50] Hydrogen abstraction to form OH is also believed to be a major reactive pathway for $O(^3P)$ reactions with simple alkane derivatives, such as alcohols, haloalkanes, and aldehydes.[47] Electronically-excited $O(^1D)$ might abstract a hydrogen atom from an alkane, but it more likely inserts, with no barrier, into a C–H bond to form an excited alcohol.[53-56] The excited alcohol can dissociate either by C–O or C–C bond fission, or it can be stabilized by collisions if the pressure is sufficiently high.[53] $O(^3P)$ might react effectively as $O(^1D)$ under conditions where the reaction time is long enough to allow for crossing between the triplet and singlet states of atomic oxygen. Such triplet–singlet crossing has been proposed in order to explain stable alcohol products obtained in reactions of $O(^3P)$ with cyclohexane clusters.[57]

Ground state $O(^3P)$ reacts with alkenes in the gas phase by adding nonstereospecifically to the double bond, thereby creating a vibrationally

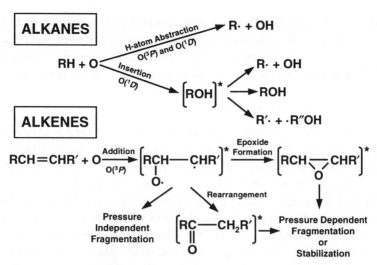

Fig. 4. Summary of probable mechanisms for the reaction of atomic oxygen with saturated and unsaturated hydrocarbons in the gas phase. The reaction of $O(^1D)$ with an alkene is believed to form the expoxide directly, thus skipping the formation of a triplet biradical.

excited triplet biradical intermediate, which can subsequently form an excited epoxide or other rearrangement products.[49,58,59] Vibrationally-excited products may be stabilized through collisions, or they may undergo further fragmentation. Several rearrangements may occur, such as a 1,2 hydrogen-atom migration, which will lead to the formation of vibrationally-excited carbonyl compounds. Pressure-independent fragmentation (PIF) and pressure-dependent fragmentation (PDF) of the intermediates can also be significant reaction pathways.[58,59] At high pressures or in condensed media, PDF is suppressed. The addition of $O(^3P)$ to the double bond of an alkene is much more rapid than hydrogen atom abstraction at low collision energies or temperatures.[60] However, H-atom abstraction becomes competitive at high incident energies or temperatures.[61] Excited-state $O(^1D)$ will also react with an alkene, but it is believed to react stereospecifically to form an epoxide directly, skipping the formation of the triplet biradical.[59]

Reaction mechanisms that are analogous to these gas-phase mechanisms have been reported for the relatively few studies of atomic-oxygen reactions with hydrocarbon surfaces. In this chapter (Sec. 3.3), it is shown that $O(^3P)$ may abstract a hydrogen atom from a hydrocarbon surface. Another study of $O(^3P)$ reactions with a liquid saturated hydrocarbon found a stable alcohol product,[62] which was apparently formed through triplet–singlet crossing. Reactions of $O(^3P)$ with an unsaturated hydrocarbon polymer, polybutadiene, were studied by Golub and coworkers, who interpreted the results largely in terms of analogous gas-phase mechanisms.[63]

1.4. *Gas–Surface Interactions*

Unique interactions between a gas-phase oxygen atom and a hydrocarbon surface complicate the picture that is developed from consideration of gas-phase reactions alone. Gas–surface reactions may be described in terms of two limiting cases, one thermal and the other nonthermal (Fig. 5). In a thermal, or Langmuir–Hinshelwood, reaction, a gas-phase reactant species becomes trapped on the surface and reacts with another species in thermal equilibrium with the surface.[64,65] The product then desorbs from the surface at energies dictated by a Maxwell–Boltzmann distribution at the surface temperature. In a nonthermal, or Eley–Rideal, reaction, the incident reactant species reacts with a collision partner on the surface on a time scale too short for thermal equilibrium to be established, and the product leaves the surface with hyperthermal translational energy.[64–66] Analogous nonreactive processes are trapping desorption (thermal) and direct inelastic scattering (nonthermal).[67–70]

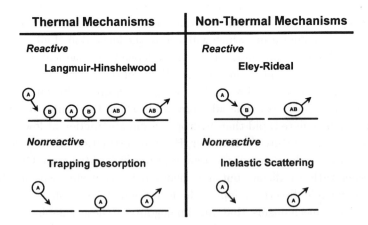

Fig. 5. A pictorial representation of the limiting cases for the interaction of a gaseous atom with a surface. Both thermal and nonthermal, reactive and nonreactive processes are shown. A process that occurs in thermal equilibrium with the surface is referred to as thermal.

1.5. *Scope of this Chapter*

This chapter focuses on the interactions of atomic oxygen with hydrocarbon polymer surfaces that are relevant to the low Earth orbital environment. While many materials other than hydrocarbon polymers are used on the external surfaces of spacecraft, hydrocarbon polymers are vitally important as thermal control, structural, and optical materials. Because of their importance and susceptibility to atomic-oxygen attack, the degradation mechanisms of hydrocarbon polymers in LEO have been relatively well studied. The relative abundance of information about polymer degradation mechanisms and the rich literature on gas-phase reactions of atomic oxygen with hydrocarbon molecules provide a basis for choosing the collective interactions of atomic oxygen with various hydrocarbon surfaces as a model system for describing the erosion and oxidation of materials in the LEO environment. The aim of this chapter is to build a picture of this model system by bringing together elements from the myriad studies that have attempted to uncover the atomic-oxygen-induced degradation mechanisms on polymer surfaces in LEO.

2. Experimental Methods

2.1. *Exposure Environments*

2.1.1. *The Low Earth Orbital Environment*

Materials intended for use in LEO must resist deterioration in their function resulting from atomic-oxygen attack, acting either alone or in synergy with other elements of the space environment.   Given the complexity of the space environment,[3,71–76] a common and intuitively satisfying approach to evaluating the durability of these candidate materials is simply to expose them to the actual LEO environment. Accordingly, many space-based exposure studies have been conducted on the Space Shuttle,[77–83] the Mir Space Station,[84,85] and other retrievable[17,86,87] and nonretrievable[88] satellites. The majority of exposures are considered "passive," where a material is exposed to the space environment and is later returned to Earth for postflight analysis. Occasionally, material exposures are designed with an active monitoring scheme that allows data collected *in situ* to be telemetered to Earth[88] or to be sent to a recording device that can be returned to Earth.[79] In some cases, the active part of the material exposure may be a device or devices which control the exposure in some way.[79,83]

Experiments involving space-based material exposures have the advantage that the exposure environment may be a faithful representation of the mission environment for the candidate material; however, control over exposure parameters is difficult, if not impossible, to ensure. Sample materials and associated flight hardware must typically be delivered to a site for integration into a flight package months before the actual launch. After return of the samples to Earth, there might be a delay of weeks or months before they are delivered to the original sample providers. During these pre- and post-flight periods, the samples may be handled by numerous persons, and even though they may be housed in clean-room facilities, there is always the possibility for a sample to be damaged, contaminated, or mounted incorrectly. Furthermore, the surface chemistry of an exposed sample may change considerably between the time that the exposure ends and the time that a sample is analyzed.[89] Contamination during a space-based exposure is common, as a variety of materials used on spacecraft outgas in the vacuum of space.[90-93] A particularly undesirable contaminant for studies of the deleterious effects of atomic-oxygen is silicone, which is ever present as a result of silicone-based adhesives that are used on spacecraft. Surface silicone contamination will react with atomic oxygen to produce an SiO_x layer, which reduces the erosion rate of the underlying material to some unknown and uncontrollable extent. Other common sources of contamination are thrusters and waste dumps. In addition to the uncertainties caused by contamination, the difficulty of defining or even knowing the actual makeup of the exposure environment leads to ambiguities in the conclusions that can be drawn from a space-based exposure. The choice of a spacecraft mission on which to expose test specimens to the LEO environment is typically opportunistic. The test material may therefore be exposed to a unique set of environmental conditions, including orbital altitude, orbital inclination, spacecraft attitude, and spacecraft contamination, that make it difficult to extrapolate the observed effects to what might occur on another mission. While there are environmental models that allow an estimation of the density of the various neutral (Fig. 1) and ionic species that might be present,[2,12] as well as the flux and spectrum of solar radiation, actual measurements of these exposure parameters are seldom made. Disentangling the role of all the environmental factors on the degradation of a sample can thus become intractable. The myriad possibilities that can blur the conclusions about the degradation mechanisms of materials in the space environment are too many to recount or predict here. Nevertheless, the sheer number of exposures that have been carried out in LEO provide a wealth of phenomenological data on a large variety of materials. Many materials have been exposed in LEO on several missions, allowing for increased confidence in the conclusions about how those materials degrade in the LEO environment.

2.1.2. *Laboratory Environments*

Laboratory environments containing atomic oxygen afford the possibility of conducting controlled experiments on the atomic-oxygen-induced degradation of materials. An ideal laboratory source of atomic oxygen for studies of material degradation mechanisms in space would be able to produce a high flux ($>10^{15}$ atoms cm^{-2} s^{-1}), directed beam of energetic (kinetic energy $\sim$450 kJ mol^{-1}) oxygen atoms in the ground electronic state (^3P), with no impurities or ultraviolet light present and preferably with a large cross sectional area (>1 cm^2). This ideal has not been realized, essentially because of the high chemical reactivity of atomic oxygen. Any method that has been developed to create atomic oxygen may have as byproducts one or more of the following: light (from the infrared to the vacuum ultraviolet), ions, excited-state species, electrons, molecular oxygen, and rare-gas atoms. In addition, the kinetic energy distribution and directionality of atomic oxygen striking a sample surface may vary over a wide range. Laboratory sources of atomic oxygen may be grouped broadly into plasma and neutral beam sources. Many approaches to producing atomic oxygen for simulation of space environmental effects have been explored.[94,95]

Plasma sources are generally thought of as thermal sources and are commonly used for material testing because of their wide availability and simplicity of use. A thermal source generates atomic oxygen with a Maxwell–Boltzmann distribution of speeds at temperatures within a few hundred degrees Celsius of room temperature. A sample may be placed directly into an oxygen or air plasma, in which case the application would be referred to as a "plasma asher",[96,97] or it may be placed downstream of an oxygen plasma in the "flowing afterglow".[98] In either arrangement, the atomic oxygen kinetic energies are near thermal and impingement on a sample surface is isotropic. A common radio frequency (RF) plasma asher uses a 13.56 MHz capacitively coupled air or oxygen plasma.[99] Another type of plasma environment can be generated with an electron cyclotron resonance (ECR) plasma that uses 2.45 GHz excitation in a magnetic field to produce a somewhat directed beam of ions, oxygen atoms, and other species.[100–102] In one configuration,[103] the output of an ECR source is deflected from a quartz surface in order to reduce the exposure of the sample surface to ions and vacuum ultraviolet (VUV) light; however all directionality of the atomic oxygen is lost. Of these less directional exposure methods, the flowing afterglow and deflected ECR beam give the exposure conditions with the fewest byproducts. Their environments consist predominantly of ground state (^3P) thermal O atoms and other less reactive neutral species. The plasma environments subject sample surfaces to intense ultraviolet radiation and high energy ions (hundreds to thousands of kJ mol^{-1}), in addition to ground- and excited-state O atoms and other neutrals, whose kinetic energies are thermal.

Beam sources may be thermal or hyperthermal. A realization of a thermal beam source would be the effusive flow of gas from the nozzle of a plasma source. A hyperthermal source produces atomic oxygen with average kinetic energies many times the average energy of a Maxwell–Boltzman distribution at 300 K ($\sim$4 kJ mol^{-1}). Hyperthermal atomic oxygen can be produced in a variety of ways, and the resulting kinetic energy distribution may be Maxwell–Boltzmann, but it is usually much narrower. The roughly 450 kJ mol^{-1} atomic oxygen collision energy encountered on materials in LEO has motivated the development and use of hyperthermal sources, although many of these sources are routinely operated with atomic oxygen kinetic energies much less than 450 kJ mol^{-1}. The general approaches to producing a beam containing hyperthermal oxygen atoms are enumerated below.

Laser detonation. The laser detonation source was originally developed by Physical Sciences, Inc.,[104–106] and sources based on this design are now used in several other laboratories.[24,107–109] The key elements of this source, shown in Fig. 6, are a pulsed molecular beam valve, coupled to an expanding conical nozzle, and a high energy (5–10 J pulse^{-1}, typically) CO_2 TEA laser. The pulsed valve is used to introduce a surge of O_2 gas into the conical nozzle. As the gas begins to expand into the nozzle, the CO_2 laser is fired, and the light pulse is focused down into the cone, where it initiates a plasma and heats it to more than 20 000 K. The high-temperature, high-density plasma expands rapidly into the diverging cone following detonation and engulfs the remaining cold gas. The local densities in the nozzle are

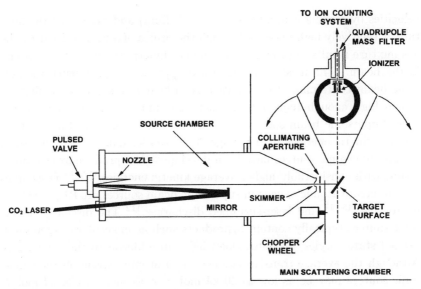

Fig. 6. Schematic diagram of the laser-detonation hyperthermal beam source, the differentially pumped scattering region containing the target surface, and the rotatable mass spectrometer detector.

sufficient to allow for efficient electron–ion recombination, but by the time the atoms formed in the plasma have cooled enough to recombine, the termolecular collision rate has dropped so low that the atomic species are, in effect, frozen in the emerging beam. The resulting beam from the nozzle, therefore, consists predominantly of fast neutrals, with a small ionic component ($<<$ 1%). The beam has a nominal direction, determined by the 20° full included angle of the conical nozzle; however, the angular divergence of the beam can be significantly reduced with the use of an aperture. This source is capable of producing beams containing atomic oxygen with average kinetic energies in the range of 300–1000 kJ mol^{-1}. The energy width (full width at half maximum) is approximately 250 kJ mol^{-1}. Both atomic and molecular oxygen exit the source at similar velocities, and the O/O_2 concentration ratio can be varied from roughly 0.5 to 3. O-atom fluxes can be more than 10^{15} atoms cm^{-1} s^{-1} at a distance of 50 cm from the apex of the conical nozzle.

Supersonic beam. Supersonic beam techniques are a well established means for producing intense beams of hyperthermal species with narrow energy distributions. Atomic oxygen may be accelerated to hyperthermal velocities by seeding it in a light gas (e.g. helium) and expanding the mixture at relatively high pressure through the orifice of a nozzle. Heating the gas mixture results in proportionally higher kinetic energies of the species in the beam. The most straightforward approach to producing a supersonic beam containing atomic oxygen is to heat a mixture containing O_2 gas to high enough temperatures such that appreciable dissociation occurs. Supersonic beam sources typically produce beams of O atoms with average kinetic energies in the range 20 to 100 kJ mol^{-1};[110,111] however, unique supersonic sources have been developed that produce oxygen atom beams with considerably higher average kinetic energies.[112,113] Discharges used to create atomic oxygen in supersonic beam sources are RF,[111,114–116] microwave,[113,117] and laser-sustained discharges.[112] The discharge supersonic sources typically contain byproducts such as molecular oxygen, ions, excited state species, and ultraviolet light, in addition to the carrier gas. Although the average translational energy of atomic oxygen in the supersonic sources may be as low as 20 kJ mol^{-1} or as high as 500 kJ mol^{-1}, energies between 100 and 300 kJ mol^{-1} are typically used for studies of materials degradation mechanisms.

Ion neutralization. Three types of ion neutralization schemes are used to produce atomic oxygen for studies of space environmental effects on materials. In one scheme,[118,119] a beam of O$^-$ ions is produced and given a known translational energy around 500 kJ mol^{-1} by ion-optic control. A laser is then used to photodetach the electrons from the ions, leaving a neutral O-atom beam with a known translational energy. A large magnetic field surrounds the source in order to counteract the divergence that limits the flux of the beam of relatively low-energy ions. Any ions that remain after photodetachment are deflected by an electric field. This source has been reported to produce a collimated beam of atomic oxygen, with a flux as high as 10^{14} atoms cm^{-2} s^{-1}. In principle, this source should produce pure ground state atomic oxygen with no byproducts. Another type of ion neutralization scheme employs a plasma chamber to create O$^+$ ions.[120] The ions are extracted from the plasma by an electric field and directed onto a metallic plate, where the ions are presumed to be neutralized and the resulting atoms are believed to be scattered primarily in the direction toward a sample surface. Little is known about the details of the exposure environment afforded by this source, because there have been few measurements to characterize it. A third ion neutralization source uses a Penning ionization gauge ion source to create an oxygen plasma.[121] The plasma is adjusted to maximize the production of O_2^+. The O_2^+ ions are extracted, the velocities are adjusted, and the O_2^+ ions are neutralized by dissociative electron recombination. The flux of atomic oxygen at 500 kJ mol^{-1} is estimated to be approximately 10^{13} atoms cm^{-2} s^{-1}. A wide range of translational energies are accessible with this type of source, and the full width at half maximum in the energy distribution is on the order of 100 kJ mol^{-1}.

Electron-stimulated desorption. An atomic-oxygen source based on electron-stimulated desorption of oxygen atoms from the surface of a silver foil is now available commercially. O_2 gas is continuously in contact with one side of the silver foil, where the O_2 molecules dissociatively chemisorb on the surface and the resulting O atoms diffuse through to the other side.[122] A 1650 eV electron beam is directed at the atomic-oxygen-covered silver surface, and O atoms are released. The kinetic energy distribution of ions leaving the surface has been measured and is assumed to represent the distribution of neutrals leaving the surface. The ions have an average energy of approximately 500 kJ mol^{-1}. The flux of O atoms has been reported to be as high as 10^{14} atoms cm^{-2} s^{-1} at a distance of 2.5 cm from the source; however, this value is an estimate. The exact makeup of the beam from this source is open to question because the neutral flux has not been measured. Presumably, the beam consists of atomic and ionic oxygen, but relative abundances of these species are unknown.

Photodissociation. Photodissociation of oxygen-containing species, such as N_2O or O_3, has been used to produce beams or environments containing atomic oxygen.[123-125] Excited-state $O(^1D)$ dissociation products may predominate in the exposure environment, or they may be quenched to the ground state $O(^3P)$. Photodissociation in a gas cell can be a means to produce a diffuse supply of O atoms for exposing a sample,[125] or photodissociation can be used in conjunction with a nozzle to produce O atoms for a supersonic source.[124] If photodissociation were used after the expansion of a beam of precursor gas into vacuum,[123] the photoproducts would be scattered out of the original beam, and a sample could be placed where it would be bombarded by the photoproducts. The polarization of the photodissociating light (usually a laser) may be adjusted to enhance the recoil of products into a certain direction, and the photon energy can be adjusted to control the energy of the products. Photodissociation has the potential to allow a high degree of control over the sample exposure environment, and a photodissociation-based technique could yield hyperthermal O-atoms with energies of hundreds of kJ mol^{-1}. However, the atomic-oxygen flux would be low, making it difficult to conduct sample exposures where degradation effects on the sample could easily be observed. For this reason, the photodissociation technique has received little attention from the space environmental effects community.

All exposure environments designed to aid in the study of materials degradation in LEO have drawbacks that must be taken into account when trying to interpret either post-exposure analysis data or data collected *in situ*. Because of the difficulty of controlling an exposure environment, many conclusions about the mechanisms by which atomic oxygen attacks a surface and assists in the oxidation and erosion of materials have been reached less systematically than would be desired.

2.2. *Material Loss Measurements*

The quantity that is frequently used to quantify the susceptibility of a material to erosion by atomic oxygen is the erosion yield, R_e, which is defined as follows:

$$R_e = \frac{\text{volume of material lost}}{\text{total number of incident O atoms}} \ (\text{cm}^3 \ \text{atom}^{-1}).$$

The erosion yield of a polymer is typically measured by two methods: (1) recession and (2) mass loss. Recession measurements are made by masking an area of a sample surface from attack and measuring the step height difference between exposed and unexposed areas. The thickness loss divided by the exposure fluence is the erosion yield. Mass loss measurements are made either by weighing a sample before and after exposure[126] or by monitoring the mass loss *in situ* of a material that was coated onto a quartz crystal microbalance.[89,127,128] Care must be taken in mass loss measurements to ensure that outgassing from the material in the vacuum of the exposure environment does not affect the results. The calculation of the erosion yield from a mass loss measurement requires knowledge of the density of the material and the surface area exposed, as well as the exposure fluence.

Very few exposures are done under conditions where the absolute atomic-oxygen flux is known. If absolute flux is measured, it is usually done by monitoring the resistance change of a thin silver strip.[129–131] In general, the integrated flux (or fluence) of an exposure is measured with the use of a standard, which is Kapton HN polyimide. The agreed upon erosion yield of room temperature Kapton in the LEO environment is 3.00×10^{-24} cm^3 $atom^{-1}$, and the density is 1.42 g cm^{-3}. The use of the Kapton standard allows the definition of a "Kapton-equivalent fluence" for any type of exposure environment, from plasmas to neutral beams. The use of Kapton as an exposure fluence standard is so ubiquitous that it is generally understood in the space environmental effects community that reported erosion yields for a given environment are based on the Kapton-equivalent fluence in the same environment.

The erosion yields that are reported in the literature[132,133] almost always come from exposures where the Kapton-equivalent exposure fluence was $>10^{19}$ atoms cm^{-2}. These erosion yields represent the material removal process that occurs under steady-state erosion conditions. Mass loss measurements on hydrocarbon materials that were deposited on quartz crystal microbalances have shown that there is an induction period, sometimes involving an initial mass gain,[134] before the erosion yield of a polymer becomes linear with fluence.[127] This linear, or steady-state, behavior is typically reached before a fluence of 10^{18} atoms cm^{-2}, so reported erosion yields are only sensitive to the surface chemistry that is occurring after the induction period.

2.3. *Scattering Dynamics*

The dynamics of the interactions that lead to polymer degradation in space are often studied with beam–surface scattering techniques in which the dynamical behavior of gas-phase products (both reactive and nonreactive) is examined. Although infrared emission has been used to detect reaction products,[135] the majority of scattering dynamics studies to date have employed mass spectrometric detection.[107,108,136–141] The most detailed dynamical studies on gas–surface interactions relevant to polymer degradation in space have been done in the laboratory of the authors.[107,136–141] The results from these studies will be summarized in Secs. 3.3 and 3.4, while the experimental methods follow in this section.

The scattering dynamics experiments done in our laboratory utilize the coupling of a laser detonation source (described above) with a crossed molecular beams apparatus (Fig. 6).[142–144] A pulsed beam containing energetic species (oxygen atoms or inert species, such as Ar and N_2) is produced in a differentially pumped chamber and is directed at a target surface, which is placed 92 cm from the apex of the conical nozzle of the source. The flux of energetic species of interest at the target surface is estimated to be in the range 10^{14}–10^{15} atoms cm^{-2} pulse^{-1}, and this flux may be reduced by as much as an order of magnitude when a velocity-selecting chopper wheel is used (discussed below). A rotatable mass spectrometer detector can be positioned such that the detector axis coincides with the beam axis in order to determine the species in the beam and their energy distributions. Because the beam source is pulsed and emanates essentially from a point in space, the energy distributions of the species in the beam may be determined by measuring their flight-time distributions between the nozzle and the detector. The beam pulses produced in the source have broad energy distributions (full width at half maximum ranging from 250 to 700 kJ mol^{-1}, depending on the species of interest and the source operating conditions). Therefore, a synchronized chopper wheel (slotted disk) is usually used to select a relatively small portion of the overall beam pulse. With the use of the chopper wheel, the ratio of the average kinetic energy of any given species in the incident beam distribution to the energy width (full width at half maximum) is typically $\langle E_i \rangle / \Delta \langle E \rangle \approx 4$. The target surface and the detector can be rotated about the same axis, allowing scattered species to be detected as a function of incident and final scattering angles θ_i and θ_f with respect to the surface normal. All raw data, corresponding both to beam source characterization and to detection of species scattered from the surface, consist of time-of-flight (TOF) distributions based on the flight time of neutral species from a modulated source (the chopper wheel or the laser detonation source itself) to the electron-bombardment ionizer of the detector. The TOF distributions are number density distributions, $N(t)$, and they may be converted into probability density distributions as a function of energy, $P(E)$, by taking into account the proportionality $P(E) \propto t^2 N(t)$ where t is the relevant flight time. The probability density distributions, or translational energy distributions, are proportional to the flux of the detected species.

An additional effusive beam source, operated with pure O_2 as the precursor gas, is used for some studies.[137] This source, shown schematically in Fig. 7, is a low pressure ($\sim$100 mTorr) RF discharge tube, with a small hole (0.5 mm) in the end. The inductively coupled plasma is typically operated at a power of 50 W. The effluent (beam) from this source has been characterized by aligning the detector with the plasma source and placing

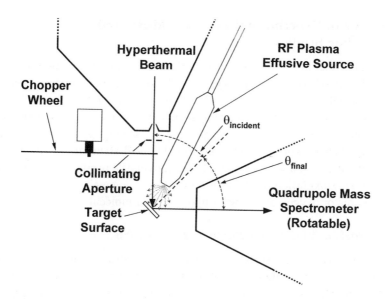

Fig. 7. Schematic diagram of the interaction region, showing the orientation of the effusive plasma and hyperthermal beam sources, chopper wheel, target surface, and detector.

a chopper wheel in front of the detector to modulate the beam. The translational energy distributions of the O and O_2 components in the effusive beam are fit well by Maxwell–Boltzmann distributions with temperatures of 465 K and 485 K, respectively. The fraction of O atoms in the beam is roughly 30 percent, and with the orifice a distance of 2.8 cm from the target surface, the flux of O atoms on the surface is approximately 10^{15} atoms cm^{-2} s^{-1}. With the filament of the electron-bombardment ionizer turned off, the ion signal in the O/O_2 beam is undetectable; thus, the plasma source provides a continuous neutral beam containing only atomic and molecular oxygen. This beam is used to oxidize a target surface continuously while energetic pulses of an inert gas (e.g. Ar and N_2) are allowed to bombard the surface.

Hyperthermal pulsed beams have thus far been directed onto two types of target surfaces, solid polymer films and low-vapor-pressure liquid hydrocarbon films, such as squalane ($C_{30}H_{62}$). The liquid surface is produced by forming a film on a stainless steel wheel that rotates through a liquid reservoir, providing a continuously refreshed surface.[145]

3. Survey of Experimental Results and Related Modeling Studies

Although it has been well established that exposure of polymers to thermal and hyperthermal atomic oxygen leads to significant erosion, uncovering the details of the chemistry behind the process of erosion has been a formidable task. Understanding the chemistry occurring at a polymer surface during exposure to atomic oxygen may be accomplished by estimating mass loss during exposure, studying the chemical structure of the surface before, during, and after exposure, detecting volatile products that are produced during oxidation, and considering the analogous gas-phase reaction mechanisms of oxygen atoms with hydrocarbon molecules.

3.1. *Erosion Yields and Surface Topography*

The erosion yields and surface topography of hydrocarbon polymers that are exposed to environments containing atomic oxygen have provided a basis for much insight, as well as speculation, towards understanding the interactions of oxygen atoms with hydrocarbon surfaces. Pioneering studies of the effect of atomic oxygen on polymers were performed by Rentzepis and coworkers,[146] in which they exposed a variety of polymers to an oxygen plasma and monitored their relative etch rates. After the erosion of polymers on early space shuttle flights was attributed to atomic oxygen,[10,16,81] many space- and ground-based exposure studies provided quantitative erosion yields for a variety of polymers.[78,83,147,148]

The polymer whose erosion yield has become the most well-characterized is Kapton, a polyimide which is used as a component of thermal blankets on spacecraft. The structure of Kapton is shown below.

The erosion yield (at 300 K) of Kapton in the space environment is now agreed to be 3.00×10^{-24} cm^3 O-atom^{-1}. Because of its well known erosion yield in space, Kapton is frequently used as a calibration standard for all exposure environments that contain atomic oxygen. With very few exceptions,[33,34,37,38] all hydrocarbon-based polymers have erosion yields in atomic-oxygen environments that are within a factor of three of the Kapton erosion yield (i.e. R_e in the range of approximately 1×10^{-24} to 6×10^{-24} cm^3 O-atom^{-1} for exposures in the LEO environment). When bombardment of O atoms on a polymer surface is hyperthermal and highly directional (in space and in directed beam environments), the eroded surface becomes rough, with rod- and cone-like structures pointing from the surface (Fig. 3). These structures evolve until they reach steady-state dimensions on the order of one micrometer. Polymers that are exposed in plasmas or in other environments that expose a surface to near isotropic attack from O atoms develop relatively minor roughness, with feature sizes much less than 0.1 μm.

The erosion rates of polymers in an oxygen plasma, measured by Rentzepis and coworkers,[146] provided the first evidence that polymer degradation in an environment containing atomic oxygen is qualitatively different than thermal oxidation of a polymer by molecular oxygen. In thermal oxidation with O$_2$, initiation must begin with spontaneous dissociation of a C–H or C–C bond, which forms a radical that can then react with molecular oxygen:

$$R - H \rightarrow R \cdot + H \cdot \text{ or } R \cdot' + R \cdot'' \tag{1}$$

$$R \cdot + O_2 \rightarrow ROO \cdot . \tag{2}$$

Propagation of thermal oxidation depends on the interaction of peroxy radicals with the polymer to form new alkyl radicals and hydroperoxides:

$$ROO \cdot + R'H \rightarrow ROOH + R \cdot' \tag{3}$$

$$ROOH \rightarrow RO \cdot + OH \cdot . \tag{4}$$

Steps (1) and (4) are slow, so thermal oxidation is usually associated with an induction period. Additionally, thermal oxidation with O$_2$ can be slowed with the use of an antioxidant, which competes with R'H for hydroperoxy radicals. When atomic oxygen is present, however, oxidation begins immediately, without evidence for an induction period, and the erosion rate is unaffected by the addition of antioxidants. Therefore, the initiation of degradation was proposed to involve reaction with atomic oxygen:

$$R - H + O \cdot \rightarrow R \cdot + OH \cdot \text{ or } R \cdot ' + R''O \cdot . \qquad (5)$$

Besides direct reaction of atomic oxygen with a stable hydrocarbon, another path to an oxide radical on the surface was thought to be reaction of O atoms, from the gas-phase or on the surface, with hydrocarbon radicals formed from the initial H-atom abstraction reaction:

$$R \cdot + O \cdot \rightarrow RO \cdot . \qquad (6)$$

If molecular oxygen is present, either in the exposure environment or on the surface as a result of O-atom recombination, then it may also add to the hydrocarbon radical site as in Reaction (2). Further reactions that ultimately lead to removal of carbon from the surface were not postulated in the study by Rentzepis and coworkers. They did note, however, that the erosion rates in the oxygen plasma varied with polymer chemistry. Polymers that contained significant hydrocarbon content eroded at similar rates, while halocarbon and vulcanized polymers eroded at much lower rates. In addition, they found that crosslinking of a polymer did not greatly affect erosion rate, and they therefore concluded that crosslinking is not a general method for improving resistance of a polymer to attack by atomic oxygen.

The initiation of polymer erosion in the LEO environment is believed to result from the same kinds of O-atom interactions that initiate polymer erosion in an oxygen plasma.[97,149] However, the striking differences in erosion yield and surface roughness between polymers exposed in these two environments suggest that the processes that govern erosion depend sensitively on the nature of the exposure environment. Erosion yields from LEO exposures indicate that the efficiency with which energetic oxygen atoms react with a polymer surface is on the order of 10 percent or greater,[150,151] while the probability of reaction when an O atom encounters a surface in an environment containing thermal oxygen atoms is in the range 10^{-3}–10^{-4}.[97]

The erosion of polymers by atomic oxygen thus seems to be an activated process, but whether a single rate-limiting step or multiple activated processes govern the erosion yield is difficult to determine. Koontz *et al.* compared the erosion yields of polyethylene and perdeutero polyethylene when exposed to a flowing afterglow from an oxygen plasma and to a supersonic beam containing O atoms with an average translational energy of 77 kJ mol^{-1}.[98] They observed a kinetic isotope effect for the thermal O-atom exposure but not for the hyperthermal beam exposure, which led them to conclude that H-atom abstraction limits the erosion rate when incident O-atom translational energies are low and that other steps in the mass loss process become rate limiting as the O-atom translational energy increases. In another study of the polyimide, Kapton, the sample temperature was varied in different thermal O-atom environments and an Arrhenius form $[R_e = A \exp(E_a/RT)]$ for the dependence of Kapton erosion yield on temperature was assumed. Activation energies, E_a, of less than 28 kJ mol^{-1} were obtained.[97] An analogous analysis of Kapton erosion yields from exposures at different incident O-atom energies in different environments (where RT was replaced by O-atom translational energy in the Arrhenius equation) was used to derive an activation energy E_a of 37 kJ mol^{-1}.[97] These activation energies are similar to or slightly larger than those observed for H-atom abstraction from various types of hydrocarbon molecules in the gas phase.[50,97] In hydrocarbon polymers that have a high aromatic or olefinic content, such as Kapton, erosion yields tend to be lower than those with more aliphatic content. The relatively low erosion yields of highly unsaturated hydrocarbon polymers have been attributed to the necessity for destruction of the aromatic ring or double bond (e.g. through triplet biradical formation and rearrangement or fragmentation) before mass loss processes can proceed through hydrogen abstraction.[63,97,98] Activation energies derived from temperature dependent sample exposures in LEO are considerably smaller than those mentioned above.[150] In fact, no temperature dependence was observed for Kapton erosion in LEO.[83,152,153] It thus appears that the high O-atom collision energies encountered in LEO overcome any reaction barriers, such as H-atom abstraction, that limit the erosion rate when oxygen atoms bombard a surface at low incident translational energies.

Empirical models[151,154,155] have suggested that production of CO or CO_2 limits the erosion rate during steady-state erosion by hyperthermal oxygen atoms. One model relates the carbon content of a polymer to its erosion yield.[154,155] A correlation between carbon content and erosion yield is found for erosion by hyperthermal oxygen atoms but not for erosion by thermal oxygen atoms, suggesting that the erosion rate in the LEO exposure environment is governed by oxidation of carbon and/or removal of carbon-containing species from the surface, as opposed to an initiation step such as H-atom abstraction. A second model[151] uses the chemical formula of a polymer in a stoichiometric reaction with atomic oxygen to form simple oxides

(CO, CO_2, H_2O, and NO) and deduces erosion yields under the assumption that all O atoms impinging upon on a surface react. They also find a correlation between polymer structure and erosion yield by hyperthermal oxygen atoms, again suggesting that an initiation step does not limit the erosion rate in the LEO exposure environment.

The conclusion that processes other than an initiation step limit the erosion rate of a polymer under attack by hyperthermal oxygen atoms seems at first to be at odds with a variety of data implying that trapping and thermalization of incident O atoms precede their reaction at the surface. This difficulty is reconciled in a reaction scheme proposed by Gregory and Peters,[150] which is based on a Langmuir–Hinshelwood mechanism. They related the dependency of erosion rate on kinetic energy only to the sticking of high-energy O atoms on the surface:

$$O_g + S \rightarrow O_s^* \qquad \text{activated adsorption} \qquad (7)$$

$$O_s^* + S \rightarrow O_s + S \qquad \text{deexcitation} \qquad (8)$$

$$O_s \rightarrow O_g \qquad \text{desorption} \qquad (9)$$

$$O_s + RH \rightarrow \text{products} \quad \text{reaction}. \qquad (10)$$

According to their model, the rate limiting step is adsorption into a reactive precursor state that is mobile, so the actual reaction of an O atom with a moiety on the surface does not depend on the incident energy that has already been dissipated. This model unifies the erosion yield measurements in thermal and hyperthermal O atom environments by assuming that the activation barrier to reaction of both thermal and hyperthermal oxygen atoms with a polymer surface is the adsorption into the precursor state. The reactions subsequent to adsorption will thus follow essentially the same path to products regardless of incident O atom translational energy. This model was justified by the small activation energies determined for carbonaceous materials eroded in LEO (which were determined from the temperature dependence of the erosion yield and were assumed to be representative of analogous thermal reactions in the gas phase) and by the near cosine angular distribution of O atoms scattered from a carbon surface flown in LEO, both of which were consistent with substantial accommodation of O atoms at the surface. The proposal of a mobile precursor state arose from the observation that O atoms are not completely accommodated on a carbon surface.

This general model has been supported by several subsequent studies. Pippin[156] pointed out that an activation energy of $\sim$28 kJ mol^{-1} (which he interprets as a global activation energy, referring to the overall process of thermal oxidation) for the reaction of O atoms with Kapton would require a temperature of approximately 500 K in order for a thermal process to occur, which is consistent with the ambient temperatures in the LEO environment of 900–1200 K, yielding average surface temperatures that vary between 200 and 400 K. He further suggested that at the relatively low fluxes of O atoms onto a surface in LEO, the rate of reaction is limited by the rate at which O atoms strike the surface. As the flux of O atoms onto a surface would be proportional to the rate of adsorption into a reactive precursor state, Pippin's view of the rate limiting step seems to be effectively the same as that of Gregory and Peters. Cross and Blais observed direct evidence for thermal accommodation of hyperthermal O atoms incident on a polymer surface and for subsequent reactions that occur in thermal equilibrium with the surface.[112] They directed a supersonic beam containing O atoms, with an average translational energy of 145 kJ mol^{-1}, at an acrylic surface and observed that unreacted O atoms scattered from the surface in a cosine distribution with respect to the surface normal, and they reported that the translational energy distributions of scattered O atoms and of reactive H_2O, CO, and CO_2 products were characteristic of the surface temperature. Experiments by Nikiforov and Skurat[115] also support the viewpoint that the erosion reactions occur in thermal equilibrium with the surface. They directed a supersonic beam containing O atoms with energies in the range 370–425 kJ mol^{-1} at a Kapton surface, and they were able to deduce the extent of energy accommodation at the surface from the drag coefficient measured with the use of a torsion balance. They found that thermal accommodation of atomic oxygen was initially incomplete and later became complete. Concomitant with an increase in thermal accommodation was a rise in the erosion yield. They attributed both increases to escalation of surface roughness during exposure. The roughness was assumed to aid in trapping of the incident O atoms, presumably through multiple collisions at the surface. The increase in surface roughness was believed to lead to enhanced surface reactivity by (1) increasing the effective density of active surface sites, (2) promoting thermal accommodation of incident O atoms, and (3) permitting multiple energetic collisions of an oxygen atom with a polymer surface. Tagawa and coworkers[134] recently reported an *in situ* study of the mass change of a Kapton film on a quartz crystal microbalance during exposure to low fluxes of energetic O atoms. At O-atom incident energies less than 1000 kJ mol^{-1}, they observed a mass gain of the film, indicating that O atoms may incorporate into a Kapton surface even before it becomes rough.

The model of Gregory and Peters is generally consistent with the erosion-yield data. In this model, incident O atoms must overcome a barrier to adsorption into a reactive precursor state. If the collision energy between the incident O atoms and the polymer surface were comparable to the barrier to adsorption into the reactive precursor state and if the barrier varied from one polymer to another, then adsorption into the precursor state would strongly influence, if not control, the relative erosion yield. A distribution of barrier heights could explain why no correlation was found (in the empirical modeling studies) between carbon content and erosion yield for polymers subjected to thermal O-atom environments. At high incident energies, adsorption into the precursor state would become facile and a sequence of reactions characteristic of thermal oxidation would follow, starting perhaps with H-atom abstraction and ending with complete, or near complete, oxidation of the atoms that make up the polymer. Because the adsorption rate of O atoms would be proportional to their flux onto a surface, the structure of the polymer would then ultimately control the erosion yield, as has been assumed in the empirical models. Even though Gregory and Peters' mechanism appears to have some compelling support from the data, the nature of the reactive precursor state, the mechanism by which incident O atoms enter the reactive precursor state (directly or through multiple collisions on a rough surface), and the subsequent reaction sequence that leads to material removal remain unclear. Recent studies of gas–surface scattering dynamics (Secs. 3.3 and 3.4) reveal the limitations of the proposed mechanism of Gregory and Peters and suggest a more detailed picture that is also consistent with all the erosion-yield data and the empirical models (Sec. 4).

3.2. *Surface Chemistry*

A polymer surface that is exposed to atomic-oxygen is chemically modified as it is eroded. The chemical modification is typically in the form of various nonvolatile oxides. The degree of oxidation and the nature of the oxygen-containing functional groups on the surface have been probed with X-ray Photoelectron Spectroscopy (XPS). Surface chemistry measurements probe the processes that ultimately lead to formation of volatile carbon-containing products. While much effort has gone into surface chemistry modifications of polymers in plasmas,[157-165] relatively few detailed studies have been done to understand the chemistry of polymers that are modified in LEO or simulated LEO environments. The studies that have been reported in the literature examine the chemistry of a surface before and after it is exposed to an atomic-oxygen environment, so the possibility of post-exposure reactions of the surface with air is ever present. A recent study by Tagawa and coworkers[89] has shown that a Kapton surface reaches a saturation level of surface oxygen content during exposure to a beam of hyperthermal oxygen atoms, and upon exposure to air, the surface oxygen atomic concentration decreases with time. Another XPS study of polystyrene surfaces that had been oxidized in a beam of hyperthermal oxygen atoms has shown the existence of unstable surface oxides that are depleted from the surface over a period of several weeks of exposure to air.[166]

Hydrocarbon polymers that were exposed in the LEO environment invariably showed an increase in surface oxide content. For example, the oxygen atomic concentration on a polyethylene surface increased by roughly 10 percent during exposure in LEO.[77] And similar exposures of polystyrene and Kapton led to an increase in surface oxygen atomic concentration of approximately 20 percent and three percent, respectively.[77,167]

The surface oxidation likely proceeds rapidly to a steady-state. An XPS study of polystyrene samples that were exposed for varying durations to a hyperthermal beam containing oxygen atoms indicates that surface oxidation approaches a steady-state during attack from the first few monolayers of impinging O atoms.[168] The surface oxygen atomic concentrations on these samples increased by less than 50 percent when the incident atomic-oxygen fluence increased by three orders of magnitude, from approximately 10 monolayers to 10 000 monolayers. Roughness on these samples is only visible with scanning electron microscopy on the sample exposed to the highest fluence of atomic oxygen.[166] The oxygen atomic concentration on this sample was 22.9 percent, which is similar to that observed on polystyrene samples that were eroded several microns and severely roughened in the LEO environment.[83] It thus appears that surface chemistry remains essentially constant as a polymer is eroded and roughened under steady-state exposure conditions in LEO. A similar conclusion was reached in a report on Kapton erosion in an oxygen plasma, where it was found that the surface atomic composition was invariant with the extent of erosion, so long as the exposure conditions were not altered.[169]

The carbon–oxygen functional groups on the surfaces of exposed polymers are inferred from high resolution XPS spectra of the C1s peak. The spectra of exposed polymers typically exhibit enhanced signals toward higher binding energies relative to the spectra of unexposed control surfaces,[77,134,166–168,170] indicating an enhancement in carbon–oxygen bonding environments on the exposed surfaces. The range of binding energies suggests the presence of all possible functional groups, including ethers, alcohols, epoxides, ketones, aldehydes, esters, acids, and carbonates. However, the shapes of the C1s curves for surfaces exposed to different environments are not the same, which reflects the strong dependence of the resulting distribution of functional groups on exposure environment. Apparently, different steady-state distributions of carbon–oxygen functional groups are formed under different exposure conditions. However, the data are too sparse to permit a meaningful correlation of these functional group distributions with reaction mechanisms or erosion rates. Furthermore, the data available may be misleading because of reactions of species other than atomic oxygen in the exposure environments and by the reactions of ambient air with a freshly exposed polymer surface. Both Golub[171] and Skurat[125] have reported evidence for reactions of molecular oxygen with radical sites formed during etching (Reaction 2) in environments where O_2 is present, leading ultimately to a surface oxide content that is higher than that formed during exposure in LEO, where the molecular oxygen density is very low.

XPS spectra provide a clue to the depth of penetration of oxygen in a polymer that is under O atom attack. Polymers with aromatic groups exhibit a $\pi \to \pi^*$ shakeup peak (the result of electronic excitation of the π-electron system on an aromatic ring by a photoelectron) in the high binding-energy tail of the C1s spectrum. This peak is typically present in the XPS spectra of polymers that have been exposed to LEO and hyperthermal beam environments.[166,168–170] On the other hand, spectra of aromatic polymers that have been exposed to oxygen plasmas exhibit little or no $\pi \to \pi^*$ shakeup peak. Apparently, the plasma treatment destroys the aromatic character throughout the surface layer probed by XPS (5–8 nm deep). The preservation of aromatic character in the surfaces exposed to LEO or a hyperthermal beam suggests a reduced reactivity, relative to the plasma, of the species in these environments with the aromatic ring (as has been inferred, see Sec. 3.1) or a more surface-specific interaction. An angle-dependent XPS study of Kapton that was exposed to a hyperthermal beam, containing atomic and molecular oxygen, indicated an oxidized layer that was more than 5 nm deep,[125] while another angle-dependent XPS study of polystyrene that was exposed to an oxygen plasma found an oxide layer just 1–2 nm thick.[165] The Kapton sample that was exposed to the beam was probably roughened, yet it is not clear how surface roughness was taken into account in the angle-dependent XPS study. In plasma exposures, the thickness of the oxide layer likely depends on the operating conditions. An additional reason for the destruction of the π-electron system in the plasma exposures is photoinduced degradation from VUV light that is more intense in the plasma environment than in beam and LEO environments. The information at hand is insufficient to draw firm conclusions about the penetration of oxygen into a polymer that is bombarded by oxygen atoms in LEO. While the angle-dependent XPS studies suggest that atomic-oxygen might penetrate as much as a few nanometers beneath the surface, the rapid approach to steady-state oxidation in a hyperthermal beam environment and the retention of the $\pi \to \pi^*$ shakeup peak in XPS spectra of LEO-exposed polymer surfaces provide evidence that oxidation takes place closer to the surface.

3.3. *Dynamics of Initial Interactions*: *O + Squalane*

Measurements of erosion yields and surface chemistry provide information about the net result of the complex reactions that occur at a surface during extensive exposure to atomic-oxygen and to any other elements of the exposure environment. While inferences may be made about the chemical and physical interactions at the surface, such studies are an insensitive probe of the individual reaction and interaction mechanisms that accumulatively contribute to the net results. Experiments can, however, be designed to study the various steps of the overall erosion process. These steps involve initial interactions of O atoms with a surface (initiation), oxidation of carbon and scission of the hydrocarbon backbone (propagation), and removal of volatile carbon-containing species (material loss). In this section, studies of the initial interactions of O atoms incident on a hydrocarbon surface[136,138,140] will be summarized.

The initial interactions/reactions that occur when an O atom encounters a hydrocarbon surface have been studied by directing a beam containing oxygen atoms at the continuously refreshed surface of the liquid, squalane (2,6,10,15,19,23-hexamethyltetracosane), which has a room temperature vapor pressure of 2×10^{-8} Torr. This surface was prepared by rotating a polished stainless steel wheel through a liquid reservoir[145] held at 25°C. The film on the wheel was cleaned by passing the wheel by a sapphire scraper, leaving a fresh 100 μm thick film on the wheel prior to exposure to the O atom beam. These studies employed the molecular beam-scattering technique described in Sec. 2.3 and seen schematically in Fig. 6.[140,142] Supersonic and laser detonantion sources were used to produce beams containing O atoms with average translational energies of 21 and 47 kJ mol^{-1} (supersonic) and 297 and 504 kJ mol^{-1} (laser detonation).

The dominant reactive product observed following O atom collisions with a squalane surface was OH. Figure 8 shows representative TOF

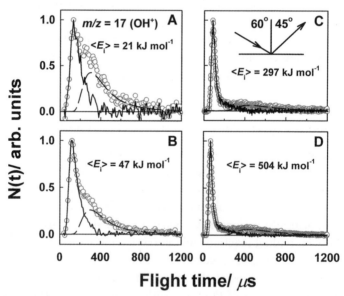

Fig. 8. Representative time-of-flight distributions for reactively scattered OH ($m/z = 17$) following reaction of four different incident oxygen-atom beams, $\langle E_i \rangle = 21$ (A), 47 (B), 297 (C), and 504 (D) kJ mol^{-1}, with a squalane surface. The angle of incidence for all four beams was 60° and the angle of detection was 45°. The distributions have been deconvoluted into hyperthermal (shorter flight times) and thermal (longer flight times) components.

distributions of OH resulting from four different O atom incident energies. All distributions contain a thermal component, corresponding to longer flight times, and a hyperthermal component, corresponding to shorter flight times. The two components in the TOF distributions were deconvoluted by assuming that the slow component corresponded to products that were in thermal equilibrium with the surface and therefore exited the surface with a Maxwell–Boltzmann (M-B) distribution of velocities. The difference between the overall TOF distribution and the assumed M-B component is taken to be the TOF distribution of the second, hyperthermal component. Figure 8 illustrates the trend that is generally observed for any given combination of incident and final angle, where the thermal OH product decreases in significance relative to the hyperthermal OH product as the incident O atom energy is increased.

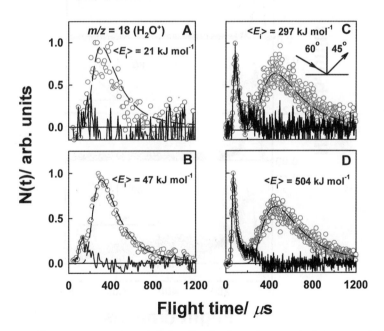

Fig. 9. Representative time-of-flight distributions for reactively scattered H_2O ($m/z = 18$) following reaction of four different incident oxygen atom beams, $\langle E_i \rangle = 21$ (A), 47 (B), 297 (C), and 504 (D) kJ mol^{-1}, with a squalane surface. The angle of incidence for all four beams was 60° and the angle of detection was 45°. The distributions have been deconvoluted into hyperthermal (shorter flight times) and thermal (longer flight times) components.

The only other reactive product observed was H_2O. Figure 9 shows representative TOF distributions of H_2O produced from O atoms colliding with a squalane surface at four different incident energies. Unlike the OH product, H_2O can only be formed through a multiple collision process at the surface. It is therefore not surprising that the H_2O product TOF distributions have a relatively large thermal component, as each successive collision at the surface brings a reactant or product closer to thermal equilibrium

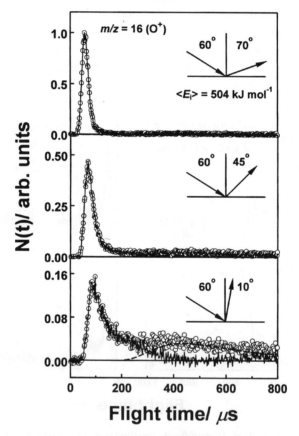

Fig. 10. Representative time-of-flight distributions for inelastically scattered oxygen atoms ($m/z = 16$) following impact with a squalane surface. The incident atomic-oxygen beam had an average energy of 504 kJ mol^{-1} and impinged on the surface at an incident angle of 60°. The scattered O atoms were detected at three final angles: 70°, 45°, and 10°. Two populations of scattered products are identified, those with thermal (dashed line) and hyperthermal (solid line) energies.

with the surface. The hyperthermal signal, which becomes intense when O atoms with high incident energies collide with the surface, indicates a sequential process that occurs on a time scale too short for thermal equilibrium to be reached. The simplest such sequential process is a double direct-abstraction mechanism where an initially-formed OH radical abstracts a second hydrogen atom from another surface hydrocarbon segment.

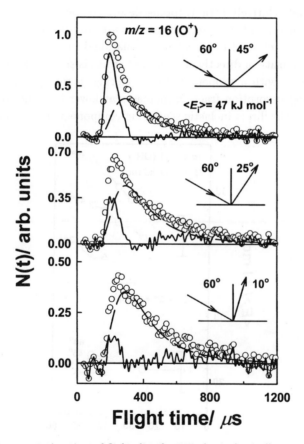

Fig. 11. Representative time-of-flight distributions for inelastically scattered oxygen atoms following impact with a squalane surface. The incident oxygen atoms approached the surface with an average energy of 47 kJ mol^{-1} and with an incident angle of 60°. The scattered O atoms were detected at three final angles: 45°, 25°, and 10°. Two populations of scattered products are identified, those with thermal (dashed line) and hyperthermal (solid line) energies.

    The most probable interaction of an oxygen atom (even with hundreds of kJ mol^{-1} of translational energy) with a hydrocarbon surface is nonreactive, or inelastic, scattering. Figure 10 shows representative TOF distributions of O atoms that scattered from the squalane surface after impinging at an incident angle, θ_i, of 60° and an incident translational energy of 504 kJ mol^{-1}. These TOF distributions were collected at three different final angles, θ_f, as shown. The small thermal (slower) component in these distributions indicates that relatively little trapping desorption occurs. The majority of the signal corresponds to direct inelastic scattering (faster component), in which only a fraction of the incident translational energy is lost to the surface. The intensity of directly scattered O atoms is seen to depend strongly on final angle, with a propensity for scattering in the specular direction. The trapping desorption fraction increases at lower incident energies (see Fig. 11), but the direct inelastic scattering component still remains strong.

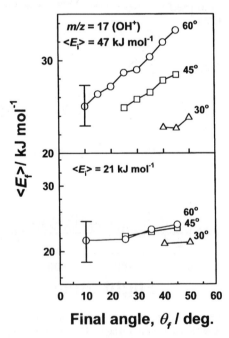

Fig. 12. Average final translational energies of *hyperthermal* OH products as a function of exit angle, corresponding to $\langle E_i \rangle = 47$ kJ mol^{-1} (top panel) and $\langle E_i \rangle = 21$ kJ mol^{-1} (bottom panel) and to three incident angles, 60°, 45°, and 30°. Solid lines connect the data points. Representative error bars are shown.

The observation of nonthermal reactions leading to production of OH and H₂O suggests that the hyperthermal products are the result of direct-abstraction, or Eley–Rideal, reactions at the surface. An examination of the dynamical behavior of the hyperthermal signal confirms this conclusion. Figure 12 shows a plot of average final translational energies of the hyperthermal OH product as a function of final angle, corresponding to the two lower incident energies used and to three different angles of incidence. There is some dependence on final angle, but more importantly, the final energies are higher when the incident energy is higher. The final energy of the hyperthermal OH product also increased with increasing energy in the experiments done with the higher-incident-energy oxygen atoms (see Fig. 13). In these experiments, the average final energies clearly increased with increasing final angle. Further manifestations of the dynamical behavior are

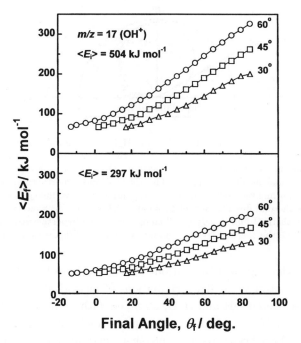

Fig. 13. Average final translational energies of *hyperthermal* OH products as a function of exit angle, corresponding to $\langle E_i \rangle = 504 \text{ kJ mol}^{-1}$ (top panel) and $\langle E_i \rangle = 297 \text{ kJ mol}^{-1}$ (bottom panel) and to three incident angles, 60°, 45°, and 30°. Solid lines connect the data points. Error bars are comparable to the symbol size.

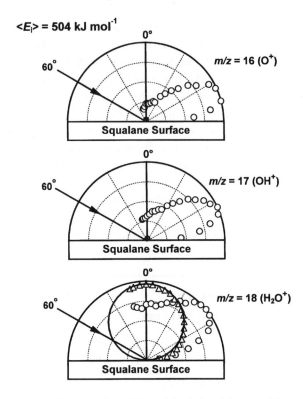

Fig. 14. Angular distributions of O, OH, and H_2O flux following O-atom impact with squalane at $\langle E_i \rangle = 504$ kJ mol^{-1} and $\theta_i = 60°$. Total fluxes were separated into thermal and hyperthermal components. The hyperthermal component (circles) is offset from the surface normal in all three plots. The thermal and hyperthermal components in each plot are normalized such that the corresponding relative fluxes may be compared directly. The thermal flux component for O and OH is shown as a circle centered at the origin, as the thermal flux is very small compared to the hyperthermal component. In the case of H_2O, the thermal (triangles) and hyperthermal fluxes were comparable.

reflected in the angular distributions of scattered product flux. Figure 14 shows representative angular distributions of inelastically-scattered O and reactively scattered OH and H_2O, obtained with an average O-atom incident energy of $\langle E_i \rangle = 504$ kJ mol^{-1} and an incident angle of 60°. These distributions have been separated into their thermal and hyperthermal components. The thermal components follow a $\cos \theta_f$ distribution, while the distributions for the hyperthermal components are far from cosine, with maxima in the vicinity of the specular angle.

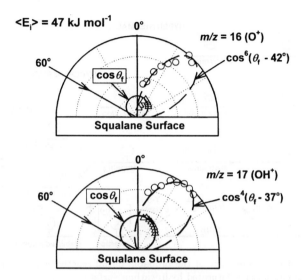

Fig. 15. Angular distributions of O and OH flux following O-atom impact with squalane $\langle E_i \rangle = 47$ kJ mol^{-1} and $\theta_i = 60°$. The thermal (triangles) and hyperthermal (circles) components in each plot are normalized such that the corresponding relative fluxes may be compared directly.

The corresponding angular distributions derived from the experiments with the lower energy incident O atoms show the same general trends (see Fig. 15). The observation of an OH product that exits the surface at hyperthermal translational energies proportional to the incident O-atom energy and that scatters preferentially toward the specular direction confirms that it is formed through a direct-abstraction mechanism.[65] The hyperthermal H$_2$O product cannot correspond to a single direct-abstraction reaction; however, it may be the result of two sequential direct H-atom abstraction reactions. The good match of the angular distributions for the slow components to a $\cos\theta_f$ distribution validates the initial assumption that the slow components in the TOF distributions correspond to products that desorb in thermal equilibrium with the surface.

Analysis of all the data on O-atom reactions with squalane and consideration of the various possible reactions of ground state O(^3P) and electronically-excited O(^1D) lead to a qualitative summary (Fig. 16) of the initial reactions between atomic oxygen and a saturated hydrocarbon surface. The first step leading to the production of volatile reaction products is direct H-atom abstraction by O(^3P). The initial OH product may

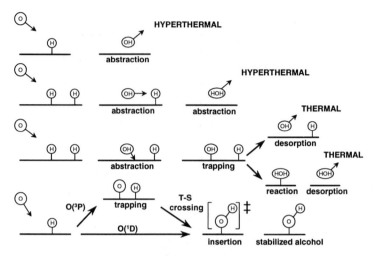

Fig. 16. A pictorial summary of the probable initial reactive events when gas phase oxygen atoms encounter a saturated hydrocarbon surface.

scatter immediately back into the gas phase with hyperthermal energies, or it may abstract a second H atom to form H_2O. The initially formed OH may also become trapped, at which point it may desorb thermally or abstract an H atom to form H_2O, which then desorbs in thermal equilibrium with the surface. If an incident O atom becomes trapped on a 300 K surface, it is unlikely to react if it is in the ^3P state, because the reaction barrier is too high. However, with a long residence time, the $O(^3P)$ can undergo a triplet–singlet crossing[56,57] and insert into a C–H bond to form an excited alcohol that would be stabilized on the surface. If the incident trapped atom were $O(^1D)$, it would insert, and the excited alcohol would also be stabilized on the surface.

Direct H-atom abstraction to form OH is a gas-phase-like process that does not occur in thermal equilibrium with the surface. In the gas phase, an oxygen atom may strike a hydrocarbon molecule from any direction with any impact parameter. Although the collision geometry is more restricted when O atoms impinge on a surface, the atom–surface collisions may resemble gas-phase collisions if the surface is rough on a molecular scale. The squalane surface is believed to consist predominantly of chain ends (CH_3-CH-CH_3) protruding from the surface.[172] Therefore, an atom incident on the liquid surface is most likely to encounter a protruding piece of hydrocarbon. Scattering of atoms from a rough surface, such as squalane, has been described by a hard-sphere collision model,[173] implying gas-phase-like collisions between an incident atom and a finite mass on the surface. While hard-sphere scattering assumes a collision that is elastic in the center-of-mass (c.m.) reference frame, it is likely that a collision between an incident atom and a section of hydrocarbon chain will be inelastic, resulting in internal excitation of the local region of the hydrocarbon chain where the collision occurs. A more detailed picture than that afforded by the hard-sphere scattering model is achieved with the use of a Newton diagram.[174] Not only can the Newton diagram be used to describe nonreactive collisions, which is where the utility of the hard-sphere model is found, but it can also be used to describe reactive collisions. Both the hard-sphere model and Newton diagram pictures pertain to direct scattering events, so they are related only to the hyperthermal components of the reactive and nonreactive signals. The use of Newton diagrams to describe the nonthermal scattering processes is described below.

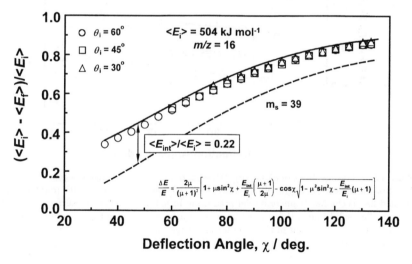

Fig. 17. Average fractional energy transfer of directly scattered oxygen atoms as a function of deflection angle, $\chi = 180 - (\theta_i + \theta_f)$. Atomic oxygen was incident on a squalane surface with $\langle E_i \rangle = 504$ kJ mol^{-1} and $\theta_i = 60°$ (circles), 45° (squares), and 30° (triangles). The dashed line is the hard-sphere model prediction based on the effective surface mass, m_s, shown. The solid line is the revised prediction after the hard-sphere model is corrected for the internal excitation of the interacting surface fragment. The correction involves the ratio, E_{int}/E_i, between the internal excitation energy in the c.m. reference frame and the incident O-atom translational energy in the lab frame. The corrected hard-sphere model equation is given in the figure.

Figure 17 shows the average fractional energy transfer for direct inelastically scattered O atoms as a function of deflection angle, $\chi = 180° - (\theta_i + \theta_f)$. The data plotted in this figure show that the average fractional energy transfer depends on the deflection angle and not on the incident or final angles alone. Such behavior is characteristic of scattering from a rough surface[175–179] and has been successfully described in terms of a hard-sphere model.[180] As seen in Fig. 17, however, the hard sphere model prediction gives a shape that roughly matches the data when an effective surface mass of 39 amu is used, but the magnitude of the predicted energy transfer grossly underestimates what is observed. Figure 18 shows the similar picture that emerges even when the incident O-atom energy is lower by a factor of more than ten.

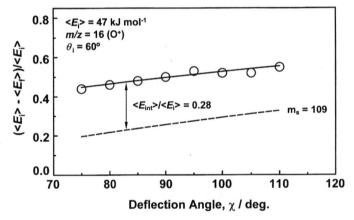

Fig. 18. Average fractional energy transfer of diretly scattered oxygen atoms as a function of deflection angle, χ, for $\langle E_i \rangle = 47$ kJ mol^{-1} and $\theta_i = 60°$ (circles). The dashed line is the hard-sphere model prediction based on the effective surface mass, m_s, shown. The solid line is the revised prediction after the hard-sphere model is corrected for the internal excitation of the interacting surface fragment. The correction is derived from a kinematic analysis of scattering in the c.m. reference frame.

The discrepancy between the observed average fractional energy transfer and the hard-sphere model prediction can be explained by a kinematic analysis (Newton diagram), which provides a view of the scattering dynamics in the c.m. frame. Figure 19 shows an example of a Newton diagram for direct inelastic scattering of O atoms from squalane at an average incident energy of 504 kJ mol^{-1} and an incident angle of 60°. Assuming a surface collision partner whose mass is finite and whose thermal velocity is negligible

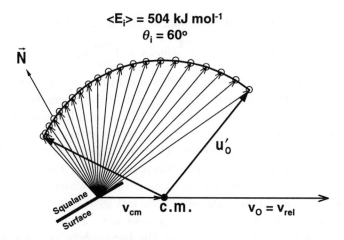

Fig. 19. Newton diagram for direct inelastic scattering of O atoms following impact of O atoms at $\langle E_i \rangle = 504$ kJ mol^{-1} and $\theta_i = 60°$ on a squalane surface.

compared to the velocity of the incident atom, then the c.m. is moving in the laboratory reference frame in the same direction as the incoming beam. Because the surface collision partner is tethered to the surface in the laboratory frame, the relative velocity v_{rel} between the incident atom and the surface is simply the velocity of the incident atom, which is the measured beam velocity. There is no *a priori* knowledge of the c.m. velocity, because the effective mass of the surface collision partner is unknown. It is found, however, that vectors corresponding to the average final velocities of the O atoms lie on a circle about a point on the relative velocity vector. This point corresponds to the tip of the c.m. velocity vector, $\mathbf{v}_{\text{cm}}$. The effective surface mass m_s follows directly from v_{cm}, because the mass of the incident atom m_O and its velocity v_O are known:

$$m_s = m_O \left(\frac{v_O}{v_{\text{cm}}} - 1 \right). \tag{11}$$

The c.m. collision energy can then be calculated as follows:

$$E_{\text{coll}} = \frac{1}{2} \mu v_{\text{rel}}^2, \tag{12}$$

where μ is the reduced mass between the incident oxygen atom and the surface collision partner. The total c.m. energy that goes into translation of the scattered atom and the recoiling surface fragment is

$$E_T = \frac{1}{2}\mu' v_{rel}'^2 = \frac{1}{2}(m_O + m_s)\frac{m_O}{m_s}u_O', \tag{13}$$

where the primes refer to post-collision quantities and u_O' refers to the c.m. velocity of the scattered atom. Implicit in the kinematic analysis is the assumption that the average final velocity is not a function of the c.m. scattering angle over the angular range of detected atoms. This assumption is supported by the almost perfect fit of a circle to the average final velocities over an angular range of $97°$. The difference between the c.m. collision energy and the total energy in translation is the remaining internal energy after the collision. In the example shown in Fig. 19, the internal energy is $E_{int} = E_{coll} - E_T = 357 - 244 = 113$ kJ mol^{-1}, which is 0.22 of the average incident energy, $\langle E_i \rangle$. When the hard-sphere model prediction is corrected by the addition of this fraction, it falls into good agreement with the data (Fig. 17). The hard-sphere model predicts fractional energy transfers that are slightly higher than those that were observed. The data are derived from average translational energies ($E = m\langle v^2 \rangle / 2$), while the hard-sphere model energies are effectively derived from average velocities ($E = m\langle v \rangle^2 / 2$). As $\langle v \rangle^2$ is lower than $\langle v^2 \rangle$, the apparent energy transfer calculated from the hard-sphere model is slightly higher than the observed energy transfer. A similar correction of the hard-sphere model prediction for internal excitation in the c.m. frame improves the agreement between the calculated curve and the data from the experiment that utilized a lower energy beam (Fig. 18). The perfect agreement between the corrected hard-sphere model prediction and the observed fractional energy transfer in this case must be regarded as a fortuitous consequence of the uncertainty in the data. Although correcting the hard-sphere model to allow for internal excitation of the interacting surface fragment significantly improves the agreement with the data, the corrected model remains an approximation, because it does not take into account multiple-bounce effects and the fact that the surface collision partner is not a sphere.

The Newton diagram can also be applied to the reactive OH products that exit the surface with hyperthermal translational energies (Fig. 20). For the hydrogen-abstraction reaction, Eq. (13) is modified by replacing m_O with m_{OH}, m_s with $(m_s - 1)$, and u_O' with u_{OH}', while Eqs. (11) and (12) remain unchanged. The effective surface mass and other dynamical quantities are almost the same as those found for inelastic scattering of

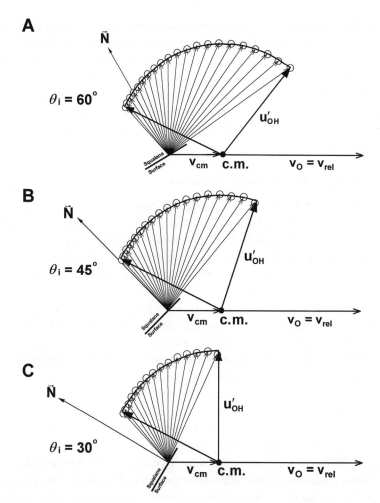

Fig. 20. Newton diagrams for reactive scattering (hyperthermal products only) of OH following impact of O atoms at $\langle E_i \rangle = 504$ kJ mol^{-1} and $\theta_i = 60°$, 45°, and 30° on a squalane surface.

O atoms. This result is expected because the relatively heavy incident O atom will determine the dynamics of an abstraction mechanism that involves the transfer of a light hydrogen atom, especially for this reaction which has a low exoergicity.

The Newton diagram reveals the center-of-mass angular distributions of scattered O and OH flux. In the laboratory reference frame, hyperthermal

O and OH products are predominantly scattered in the specular direction (Fig. 14). However, the specular direction in the laboratory frame corresponds to sideways scattering, with respect to the direction of the incident O atom, in the c.m. frame. Sideways scattering suggests a strong interaction between the incident atom and a hydrocarbon group on the surface. In the case of an abstraction reaction, these dynamics indicate a mechanism where the incident O atom reacts through a relaxed collinear O–H–C transition state.[138]

3.4. *Steady-State Material Removal*

The results discussed in Sec. 3.3 strongly suggest that the first step toward the erosion of a hydrocarbon polymer surface that is under atomic-oxygen attack is direct hydrogen-atom abstraction. The following steps become varied and complex, and although mechanisms have been proposed,[63,97,149] it is difficult to track the overall mechanism that ultimately leads to significant material loss from a polymer surface. Initial radical products, such as OH, may undergo secondary reactions on the surface. Radical sites on the hydrocarbon backbone will be susceptible to further reactions with incoming O atoms, neighboring parts of the polymer, or various other species present in the exposure environment. Ultimately, volatile carbon-containing products must carry mass away from the surface. CO and CO_2 are presumed to be important, if not dominant, carbon-containing products when a polymer surface is continuously bombarded with O atoms.[112,135,151,154,155,181] However, long-chain hydrocarbon fragments may desorb from the surface after enough chain scission reactions have occurred.[108] Little work has been done to quantify the fraction of carbon that leaves a surface in the form of hydrocarbon fragments, as opposed to CO or CO_2 molecules. Nevertheless, several experiments have confirmed the production of CO and CO_2 during O-atom bombardment of a polymer, and these products have been detected as important products even in experiments where polymer fragments were observed. The removal mechanisms for CO and CO_2 are themselves complex, as scattering dynamics experiments have detected the presence of both thermal and hyperthermal CO and CO_2 products when polymer surfaces are exposed to beams containing O atoms with translational energies from 50 to 600 kJ mol^{-1}.[107,139–141,181]

Assuming the bulk of the mass loss during continuous atomic-oxygen bombardment is the result of removal of carbon in the form of CO and

CO_2, then the erosion rate might depend on the formation and/or removal rates of these species. A kinetic barrier to the removal of CO and CO_2 from the surface would then impede erosion, with the erosion rate being inversely related to the barrier height. If the rate of product removal from the surface is slower than the rate of formation of these products, then the material removal mechanisms, not the reaction mechanisms, could dominate the erosion yield of a polymer. The possibility of collision-induced release of CO or CO_2 from an oxidized surface has been investigated in our laboratory, and the results are summarized in this section.

The experimental approach was outlined in Sec. 2.3. Hyperthermal beams of energy selected Ar atoms or N_2 molecules were directed at polymer surfaces that were continuously bathed with the effluent of a low pressure RF plasma source of oxygen atoms (see Fig. 7). Interaction of the effusive O-atom beam with polymer samples produced both CO and CO_2 continuously, and data were collected under conditions of steady-state oxidation, as verified by the unchanging signals from CO and CO_2 in the mass spectrometer. Product TOF distributions were collected at $m/z = 28(CO^+)$ and $44(CO_2^+)$, following impingement of hyperthermal beam pulses on polymer surfaces that had reached a steady state of oxidation.

Figure 21 shows TOF distributions of CO_2 following impingement of five different Ar beams on an oxidized Kapton surface. Time zero in these distributions corresponds to the time at which the peak of the beam pulse impinged on the surface. The product TOF distributions have been normalized to the integrated intensities of the respective incident beams. The TOF distributions are bimodal, indicating at least two mechanisms through which CO_2 leaves the surface following impact of the hyperthermal beam pulse. The slower component corresponds to CO_2 molecules that exit the surface with a Maxwell–Boltzmann distribution of velocities, corresponding to the surface temperature ($\sim$320 K), while the faster, or hyperthermal, component corresponds to CO_2 molecules that exit the surface with approximately 70 kJ mol^{-1} of kinetic energy. These data are representative of what has been seen in analogous experiments with hyperthermal Ar atoms striking an oxidized polystyrene surface and with hyperthermal N_2 molecules striking oxidized Kapton and polystyrene surfaces.[107,137] Figure 22 shows the relationship between the total integrated intensity of the CO_2 signal (correlated with hyperthermal beam impingement) and the incident kinetic energy of the impinging species for several different experiments. The

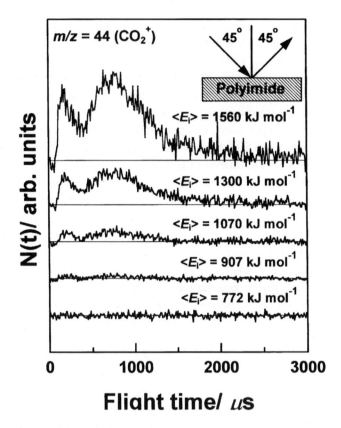

Fig. 21. Time-of-flight distributions of CO_2 exiting a continuously oxidized Kapton surface following exposure to pulses of five hyperthermal argon beams whose average translational energies are shown. The distributions are normalized with respect to the respective incident beam intensity.

curves through each set of data points have the form, $I = Ae^{-b/\langle E_i \rangle}$, underscoring the strong dependence of incident energy on CO_2 release from a continuously oxidized surface. Similar results have been obtained for the CO signal (which is comparable in magnitude to the CO_2 signal).

The data demonstrate the existence of a collisional process that leads to ejection of CO_2 (and CO) from an oxidized polymer surface upon impingement of hyperthermal inert atoms or molecules and show that the rate of release of these carbon oxide molecules from the surface rises dramatically with incident kinetic energy above ~800 kJ mol^{-1}. The fact

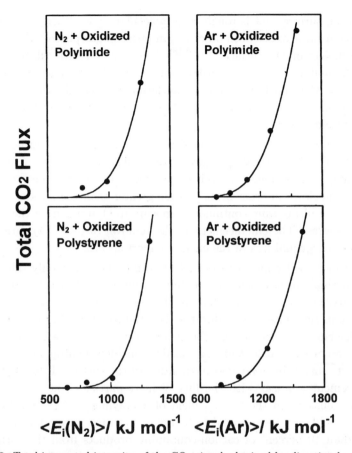

Fig. 22. Total integrated intensity of the CO_2 signal, obtained by directing hyperthermal beams of Ar or N_2 at continuously oxidized Kapton and polystyrene surfaces, as a function of incident Ar or N_2 average translational energy. The solid lines represent fits that have the functional form, $I = Ae^{-b/\langle E_i \rangle}$.

that bombarding species of different mass give rise to the same energy threshold of ~ 800 kJ mol^{-1} suggests that collision energy (as opposed to momentum) governs the collisional production of reaction products from the surface. Although the data are not shown, the same dramatic rise from a roughly 800 kJ mol^{-1} threshold is seen in the CO_2 signal intensity when the incident angle of the hyperthermal beam is 30° or 60°, instead of 45°. Thus, it appears that total energy, rather than normal energy, governs the collisional enhancement in the signal. This observation is consistent with an eroding surface that is rough on a molecular scale. Energy transfer on such rough surfaces has been described in terms of gas-phase-like scattering from a moiety on the surface.[138,180,182] The observed data exhibit distinct hyperthermal and thermal translational energy distributions, as has been seen in the OH and H_2O product translational energy distributions when hyperthermal atomic oxygen bombards a hydrocarbon surface that is rough on a molecular scale.[140,141]

The enhancement in the rate of CO_2 release from the surface is significant. The CO_2 signals produced from collisions of energetic atoms or molecules with continuously oxidized polymer surfaces have been observed to be as much as an order of magnitude greater than the background signal levels resulting from continuous exposure to O atoms from the effusive plasma source. The enhancement in the removal rate of CO_2 varies with incident energy and flux of the energetic inert species.

The increased removal rates of CO_2 (and CO) occur only during the hyperthermal pulse, which is estimated to contain somewhat less than 10^{14} inert particles cm^{-2}, or somewhat less than one tenth of a monolayer. Given this flux in the pulse, whose duration is ~ 100 μs, it would take 10^{18} inert particles cm^{-2} s^{-1} to increase the steady-state removal rate of CO_2 (and CO) by an order of magnitude, provided the incident O-atom flux were high enough to maintain steady-state oxidation of the surface. If the collisional enhancement is linear with the flux of inert particles, the effect of high energy collisions on the erosion yield of a polymer under atomic-oxygen attack would only be expected to be important (accounting for ejection of more than 10 percent of carbon-containing products from the surface) if the flux of energetic species onto the surface were on the order of 10^{16} inert particles cm^{-2} s^{-1}, or approximately 10 monolayer per second. Thus, it appears from the scattering experiments that the flux of energetic species onto a continuously oxidized surface would have to be relatively high in order to enhance significantly the erosion rate.

This conclusion was tested with a practical experiment in which the erosion yield of polystyrene under exposure to the effusive plasma source was measured with and without the addition of the pulsed hyperthermal Ar beam. In this experiment, the Ar beam pulse was not chopped, so the entire beam pulse produced from the laser detonation source was allowed to strike the surface. The erosion yield was determined by placing a screen over the samples during exposure and measuring the etch depth of the exposed areas with a profilometer. The erosion yield when the Ar beam bombarded the sample in addition to the effluent from the effusive plasma source was approximately 30 percent higher than the erosion yield observed when only the effusive source was used to expose the surface. Presumably, only the high-energy tail of the Ar beam contributed significantly to the observed enhancement in the erosion yield. The flux of this part of the beam is estimated to be on the order of 10^{14} atoms cm^{-2} s^{-1}. Apparently, the conclusion that a practical erosion yield would be unaffected by this average flux of energetic Ar atoms was erred. More work is needed to understand the nature of the collisional mechanisms, in order to make an accurate assessment of the role of energetic collisions in the erosion of a polymer surface that is under simultaneous attack by atomic oxygen.

The collisional process that has been identified may be an important factor in material erosion in LEO. The second most dominant component of the LEO environment (after atomic oxygen) is molecular nitrogen, which can have densities from a few percent to more than 50 percent those of atomic oxygen, depending on orbital altitude. At common orbital altitudes of 300–400 km, the N_2 density is typically 5–10 percent that of atomic oxygen, or roughly 2×10^7 molecules cm^{-3}. Combining this density with an impact velocity of 7.4 km s^{-1} results in a flux of N_2 molecules on ram surfaces of 1.4×10^{13} molecules cm^{-2} s^{-1}, about one tenth the estimated average flux of energetic Ar atoms in our erosion experiment. The average collision energy of N_2 on the ram surfaces of a spacecraft is $\sim$790 kJ mol^{-1}, and the collision energy distribution can reach up to 1000 kJ mol^{-1}. In fact, approximately half the N_2 flux on a spacecraft collides with ram surfaces at impact energies above the 800 kJ mol^{-1} threshold, where we have observed significant enhancement in the production of CO_2 and CO, although the majority of N_2 collisions will be close to the threshold energy. Given the relatively low N_2 flux at altitudes of 300 km and above, coupled with the fact that most collisions are expected to be near the energy threshold for assisting in production of CO and CO_2 from a surface, the role of collisions in assisting atomic oxygen in material removal from a polymer is expected to be minimal. However, collision-assisted erosion may become significant on the ram surfaces of a spacecraft at altitudes lower than 300 km. For example, the N_2 flux is approximately 3×10^{14} molecules cm^{-2} s^{-1} on the ram surface of a spacecraft at an altitude of 250 km. The average flux of hyperthermal Ar atoms in the erosion experiment on polystyrene was on the same order as or lower than the N_2 flux on ram surfaces of spacecraft below 300 km. As there was a 30 percent enhancement in erosion yield when the continuously oxidized surface was simultaneously bombarded with hyperthermal Ar atoms, it is likely that N_2 in LEO also assists atomic oxygen in the erosion of spacecraft surfaces at orbital altitudes below 300 km, where the flux of N_2 collisions above 800 kJ mol^{-1} is greater than 10^{14} molecules cm^{-2} s^{-1}.

Ground-based simulations of atomic-oxygen effects in LEO may also include collisional processes. A laser-detonation source similar to the one employed in these experiments is often used to expose candidate spacecraft materials. With O_2 as a precursor gas in this source, we observed CO_2 product signals whose dynamical behavior was remarkably similar to the behavior of the signals that were generated with the combination of hyperthermal Ar or N_2 and a continuously oxidized surface. It is therefore possible that O_2 in the hyperthermal O/O_2 beam acts to remove material through similar collisional processes. Perhaps, then, the heretofore unwanted molecular oxygen byproduct in the laser-detonation source is actually a desirable component that plays the role of N_2 collisions in ground-based simulations.

4. Relevance of Gas–Surface Interaction Dynamics to the Study of Space Environmental Effects

The mechanisms through which atomic oxygen and other species in the LEO environment attack a polymer surface and eventually erode it are reflected in the gas–surface interaction dynamics. In the absence of direct dynamical probes of the gas–surface interactions, degradation mechanisms have been inferred from gas-phase-like mechanisms and measurements of erosion yields and surface chemistry under conditions that were controlled to varying degrees. Although much remains to be learned about the detailed mechanisms, the rare studies that are sensitive to the dynamical interactions (e.g. those described in Secs. 3.3 and 3.4) are beginning to bring unity to the disparate conclusions and proposals that have been advanced in the literature. The driving force behind all attempts to understand the basic mechanisms has been to develop the knowledge to predict the performance of materials in LEO and to design materials that will be durable. The reductionist approach that is implicit in the attempts to unveil the underlying dynamics is to break down the degradation process into its components and then to put the components together to build a picture of polymer degradation under a specified set of exposure conditions.

The dynamics of the initial interactions when a hyperthermal oxygen atom encounters a surface supports the view that initial H-atom abstraction limits the erosion rate of a polymer in LEO. As shown in Fig. 16, the first step toward polymer degradation is H-atom abstraction to form OH. The dynamics of this reaction suggest a gas-phase-like process, in which the surface plays a minimal role. No evidence was observed for a reactive precursor state of the type proposed by Gregory and Peters [see Eqs. (7)–(10)]. Initial H-atom abstraction is sufficient to account for the large difference in erosion yields for polymers subjected to environments containing thermal and hyperthermal oxygen atoms, because the reaction of ground state $O(^3P)$ has a barrier to H-atom abstraction of approximately 13 to 29 kJ mol^{-1}, depending on which type of hydrogen atom is abstracted. When the O-atom incident energy is significantly above the barrier, the erosion rate would thus be expected to become essentially energy and temperature independent, as has been reported (see Sec. 3.1).

The postulation of a reactive precursor state is not required if a rate limiting abstraction reaction is followed by thermal accommodation of the products on a rough surface. An oxygen atom incident on a rough polymer surface is most likely to encounter the steeply sloping surface of a microscopic cone-like feature (Fig. 3). As seen from the angular distributions in Figs. 14 and 15, grazing incidence collisions will result in reactive OH products and unreacted O atoms that scatter preferentially into the specular (forward) direction. Figures 12 and 13 show that grazing collisions transfer the least energy to the surface. (The dependence of final translational energy on exit angle for inelastically scattered O atoms follows similar behavior to that seen in Figs. 12 and 13.) Therefore, oxygen atoms or OH radicals will tend to scatter from the sloping sidewall of a feature on the rough surface down into the pit below, where they will collide with relatively high energies. The few O or OH species that do not react immediately will likely be thermally accommodated following multiple bounces in the confined region in the pit. Once the initial reactive and inelastic products accommodate to the surface, further reactions will take place in thermal equilibrium with the surface,[183] and products will eventually be released with thermal energies, as described by a Maxwell–Boltzmann distribution. Although many studies of scattering dynamics have shown that a fraction of the reaction products may exit the surface with hyperthermal translational energies, these studies employed surfaces that were not significantly roughened. Unpublished results from our laboratory suggest that most products exit a polymer surface at thermal energies when the surface develops a rough texture similar to surfaces of polymers that are eroded in space. In summary, the reaction of hyperthermal oxygen atoms with a rough surface can give the impression that all reactive steps occur in thermal equilibrium with the surface even when the initial, activated, step is a nonequilibrium reaction.

The scattering dynamics suggest a reaction scheme that is based on an initial Eley–Rideal mechanism, as opposed to the Langmuir–Hinshelwood mechanism proposed by Gregory and Peters. In the new scheme, H-atom abstraction limits the erosion rate at near-thermal O-atom/surface collision energies. At high collision energies, H-atom abstraction occurs readily. Reactions during an induction period of $\sim 10^{19}$ O atoms cm^{-2} will lead to a rough surface (see Fig. 3), and initially scattered reactive and inelastic species will become thermally accommodated on this rough surface. Oxidation reactions in thermal equilibrium with the surface will then proceed. During steady-state oxidation/erosion, incident O atoms may react with radical sites and a variety of other functional groups. Highly energetic species that strike the continuously oxidized surface at collision energies greater than 800 kJ mol^{-1} may assist in the production of oxidation products from the surface.

A refined picture of the interaction dynamics can be incorporated into models whose aim is to predict the extent of polymer erosion in LEO. For example, the empirical models mentioned in Sec. 3.1 have the possibility to make quantitative predictions of erosion yields, but they have suffered from a lack of information about the initiation steps and the reaction products. Both models[151,155] have assumed that CO and CO_2 carry carbon away from the surface, but quantitative predictions depend on the branching between these two products. The models can be extended to include contributions from all oxidation products that may carry mass away from the surface, as long as their identity and relative yields are known. In addition, the accuracy of the model predictions would be improved by knowledge of the initial trapping probability of incident oxygen atoms. Dynamical data, such as those described in Secs. 3.3 and 3.4, can bolster erosion models and thus expand their acceptance by the community.

Gas–surface interaction dynamics govern the development of surface topography during erosion in the LEO environment. Surface roughening alters thermal and optical properties of polymers because of increased scattering and trapping of light on rough surfaces relative to smooth surfaces. Not only does surface roughening alter the functional properties of a polymer, but it also changes the nature of the interaction mechanisms. If a surface remained smooth during erosion, then bombarding atoms or molecules and initial reaction products would have less chance of becoming trapped. Therefore, nonthermal gas–surface interactions may dominate the degradation processes on a relatively smooth surface. In addition, the evolution of etch pits at coating defects will limit the efficacy of protective coatings for polymers.[5] Whether surfaces erode around coating defects, areas of contamination,[83] or regions where the molecular structure of a surface makes it relatively unreactive,[151,184] scattering dynamics will govern the resultant surface topography. Both inelastic and reactive scattering dynamics play a critical role. The angular and translational energy distributions of inelastically scattered O atoms will control the location and reactivity of these atoms on subsequent collisions. The site-dependent reactivity of scattered reaction products, such as OH, will also depend on the directions and translational energies with which these products exit the local surface. Furthermore, the reactivity of a gas–surface collision is a function of the incident angle (with respect to the local surface normal). The close link between scattering dynamics and etch profile evolution was demonstrated in a study that applied a Monte Carlo approach to a model of silicon etching by atomic fluorine.[185] Only two parameters were needed because experimental data on atom–surface scattering dynamics[180] was incorporated into the model. The validity of the model was shown in its ability to predict etch profiles achieved under previously untried exposure conditions. A Monte Carlo model has also been developed to describe atomic-oxygen-induced erosion profiles at defect sites on protected polymers.[96] Although this model is successful at predicting observed erosion profiles, it employs assumptions about the dynamics that have been shown to be false for interactions of hyperthermal oxygen atoms with a hydrocarbon surface. The apparent success of the model is presumably a result of the number of adjustable parameters used. It would be straightforward to couple the known dynamics with the existing model to reduce the number of assumptions. Modeling approaches for predicting the topographical development of polymer surfaces exposed to the LEO environment currently exist; however, the models will become more grounded in physical reality as they continue to embrace new data on the dynamics of the degradation reactions.

All models that have been proposed are empirical and require some form of test data, either for their validation or for their use as predictive tools. Testing of candidate spacecraft materials in laboratory exposure environments is widely used as an economical and time efficient alternative to in-space experiments. However, direct simulation of the LEO environment is essentially impossible, and a variety of exposure environments are used (see Sec. 2). An attempt to improve and standardize ground-based materials testing in atomic-oxygen environments has resulted in a test protocol.[186] Although the knowledge upon which this protocol is based is largely phenomenological, it provides a framework for more consistent testing and identifies the exposure parameters that must be considered for a test. The phenomenological understanding of materials degradation in LEO has tended to make the specification of test parameters uncertain and arbitrary. A more concrete understanding of the degradation mechanisms will enhance confidence in test protocols.

The conclusions afforded by the dynamical studies presented here point to specific areas where test parameters should be evaluated. The conclusion that the rate limiting step is H-atom abstraction suggests that O-atom translational energies need only be significantly above $\sim$29 kJ mol^{-1} to give results that are representative of reactions in LEO. However, the distributions of inelastically scattered O atoms exhibit a strong energy dependence between 20 and 500 kJ mol^{-1}, so the evolution of surface roughness might not be representative of what occurs in LEO even if the incident O-atom energy in the test environment is high enough to overcome the barrier to H-atom abstraction. The dependence of surface roughness on incident O-atom translational energy is not known at this time, and further investigations of this dependence would help narrow the acceptable range of O-atom energies in test environments. Collision-assisted material removal may enhance the material removal rate under certain exposure conditions, especially when a laser-detonation source is used to produce atomic oxygen. If absolute O-atom-induced erosion rates are required, then the effect of collisions must be considered. Nevertheless, because the collisional phenomenon appears to be general, different polymers may show the same degree of enhancement, so the Kapton equivalent erosion yield of a test specimen may be similar to what would be measured in the LEO environment.

5. Conclusions and Future Directions

The degradation mechanisms of polymers in the LEO environment are complex, and they involve the combined effects of atomic oxygen and other species in conjunction with the unique set of exposure parameters encountered in LEO. This chapter has focused on the role of atomic oxygen in the erosion of a hydrocarbon polymer. Atomic-oxygen interactions with hydrocarbon polymers form a model system upon which to base descriptions of degradation mechanisms of other materials subjected to the LEO environment. The in-depth study of the gas–surface interaction dynamics associated with this model system is expected to build a knowledge base of fundamental gas–surface processes in a unique range of collision energies, to strengthen predictions of materials durability in LEO, and to influence the development of surface modification technologies.

Studies of hyperthermal atomic-oxygen interactions with polymer surfaces have revealed the importance of nonthermal (nonequilibrium) processes at these surfaces. Direct inelastic and reactive scattering events dominate the initial interactions. Hyperthermal product signals from CO and CO_2 indicate the occurrence of additional nonthermal processes. All these nonthermal processes become increasingly important as the incident O-atom translational energy increases from near-thermal to hundreds of kJ mol^{-1}. The existence of these nonthermal interactions offers the possibility to discover interesting new reactive pathways at the gas–surface interface.

The model system comprising the reaction sequence from initial O-atom attack to steady-state erosion of a hydrocarbon surface can serve as a benchmark for fundamental atom–surface interactions at hyperthermal collision energies and for etching mechanisms of materials. Within this model system, there is still much to learn. It is likely that when a hyperthermal oxygen atom strikes a saturated hydrocarbon surface, it will either abstract a hydrogen atom or it will scatter inelastically. The subsequent reaction sequence becomes murky. Very little is known about the mechanisms of oxidation, surface roughening, or material loss. In fact, even the sticking probability of hyperthermal atomic oxygen on a hydrocarbon surface is not known. Experiments that probe various aspects of the dynamics of initial attack and subsequent erosion are needed to complete the model picture. Currently, this picture relies on speculation to fill the large gaps between the studies of the initial interaction dynamics and the results of erosion-yield studies.

The detailed understanding of how a hydrocarbon polymer erodes under atomic-oxygen attack in LEO will validate models and lend credence to test protocols. The approach toward prediction of a material's durability in LEO has been somewhat haphazard, with a variety of materials being exposed to different environments. In most cases, the exposure parameters were not well controlled. Nevertheless, the space environmental effects community has a general idea of the relative erosion yields of a variety of polymers. However, the details of the erosion mechanisms are still lacking, including knowledge of which processes are common to all polymer erosion in LEO and which are specific to a particular polymer structure. The fact that the erosion yields for all hydrocarbon-based polymers in the LEO environment have the same order of magnitude suggests that some processes are common to erosion of all polymers in an atomic-oxygen environment. There must also be some dependence on chemical structure, because polymers do differ in their erosion yields. Understanding the model system will provide the first step in linking the erosion mechanisms of a wide variety of polymers, and this understanding will provide valuable input to empirical models. Eventually, as more experiments are done and models become more secure, a spacecraft designer should need only the mission profile in LEO, the chemical structure of a material, and/or a set of test results on the material in order to predict the material's performance during the mission.

The best predictions will ultimately account for both the chemical mechanisms of degradation and the concomitant physical roughening of the eroded surface. In this chapter, roughness has been considered insofar as it tends to drive the relevant gas–surface interactions toward thermal equilibrium with the surface. Yet roughness must also be coupled to the initial reactivity by determining the relative fraction of grazing incidence versus normal incidence collisions between an incident O-atom (or scattered product) and the local region of the surface. The scattering dynamics ultimately determines both chemistry and topography. Perhaps by controlling the chemical structure of a polymer, the dynamics will be altered and the surface will erode with less roughness. Reduced roughness may increase the importance of nonthermal reactions and significantly alter the sequence of reactions that lead to erosion. Only a deep understanding of the dynamics will make the link between chemistry and roughness apparent.

The new observations of nonequilibrium processes at the gas–polymer interface elicit thoughts of heretofore unfamiliar reaction dynamics driving new materials technologies in etching and surface chemistry modification. Such nonequilibrium dynamics dominate when collision energies are high (hundreds of kJ mol^{-1}). In principle, the fact that a regime exists where nonequilibrium processes dominate means that the outcome of a process may be controlled by the reaction conditions. The appropriate exercise of this control may result in the development of new or improved processes for topographical patterning or chemical modification of polymer surfaces. At the very least, studies of the unique nonequilibrium interactions of gaseous species with surfaces in such "extreme environments" as LEO, should enhance understanding of materials etching and chemical modification in other processing environments.

Acknowledgments

We are grateful to J. Seale, A. Frandsen, T. Thompson and J. Zhang, who contributed to the work done in the authors' lab that was summarized in this chapter. Special thanks are extended to P. Casavecchia and G. Nathanson, from whom we have learned much about gas-phase and gas–surface dynamics. We also wish to thank R. Dressler, E. Murad and M. Tagawa for their critical reviews of the manuscript. This work was supported in part by grants from the Department of Defense Experimental Program for the Stimulation of Competitive Research (DEPSCoR), administered by the Air Force Office of Scientific Research (Grant Nos. F49620-96-0276 and F49620-99-1-0212), and by a grant from Ionwerks.

References

1. K. S. W. Champion, A. E. Cole and A. J. Kantor, "Standard and Reference Atmospheres", *Handbook of Geophysics and the Space Environment*, Ed. A. S. Jursa (United States Air Force, Air Force Geophysics Laboratory, 1985) Chap. 14.
2. *U.S. Standard Atmosphere*, Ed. A. Jursa (U.S. Government Printing Office, Washington, D.C., 1976).

3. R. G. Roble, "Energetics of the Mesosphere and Thermosphere", *The Upper Mesosphere and Lower Thermosphere: A Review of Experiment and Theory*, Eds. R. M. Johnson and T. L. Killeen (American Geophysical Union, Washington, D.C., 1995) pp. 1–21.

4. J. T. Visentine, "Environmental Definition of the Earth's Neutral Atmosphere," *NASA/SDIO Space Environmental Effects on Materials Workshop held in Hampton, VA, 28 June–1 July 1988*, NASA Conference Publication 3035, Part 1, Eds. L. A. Teichman and B. A. Stein (NASA, Washington, D.C., 1989) pp. 179–195.

5. B. A. Banks, B. Auer and F. J. DiFilippo, "Atomic-Oxygen Undercutting of Defects on SiO$_2$ Protected Polyimide Solar Array Blankets," *Materials Degradation in Low Earth Orbit: Proceedings of a Symposium Sponsored by the TSM-ASM Joint Corrosion and Environmental Effects Committee held at the 119th Annual Meeting of the Minerals, Metals and Materials Society in Anaheim, CA, 17-22 February 1990*, Eds. W. Srinivasan and B. A. Banks (Minerals, Metals and Materials Society, Warrendale, PA, 1990).

6. E. Murad, *J. Spacecraft Rockets* **33**, 131 (1996).

7. B. A. Banks, K. K. de Groh, S. L. Rutledge and F. J. DiFilippo, *Prediction of In-Space Durability of Protected Polymers Based on Ground Laboratory Thermal Energy Atomic-Oxygen*, NASA Technical Memorandum 107209 (NASA, Washington, D.C., 1996).

8. A. E. Hedin, J. E. Salah, J. V. Evans, C. A. Reber, G. P. Newton, H. W. Spencer, D. C. Kayser, D. Alcaydé, P. Bauer, L. Cogger and J. P. McClure, *J. Geophys. Res.* **82**, 2139 (1977).

9. A. E. Hedin, C. A. Reber, G. P. Newton, N. W. Spencer, H. C. Brinton, H. G. Mayr and W. E. Potter, *J. Geophys. Res.* **82**, 2148 (1977).

10. L. J. Leger and J. T. Visentine, *J. Spacecraft* **23**, 505 (1986).

11. A. E. Hedin, *J. Geophys. Res.* **88**, 10170 (1983).

12. A. E. Hedin, *J. Geophys. Res.* **92**, 4649 (1987).

13. *Natural Orbital Environmental Guidelines for Use in Aerospace Vehicle Development*, Eds. B. J. Anderson and R. E. Smith, NASA Technical Memorandum 4527 (NASA, Washington, D.C., 1994).

14. L. K. English, *Mat. Eng.* August, 39 (1987).

15. D. E. Hunton, *Sci. Am.* **261**, 92 (1989).

16. L. J. Leger, *Oxygen Atom Reaction with Shuttle Materials at Orbital Altitudes*, NASA Technical Memorandum 58246 (NASA, Houston, TX, 1982).

17. L. E. Murr and W. H. Kinard, *Am. Sci.* **81**, 153 (1993).

18. *NASA/SDIO Space Environmental Effects on Materials Workshop held in Hampton, VA, 28 June–1 July 1988*, NASA Conference Publication 3035, Eds. L. A. Teichman and B. A. Stein (NASA, Washington, D.C., 1989).

19. *Proceedings of the NASA Workshop on Atomic-Oxygen Effects, held in Pasadena, CA, 10–11 November 1986*, NASA Contractor Report 181163, JPL Publication 87-14, Ed. D. E. Brinza (NASA, Pasadena, CA, 1987).

20. S. Koontz, L. Leger, K. Albyn and J. Cross, *J. Spacecraft* **27**, 346 (1990).

21. A. E. Stiegman, D. E. Brinza, E. G. Laue, M. S. Anderson and R. H. Liang, *J. Spacecraft Rockets* **29**, 150 (1992).
22. S. K. Rutledge, B. A. Banks and M. Kitral, *A Comparison of Space and Ground Based Facility Environmental Effects for FEP Teflon*, NASA Technical Memorandum 1998-207918/REV1 (NASA, Cleveland, OH, 1998).
23. H. Tahara, L. Zhang, M. Hiramatsu, T. Yasui, T. Yoshikawa, Y. Setsuhara and S. Miyake, *J. Appl. Phys.* **78** (1995).
24. M. Van Eesbeek, F. Levadou and A. Milintchouk, "A Study of FEP Behaviour in the Space Environment," *25th International Conference on Environmental Systems, held in San Diego, CA, 10–13 July 1995* (Society of Automotive Engineers, 1995) Paper No. 951640.
25. E. Grossman, Y. Noter and Y. Lifshitz, "Oxygen and VUV Irradiation of Polymers: Atomic Force Microscopy (AFM) and Complementary Studies," *7th International Symposium on Materials in Space Environment held in Tolouse, France, 16–20 June 1997*, Ed. T. D. Guyenne (European Space Agency, Paris, France, 1997) pp. 217–223.
26. S. L. Koontz, L. J. Leger, J. T. Visentine, D. E. Hunton, J. B. Cross and C. L. Hakes, "An Overview of the Evaluation of Oxygen Interactions with Materials III Experiment: Space Shuttle Mission 46, July–August 1992," *LDEF, 69 Months in Space, Third Post-Retrieval Symposium: Proceedings of a Symposium Sponsored by the National Aeronautics and Space Administration, Washington, D.C., and the American Institute of Aeronautics and Astronautics, Washington, D.C., and held in Williamsburg, VA, 8–12 November 1993*, NASA Conference Publication 3275; Part 3, Ed. A. S. Levine (NASA, Washington, D.C., 1993) pp. 869–902.
27. F. Levadou and M. Van Eesbeek, "Follow-up on the Effects of the Space Environment on UHCRE Thermal Blankets," *LDEF Materials Results for Spacecraft Applications: Proceedings of a Conference held in Huntsville, AL, 27–28 October 1992*, NASA Conference Publication 3257, Eds. A. F. Whitaker and J. Gregory (NASA, Washington, D.C., 1994) pp. 73–84.
28. B. D. Green, G. E. Caledonia and T. D. Wilkerson, *J. Spacecraft* **22**, 500 (1985).
29. J. B. Cross, J. A. Martin, L. E. Pope and S. L. Koontz, *Surface and Coating Technology* **42**, 41 (1990).
30. H. B. Garrett, A. Chutjian and S. Gabriel, *J. Spacecraft* **25**, 321 (1988).
31. U. von Zahn and E. Murad, *Nature* **321**, 147 (1986).
32. T. G. Slanger, *Geophys. Res. Lett.* **10**, 130 (1983).
33. J. W. Connell, J. V. Crivello and D. Bi, *J. Appl. Polymer Sci.* **57**, 1251 (1995).
34. I. Yilgor, E. Yilgor and M. Spinu, *Polymer Preprints* **28**, 84 (1987).
35. R. L. Kiefer, R. A. Orwoll, E. C. Aquino, A. C. Pierce, M. B. Glasgow and S. A. Thibeault, *Polymer Degradation and Stability* **57**, 219 (1997).
36. L. L. Fewell and L. Finney, *Polymer Comm.* **32**, 393 (1991).

37. J. G. Smith, Jr., J. W. Connell and P. M. Hergenrother, *Polymer* **35**, 2834 (1994).
38. J. W. Connell, J. G. Smith, Jr., C. G. Kalil and E. J. Siochi, *Polymers Adv. Tech.* **9**, 11 (1998).
39. E. H. Lee, M. B. Lewis, P. J. Blau and L. K. Mansur, *J. Mat. Res.* **6**, 610 (1991).
40. Z. Iskanderova, J. Kleiman, Y. Gudimenko, W. D. Morison and R. C. Tennyson, *Nucl. Instr. Meth. Phys. Res.* **B127/128**, 702 (1997).
41. B. A. Banks and C. LaMoreaux, "Performance and Properties of Atomic-Oxygen Protective Coatings for Polymeric Materials," *Advanced Materials — Meeting the Economic Challenge: Proceedings of the 24th International SAMPE Technical Conference held in Toronto, Canada, 20–22 October 1992*, Eds. T. J. Reinhart and F. H. Froes (SAMPE, Covina, CA, 1992) pp. T165–T173.
42. M. Hanichak and M. Finckenor, "Effects of Simulated Orbital Atomic-Oxygen on Germanium-Coated Kapton Flims," *Technology Transfer in a Global Community: Proceedings of the 28th International SAMPE Technical Conference held in Seattle, WA, 4–7 November 1996*, Ed. J. T. Hoggatt (SAMPE, Covina, CA, 1996) pp. 1148–1157.
43. R. Mutikainen, *Thin Solid Films* **238**, 248 (1994).
44. M. McCargo, R. A. Dammann, T. Cummings and C. Carpenter, "Laboratory Investigation of the Stability of Organic Coatings for Use in a LEO Environment," *Third Symposium on Spacecraft Materials in a Space Environment: Proceedings of a Symposium held at ESTEC, Noordwijk, The Netherlands, 1–4 October 1985*, ESA Special Publication 232 (European Space Agency, Paris, France, 1985) pp. 91–97.
45. R. C. Tennyson, *Surf. Coatings Tech.* **68/69**, 519 (1994).
46. Y. Gudimenko, Z. Iskanderova, J. Kleiman, G. Cool, D. Morison and R. Tennyson, "Erosion Protection of Polymer Materials in Space," *7th International Symposium on Materials in a Space Environment, held in Tolouse, France, 16–20 June 1997*, ESA Special Publication 399, Ed. T. D. Guyenne (European Space Agency, Paris, France, 1997) pp. 403–410.
47. J. T. Herron, *J. Phys. Chem. Ref. Data* **17**, 967 (1988).
48. *Combustion Chemistry*, Ed. W. C. Gardiner (Springer-Verlag, New York, 1984).
49. M. C. Lin, "Dynamics of Oxygen Atom Reactions", *Advances in Chemical Physics: Potential Energy Surfaces*, Ed. K. P. Lawley (John Wiley & Sons, Ltd., Chichester, Great Britain, 1980) pp. 113–167.
50. P. Andresen and A. C. Luntz, *J. Chem. Phys.* **72**, 5842 (1980).
51. N. J. Dutton, I. W. Fletcher and J. C. Whitehead, *Mol. Phys.* **52**, 475 (1984).
52. G. M. Sweeney, A. Watson and K. G. McKendrick, *J. Chem. Phys.* **106**, 9172 (1997).
53. H. Yamazaki and R. J. Cvetanović, *J. Chem. Phys.* **41**, 3703 (1964).

54. A. C. Luntz, *J. Chem. Phys.* **73**, 1143 (1980).

55. C. R. Park and J. R. Wiesenfeld, *J. Chem. Phys.* **95**, 8166 (1991).

56. Y. Rudich, Y. Hurwitz, G. J. Frost, V. Vaida and R. Naaman, *J. Chem. Phys.* **99**, 4500 (1993).

57. Y. Rudich, Y. Hurwitz, S. Lifson and R. Naaman, *J. Chem. Phys.* **98**, 2936 (1993).

58. R. J. Cvetanović, *J. Chem. Phys.* **23**, 1375 (1955).

59. R. J. Cvetanović, *Adv. Photochem.* **1**, 115 (1963).

60. R. J. Cvetanović, "Kinetics and Mechanisms of Some Atomic-Oxygen Reactions," *Proceedings of the NASA Workshop on Atomic-Oxygen Effects held in Pasadena, CA, 10–11 November 1986*, NASA Contractor Report 181163, JPL Publication 87-14, Ed. D. E. Brinza (NASA, Pasadena, CA, 1987) pp. 47–54.

61. R. A. Ferrieri and A. P. Wolf, *J. Phys. Chem.* **96**, 4747 (1992).

62. P. Patiño, F. E. Hernández and S. Rondón, *Plasma Chem. Plasma Process.* **15**, 159 (1995).

63. M. A. Golub, N. R. Lerner and T. Wydeven, *Polymer Preprints* **27**, 87 (1986).

64. W. H. Weinberg, "Kinetics of Surface Reactions", *Advances in Gas-Phase Photochemistry and Kinetics: Dynamics of Gas–Surface Interactions*, Eds. C. T. Rettner and M. N. R. Ashfold (The Royal Society of Chemistry, North Yorkshire, Great Britain, 1991) pp. 171–219.

65. C. T. Rettner and D. J. Auerbach, *Science* **263**, 365 (1994).

66. W. H. Weinberg, *Acc. Chem. Res.* **29**, 479 (1996).

67. D. J. Auerbach, in *Atomic and Molecular Beam Methods*, Chap. 14 of Vol. I and Chap. 15 of Vol. II, Ed. G. Scoles (Oxford University Press, New York, 1988 and 1992).

68. J. E. Hurst, C. A. Becker, J. P. Cowin, K. C. Janda, L. Wharton and D. J. Auerbach, *Phys. Rev. Lett.* **43**, 1175 (1979).

69. C. T. Rettner, E. K. Schweizer and C. B. Mullins, *J. Chem. Phys.* **90**, 3800 (1989).

70. C. T. Rettner, D. J. Auerbach, J. C. Tully and A. W. Kleyn, *J. Phys. Chem.* **100**, 13021 (1996).

71. *The Natural Space Environment: Effects on Spacecraft*, Eds. M. B. Alexander, B. F. James and O. W. Norton, NASA Reference Publication 1350 (NASA, Washington, D.C., 1994).

72. D. R. Peplinski, G. S. Arnold and E. N. Borson, "Satellite Exposure to Atomic Oxygen in Low Earth Orbit," *13th Space Simulation Conference, The Payload — Testing for Success: Proceedings of a Symposium held in Orlando, FL, 8–11 October 1984*, NASA Conference Publication 2340 (NASA, Washington, D.C., 1984) pp. 133–145.

73. A. C. Tribble, *The Space Environment: Implications for Spacecraft Design* (Princeton University Press, New Jersey, 1995).

74. *LDEF, 69 Months in Space, First Post-Retrieval Symposium, Kissimmee, FL, 2–8 June 1991*, NASA Conference Publication 3134, Ed. A. S. Levine (NASA, Washington, D.C., 1991).

75. P. N. Peters, R. C. Sisk and J. C. Gregory, *J. Spacecraft* **25**, 53 (1988).

76. D. Hastings and H. Garrett, *Spacecraft-Environment Interactions* (Cambridge University Press, United States, 1996).

77. R. H. Liang, A. Gupta, S. Y. Chung and K. L. Oda, *Mechanistic Studies of Polymeric Samples Exposed Aboard STS VIII*, NASA Contractor Report 182364, JPL Publication 87-25 (NASA, Pasadena, CA, 1987).

78. *Atomic-Oxygen Effects Measurements for Shuttle Missions STS-8 and 41-G*, NASA Technical Memorandum 100459, Ed. J. T. Visentine (NASA, Houston, TX, 1988).

79. S. L. Koontz, L. J. Leger, J. T. Visentine, D. E. Hunton, J. B. Cross and C. L. Hakes, *J. Spacecraft Rockets* **32**, 483 (1995).

80. S. L. Koontz, L. J. Leger, S. L. Rickman, C. L. Hakes, D. T. Bui, D. E. Hunton and J. B. Cross, *J. Spacecraft Rockets* **32**, 475 (1995).

81. L. J. Leger, "Oxygen Atom Reactions with Shuttle Materials at Orbital Altitudes — Data and Experimental Status," *AIAA 21st Aerospace Sciences Meeting held in Reno, NV, 10–13 January 1983*, (AIAA, New York, 1983) Paper No. 83-0073.

82. H. Hamacher, H. E. Richter and F. Joo, *Surface Erosion in Low Earth Orbit: Proceedings of the Norderney Symposium on Scientific Results on the German Spacelab Mission D2, held in Norderney, Germany, 14–16 March 1994* (Wissenschaftliche Projektfeuhrung DI c/o DFVLR, Keöln, Germany, 1995) pp. 159–170.

83. D. E. Brinza, S. Y. Chung, T. K. Minton and R. H. Liang, *Final Report on the NASA/JPL Evaluation of Oxygen Interactions with Materials — 3 (EOIM-3)*, NASA Contractor Report 198865, JPL Publication 94-31 (NASA, Pasadena, CA, 1994).

84. D. A. Jaworske, K. K. de Groh, T. J. Skowronski, T. McClooum, G. Pippin and C. Bungay, *Evaluation of Space Power Materials Flown on the Passive Optical Sample Assembly*, NASA Technical Memorandum 209061 (NASA, Washington, D.C., 1999).

85. J. M. Zwiener, R. K. Kamenetzky, J. A. Vaughn and M. M. Finckenor, "The Passive Optical Assembly (POSA-I) Experiment: First Flight Results and Conclusions," *37th AIAA Aerospace Sciences Meeting, held in Reno, NV, 11–14 January 1999* (AIAA, Reston, VA, 1999).

86. *LDEF Materials Results for Spacecraft Applications-Executive Summary: Proceedings of a Conference, held in Huntsville, Alabama, October 27–28*, NASA Conference Publication 3261, Eds. A. F. Whitaker and D. C. Dooling (NASA, Marshall Space Flight Center, AL, 1995).

87. P. R. Young, W. S. Slemp and E. J. Siochi, "Polymer Performance in Low Earth Orbit: A Comparison of Flight Results," *Moving Forward with 50*

Years of Leadership in Advanced Materials: Proceedings of the 39th International SAMPE Symposium, held in Anaheim, CA, 11–14 April 1994 (SAMPE, Covina, CA 1994).

88. J. B. Cross, S. L. Koontz and E. H. Lan, "Atomic-Oxygen Interactions with Spacecraft Materials: Relationship Between Orbital and Ground-Based Testing for Materials Certification," *5th International Symposium on Materials in a Space Environment, held in Cannes, France, 16–20 September 1991*, Ed. C. Salmon (Cepadues-Editions, Toulouse, France, 1992).

89. M. Tagawa, K. Yokota, N. Ohmae, H. Kinoshita, M. Umeno and K. Gotoh, "Atomic-Oxygen-Induced Erosion of the Polyimide Films Studied by QCM, AFM, XPS, and Contact Angle Measurements," *Needs and Trends in Advanced Materials for the 21st Century: Proceedings of the 6th Japan International SAMPE Symposium and Exhibition, held in Tokyo, Japan, 26–29 October 1999* (SAMPE, Covina, CA, 1999) p. 1147.

90. A. P. M. Glassford, R. A. Osiecki and C. K. Liu, *J. Vac. Sci. Tech.* **A2**, 1370 (1984).

91. A. P. M. Glassford and C. K. Liu, *J. Vac. Sci. Tech.* **17**, 696 (1980).

92. A. C. Tribble, B. Boyadjian, J. Davis, J. Haffner and E. McCullough, *Contamination Control Engineering Design Guidelines for the Aerospace Community*, NASA Contractor Report 4740 (NASA, Washington, D.C., 1996).

93. Y. Haruvy, *ESA J.* **14**, 109 (1990).

94. A. Chutjian and O. J. Orient, "Fast Beam Sources", *Atomic, Molecular, and Optical Physics: Atoms and Molecules*, Vol. 29B, Eds. F. B. Dunning and R. G. Hulet (Academic Press, New York, 1996) pp. 49–66.

95. D. Dooling and M. M. Finckenor, *Material Selection Guidelines to Limit Atomic-Oxygen Effects on Spacecraft Surfaces*, NASA Technical Publication 209260 (NASA, Marshall Space Flight Center, AL, 1999).

96. B. A. Banks, B. M. Auer, S. K. Rutledge, K. K. de Groh and L. Gerbauer, "The Use of Plasma Ashers and Monte Carlo Modeling for the Projection of Atomic-Oxygen Durability of Protected Polymers in Low Earth Orbit," *Terrestrial Test for Space Success: Proceedings of the 17th Space Simulation Conference, held in Baltimore, MD, 9–12 November 1992*, Ed. J. Stecher (NASA, Washington, D.C., 1992).

97. S. L. Koontz, K. Albyn and L. J. Leger, *J. Spacecraft* **28**, 315 (1991).

98. S. L. Koontz and P. Nordine, "The Reaction Efficiency of Thermal Energy Oxygen Atoms with Polymeric Materials", *Materials Degradation in Low Earth Orbit: Proceedings of a Symposium Sponsored by the TSM-ASM Joint Corrosion and Environmental Effects Committee, held at the 119th Annual Meeting of the Minerals, Metals and Materials Society in Anaheim, CA, 17–22 February 1990*, Eds. V. Srinivasan and B. A. Banks (Minerals, Metals and Materials Society, Warrendale, PA, 1990) p. 189.

99. E. J. H. Collart, J. A. G. Baggerman and R. J. Visser, *J. Apply, Phys.* **78**, 47 (1995).

100. R. A. Synowicki, J. S. Hale and J. A. Woollam, *J. Spacecraft Rockets* **30**, 116 (1993).

101. H. Tahara, T. Yasui, K. Onoe, Y. Tsubakishita and Y. Yoshikawa, *Japan. J. Appl. Phys.* **32**, 1298 (1993).

102. H. Tahara, K. Minami, T. Yasui, K. Onoe, Y. Tsubakishita and T. Yoshikawa, *Japan. J. Appl. Phys.* **32**, 1822 (1993).

103. B. A. Banks, S. K. Rutledge, K. K. de Groh, C. R. Stidham, L. Gerbauer and C. M. LaMoreaux, "Atomic-Oxygen Durability Evaluation of Protected Polymers Using Thermal Energy Plasma Systems," *Plasma Synthesis and Processing of Materials: Proceedings of a Symposium Sponsored by the Structural Materials Division of TMS, held during the 1993 TMS Annual Meeting in Denver, CO, 22–25 February 1993*, Materials Degradation in Low Earth Orbit (LEO), Ed. K. Upadhya (Minerals, Metals and Materials Society, Warrendale, PA, 1993) pp. 61–75.

104. G. E. Caledonia, "Laboratory Simulations of Energetic Atom Interactions Occurring in Low Earth Orbit," *Rarefied Gas Dynamics — Space-Related Studies: Proceedings of the 16th International Symposium on Rarefied Gas Dynamics, held in Pasadena, CA, 10–16 July 1988*, Progress in Astronautics and Aeronautics, Vol. 116, Eds. E. P. Muntz, D. P. Weaver and D. H. Campbell (AIAA, Washington, D.C., 1989) pp. 129–142.

105. G. E. Caledonia, R. H. Krech and B. D. Green, *AIAA J.* **25**, 59 (1987).

106. G. E. Caledonia, R. H. Krech, B. D. Green and A. N. Pirri, *Source of High Flux Energetic Atoms*, Physical Sciences, Inc., US Patent No. 4 894 511, 1990.

107. T. K. Minton, J. W. Seale, D. J. Garton and J. Zhang, "Mechanisms of Polymer Erosion in Low Earth Orbit: Implications for Ground-Based Atomic-Oxygen Testing," *Evolving and Revolutionary Technologies for the New Millennium: 44th International SAMPE Symposium and Exhibition, Long Beach, CA, 23–27 May 1999*, Science of Advanced Materials and Process Engineering Series, Vol. 44, Ed. L. J. Cohen (SAMPE, Covina, CA, 1999) pp. 1051–1063.

108. B. Cazaubon, A. Paillous and J. Siffre, *J. Spacecraft Rockets* **35**, 797 (1998).

109. M. Tagawa, H. Kinoshita, Y. Ninomiya, Y. Kurosumi, N. Ohmae and M. Umeno, "Volume Diffusion of Atomic Oxygen in a-SiO_2 Protective Coating at Various Temperatures," *Evolving and Revolutionary Technologies for the New Millennium: 44th International SAMPE Symposium and Exhibition, Long Beach, CA, 23–27 May 1999*, Science of Advanced Materials and Process Engineering Series, Vol. 44, Ed. L. J. Cohen (SAMPE, Covina, CA, 1999) pp. 402–411.

110. M. Alagia, V. Aquilanti, D. Ascenzi, N. Balucani, D. Cappelletti, L. Cartechini, P. Casavecchia, F. Pirani, G. Sanchini and G. G. Volpi, *Israel. J. Chem.* **37**, 329 (1997).

111. S. J. Sibener, R. J. Buss, C. Y. Ng and Y. T. Lee, *Rev. Sci Instr.* **51**, 167 (1980).

112. J. B. Cross and N. C. Blais, *Progress in Astronautics and Aeronautics* **116**, 143 (1989).

113. R. C. Tennyson and W. D. Morison, "Atomic-Oxygen Effects on Spacecraft Materials," *Materials Degradation in Low Earth Orbit: Proceedings of a Symposium Sponsored by the TSM-ASM Joint Corrosion and Environmental Effects Committee, held at the 119th Annual Meeting of the Minerals, Metals and Materials Society in Anaheim, CA, 17–22 February 1990*, Eds. V. Srinivasan and B. A. Banks (Minerals, Metals and Materials Society, Warrendale, PA, 1990) pp. 59–75.

114. H. G. Pippin and R. Curruth, "Materials Screening Chamber for Testing Materials Resistant to Atomic-Oxygen," *34th International SAMPE Symposium, held in Reno, NV, 8–11 May 1989*, Ed. G. A. Zakrzewski (SAMPE, Covina, CA, 1989) pp. 2101–2106.

115. A. P. Nikiforov and V. E. Skurat, *Chem. Phys. Lett.* **212**, 43 (1993).

116. J. E. Pollard, *Rev. Sci. Instr.* **63**, 1771 (1992).

117. G. S. Arnold, D. R. Peplinski and F. M. Cascarano, *J. Spacecraft* **24**, 454 (1987).

118. O. J. Orient, A. Chutjian and E. Murad, "Experimental Investigations of Low-Energy (4–40 eV) Collisions of $O(^2P)$ Ions and $O(^3P)$ Atoms with Surfaces," *Materials Degradation in Low Earth Orbit: Proceedings of a Symposium by the TSM-ASM Joint Corrosion and Environmental Effects Committee, held at the 119th Annual Meeting of the Minerals, Metals and Materials Society in Anaheim, CA, 17–22 February 1990*, Eds. V. Srinivasan and B. A. Banks (Minerals, Metals and Materials Society, Warrendale, PA, 1990) pp. 87–95.

119. O. J. Orient, A. Chutjian and E. Murad, *Phys. Rev.* **A41**, 4106 (1990).

120. J. V. Vaughn, R. Kamenetzky, M. Finckenor and D. Edwards, "Space Environmental Effects Testing Capabilities at MSFC," *4th Annual Workshop on Space Operations Applications and Research (SOAR '90): Proceedings of a Workshop Sponsored by the National Aeronautics and Space Administration, Washington, D.C., the U.S. Air Force, Washington, D.C., and Cosponsored by the University of New Mexico, Albuquerque, NM, held in Albuquerque, NM, 26–28 June 1990*, NASA Conference Publication 3103, Vol. 2, Ed. R. T. Savely (NASA, Albuquerque, NM, 1996) pp. 734–741.

121. M. Tagawa, M. Tomita, M. Umeno and N. Ohmae, *AIAA J.* **32**, 95 (1994).

122. G. B. Hoflund and J. F. Weaver, *Meas. Sci. Tech.* **5**, 201 (1994).

123. Y. Kinugawa, T. Sato and T. E. A. Arikawa, *J. Chem. Phys.* **93**, 3289 (1990).

124. D. F. Plusquellic, M. Casassa and J. Stephenson, *Rev. Sci. Instr.* (2000).

125. V. E. Skurat, E. A. Barbashev, Y. I. Dorofeev, A. P. Nikiforov, M. M. Gorelova and A. I. Pertsyn, *Appl. Surf. Sci.* **92**, 441 (1996).

126. A. F. Whitaker and B. Z. Jang, *J. Appl. Polymer Sci.* **48**, 1341 (1993).

127. R. C. Tennyson, "Atomic-Oxygen and its Effects on Materials in Space," *Rarefied Gas Dynamics: Technical Papers from the Proceedings of the 18th International Symposium on Rarefied Gas Dynamics, held in Vancouver, B.C., Canada, 26-30 July 1992*, Progress in Astronautics and Aeronautics, Eds. B. D. Shizgal and D. P. Weaver (AIAA, Washington, D.C., 1992) pp. 461-477.

128. E. B. D. Bourdon, R. H. Prince, W. D. Morison and R. C. Tennyson, *Surf. Coating Tech.* **52**, 51 (1992).

129. W. R. Henderson, *J. Geophys. Res.* **79**, 3819 (1974).

130. J. C. Gregory, G. P. Miller, P. J. Pettigrew, G. N. Raikar, J. B Cross, E. Lan, C. L. Renschler and W. T. Sutherland, "Atomic-Oxygen Dosimetry Measurements Made on STS-46 by CONCAP-II," *LDEF, 69 Months in Space, Third Post-Retrieval Symposium: Proceedings of a Symposium Sponsored by the National Aeronautics and Space Administration, Washington, D.C., and the American Institute of Aeronautics and Astronautics, Washington, D.C., held in Williamsburg, VA, 8-12 November 1993*, NASA Conference Publication; 3275, Part 3, Ed. A. S. Levine (NASA, Washington, D.C., 1993) pp. 957-964.

131. M. R. Curruth, R. F. DeHaye, J. K. Norwood and A. F. Whitaker *Rev. Sci. Inst.* **61**, 1211 (1990).

132. *LDEF Materials Results for Spacecraft Applications: Proceeding of a Conference held in Huntsville, AL, 27-28 October 1992*, NASA Conference Publication 3257, Eds. A. F. Whitaker and J. Gregory (NASA, Washington, D.C., 1994).

133. M. R. Reddy, *J. Mat. Sci.* **30**, 281 (1995).

134. H. Kinoshita, M. Tagawa, M. Umeno and N. Ohmae, *Trans. Japan Soc. Aero. Space Sci.* **41**, 94 (1998).

135. K. W. Holtzclaw, M. E. Fraser and A. Gelb, *J. Geophys. Res.* **95**, 4147 (1990).

136. J. Zhang, D. J. Garton, J. W. Seale, A. K. Frandsen and T. K. Minton, submitted to *J. Phys. Chem.* **B** (2000).

137. T. K. Minton, J. Zhang, D. J. Garton and J. W. Seale, *High Performance Polymers* **12**, 27-42 (2000).

138. D. J. Garton, T. K. Minton, N. Balucani, P. Casavecchia and G. G. Volpi, *J. Chem. Phys.* **112**, 5975-84 (2000).

139. T. K. Minton, J. W. Seale, D. J. Garton and A. K. Frandsen, "Dynamics of Atomic-Oxygen-Induced Degradation of Materials," *Proceedings of the 4th International Conference on the Protection of Materials and Structures from Low Earth Orbital Space Environment, held in Toronto, Ontario, 23-24 April 1998* (Kluwer Academic Publishers, Boston, 2000).

140. D. J. Garton, T. K. Minton, M. Alagia, N. Balucani, P. Casavecchia and G. G. Volpi, *Faraday Discuss.* **108**, 387 (1997).

141. D. J. Garton, T. K. Minton, M. Alagia, N. Balucani, P. Casavecchia and G. G. Volpi, "Atomic-Oxygen Interactions with Saturated Hydrocarbon Surfaces," *A Bound Collection of Papers: AIAA Defense & Space Programs Conference & Exhibit, held in Huntsville, AL, 23–25 September 1997* (AIAA, Reston, VA, 1997) Paper No. 97-3947.

142. Y. T. Lee, J. D. McDonald, P. R. LeBreton and D. R. Herschbach, *Rev. Sci. Instr.* **40**, 1402 (1969).

143. R. K. Sparks, Ph.D. Thesis, University of California, Berkeley, 1980.

144. M. J. O'Laughlin, B. P. Ried and R. K. Sparks, *J. Chem. Phys.* **83**, 5647 (1985).

145. M. E. Saecker, S. T. Govoni, D. V. Kowalski, M. E. King and G. M. Nathanson, *Science* **252**, 1421 (1991).

146. R. H. Hansen, J. V. Pascale, T. DeBenedicts and P. M. Rentzepis, *J. Polymer Sci.* **A3**, 2205 (1965).

147. LDEF, *69 Months in Space, Third Post-Retrieval Symposium: Proceedings of a Symposium Sponsored by the National Aeronautics and Space Administration, Washington, D.C., and the American Institute of Aeronautics and Astronautics, Washington, D.C., held in Williamsburg, VA, 8–12 November 1993*, NASA Conference Publication; 3275, Part 3, Ed. A. S. Levine (NASA, Washington, D.C., 1993).

148. R. H. Krech, "Determination of Oxygen Erosion Yield Dependencies Upon Specific LEO Environments", NASA Contractor Report PSI-1104/TR-1176 (NASA Lewis, Cleveland, OH, 1993).

149. J. B. Cross, S. L. Koontz, J. C. Gregory and M. J. Edgell, "Hyperthermal Atomic-Oxygen Reactions with Kapton and Polyethylene," *Materials Degradation in Low Earth Orbit: Proceedings of a Symposium Sponsored by the TSM-ASM Joint Corrosion and Environmental Effects Committee held at the 119th Annual Meeting of the Minerals, Metals and Materials Society in Anaheim, CA, 17–22 February 1990*, Eds. V. Srinivasan and B. A. Banks (Minerals, Metals and Materials Society, Warrendale, PA, 1990).

150. J. C. Gregory and P. N. Peters, "The Reaction of 5 eV Oxygen Atoms with Polymeric and Carbon Surfaces in Earth Orbit", *Atomic-Oxygen Effects Measurements for Shuttle Missions STS-8 and 41-G*, NASA Technical Memorandum 100459, Vol. 2, Ed. J. Visentine (NASA, Houston, TX, 1988) pp. 4.1–4.5.

151. V. E. Skurat, "Evaluation of Reaction Efficiencies of Polymeric Materials in Their Interaction with Fast (5 eV) Atomic-Oxygen", *7th International Symposium on Material in the Space Environment, held in Toulouse, France, 16–20 June 1997*, Enseignment et espace (Cepadues-Editions, Toulouse, France, 1997) pp. 231–235.

152. P. N. Peters, J. C. Gregory and J. T. Swann, *Appl. Opt.* **25**, 1290 (1986).

153. M. J. Meshishnek, W. K. Stuckey, J. S. Evangelides, L. A. Feldman, R. V. Peterson, G. A. Arnold and D. R. Peplinski, *Effects on Advanced*

Materials: Results of the STS-8 EOIM Experiment, Report No. SD-TR-87-34 (Aerospace Corporation, Los Angeles, CA, 1987).

154. Z. A. Iskanderova, J. I. Kleiman, Y. Gudimenko and R. C. Tennyson, *J. Spacecraft Rockets* **32**, 878 (1995).

155. J. I. Kleiman, Z. A. Iskanderova, Y. I. Gudimenko, V. Lemberg, D. Talas and R. C. Tennyson, "Predictive Models of Erosion Processes in LEO Space Environment: A Basis for Development of an Engineering Software", in *Computer Modeling of Electronic and Atomic Processes in Solids*, Eds. R. C. Tennyson and A. E. Kiv (Kluwer Academic Publishers, Boston, 1997) pp. 277–287.

156. H. G. Pippin, *Surf. Coating Tech.* **39/40**, 595 (1989).

157. F. Garbassi, M. Morra and E. Occhiello, *Polymer Surfaces: From Physics to Technology* (John Wiley & Sons, New York, 1994).

158. C.-M. Chan, T.-M. Ko and H. Hiraoka, *Surf. Sci. Rep.* **24**, 1 (1996).

159. F. D. Egitto and L. J. Matlenzo, *IBM J. Res. Develop.* **38**, 423 (1994).

160. E. M. Liston, *J. Adhesion* **30**, 199 (1989).

161. M. A. Hartney, D. W. Hess and D. S. Soane, *J. Vac. Sci. Techn.* **B7**, 1 (1989).

162. O. Joubert, J. Pelletier, C. Fiori and T. A. Nguyen Tan, *J. Appl. Phys.* **67**, 4291 (1990).

163. A. S. Hoffman, *J. Appl. Polymer Sci. Appl. Polymer Symp.* **42**, 251 (1988).

164. G. S. Selwyn, *J. Appl. Phys.* **60**, 2771 (1986).

165. P. C. Schamberger, J. I. Abes and J. A. Gardella, *Colloids Surf.* **B3**, 203 (1994).

166. T. L. Thompson, M. S. Thesis, Montana State University, 1997.

167. R. H. Liang, A. Gupta, S. Y. Chung and K. L. Oda, "Mechanistic Studies of Polymeric Materials Samples Exposed Aboard STS-8", in *Atomic-Oxygen Effects Measurements for Shuttle Missions STS-8 and 41-G*, NASA Technical Memorandum; 100459, Ed. J. T. Visentine (NASA, Houston, TX, 1988) p. 6.1–6.17.

168. T. L. Thompson, B. J. Tyler and T. K. Minton, *Polymer Preprints* **38**, 1063 (1997).

169. M. A. Golub, T. Wydeven and R. D. Cormia, *Polymer Comm.* **29**, 285 (1988).

170. M. Tagawa, M. Matsushita, M. Umeno and N. Ohmae, "Laboratory Studies of Atomic-Oxygen Reactions on Spin-Coated Polyimide Films," *Proceedings of the 6th International Symposium on Materials in a Space Environment, Noordwijk, The Netherlands, 19–23 September 1994*, ESA Special Publication 368, Ed. T. D. Guyenne (European Space Agency, Paris, France, 1994) p. 189.

171. M. A. Golub, "Reactions of Atomic Oxygen [O(^3P)] with Polymer Films," *Invited Lectures Presented at Polymer 91, International Symposium on Polymer Materials — Preparation, Characterization, and Properties, held in*

Melborne, Australia, 10–15 February 1991, Makromolekulare Chemie. Macromolecular Symposia, Vol. 53 (Heuthig and Wepf, New York, 1992) pp. 379–391.

172. J. G. Harris, *J. Phys. Chem.* **96**, 5077 (1992).
173. J. Harris, "Mechanical Energy Transfer in Particle-Surface Collisions", in *Dynamics of Gas Surface Interactions*, Advances in Gas-Phase Photochemistry and Kinetics, Eds. C. T. Rettner and M. N. R. Ashfold (The Royal Society of Chemistry, Cambridge, United Kingdom, 1991) pp. 1–46.
174. J. I. Steinfeld, J. S. Francisco and W. L. Hase, *Chemical Kinetics and Dynamics*, 2nd edition (Prentice Hall, New Jersey, 1998).
175. M. E. King, G. M. Nathanson, M. A. Hanning-Lee and T. K. Minton, *Phys. Rev. Lett.* **70**, 1026 (1993).
176. E. K. Grimmelmann, J. C. Tully and M. J. Cardillo, *J. Chem. Phys.* **72**, 1039 (1980).
177. S. Cohen, R. Naaman and J. Sagiv, *Phys. Rev. Lett.* **58**, 1208 (1987).
178. C. T. Rettner, J. A. Barker and D. S. Bethune, *Phys. Rev. Lett.* **67**, 2183 (1991).
179. A. Amirav, M. J. Cardillo, P. L. Trevor, C. Lim and J. C. Tully, *J. Chem. Phys.* **87**, 1796 (1987).
180. T. K. Minton, K. P. Giapis and T. A. Moore, *J. Phys. Chem.* **A101**, 6549 (1997).
181. T. K. Minton and T. A. Moore, "Molecular Beam Scattering from ^{13}C-Enriched Kapton and Correlation with EOIM-3 Carousel Experiment," *LDEF, 69 Months in Space, Third Post-Retrieval Symposium: Proceedings of a Symposium Sponsored by the National Aeronautics and Space Administration, Washington, D.C., and the American Institute of Aeronautics and Astronautics, Washington, D.C., held in Williamsburg, VA, 8–12 November 1993*, NASA Conference Publication; 3275, Part 3, Ed. A. S. Levine (NASA, Washington, D.C., 1993) pp. 1095–1114.
182. M. E. Saecker and G. M. Nathanson, *J. Phys. Chem.* **99**, 7056 (1993).
183. H. Kinoshita, M. Tagawa, M. Umeno and N. Ohmae, *Surf. Sci.* **440**, 49 (1999).
184. J. K. Baird, *J. Spacecraft Rockets* **35**, 62 (1998).
185. G. S. Hwang, C. M. Anderson, M. J. Gordon, T. A. Moore, T. K. Minton and K. P. Giapis, *Phys. Rev. Lett.* **77**, 3049 (1996).
186. T. K. Minton, *Protocol for Atomic-Oxygen Testing of Materials in Ground-Based Facilities*, NASA Contractor Report 95-112589, JPL Publication 95-17 (NASA, Pasadena, CA, 1995).

CHAPTER 10

ATOMIC-LEVEL PROPERTIES OF THERMAL BARRIER COATINGS: CHARACTERIZATION OF METAL–CERAMIC INTERFACES

Asbjorn Christensen,* Emily A. A. Jarvis and Emily A. Carter[†]

Department of Chemistry and Biochemistry, Box 951569
University of California, Los Angeles, CA 90095-1569, USA

Contents

---

*Present address: Physics Department, Building 307, Technical University of Denmark, DK-2800 Lyngby, Denmark.
[†]To whom correspondence should be addressed.

1. Introduction and Chapter Overview

This chapter considers basic research related to the extreme environment of an aircraft engine and the use of Thermal Barrier Coatings (TBC's) to ameliorate the effects of extreme temperature cycling on metal engine components. The failure of these TBC's is a serious technological problem; one that, if solved, should greatly increase the fuel efficiency and operating lifetimes of airplane engines. These TBC's are comprised of ceramics, with favorably low thermal conductivity, deposited on the engine metals. Accordingly, we are concerned with the characterization of metal–ceramic (M/C) interfaces at a fundamental level. In this chapter, we attempt to provide an overview of experimental techniques for characterizing M/C interfaces. However, since we are theorists, much of the review is focused on providing a detailed, critical analysis of theoretical methods in use today to study such systems. We also give examples from our own modeling at the atomic level that has yielded some insights into the interfacial behavior of TBC's.

Let us outline the problem caused by the extreme combustion environment. The ideal engine would operate at very high temperatures without failure in order to have the highest fuel efficiency. Typically, the combustion gas is held at temperatures above 1370°C, while the metal superalloys that constitute the engine components have melting points ranging from 1230–1315°C! This makes it imperative to either cool the metal components (e.g. by drilling holes and flowing cool air) or to provide thermal protection from the combustion gas.[1] Ideas for optimizing the cooling techniques and engine metal alloy compositions reached a point of diminishing returns at least 10 years ago.[2] Hence, engineers looked to ceramic materials as a means of providing a thermal barrier coating that will: (1) extend the life of gas turbine components, (2) reduce cooling requirements (thereby decreasing fuel consumption), or (3) allow for an increase in gas inlet tempratures (thereby increasing thrust).[3]

Ceramics are thermal insulators that can provide the thermal barrier desired. The challenge of working with such ceramics is that typically they have completely different thermochemical properties from the material to which they are expected to adhere, namely a metallic alloy. As a result, the usual thermal cycling that such M–C interfaces undergo tends to stress these interfaces to the point of fracture and spallation (chipping-off).[1] It has been suggested that research is needed to try to connect these macroscopic phenomena to microscopic properties, in order to make progress in understanding how to design the best M/C junction.[4,5] In particular, the exact nature of the interface at the atomic level is poorly characterized; the exact mechanisms have yet to be identified by which the interface is formed, stressed, fractured, and spalled. Furthermore, it remains unclear what roles oxidation and temperature play in stabilizing or destabilizing the interface. Gaining an atomic-level understanding of these mechanisms should help elucidate ways to optimize the M/C couple: simultaneously maximizing thermal insulation and minimizing spallation.

This chapter initially explores the TBC from an experimental point of view. In Sec. 2 of this chapter, we present a detailed description of the TBC structure and composition. We explore the typical chemical make-up of a TBC, the formation of the thermally grown oxide, and the physical methods used in TBC fabrication. Section 3 lists a number of the experimental techniques used to characterize functional TBC's and explains some experimental characterization of ideal interfaces. Section 4 investigates the problem of TBC spallation. This includes identifying likely perpetrators of spallation, describing proposed spallation mechanisms, looking at some features that complicate spallation studies, and expounding on the issue of adhesion. We also mention some of the techniques used for measuring stress, fracture, and spallation at the end of this section. After looking at

TBC's from an experimental standpoint, we turn our discussion to present applicable theory. We describe atomistic modeling approaches in Sec. 6 and how those can be applied to M/C interfaces. This includes exploring the use of cluster and slab models, the idea of interface stoichiometry, and the problem of lattice misfit. We then consider the available *ab initio* techniques for modeling M/C interfaces. This is broken down to more specific methods including quantum chemical approaches, density functional theory, the approximation of the Harris functional, and tight binding schemes. We also present results obtained from a number of density functional theory studies, including our own, as well as tight binding predictions. Sections 7 and 8 review theory designed to handle larger systems than are computationally feasible with the *ab initio* techniques. Finally, we offer brief conclusions that attempt to look to the future for ways of enhancing understanding and optimization of thermal barrier coatings.

2. Creating a TBC System

2.1. *Commonly Used Materials*

When creating a TBC, generally a top coat and a bond coat layer must be deposited. The top coat serves as the insulator and the bond coat mediates contact between the top coat and metal–alloy substrate. The nickel-based "superalloy" substrate is a real pot pourri of elements, consisting of Ni, Co, Cr, Mo, Al, Ta, Ti, C, Zr, and B (where the non-Ni elements are present at the few to hundredths a weight percent level.)[6] Yttria-Stabilized Zirconia (YSZ) is a favored top coat material. Pure zirconia has relatively high strength, wear resistance, and fracture toughness. Likewise, it exhibits an extremely low thermal conductivity. In fact, excluding Pyrex glass, the thermal conductivity of zirconia is lower than any other engineering ceramic by over an order of magnitude.[6] In practical terms, this property allows even a thin zirconia film (less than one millimeter thickness) to potentially reduce the temperature of the underlying alloy several hundred degrees Celsius.[7] Furthermore, the linear thermal expansion coefficient and elastic modulus of zirconia (especially the tetragonal phase) are well-matched to several popular nickel-based superalloys, compared to possible alternative ceramics. These properties are essential to coating survival during thermal cycling.[7–9] Unfortunately, zirconia also exhibits polymorphism as a function of temperature, primarily between the monoclinic, tetragonal, and cubic phases. Although the polymorphism can be exploited to inhibit crack propagation via volume changes that occur upon transformation,[7,10,11] unregulated polymorphism can be detrimental to TBC functionality. As a result, other cubic oxides are added to control or eliminate polymorphism. Partial stabilization of the tetragonal phase across the relevant temperature range affords some phase control yet preserves the valuable inhibition of stress-induced micro-crack propagation. Greater than 8.5 mole percent Y_2O_3 dopant produces fully stabilized cubic zirconia while a 2–8.5 percent Y_2O_3 concentration creates a partially stabilized (tetragonal) form;[12] similarly, adding CeO_2, CaO, or MgO can generate stabilized zirconia.[13–16] Conversely, using TiO_2 as the stabilizing oxide is less effective than the other oxides studied.[17,18] However, TiO_2 added to Y_2O_3 or CeO_2 stabilized ZrO_2 has been shown to produce zirconia polycrystals with favorable properties.[19,20] Although a percentage of the partially stabilized ZrO_2 tends to remain in the monoclinic phase throughout thermal cycling, the initially formed cubic phase converts to tetragonal after thermal treatment in air.[21,22] Despite the favorable properties of a YSZ top coat, the difference in thermomechanical properties between a YSZ top coat and a metal–alloy substrate is enough to require the introduction of an intermediate layer. This bond coat is important for adhesion and grading the thermal expansion mismatch between the top coat and substrate. A typical bond coat contains nickel, chromium, aluminum, and yttrium, with nickel as the primary element for nickel–alloy substrate applications. With other alloys, it is sometimes desirable to use iron or cobalt in place of the nickel in the bond coat.[23,24]

2.2. *Formation of the Thermally Grown Oxide*

A third layer present in a TBC is the Thermally Grown Oxide (TGO). Bond coat oxidation can reduce TBC adhesion to the substrate. In most cases, the oxidation products begin to form even prior to top coat deposition. It is standard practice, in zirconia film creation chambers, to supply oxygen to ensure stoichiometry of the zirconia layer. Moreover, the bond coat continues to be oxidized once the zirconia layer is in place, since zirconia readily conducts oxygen ions at high temperature.[25] Accordingly, the TGO layer between the bond coat and top coat thickens with thermal cycling. Although this layer is generally thin compared to the TBC layer, it can lead to large stresses in the system due to significant thermal expansion mismatch of the TGO and bond coat.

The growth of the TGO falls into two primary regimes. Fast initial growth of nonprotective oxides is followed by continuous protective scale growth, largely controlled by diffusion. The primary protective scale formed with bond coat oxidation is alumina, Al_2O_3. For a bond coat composed of NiCrAlY, the oxidation also produces $Ni(Cr,Al)_2O_4$ spinels, Y_2O_3, NiO, $AlYO_3$, and $Al_5Y_3O_{12}$ bands oriented perpendicularly to the bond coat.[23] In the initial oxidation stages, X-ray diffraction of a FeCrAlY bond coat indicate that both $FeCr_2O_4$ and Cr_2O_3 form in that system.[24]

Oxide scale studies are complicated further by the coexistence of various oxide phases. For instance, at early times, both Θ-alumina, a transient phase formed when γ-alumina is heated, and α-alumina are present. Amorphous alumina transforms through a series of metastable phases until the more stable α-alumina dominates at long oxidation times.[26] Although thicknesses of the layers vary, some fairly typical values are 250–500 μm for the top coat, 100–150 μm for the bond coat (this varies since some bond coats are designed for surface roughness and with several gradation layers), and about an order of magnitude less than these for the thermally grown oxide.[23] Figure 1 displays a schematic cross-section of a TBC.

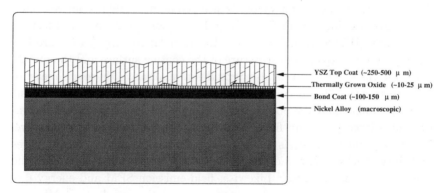

Fig. 1. Cross-section of a TBC (not pictured to scale).

2.3. *Depositing the TBC and Bond Coat*

Several methods effectively apply thin coatings to substrates. One process commonly employed for TBC fabrication is thermal plasma spraying. For this technique, the bond coat is deposited with low-pressure plasma spraying, and the ZrO_2 top coat is then created using atmospheric plasma spraying.[27] A potential problem with this method is coating fracture due to weak interlamellar adhesion, resulting from discontinuities within the solidifying process.[28] Accordingly, plasma-sprayed coatings tend to fail within the TBC itself. Alternatively, electron-beam physical-vapor deposition can be used for TBC fabrication. One drawback of this method is possible spallation, via separation at the bond coat interface during cooling.[26] Nevertheless, successful coatings of this type possess columnar micro-structure, supplying some of the stress compliance required by the TBC. Furthermore, this method is suited for creating a graded Al_2O_3–ZrO_2 layer. It is hoped that a gradual transition from the bond coat oxide to the top coat would improve the life of the TBC;[29] however, more work is needed to understand the practical effectiveness of such a layer.[30] Sputter deposition is another useful means for producing thin coatings.[31,32] Each deposition method has its own set of pros and cons, hence this variety in fabrication techniques continues.

3. Characterizing the TBC

3.1. *Characterization Techniques*

Advanced characterization methods are required to gain a microscopic-level understanding of TBC's. High-Resolution Transmission Electron Microscopy (HRTEM) and Scanning Electron Microscopy (SEM) are frequently used to explore the structure of the coating surface and the metal–oxide interface.[22,33,34] SEM also provides a qualitative thickness measure of oxide growth. The coating's porosity can be measured by image analysis techniques and mercury porosimetry. Use of the latter can pose difficulties since large pores are filled with mercury prior to applying pressure. Conversely, image analysis is ineffective for measuring very small pores and microcracks.[35] X-ray diffraction provides a means to determine the oxide species and phases.[22] Likewise, field-emission SEM and energy dispersive spectroscopy aid in characterizing oxidation products.[23] Microindentation tests determine the elastic modulus of the prepared coatings; however, due to inherent inhomogeneity in the coatings, results from this method must be appropriately averaged.[35] Thermal conductivity measures have been provided by laser flash methods that measure diffusivity and specific heat measurements from differential scanning calorimetry.[36,37] Alternatively, diffusivity values can be obtained with a multiproperty apparatus that measures temperature gradients, sample geometry, and heat flux.[38] The single-wavelength pyrometer method gathers spectral effective emissivities. Although thermocouples could also be used, the single-wavelength pyrometer technique has the superior attributes of not requiring surface contact during temperature measurements, immunity to electromagnetic interference from the surroundings, and no system perturbation during measurements.[39] The use of Auger Electron Spectroscopy (AES), X-ray photoelectron spectroscopy (XPS), and Auger parameter (α') analysis provide means to explore the nickel–alumina interface formation. Furthermore, these techniques are suited for determining ionicity and growth mechanisms at metal–oxide interfaces.[40]

3.2. *Characterization of Ideal Interfaces*

At high temperatures, interdiffusion leads to the formation of new phases at the M/C interface.[4] For example, Qin and Derby[41] used optical and electron microscopy to characterize the strength of Ni/ZrO_2 and $Ni/NiO/ZrO_2$ interfaces. They found that the strongest interfaces were formed by annealing Ni/ZrO_2 in air, which allowed a thin layer of NiO to form that helped adhesion to the ceramic. Qin and Derby also studied the formation of an interface between ZrO_2 and a Ni(Cr) alloy, where a mixed oxide with a spinel structure was formed at the interface.[42] The reaction at the interface appeared to be accompanied by local melting after interdiffusion, although it is not clear which elements were actually diffusing. Wagner *et al.*[43] examined charge flow in a $Ni–ZrO_2–Zr$ cell and showed that at 1273 K with no applied voltage, oxygen ions move from the Ni to the Zr electrode, forming Ni_5Zr and Ni_7Zr_2 at the Ni electrode while reducing ZrO_2 to Zr. At the Zr electrode, monoclinic ZrO_2 forms. Thus, oxygen–ion diffusion appears to be an important process by which adhesion takes place.

As mentioned earlier, Al_2O_3 (TGO) also forms at the TBC-bond coat–superalloy junction. Trumble and Rühle[44] showed by Transmission Electron Microscopy (TEM) that at 1390°C, pure Ni does not form a spinel ($NiAl_2O_4$) at the interface between Ni and Al_2O_3. However, even 0.07 percent oxygen in the Ni will induce formation of a thin spinel layer at the interface (with no NiO intermediate required), where the kinetics appear to be controlled by oxygen diffusion. Shear strength measurements by Loh *et al.*[45] determined that, under conditions where the spinel $NiAl_2O_4$ is formed, fracture occurs along the spinel-Ni interface. It is still controversial as to whether formation of the spinel compound at these interfaces actually helps or hinders adhesion. Zhong and Ohuchi,[46] using X-ray photoelectron spectroscopy, and later Brydson *et al.*,[5] using spatially resolved transmission Electron Energy Loss Spectroscopy (EELS) and High-Resolution Electron Microscopy (HREM), determined that the interface between Ni and Al_2O_3 forms direct Ni–Al bonds under reducing conditions at high temperatures. They presented evidence for a Ni_3Al phase at the interface, which they suggested was formed by Al diffusion into the Ni, and that this phase provides a driving force for the formation of Ni–Al bonds rather than Ni–O bonds. They suggested that formation of any spinel phase containing Ni–O bonds needs to be minimized because the spinel is brittle, an exactly opposite conclusion to that reached by Qin and Derby for the Ni–ZrO_2 interface.

In addition to M/C interactions, characterization of ZrO_2–Al_2O_3 is interesting as well, because of the Al_2O_3 that forms between the bond coat and the TBC. Aita and coworkers have grown and characterized with HREM nanolaminate films that alternate between polycrystalline ZrO_2 and Al_2O_3 layers. They have shown that ZrO_2 grows with the close-packed tetragonal(111) or monoclinic(11$\bar{1}$) surfaces parallel to the substrate.[47] Interestingly, we recently calculated those surfaces to be the most stable for each of these two phases.[48] They also observed that the amount of t-ZrO_2 increases with decreasing ZrO_2 film thickness and that the crystallites grow in this tetragonal phase up to a critical thickness of about 6.0 nm, at which point additional ZrO_2 converts the crystallite into the monoclinic phase.[49] In other work, they showed that the stress-induced transformation of tetragonal to monoclinic zirconia is limited to nanometer-scale regions in these nanolaminates.[50] In Secs. 6.2.2.2, we discuss our own theoretical rationalization of these observations. The role of alumina seems merely to confine the size of ZrO_2 crystallites formed. Our own calculations on ZrO_2–Al_2O_3 (discussed in Secs. 6.2.2.2) are consistent with these observations.

4. TBC Failure

Differences in crystal structure, size, and thermal expansion coefficients between the top coat, bond coat, and substrate introduce strain into the TBC. The thermal expansion coefficient of the top coat is generally lower than that of the bond coat or the alloy substrate. As mentioned earlier, a properly designed bond coat serves as a graded thermal expansion layer to reduce the strain caused by thermal expansion mismatch. Unfortunately, thermal cycling in air enhances the mismatch and increases strain resulting from interfacial damage during oxidation.[24] With repeated cycling, this strain contributes to coating failure.

4.1. Likely Culprits

TBC failure involves a number of contributing factors. Obviously, the strains introduced with thermal cycling are a major area of concern. The buildup of the TGO, migration and segregation of the components, and phase transitions of the TGO and top coat may each contribute to the coating failure. Even the type of porosity in the coating, largely dependent on the coating procedure used, can affect the coating lifetime. A great deal of experimental work connected with TBC's exists in the literature. Ernst published an excellent review that covers many experiments performed before 1995.[4] Due to inherent complexities of TBC's, studies to elucidate the failure mechanisms generally concentrate on only certain aspects contributing to failure. The subtle relationships of all the factors are not fully understood; hence, further modeling and experiments are needed to gain a more thorough understanding of present weaknesses and possible future improvements.

4.2. Spallation Mechanisms

Spallation is the process by which the TBC peels off the substrate; and naturally, after the coating has spalled, the continuous thermal protection layer no longer exists. Likewise, the spalled fragments can block gas flow, contaminate products, and permit corrosion. The failure of the TBC leading to spallation appears closely related to the damage process within the TGO layer. Unfortunately, due to the highly complex nature of the problem, there is limited understanding of how growth kinetics of the TGO and micro-crack damage affect TBC lifetime. Three primary scenarios have been proposed for the spallation mechanism.[51] The first of these states that a buckling effect can result from planar compressive stresses within the ceramic layer. It seems this is a plausible mechanism provided there exists an interfacial delamination crack, formed due to local conditions creating both out-of-plane and shear tensions. Recent microscopy data indicate spallation can proceed via this buckling mechanism when the delamination crack is at least sixteen times the TBC thickness.[51,52] However, for thick oxides, the buckling mode is not viable. Those systems may undergo a surface wedging effect.[53] This second model relies on the development of a through-thickness shear crack in the TBC due to compression.[54] Finally, a more recently proposed mechanism depends upon a void formation under the TBC and subsequent folding effects which may lead to cracking.[55]

This effect is also known as "wrinkling" or "rumpling." Naturally, in addition to voids formed through thermal cycling, the pore types initially created in the TBC may influence the eventual spallation mode. Although each model appears plausible in its description of the final spallation effect, an understanding that would elucidate dynamic evolution conditions leading to spallation is desired. Figure 2 displays schematic diagrams of the three spallation models.

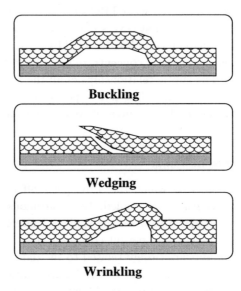

Buckling

Wedging

Wrinkling

Fig. 2. Spallation Mechanisms of the TBC.

4.3. *Complicating Features of Spallation*

Although an understanding of spallation mechanisms is useful, it seems TBC and substrate delamination results from a highly complex relationship, which includes bond coat oxidation, micro-crack evolution, and progressive buckling.[51] Fracture surfaces created by TBC spallation suggest that failure occurs in the TBC-bond coat interfacial region.[56] A piezospectroscopic study shows large failure regions may follow the oxide grain impressions in the bond coat, indicating failure where the TGO has grown into the bond coat grain boundaries.[26] Certain morphological instabilities of the oxide–bond coat interface may cause the fracture that joins the failure regions. These instabilities could result in local normal forces across the interface leading to interfacial fracture. A similar phenomenon occurs with spontaneous spallation after cooling a TBC to room temperature. Consequently, subcritical crack growth may be hastened by environmental factors. Propensity for moisture to enhance such crack growth in metal–alumina interfaces further supports these ideas.[26,57,58] Using thermal plasma spraying for bond coat deposition may result in a layered bond coat with irregular thickness. Samples possessing this discontinuity in bond coat thickness failed after thermal treatment under an argon atmosphere, indicating that at least one factor in thermal spallation is increased residual stress on the YSZ after thermal treatment.[21] Thus, although the buildup of the TGO may play a large role in the eventual TBC failure, an overall spallation mechanism encompasses a variety of complicating features.[51]

4.4. *Adhesion*

Composition of the substrate and bond coat affects adhesion of alumina formed during bond coat oxidation. Of course, when the alumina spalls, the zirconia layer is not maintained; so the TBC fails. Migration, segregation, and stress generation are key areas of concern in the thermally grown oxide region. Migration of aluminum ions from bond coat to metal–alloy substrate occurs due to a concentration gradient. Likewise, migration from the substrate and bond coat into the thermally grown oxide region is also expected. Actually, it appears that limited yttrium ion migration to the alumina improves adhesion;[25] however, contaminants in the alumina generally lead to additional stresses, which decrease adhesion during thermal expansion.[59] The high-temperature diffusion of bulk yttria ions to the YSZ surface destabilizes the top coat.[60,61] This permits zirconia phase transformation to monoclinic,[62] resulting in an undesirable volume expansion that may contribute to spallation. Furthermore, an SEM study shows bond coat aluminum can diffuse into the YSZ layer.[39] Previously, it has been suggested that both neutral and ionic aluminum diffusion into the YSZ layer induces spallation.[63] Accordingly, diffusion-controlled migration in the TBC may promote harmful phase transitions and enhance thermal stresses.

Although a concentration gradient allows migration via diffusion, the oxygen chemical potential gradient, arising during oxidation, supplies another driving force. Studies show that this chemical potential gradient penetrates the top coat, bond coat and into the metal substrate. It affects oxygen-reactive species in all these layers, including those initially in the form of stable oxides, nitrides, carbides and sulfides.[64–67]

Segregation to the metal-TGO interface also affects oxide adhesion. For instance, sulfur segregates to this interface, which can prove detrimental to TBC lifetime. The interfacial sulfur increases the thickness of the oxidation layer, decreases adhesion of the oxide layer to the metal, increases transformation of metastable alumina to the alpha phase, and enhances pore formation at the interface and within the oxide layer. Increased interfacial roughening and void formation results. The formation of voids is problematic since voids act as stress concentration sites within the oxidation layer.[65,68–72] De-sulfurization of the sample to less than 1 ppm may help prevent spallation.[73–76] However, it is possible that the presence of other species (such as Y and Zr) in the bond coat and substrate may supress the harmful effects of the sulfur and render de-sulfurization unnecessary.[25,65,77] Finally, scale stresses can result from isothermally generated growth stress during the oxide formation or from cooling stress due to thermal expansion mismatch during thermal cycling.[78]

4.5. *Measures of Stress, Fracture and Spallation*

In addition to experimental means for initially characterizing TBC systems, a number of techniques are useful for investigating stress, fracture, and spallation. Cr^{3+} piezospectroscopy is an optical method, sensitive to Cr^{3+} dopants in the alumina, that permits study of oxidation stresses through the TBC. Prior to this technique, non-destructive study of the oxide layer had proven difficult due to the physical location of the alumina. Fortunately, zirconia is fairly transparent at the frequencies of interest. As a result, this piezospectroscopic technique is able to detect stresses and some phase differences within the oxide layer.[79] Techniques for determining fracture energy include double cantilever beam experiments,[80,81] four-point bending tests,[82–84] and wedge-loaded peel tests.[83] Furthermore, laser spallation provides a means to measure tensile strength of the metal–oxide interface.[85–87] Pulsed lasers, used as high shock generators, allow exploration of the spallation process through simulated high-pressure loading.[88,89] A method based upon Thin Layer Activation (TLA) attempts to directly measure spallation. TLA relies on the creation of radionuclides in the surface layer after exposure to a high-energy beam of charged particles.[90,91] Loss of activated material due to spallation results in a decreased γ-activity signal. Furthermore, it is possible to collect the spalled material, allowing mass quantification that provides an additional sensitive spallation measure.[92] Recently, effective measurements of amplitude and profile of laser generated stress pulses have been made using a Doppler velocity interferometer system for any reflector. An advantage of time resolved interferometry is that it allows the researcher to investigate dynamic behavior of the material by looking at damage effects on wave propagation.[93] Dilatometry provides a quantitative means to explore interfacial damage due to oxidation, allowing thermal expansion experiments.[24] The strain introduced by thermal expansion mismatch plays a major role in TBC failure, so techniques for exploring thermal expansion are very important.

5. Theoretical Prediction of Interface Stability

Theoretical methods offer the opportunity to explore structure-property relationships in ideal metal–ceramic interfaces. Ultimately, improved understanding of the causal sequence leading to a particular interface structure and set of properties would enable further optimization of manufacturing parameters. Atomistic modeling constitutes the perfect laboratory in this respect. Within the limits of the specific approximations used for interatomic interactions, physical properties may be resolved to arbitrary accuracy and competing effects may be separated.

Here we will be concerned mainly with the interface structural stability and electronic structure. One of the most important factors determining the interface stability is the interface cohesive strength. Other factors may be equally important for stability of the interface, depending on the actual situation, e.g., corrosion resistance, thermal expansion and elastic matching of the metal and ceramic, the flexibility of the structure to release stress buildup during thermal cycling, stability towards structural degradation by unwanted interface segregation of certain elements and undesirable mixed phase formation at the interface. In this section, we give an overview of strategies by which theory can play an important role in helping characterize such interfaces. A recent and excellent review on the theoretical aspects of the M/C interface was given by Finnis[94] in 1996; we will therefore concentrate on later developments in the understanding of M/C interfaces.

6. Atomistic Approaches

The discipline of atomistic modeling has proliferated tremendously over the past two decades, due to the increased capacity of modern computers. The basic trade-off in atomistic modeling is always between accuracy of calculated energies and forces and, on the other hand, size of the atomic ensemble meant to model a macroscopic system. The larger the ensemble size, the smaller is the influence of miscellaneous finite-size effects on the physical properties. However, the larger the system size, the more approximate the atomistic description necessarily becomes.

Which trade-off to choose depends on the situation: some physical properties are insensitive to details in the interatomic potentials. In such cases, a model potential will provide essentially identical results to the exact, as long as the model potential reproduces certain characteristic quantities like, e.g., elastic constants and bulk cohesive energies.

Physical properties may be insensitive to details of the interatomic potential for a variety of reasons. For example at low temperatures, the atomic system may probe a bounded region of phase space, most often the elastic regime around the equilibrium state; and in this bounded region, the interatomic interactions may be represented well by a suitable model potential. Some classes of physical properties, like critical phenomena, are intrinsically insensitive to many details in the interatomic potentials, as they are controlled by collective behavior. For other physical phenomena, like melting, effects of fine details in the interatomic potentials tend to average out.

However, in other situations, there will be a strong sensitivity to details in the interatomic potential. This is the case when the physics and chemistry is governed by rare events, like crossing activation barriers and the breaking and forming of chemical bonds. If the transition state is not well-characterized and included carefully in the parameterization data base for the model potential, the activation barrier is very likely to be described erroneously by the model potential, leading to wrong transition rates. This is where *ab initio* methods are needed.

In the sections below, the particular aspects associated with atomistic modeling of M/C interfaces will be reviewed along with applicable recent work done in this field.

6.1. *Structural Models for M/C Interfaces*

Atomistic approaches need a structural model to represent the real system. In this section, we will briefly discuss the most common choices of structural models along with advantages and disadvantages, in order to enable readers unfamiliar with details to critically judge results presented in the literature. We will then discuss issues involved in choosing the structure of the interface which involves lattice stoichiometry and misfit. The most common structural models of M/C interfaces fall into two main classes: cluster and slab representations.

6.1.1. *Cluster Models*

These models are usually constructed by scooping out a small representative region (see Fig. 3A) containing 5–30 atoms from the real M/C interface. Cluster models emphasize chemistry as local by only considering a small "active" region. Unfortunately, some early cluster studies used unrealistically small clusters, which were poor representations of the real interface they attempted to model.[95] In such cases, important long-range effects are missed.

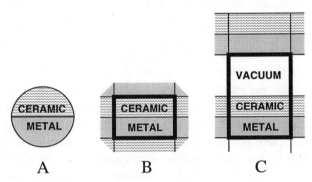

Fig. 3. Schematic picture of M/C interface structural models. A: Cluster model with vacuum around it. B: Dense M/C interface unit cell, physically corresponding to a superlattice or sandwich structure. C: Slab model with vacuum between periodic images perpendicular to the interface. Physically, the structure corresponds to M/C thin films. The borders of the periodically replicated unit cells in B and C are shown with bold lines.

Sometimes the ceramic side of the cluster is embedded in an array of the nominal anion and cation point charges in the proper structure, to emulate the real Madelung potential in the ceramic — this definitely improves the realism of the cluster model.[96–100] A Green's function constructed from the perfect host crystal has also been used to embed a ceramic cluster.[101] An alternative to anion and cation point charges is embedding the cluster into an array of overlapping anion and cation pseudopotentials,[102,103] thus trying to capture some of the electron–electron interaction with the surrounding ceramic medium. A supplement to this is mechanical embedding of the cluster via force field interactions with the surrounding substrate lattice,[104] so-called molecular mechanics. The latter neglects the perturbations in interatomic force constants caused by metallic adsorption and is likely to be rather inaccurate in the case of reconstruction or extensive relaxations in the interface region.

Cluster models have other generic finite-size effects. Interfaces are often geometrically frustrated: for all metal atoms, there is a preferred adsorption site on the ceramic surface. When one metal atom occupies that site, the neighboring metal atom generally will not be able to occupy an equally favorable site, due to the lattice mismatch between metal and ceramic. This frustration effect is not modeled properly, if the metal side of the cluster is too small. Another finite-size effect is that the electronic density of states oscillates significantly with the cluster size.[105] This is critical near the Fermi level, because such oscillations in the density of states available for bonding to the ceramic is likely to produce predictions that converge slowly with cluster size.[106] This effect may be countered by "state preparation"[105] or by embedding the metal cluster in a periodic metal slab.[107] Unfortunately, since many finite-size effects converge slowly with the cluster size, unmanageably large clusters may be required if not embedded properly. Furthermore, charged clusters are often considered, due to the nonstoichiometry of the ceramic piece of the cluster. Otherwise, the anions and cations will not be in the proper charge state, corresponding to the overall neutral bulk ceramic. A back-of-the-envelope estimate suggests that the nonphysical polarization induced by charged fragments may substantially affect the chemical predictions. The one significant advantage of cluster models is that the power of *ab initio* quantum chemical methods can be applied; and therefore, systematically converged results independent of experiment can be obtained, albeit for a small piece of the macroscopic system. This will be discussed in more detail shortly.

6.1.2. *Slab Models*

These models are the preferred geometry in electronic structure methods originating from solid state physics. Periodic boundary conditions are applied to a unit cell representing the M/C interface. Therefore, slab models are restricted to coherent interfaces, which means that periodicity parallel to the interface is present — this corresponds to a "locked in" interface structure. Of course, the periodicity may have a long repetition length, which may not be modeled properly in some periodic slab calculations. There are two types of slab cells: those that do not include a vacuum layer and those that do. Dense unit cells (Fig. 3B) physically correspond to sandwich structures with two M/C interfaces per unit cell. From an energetic point of view, this geometry is rather restrictive, because it requires sufficient symmetry such that the interfaces are identical (in order for the interface energy to be uniquely defined). Unfortunately, this requirement often restricts transverse relaxation, because the symmetry "locks" the M/C interface. On the other hand, if symmetry is enforced, this unit cell has no net perpendicular dipole moment, so that unphysical electrostatic coupling between the interfaces is avoided. It is also necessary to make both the metal and ceramic layer thick enough so that the interfaces do not interact via electronic structure perturbations or strain, in order to have such a model realistically portray a single M/C interface.

A more general interface geometry is shown in Fig. 3C. Physically this corresponds to an infinite array of M/C thin film couples separated by vacuum. A salient point is that the vacuum layers should be thick enough that adjacent M/C slabs do not interact. Interaction is possible in two ways: either via electronic wavefunction overlap in the vacuum or via Coulombic multipoles. The former interaction is usually vanishing, if more than ~ 10 Å of vacuum is present. The latter interaction is rather long ranged, but fortunately methods have been devised to electrostatically decouple the slabs.[108-110] Of course, it is required that both the metal and ceramic layers are thick enough that the interface and surfaces do not interact.

The slab geometry also suffers from other finite-size effects. If the extent of the unit cell parallel to the interface is too small, artificial strain effects are introduced, because the metal and ceramic are forced to be coherent by the periodic boundary conditions. Of course, this may be eliminated by enlarging the unit cell, which unfortunately leads to very computer-intensive calculations, as is the case with the cluster models. However for the slab model, the oscillations in the electronic density of states are not as dramatic when varying the number of atoms as in the case with clusters. This is because the slab is infinite parallel to the interface. This implies the spectrum is continuous, and the metal slab does not have an artificial band gap, unlike the metal cluster.

The artificial requirement of the unit cell periodicity perpendicular to the interface may be avoided, if a Green's function technique is used. Such techniques have been used for metal–metal interfaces[111] with high symmetry, but no such calculations for M/C interfaces have been reported yet to our knowledge.

6.1.3. *Interface Stoichiometry*

If no experimental evidence is available concerning the orientation of the metal and ceramic crystals with respect to each other for a coherent interface, one needs somehow to produce a reasonable guess as to how the parent crystals may match up at the interface. The possibilities are immense, especially if the ceramic has low symmetry: first one needs to consider which faces from each parent crystal will match up. Symmetry considerations are often helpful at this point. Free energy consideration might also be guiding: weakly interacting M/C pairs are likely to match up on their low-energy surfaces. Strongly interacting M/C pairs are more likely to match up on their high-energy surfaces, because these generally contain more unsaturated atoms ready for bonding. However, if a new reaction phase forms at the interface, the situation becomes more complicated. If the bonding mechanism is dominantly electrostatic, image theory, which we return to later, predicts the most polar ceramic surface to bind most tightly to the metal, i.e. it has the largest work of separation W. However, this does not mean that polar interfaces are most stable from a thermodynamical point of view; the most stable M/C interface that can be formed between the pair M and C is the one with smallest free energy, i.e. smallest interface tension γ^{MC}:

$$\gamma^{\text{MC}} = \gamma^M + \gamma^C - W - \sum_i \mu_i \Delta N_i. \tag{1}$$

This is the Dupré relation, generalized to allow mass exchange, e.g. with the ambient gas or with the bulk, at chemical potentials μ_i. ΔN_i is the change in abundance of species i in the interface region on forming the interface, and γ^M, γ^C are surface tensions of bare metal and ceramic, respectively. Thus, because a polar surface is energetically very unfavorable (theoretically γ^C diverges for a perfect, infinite polar ceramic surface), it may not be favorable to form an interface with a polar ceramic termination, even though the interface binding is relatively strong. Another approach for guessing the ceramic termination at the interface is to estimate anion/cation/metal bond combinations from the appropriate bulk phases. But no foolproof rules can be formulated on how the ceramic is terminated at the M/C interface. Stoichiometry is a complicating aspect in this context, because it depends on the chemical potentials present; and the interface chemistry changes dramatically with stoichiometry, as we will discuss later.

6.1.4. *Lattice Misfit*

Having settled on some metal and ceramic surfaces that we think will match, we must determine the relative orientation and translation of these surfaces with respect to each other. Certain directions in the metal and ceramic are likely to be aligned for a stable interface. There will often be multiple minima, corresponding to different lock-in possibilities for the coherent interface. Lastly, the size and shape of the interface unit cell needs to be determined, if we assume a coherent interface, which is implicit if periodic boundary conditions are applied. A realistic unit cell will of course correspond to low strain on both the metal and ceramic side.

This vast number of possibilities calls for a systematic procedure to identify a subset of the most likely interface matchings of the parent crystals. This subset will then be the starting point for atomistic modeling. The question about unit cell size and shape is relatively simple to address. Many related procedures based on linear elasticity theory and lattice strain estimates may be adopted. The basic situation is sketched in Fig. 4: an overlayer unit cell A needs to be matched together with a substrate unit cell B. Matching pairs of unit cells are, in general, multiples of primitive cells in the interface plane for the metal and ceramic, respectively.

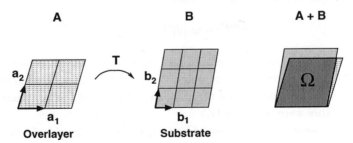

Fig. 4. Schematic picture of M/C interface lattice matching. A: Arbitrary overlayer unit cell constituted of many primitive cells spanned by the vectors a_1, a_2. B: Arbitrary substrate unit cell constituted of many primitive cells spanned by the vectors b_1, b_2. T: Linear transformation between A and B. A+B: The "overlap" Ω between cells A and B.

The most crude approach to quantify the commensurability of the two given unit cells is to assign a mismatch factor based on the overlap between the cells A and B, i.e.

$$\mu = 1 - \frac{2|\Omega|}{|A| + |B|} \tag{2}$$

where $|A|, |B|$ and $|\Omega|$ signify the area of cell A, B and the overlap area, respectively. This quantity vanishes, if $A \equiv B$ and increases with decreasing overlap Ω, as sketched in Fig. 4. Thus one would like to choose an interface matching that minimizes μ.

Another geometrical approach was taken by Bolding and Carter,[112] who considered the matching of orthogonal unit cells. By taking multiples of each unit cell a suitable number of times in each direction, an arbitrarily low misfit might be achieved, in the same way that an arbitrary irrational number might be approximated by a rational number, if the numerator and denominator are sufficiently large. More specifically, they defined the strain variables

$$\delta_1^{nl} = \frac{la_1 - nb_1}{nb_1}$$

$$\delta_2^{mk} = \frac{ka_2 - mb_2}{mb_2} \tag{3}$$

where a_1, a_2, b_1, b_2 are the length of the basis vectors spanning the primitive unit cells of which A and B are multiples, see Fig. 4. (n, m, l, k) are integers giving the unit cell enlargement along each basis vector. Then Bolding and Carter proposed minimizing the quantity

$$\mu_{BC}^{nmlk} = w_1|\delta_1^{nl}| + w_2|\delta_2^{mk}| \tag{4}$$

which obviously is positive definite and vanishes at perfect coherency between A and B. Here w_1, w_2 are weight factors that might be related to the elastic properties of metal and ceramic. In Bolding and Carter's case, these weight factors were set to unity.

A more general approach would be to directly consider the elastic energy associated by deforming cell A into cell B so as to achieve coherency. If this deformation is designated by the 2×2 linear transformation matrix T

$$B = TA \tag{5}$$

where the matrices A and B contain the vectors spanning the (almost) matching pair of metal and ceramic unit cells, it is easy to show that the corresponding *exact* strain tensor ε and the corresponding elastic energy u is given by

$$\varepsilon = \frac{1}{2}[T^t T - I] \tag{6}$$

$$u = \frac{1}{2}\varepsilon : \gamma : \varepsilon . \tag{7}$$

It is important to use the exact strain tensor definition, Eq. (6), to achieve rotational invariance with respect to lattice rotation; the conventional linear strain tensor only provides differential rotational invariance of u in Eq. (7).[113,114] A hierarchy of approximations may be used for the elastic tensor γ. The most rigorous approach is to transform the bulk elastic tensor c according to

$$\gamma_{pqrs} = R^t_{pi} R^t_{qj} c_{ijkl} R_{kr} R_{ls} \tag{8}$$

where a summation convention over same indices applies and R is the rotation matrix from interface to bulk crystal Cartesian frames.[113] However, it is not guaranteed that a thin film has the same ratio between elastic constants as the bulk crystal. Generally, six elastic constants are inequivalent in 2D elasticity theory, which follows from the permutational symmetry of c. As an approximation we may assume only c_{11}, c_{12} and c_{44} are nonvanishing, which is true for cubic symmetry. The lowest level of approximation is to assume isotropy, in which case the relation $c_{11} = c_{12} + 2c_{44}$ applies and only a ratio, like c_{12}/c_{11}, needs to be estimated.

The approach assumes a fixed substrate; the most sophisticated procedure would be to minimize the elastic energy $u = u^{\text{substrate}} + u^{\text{overlayer}}$. This properly results in a small overall rescaling of the elastic energies. However, due to the physical simplicity of this elastic model, it is doubtful whether a better description is obtained unless both substrate and overlayer are highly anisotropic. Further, if material A is grown onto a bulk substrate B, it is unlikely that a transversal deformation of the substrate would take place in reality anyway.

It is instructive to compare this model to that of Bolding and Carter[112] discussed above, for the case where both substrate and overlayer have orthogonal lattices (in the interface plane). If we multiply the overlayer unit cell (l, k) times along a_1, a_2 and correspondingly the substrate unit cell (n, m) times along b_1, b_2, the transformation matrix T in Eq. (5) becomes

$$T^{nmlk} = \begin{pmatrix} \dfrac{nb_1}{la_1} & 0 \\ 0 & \dfrac{mb_2}{ka_2} \end{pmatrix}. \tag{9}$$

Inserting this into the strain tensor Eq. (6), we get the harmonic energy, Eq. (7), to second order in the strain variables of Bolding and Carter,

$$u^{nmlk} = \frac{1}{2}\gamma_{11}(\delta_1^{nl})^2 + \frac{1}{2}\gamma_{22}(\delta_2^{mk})^2 + \gamma_{12}\delta_1^{nl}\delta_2^{mk}. \tag{10}$$

Compared to the misfit factor of Bolding and Carter, μ_{BC}^{nmlk} [Eq. (4)], this expression contains a cross term and scales differently in the strain variables. Both expressions will be zero for perfect coherency ($\delta_1 = \delta_2 = 0$), but for competing interface lock-in possibilities, one might expect Eq. (10) to be qualitatively more accurate. In conclusion, to find the most favorable unit cell lock-ins for a given substrate and overlayer lattice, from an elasticity point of view, we simply need to scan an expression like Eq. (10) in four indices (n, m, l, k) to find a subset of likely coherent interface unit cells. This is easily done on a computer.

In addition to the elastic energy, of course, the chemical interaction energy between metal and ceramic must be accounted for. The possibility of competition between elasticity and chemistry exists such that a rather elastically strained interface combination may be more stable than an unstrained one, because the strained one may have favorable chemical bonding between metal and ceramic atoms at the interface. Therefore, other interface combinations than the least strained need to be considered in general.

We now move on to discuss how one can assess structure and energetics of such interfaces from theory, starting from the most accurate models and ending with the most approximate.

6.2. *Ab Initio Techniques*

First principles techniques can be distinguished from semiempirical methods in that they include quantum mechanical effects explicitly using exact or moderate approximations for electronic exchange and correlation. Generally, explicit inclusion of quantum mechanical effects is computationally demanding, which restricts *ab initio* methods to small systems, typically less than 50 inequivalent atoms on modern workstations (1999). Using massively parallel computers, up to 300 inequivalent main group atoms can be handled *ab initio* today.[a] In specialized cases, significantly more atoms can be handled by so-called $O(N)$ *ab initio* techniques, where the computational load scales essentially linearly with the number of atoms N. For insulating systems, tight binding calculations have shown that electronic density matrix methods[115-118] may be applied since the electronic density matrix has a finite range, and simulations with 650 atoms on a modern workstation have been reported.[119] Alternatively, divide and conquer[120] and localized wave function approaches[121-124] are well suited to achieve $O(N)$ scaling for insulating systems. If an artificial nonzero temperature is assumed for the electron system, density matrix[117,125-127] and localized wave function approaches[128] are also feasible for metallic systems, but imposing a nonphysical electronic temperature — unless it is removed by the end of the calculation[127] — is a dubious approximation for systems where the electronic density of states has much structure around the Fermi level, as in the case for transition metal/ceramic interfaces. For free-electron-like bulk systems, Orbital-free Density Functional Theory involving kinetic energy density functionals has allowed the treatment of ~6000 symmetry-inequivalent metal atoms.[129] The drawback of this method is that it is limited by the accuracy of the kinetic energy density functional, which is not known exactly for many-electron systems. Improving such functionals is, however, an active area of research.[130-144]

The restriction on the number of inequivalent atoms implies that only properties involving short length scales can be treated accurately, unless these properties are consistent with periodic boundary conditions. This restriction is rather severe in the case of a M/C interface, because the real interface is often transversely aperiodic and may contain long-ranged strain fields and defect structures perpendicular to the interface, which will require an extremely large unit cell to model. Thus only idealized M/C interfaces having small unit cells may be treated realistically by *ab initio* methods. This excludes many M/C combinations with large lattice mismatch between

---

[a]The definition of the concept *ab initio* is somewhat fluid, but usually means essentially free of empirical parameters. Being *ab initio* is generally considered a quality stamp, but one should be aware that some approaches, which are formally parameter-free, e.g. Thomas–Fermi density functional theory or *ab initio* tight binding, fail miserably outside limited application ranges.

the metal and the ceramic, because a small interface unit cell implies that either the metal or the ceramic must be strained to an unphysical extent, leading to unphysical results induced by the artificial periodic boundary conditions. Still, for M/C interfaces where the natural lattice mismatch is low, *ab initio* studies of idealized M/C interfaces are instructive, giving valuable insights into quantum effects at the interface. We now discuss what has been accomplished with *ab initio* calculations of idealized interfaces.

6.2.1. *Quantum Chemical Approaches*

A more precise description for this class is full wavefunction methods, where the basic variable is the full many-body wavefunction. The main problem with full wavefunction approaches is that the computational load increases drastically with the number of electrons N. At the Hartree–Fock level, the load increases as N^{3-4}, and the scaling with N increases steadily, the more complete the inclusion of electronic correlation. Since realistic modeling of interfaces requires quite large unit cells or large clusters, very few full wavefunction studies at and beyond the Hartree–Fock level have been reported in the literature. The interaction of a Cu[145] and a Pd[146] atom with a single MgO molecule have been studied using multireference configuration interaction. In both cases, binding of the metal to the O-atom in the MgO molecule was predicted as most stable (i.e. a linear molecule), with an "adsorption" energy of 2.28 eV and 2.65 eV for Cu and Pd, respectively. This is in accordance with the trend, observed by less accurate electronic structure methods, that metal atoms adsorb on top of the O atoms on the MgO(001) surface.

Hartree–Fock studies have been undertaken for cluster models of the Cu/MgO[147] and Ni/Al$_2$O$_3$[148] interfaces. Both studies find significant charge donation from the metal atom to the oxide conduction states. Hartree–Fock with *a posteori* electron correlation taken from a free electron gas (HF-CC)[154] in the supercell approach has been applied to Ag/α–Al$_2$O$_3$[155] and Ag/MgO(001).[155,156] Ag was found to be weakly bound to the completely O-terminated α–Al$_2$O$_3$(0001) surface, although significant charge transfer occurred. In agreement with density functional calculations, HF-CC calculations find Ag physisorbed above the O atoms in the MgO(001) surface. The adsorption energy above the O(Mg) site is $0.47(0.11)$ J/m^2, in good agreement with the experimental value, 0.45 J/m^2 for a thick Ag film.[157] However, this number is likely to be an average of adsorption of Ag over both O and Mg in separate domains, due to the (small) lattice misfit between Ag and MgO. A principal problem with the HF-CC method is that the *a posteori* free electron gas electronic correlation is not consistent with the exact exchange energy used in the underlying Hartree–Fock method — important error cancellations between exchange and correlation are lost.[158] Initial tests[154] indicated a confidence level comparable to the best density functional models for exchange and correlation, but the approach needs to be tested on more systems to see if systematic errors exist. Related in spirit to the HF-CC method is the B3LYP[159] approach, which evaluates the electronic exchange-correlation energy as a weighted average of the exact exchange energy and the electron gas results for the exchange-correlation energy. The weighting parameters for this average were determined semiempirically, which makes this approach aesthetically unpleasant. However, for many molecules and finite clusters this approach has delivered accurate results. An instructive study has appeared,[160] which compares the adhesion of Cu atoms on MgO(001) using many electronic structure methods, ranging from various forms of density functional theory to post-Hartree–Fock approaches. Here, the results of the B3LYP method agreed rather well with the results obtained by the best post-Hartree–Fock approach, which estimated the binding energy of the Cu atom on MgO(001) to 0.40 ± 0.05 eV. This is probably the most accurate theoretical estimate of this quantity today. This paper also illustrated quite dramatically the sensitivity of the results to choice of DFT functional, which should sound a note of caution regarding the general applicability and transferability of current implementations of DFT.

In summary, full wavefunction methods are the most accurate *ab initio* methods, if electronic correlation is accounted for properly and the basis set for expansion of the electronic wavefunction is sufficiently complete — this might be difficult to judge by a non-expert. Furthermore, if a cluster model is used, it needs to be of sufficient size or be properly embedded for conclusions to be representative of an extended M/C interface.

6.2.2. *Density Functional Theory*

For condensed phase systems, Density Functional Theory (DFT)[158,161,162] methods constitute the optimal compromise between accuracy and efficiency of all *ab initio* methods available today. The key point of DFT is to show that the exact quantum mechanical total energy E is a functional only of the total electron density $\rho(r)$, a 3-dimensional function, which corresponds to the intractable many-body electronic wavefunction Ψ, a $3N$-dimensional function where N is the number of electrons. The total energy within DFT, $E[\rho]$, is related to the total energy expressed in terms of Ψ, $E(\Psi)$, as

$$E[\Psi] = \langle \Psi | \hat{T}^{\text{elec}} + \hat{U} | \Psi \rangle + T^{\text{ion}}$$

$$= T_s^{\text{elec}}[\rho] + U_{\text{classical}}[\rho] + U_{xc}[\rho] + T^{\text{ion}}$$

$$= E[\rho]. \tag{11}$$

In principle, DFT is thus a formally exact mean field theory. Here T^{ion} is the ionic kinetic energy, $U_{\text{classical}}$ is the classical mean field Coulomb energy for ionic cores and charge distribution ρ, T_s^{elec} represents the kinetic energy of the (noninteracting) electrons, and U_{xc} is a term describing lowering of electron–electron repulsion by correlated electronic motion and electronic exchange, as well as a kinetic energy component due to electron–electron interactions. U_{xc} in Eq. (11) is the only term whose functional form is not known exactly: the most commonly used approximations to it today are the Local Density Approximation (LDA) and the Generalized Gradient Approximation(s) (GGA).[149] A variety of slightly different GGA parameterizations have been proposed over the years;[150–153] indeed the choice of which GGA functional to use is a nonsystematic aspect of DFT calculations.

A slightly lower level of approximation than the LDA is the Xα method, where the correlation energy is assumed to be proportional to the exchange energy, for which the LDA uniform electron gas expression is used. The Xα-approximation has been used for many pioneering metal/ceramic model studies. A consensus has developed in the literature[154,160,163–165] that the GGA on average is better than the LDA. But in certain cases, the GGA tends to overcorrect the LDA.[166,167] For ionic surfaces, a significant lowering of the surface energy going from LDA to GGA has been noticed in some cases.[168] Another major drawback of the LDA/GGA approximations in this context is that they are not able to describe strongly correlated oxides, such as many $3d$ transition metal oxides.[169] These cases with breakdown of the LDA/GGA are due to two shortcomings of the LDA/GGA. The first is the general underestimation of the band gap, which can be traced back to a discontinuity feature in the exchange-correlation potential as function of electron filling[162] not captured by the LDA/GGA. The second is the self-interaction problem in the LDA/GGA in which electrons artificially interact with themselves due to the electron gas approximation for the exchange-correlation energy. This may be remedied by considering a self-interaction correction (SIC)[170,171] scheme to the LDA/GGA, e.g. the LDA+U method.[172]

For insulating materials, such as ceramics, this normally does not pose problems, because the conduction band is empty and the band gap error is not reflected energetically. However, for M/C interfaces there may be some reason to worry, since mixing between oxide conduction bands and metal states will influence the magnitude of adhesion and charge transfer. These states may become filled, depending on the Fermi level on the metal side. Further research on this issue is warranted.

Because DFT-based techniques have the electronic density $\rho(r)$ as the basic variable, the computational load scales moderately with the number of electrons N, $O(N^{1-3})$. Thus they are able to handle significantly more atoms than traditional quantum chemical approaches that retain the full electronic many-body wavefunction as the basic variational quantity. This favorable scaling currently makes DFT-based techniques most promising for *ab initio* studies of M/C interfaces, and therefore we will emphasize this group of methods in our review.

6.2.2.1. *Applications of DFT to M/C Interfaces*

An increasing number of studies using self-consistent DFT-based techniques have been reported recently. The most popular M/C interface among theoreticians has been Ag/MgO(001), due to its small unit cell and small lattice mismatch of 3% and hence simple epitaxial character. Furthermore, this interface is well-characterized experimentally. Generally, studies based on electronic structure methods agree that this interface is characterized by weak physical rather than strong chemical adhesion, with marginal charge transfer to Ag, and the O-site as most favorable adsorption site.[155,156,173–175] The MgO(001) surface is stoichiometric (i.e. there is an equal number of Mg and O ions on the surface) and O is nominally in the O^{2-} charge state. As also noted by Finnis,[94] it is interesting to observe the scattering in adhesion energies obtained by different basis set expansions. For Ag adsorbed on top of O(Mg), the Full-Potential Linear Muffin Tin Orbital (FP-LMTO) method gives 1.59(0.78) J/m^2 for 3 layers of Ag,[176] the Full-Potential Linear Augmented Plane Wave (FLAPW) method gives

0.53(0.53) J/m² for 1 Ag layer,[174] whereas a Gaussian-based Linear Combination of Atomic Orbitals representation gives 1.9(1.08) J/m² for 1 Ag
layer.[177] All of these studies were performed within the LDA, which is no-
torious for its tendency towards overbinding.[158] This overbinding trend is
followed for the Ag/MgO system, confirmed by comparing to the experimental value 0.45 J/m² for a thick Ag film.[157] Again, this value is likely
to be an average of alternating domains with Ag adsorbed over O and Mg
respectively, due to misfit dislocations induced by strain. The scatter in the
theoretical results is not explained by the extent to which relaxations in the
interface region were included; all authors relaxed the Ag–O distance, and
none included full relaxations in the oxide. The FP-LMTO study[176] relaxed
the oxide planes perpendicular to the surface, the FLAPW study kept the
oxide frozen. The free MgO(001) surface termination displays a rumpling
of the ions; however, our own DFT calculations,[178,179] which feature full
relaxation, show that the Ag-film smooths the MgO(001) surface rumpling,
so that the oxide layer nearest to the interface is flat. Furthermore, the
oxide interlayer distances are as in bulk MgO to within 0.005 Å. Figure 5

Fig. 5. Ag/MgO(001) charge density plot cross-sectional view. There are seven Ag sites
on top of the layer of seven O sites. The Mg, with little valence charge, appear faintly
beneath the O sites. Another O layer and Mg layer appears at the bottom. The contour
lines at the top of the box are a result of periodic boundary conditions.

displays a repeated unit cell from our calculations of an Ag layer on top of a MgO(001) surface. Thus, in this particular case, the structural constraint in the FLAPW study did not affect the results. The large variations in theoretical predictions remains an unsatisfying aspect of these calculations.

Polar M/C interfaces have only been studied sparsely. The polar (111) and nonpolar (100) Cu/MgO interfaces were compared in a slab calculation within the LDA.[180] As expected from the image model theory, the polar interface displayed a considerably higher work of separation. In addition, these authors found a larger charge transfer and Cu–MgO orbital mixing for the polar interface.

Generally, *ab initio* methods have found that metals adsorb at the O-site of MgO(001) and that the Mg-site corresponds to a local maximum on the adsorption potential energy surface (PES), but the accord ends here. For Ag/MgO(001), Hong *et al.*[177] and Schönberger *et al.*[176] find a corrugation of the PES of the order of 50% of the cohesive energy whereas Li *et al.*[174] find an extremely flat PES with corrugation less than 1%. For comparison, for Pd/MgO(001), Goniakowski[181] finds a corrugation of order 20%. Thus, even though these systems appear easy to study, the disagreement upon even qualitative features of the interaction is rather disturbing.

It is an interesting question whether the GGA improves the agreement with experiments, as is often the case, or even changes trends predicted within the LDA. We investigated this by calculating the Ag/MgO(001) adhesion within both the LDA and GGA, using state-of-the-art ultrasoft pseudopotentials[178] in a converged planewave basis. For Ag adsorbed on top of O(Mg), we obtain adhesion energies of 0.62(0.27) J/m^2 within the LDA and 0.23(0.06) J/m^2 within the GGA. This indicates that the GGA overcorrects the LDA somewhat, compared to the experimental value of 0.45 J/m^2.[157] Also, a sensitivity on details of how the GGA functional is parameterized has been noticed.[160,182]

Only recently have other M/C interface studies appeared that tested the GGA in this context. Pacchioni and Rösch[183] studied Ni and Cu on MgO(001) in a cluster model. As might be expected, they found that Ni binds more strongly than Cu, due to the open *d*-shell of Ni. In both cases, the O-site is favored. They found a significant lowering of the adsorption energy for Cu when using the GGA, 0.65 J/m^2, compared to the value 2.54 J/m^2 obtained recently in a cluster calculation using the LDA.[184] Pacchioni and Rösch also noted that in these cases metal–metal bonds were stronger than metal–substrate bonds, thus predicting a 3D (cluster) growth mode as opposed to layer-by-layer growth (wetting). A more systematic study of transition metal adsorption on MgO(001) using GGA was presented by Yudanov et al.,[175] using a cluster model. Interestingly, they found no obvious correlation between cohesion and d-band filling as indicated by other studies. They found that single Cu, Ag, Au, Cr and Mo atoms exhibited weak interface bonds, whereas Ni, Pd, Pt and W atoms formed strong bonds. For Ag/MgO, they obtained 0.36 J/m^2, similar to our findings for a coherent Ag monolayer.

Square-planar M$_4$ cluster adsorption for M = (Ni, Cu, Pd, Ag) on MgO(001) has been studied within the GGA.[185] Although here open shell transition metals adsorbed stronger, metal–metal and metal–oxide bond competition were found to be complex. An extensive study of Cu growth on MgO(001) was performed for clusters with 1–13 Cu atoms.[186] Here again, the authors found the metal–metal bonds stronger than the metal–ceramic bonds, thus favoring a 3D (cluster) growth mode. Relatively weak cohesion for the Cu/MgO interface is also expected on the basis of the large lattice mismatch (13%) between the metal and ceramic.

Electronic structure calculations of metal/Al$_2$O$_3$ interfaces were initiated by Johnson and Pepper.[187] They modeled this interface with a metal–AlO$_6$$^{9-}$ cluster using the Xα-method, with the metal atom adsorbed in an O-triangle emulating a hollow site on a O^{2-}-terminated surface. This study concluded the bonding became weaker as the transition-metal adsorbate became more noble, which was explained in the conventional orbital mixing picture, where antibonding levels become filled in the right side of the transition series. This picture was essentially confirmed by later investigators. Most authors focus on the O-terminated α–Al$_2$O$_3$(0001) surface, which is nonstoichiometric, as the substrate for metallic growth. Under oxygen deficient conditions however, it is not clear that the O-termination is appropriate for growth modeling. Grazing incidence X-ray scattering data indicate that the α–Al$_2$O$_3$(0001) is Al-terminated under ultrahigh vacuum (UHV).[188,189] Furthermore, DFT calculations predict that the charge–neutral O-terminated α–Al$_2$O$_3$(0001) has a significantly higher cleavage energy[190] than Al-terminated α–Al$_2$O$_3$(0001). Experimentally, when Ni was deposited on α–Al$_2$O$_3$ in UHV, no Ni–O interaction was detected by XPS.[191] It is certainly reasonable to consider Ni–Al bonds, since NiAl is a very stable alloy. In any case, discussions about which Al$_2$O$_3$-substrate termination at the clean metal/Al$_2$O$_3$ interface is most relevant can be somewhat academic, since reaction phases, like spinels (MAl$_2$O$_4$),[192] are often formed at the interface. Our point of view is that most structural possibilities, at this stage, are interesting, since they help to furnish a general understanding of the electronic structure of the M/C interface under various conditions.

We have used DFT to study the Ni/Al_2O_3 interface. These calculations were performed within the planewave DFT code VASP using ultrasoft pseudopotentials.[178,193] Calculations with one and three layers of α–Al_2O_3 deposited onto the Ni(111) surface were performed. Both Al and O-terminations of the α–Al_2O_3(0001) surface were examined. Figure 6 displays a repeated view of a unit cell with relaxed atomic coordinates for a Ni(111)/Al_2O_3(0001) interface calculation. The lighter atoms in the top

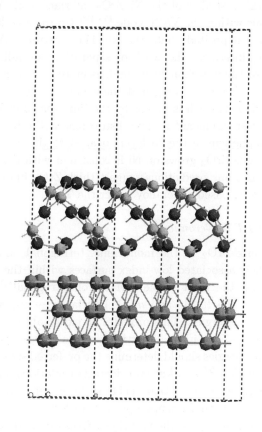

Fig. 6. Ni(111)/Al_2O_3(0001) relaxed interface geometry.

layer correspond to aluminum and the darker atoms are oxygen. In the Al-terminated α–Al_2O_3(0001) surfaces, it appears that interfacial bonding between the alumina and Ni substrate decreases with increasing Al_2O_3 thickness. This effect may contribute to the increased spallation observed experimentally with the thickening of the TGO.

Kruse *et al.*[194,195] studied Nb monolayers (and multiple Nb layers to simulate bulk-like properties) on the O-terminated α–Al_2O_3(0001) surface and found that Nb bonded to the hollow sites above the vacant octahedral sites. These authors emphasized the importance of ion relaxations when comparing competing adsorption geometries, a feature often lacking in many reported studies.

We have also undertaken studies of the Ni/ZrO_2 interface again using ultrasoft pseudopotentials within the VASP code.[178,196] Up to three layers of cubic ZrO_2(111) were "deposited" onto the Ni(111) substrate. As we found for the Ni/Al_2O_3 interface, the work of separation (or adhesion strength) is quite large for the first monolayer; and, as more ZrO_2 is added, the intra-ceramic bonds increase in strength at the expense of the interfacial bonding. The adhesion strength is halved by the time that 3 layers of ZrO_2 are present. We find significant atomic distortions at the interface and a suppression of magnetism in the Ni layers nearest to the interface. The decrease in adhesion as ZrO_2 grows on Ni is consistent with the need for a bond coat alloy, since our predictions suggest thick ZrO_2 films may not readily adhere to the Ni-based superalloy under the TBC.

6.2.2.2. *Aplications of DFT to Ceramics*

We have also studied bulk ZrO_2 in the monoclinic, tetragonal, and cubic phases as well as their associated low-index surfaces within the LDA to DFT.[48] We found that the most stable surfaces of tetragonal(t) and monoclinic(m) zirconia — the t(111) and m($\bar{1}$11) surfaces — are nearly degenerate in their surface energies. This provides an explanation for the preferential stabilization of t-ZrO_2 in small particles, where surface energies rather than bulk cohesive energies should determine the preferred structure. By using a Wulff construction[197] and known orientation relationships between the t and m phases, we showed that t(111) surfaces will transform to half m($\bar{1}$11) and half m(111) surfaces upon phase transformation. While the m($\bar{1}$11) surfaces are low in energy, the m(111) surfaces are predicted to have somewhat higher energies. Thus, we suggested that the suppression of the tetragonal to monoclinic phase transition in small particles may be due to the thermodynamically unfavorable nature of $m(111)$ surfaces that would be forced to form. This idea prompted us to consider ways to keep ZrO_2 in the tetragonal phase over a very wide temperature range. One possible solution is to embed the small ZrO_2 particles in an alumina matrix, for example, as in the idea behind the nanolaminate films comprised of ZrO_2 and Al_2O_3 multilayers.

We then undertook a DFT-GGA study (within the Projector Augmented Wave formalism)[198] of the ZrO_2–Al_2O_3 interface, in order to understand how alumina might serve this purpose of confining/stabilizing small t-ZrO_2 particles and to shed light on the nature of the interaction between the bond coat and the TBC that occurs when the bond coat is oxidized to Al_2O_3.[199]

Our main conclusions are that the alumina/zirconia interface is weakly interacting: we find negligible charge transfer, no evidence of covalent bonding, and a very low adhesion energy of 1.065 J/m^2. The low adhesion energy is probably due to the fact that these surfaces of ZrO_2 and Al_2O_3 reconstruct to obtain approximate coordinative saturation, and therefore the lack of dangling bonds on these surfaces minimizes the interaction they have between them. This suggests that the role of the Al_2O_3 in the nanolaminate coatings is simply to act as a physical barrier to growth of the ZrO_2 layer and that there is no true chemical bonding between these layers. Further, this weak interaction has broad implications for thermal barrier coatings. When the bond coat oxidizes, it is known that Al_2O_3 forms and that the lifetime of the TBC is tied to the oxidation kinetics of the bond coat. Therefore, a possible microscopic explanation is now available: once Al_2O_3 forms, there is only a weak interaction between ZrO_2–Al_2O_3, and thus the ZrO_2 de-adheres at that point. A solution to this problem may be to minimize the amount of Al in the bond coat. Al may oxidize more readily than other elements present in the bond coat, and so perhaps increasing, for example, the amount of Cr in the bond coat may help inhibit spallation.

To summarize the application of DFT methods to M/C and ceramic interfaces, a significant "noise level" is apparent in predicted energetic features, originating from different basis set expansions, different degrees of ionic relaxation allowed for, and different approximations for electron exchange-correlation effects. However, qualitative trends from different studies are most often in agreement with each other and experiments. We now turn to a more approximate DFT technique which will allow larger systems to be studied.

6.3. *The Harris Functional*

A conceptually important approximation within DFT is the Harris functional.[200,201] In the Harris functional, the electronic ground state density ρ in Eq. (11) is approximated by a sum of overlapping atom-like densities ρ_m, suitably chosen[202,203] for each species $m = 1 \cdots M$

$$\tilde{\rho}(r) = \sum_{m=1}^{M} \sum_{i=1}^{N_m} \rho_m(r - R_{mi}) \tag{12}$$

where R_{mi} designates the position of the ith atom of species m. This density ansatz makes it possible to express the cohesive energy and Hamiltonian matrix elements in terms of atomic contributions, which is very convenient for calculational and interpretive purposes. The Harris functional has been used as the starting point for deriving many approximate total energy schemes.[201,204–206] Inserting Eq. (12) into Eq. (11) we get

$$\tilde{E}[\tilde{\rho}] = \sum_{i \in \text{occup}} \varepsilon_i[v[\tilde{\rho}]] - \int v[\tilde{\rho}]\rho dr + U_{\text{classical}}[\tilde{\rho}] + U_{xc}[\tilde{\rho}] + T^{\text{ion}} \tag{13}$$

where the one-electron Schrödinger equations in the Kohn–Sham formalism[161] was used to get the first two terms and the one-electron potential v is given by

$$v[\tilde{\rho}] = \left. \frac{\delta(U_{xc} + U_{\text{classical}})}{\delta\rho} \right|_{\rho=\tilde{\rho}}. \tag{14}$$

The main problem with the Harris functional is that charge transfer and screening effects are not included properly, due to the static density ansatz, Eq. (12). This has led Finnis, on the basis of the image charge model[207,208] for M/C bonding, to object that the Harris functional — and therefore approximate total energy schemes derived from the Harris functional, such as non-self-consistent tight binding — are unsuitable for describing M/C bonding. He asserts that the attractive component of the M/C cohesion is mainly a polarization effect in the image charge model, which we will discuss later. However, despite the static density ansatz, Eq. (12), static screening effects are included via the eigenvalue sum in Eq. (13). Thus it remains to be demonstrated on a broad range of M/C systems how well the Harris functional performs. Encouraging results have been reported for Ag/MgO and Al/MgO with and without C and S impurities.[177,209]

6.4. Tight Binding

Tight binding (TB) schemes[210,211] refer to a rather ill-demarcated group of electronic structure techniques for describing interatomic interactions. The key merit of TB methods, compared to lower levels of interatomic potentials, is that they include quantum effects explicitly, although at a fairly approximate level. Their simpler nature produces a large computational speed gain over *ab initio* methods. This enables approximate studies of electronic structure effects in extended systems such as TBC's. In the chemically oriented literature, TB methods are associated with Extended Hückel Theory(EHT)[212,213] or variants like Atom Superposition and Electron Delocalization Molecular Orbital (ASED-MO) theory, which is EHT augmented by an inter-nuclear repulsion term.[214]

The archetype for the TB total energy expression, E^{TB}, for ionic coordinates $\{R_n\}$ is

$$E^{TB}\{R_n\} = \sum_{i \in \text{occup}} \varepsilon_i\{R_n\} + E_{\text{rep}}\{R_n\} \qquad (15)$$

where E_{rep} is a repulsive term balancing the propensity to gain energy in the spectral sum by increasing orbital coupling. The electronic spectrum $\{\varepsilon_i\}$ is obtained by solving a matrix eigenvalue problem

$$|\mathbf{H} - \mathbf{S}E| = \mathbf{0} \qquad (16)$$

corresponding to an implicit or explicit atomic basis, usually chosen to be minimal, i.e. one s-orbital, three p-orbitals per main group atom, and five d-orbitals per atom for transition metals. $\mathbf{H}$ and $\mathbf{S}$ are the Hamiltonian and overlap matrices corresponding to this basis. Comparing Eqs. (15) and (16) to Eq. (13), the inherent approximations of TB appear: (1) The overlap matrix $\mathbf{S}$ is often set to the unit matrix, i.e. the atomic basis is assumed orthogonal (orthogonal TB) — this reduces accuracy. Proper inclusion of the overlap matrix $\mathbf{S}$ (non-orthogonal TB) increases transferability, but is computationally more demanding. (2) The Hamiltonian and overlap matrices are often fitted to simple pairwise analytic forms from a limited set of data, instead of being appropriate matrix elements with respect to a given basis. This reduces accuracy. (3) The repulsive term E_{rep} is usually represented by an isotropic pair-potential, summed over all atomic pairs. The counterpart for E_{rep} in Eq. (13) is nonlinear in the exchange-correlation terms, therefore accurate representation of E_{rep} requires environment-dependent pair-potentials[215-217] or three-body corrections. (4) Usually there is a lack of charge self-consistency [i.e. charge redistribution, which is neglected in the density ansatz, Eq. (12)]. This is usually only done by adjusting the diagonal Hamiltonian matrix elements H_{ii}.[211,218] Charge transfer may induce long-ranged effects, i.e. they may create a Madelung potential or induce band bending. The most modern TB schemes put emphasis on a more realistic and detailed inclusion of self-consistency effects.[206,219]

The accuracy to which these points are treated determines the transferability of the TB scheme. For low-level TB schemes, at most only the qualitative trends should be trusted. A pitfall in this context is overly parameterized TB schemes, i.e. too many fitted input parameters — this strategy will tend to hide inherent limitations. In principle, carefully worked-out TB schemes can achieve almost the accuracy of self-consistent LCAO-DFT[204,206] without use of empirical fitting parameters.

Many interesting TB studies have been performed on the energetics and electronic structure of M/C interfaces. Most TB studies have focused on transition metals deposited on α–alumina. Alemany *et al.*[220] obtained band structures using TB in an extended Hückel framework to study a variety of crystalline alumina terminations, where O anions were in the O^{2-} charge state. They generally found attractive interactions between adsorbed $3d$-transition metal atoms coordinating with Al^{3+} cations but repulsive interactions when coordinating with O^{2-} anions, where the repulsiveness increased with nobleness. They found only small steric effects in this pattern, which provides support for phenomenological approaches to quantifying interface cohesion. They attributed the transition metal–Al^{3+} attraction to mixing of the adsorbate d-states and the coordinatively unsaturated Al^{3+} dangling bond (empty sp-hybrid) surface states.

When O ions were in the O^- charge state, the anions were found to be very reactive, with maximum interface adhesion for V and decreasing monotonically towards each side in the d series, but with Cu anomalously stabilized. The interface strength of O^--terminated α–$Al_2O_3(0001)$ is estimated to twice that of the Al^{3+} terminated. This picture is in some qualitative disagreement with the findings of Nath and Anderson,[221] who found the interface bond strength to O^--terminated α–$Al_2O_3(0001)$ decreasing monotonically with nobleness along the entire transition series. The Nath and Anderson study used ASED-MO theory to investigate their semiempirical cluster models where bonding was composed of covalent, charge transfer, and two-body repulsion terms. The uncertainty between different studies even at the qualitative level is a signature of the sensitivity on TB input parameters and specific approximations in the TB scheme applied. This is also seen comparing the results of Anderson *et al.*'s[222] ASED-MO cluster calculations to Ward *et al.*'s[223] approximate weighted H_{ij} extended Hückel slab calculations for Pt on α–$Al_2O_3(0001)$. Both authors agree that Pt binds to the O^{2-}-terminated surface by ~ 0.1 eV/Pt, but for the Al^{3+}-terminated surface Ward *et al.*[223] find 3.7 eV/Pt whereas Anderson *et al.*[222] obtain 2.5 eV/Pt. Interestingly, in the same study, Ward *et al.* find that Rh does not bind at all whereas Pd does, opposite to the nobleness trend found for the O^- termination.

Ohuchi and Kohyama[191] conducted empirical TB energy-band (slab) calculations for many transition metal/alumina interfaces, where the α–$Al_2O_3(0001)$ was Al^{3+} terminated. They generally found the interface cohesion to change from ionic to covalent character, when moving from left to right in the transition series. They focused especially on Nb/α–$Al_2O_3(0001)$ interfaces, where they noticed that the electronic perturbation was essentially confined to the Nb-layer nearest to the Al_2O_3 surface. This observation is in contrast to the DFT study by Kruse *et al.*,[195] which found that the electronic perturbation of the Nb layers was significant away from the nearest layer. Although the Al_2O_3 in the study by Kruse *et al.* was O-terminated, this is more likely to be a consequence of truncation of the TB Hamiltonian to nearest-neighbor contributions and limited inclusion of self-consistency in the work of Ohuchi and Kohyama.

Impurities and dopants at the M/C interface are very important in TBC systems. This issue also has been addressed by TB studies. For the Ni/Al_2O_3 system, Hong *et al.*,[224] using clusters treated with ASED-MO theory, found S impurities significantly decrease the interface strength, due to S^{-2}–O^{-2} closed-shell repulsions increasing the distance of the Al^{3+} from the Ni surface layer. Anderson *et al.*[225] found increased interface adhesion for Y dopants at the Ni/Al_2O_3 interface. Open d-shell transition metals are indeed expected to have the greatest bonding to ceramics. Further, Ni and Y form rather stable alloys, suggesting favorable Ni–Y interaction at the interface. Application of dopants for TBC's of course requires the dopants to be stable at the interface upon thermal cycling.

The moral of the TB story is to judge the results with some caution, since as we outlined above, even predicted qualitative trends vary, depending on the parametrization and particular form of the TB scheme applied.

6.5. *Semiempirical Potentials*

Many issues important for the understanding and engineering of real-life interfaces involve long length scales, such as static and dynamic macroscopic properties, interface defects, toughness, wear, corrosion resistance, complex reaction phases, phase transitions, and interface deadhesion processes. These aspects of interface modeling are far outside the reach of *ab initio* and TB methods described above, which currently — and for years to come — are limited to studying model systems mimicking authentic M/C interfaces.

Therefore, an important objective in the theory branch of M/C interface research is to envisage new types of interatomic potentials suitable for modeling M/C interfaces to bridge the length scale gap between theory and experiments. In this context, *ab initio* studies are very useful: firstly, for exploring the qualitative nature of M/C chemistry which must be captured by novel model interaction potentials; secondly, to provide a broad database for testing new interaction potentials.

For metals and ceramics considered separately, reliable interaction potentials have been developed in the past. For many metals, embedded-atom-type potentials[226–230] have proven successful. Alternatives and refinements have been developed.[231–233] For ceramics, rigid-ion and shell model potentials[234] and their refinements[235–238] have proven equally capable.

For the M/C interface, atomistic interaction potentials have developed around the idea of the image charge interaction,[239] where the ceramic ions are attracted to their electrostatic images induced in the metal (assumed perfectly conducting). This simple picture gives the right order of magnitude for the cohesion. Image charge interaction is only accurate beyond the distance where wavefunctions of metal and ceramic overlap. In that case, the ceramic acts as an external Coulombic potential on the metal and Finnis[208] derived the image charge interaction from DFT for this case. At close range, where wavefunctions overlap, orbital mixing contributes to the M/C interaction. Finnis[208] argues that this may be represented by a closed-shell repulsion term by considering the ceramic as a perturbed noble gas crystal, where protons have been transferred from anions to cations but the electron distribution is kept frozen. This point of view may be valid for weak M/C interfaces, but needs corrections for strongly interacting M/C interfaces with significant charge transfer or covalent contributions to the bonding.

An interesting implication of the image interaction picture is that metals will generally adhere more strongly to polar rather than nonpolar surfaces. This is because the net-charged ceramic surface will induce a net-charged image in the metal. By contrast, polar ceramic surfaces themselves are unstable, also for electrostatic reasons.

A refinement of the image model, the discrete classical model (DCM), was also formulated by Finnis.[208] Here the perfectly conducting halfspace representing the metal is replaced by a halfspace of conducting spheres. This discretization of the metal is necessary for atomistic modeling; and further refinement of this model gives a more realistic short-range description than the image model, where the interaction energy diverges at the image plane (i.e. the metal surface). Unfortunately, Duffy *et al.*[240] found that the DCM predicts that Ag adsorbs over Mg^{2+} instead of O^{2-} on the MgO(001) surface,[240] at variance with all electronic structure calculations.

A hybrid model for full atomistic relaxation was used recently by Purton *et al.*[241] to simulate the weakly interacting Ag/MgO(001) interface. The metal was described by an embedded-atom potential, the ceramic by a shell model and the M/C interaction by the DCM (i.e. an atom interacts via 2 different potentials). These authors found that Ag adsorbed over O^{2-} and attributed this to the fact that they used a different short-range potential in the DCM than Duffy *et al.*[240] used.

This illustrates that we are far from the point where we can develop robust and transferable semiempirical potentials for the general M/C interface. This task is fundamentally difficult because the bonding changes character at the interface, from metallic to ionic, and because of the structural complexity present. This probably requires explicit inclusion of quantum effects, at least at a low level.

Because semiempirical potentials for M/C interfaces are not yet a mature field, simulation results that rely critically on details of the metal and ceramic layer adjacent to the interface need to be judged with great caution, because of the limited transferability expected for these potentials.

For strongly coupled M/C interfaces, no reliable functional forms of atomistic interaction potentials have been proposed yet.

7. Multiscale Modeling

As mentioned before, a salient problem with the atomistic modeling of realistic M/C interfaces is the contemporary presence of effects on multiple length scales, chemical interactions in the interface region along with defect structures and strain fields extending far away from the interface region. Naturally, this is a well-recognized difficulty in many modeling applications, and numerous techniques have been introduced that attempt to overcome this problem. One useful approach is to couple an active region (in our case the interface vicinity) which is treated explicitly atomistically, together with the surrounding system within linear elasticity theory.[242,243] Course-Grained Molecular Dynamics (CGMD) is also designed to allow the treatment of larger systems than would be accessible with conventional *ab initio* techniques.[244] In CGMD, a statistical course-graining method is used to derive the equations of motion from finite temperature Molecular-Dynamics(MD). Accordingly, these equations coincide with conventional MD for atomic-scale mesh sizes, which allows a smooth coupling of the length scales. As a result, CGMD provides a means to computationally concentrate on "active regions", by creating finer meshes there, while permitting large-system simulations.

Another promising approach is the quasicontinuum method proposed by Tadmor *et al.*,[245] which combines the finite element method with adaptive mesh techniques. Atoms are partitioned into "local" and "nonlocal" atoms, according to the strain at the atom. Atoms characterized by low strain (local atoms) are typically far from the active region, and the energy of these atoms is well-approximated by atoms in a perfect crystal subjected to a similar (low) strain. This energy can be obtained from standard first principles methods. Technically, the energy of the local atoms in the large, remote regions is obtained by finite element integration on a sparse grid. The nonlocal atoms are subjected to high strain in the vicinity of an active region. The cohesive energy of a "nonlocal" atom is calculated as an explicit function of its coordination shell in real space, out to a given cutoff radius, which is more time consuming. Conventional atomistic simulations normally consider all atoms as "nonlocal", which is why the quasicontinuum method can treat of order 1000 times more atoms than feasible with conventional atomistic simulation.

A limitation of the present formulation is that it relies on the partitioning of the energy into atomic contributions, i.e.

$$E^{\text{tot}} = \sum_{i=1}^{\text{All atoms}} E_i \qquad (17)$$

where E_i is an average energy for a "representative" atom. This is a virtue of approximate schemes only, like the embedded atom method and pair-potentials; the exact total energy is a collective quantity and imposing energy separability, Eq. (17), for first principles methods introduces some degree of ambiguity, especially for the atoms treated as "nonlocal". Also due to the energy separability requirement, extension of the quasicontinuum method to multicomponent systems is less clear, if energies and forces are generated from *ab initio* methods. One may be able to get around this problem by defining E_i in the sum in Eq. (17) to be the total energy per unit cell for an associated strained reference system.

No results for M/C interfaces using these rather new multiscale modeling techniques, even with approximate total energy schemes, have emerged yet to our knowledge; but studies of weakly coupled M/C interfaces using a hybrid potential, like that of Purton *et al.*,[241] would certainly be feasible.

In addition to bridging length scales, as in the quasicontinuum method, it is equally important to consider how to bridge time scales. For example, the time scale of atomic motion is many orders of magnitude smaller than the time scale for materials creep. One route to accomplish time-scale bridging is the Kinetic Monte Carlo (KMC) technique.[246] In this method, the system is evolved stochastically according to probabilities given by ratios of rate constants k_i to k_{ref}, where k_{ref} is the fastest rate constant in the system. By first choosing randomly a species in the system and then choosing randomly from a palette of possible elementary events that may change the nature of that species, and comparing the rate constant ratio for the selected elementary event to a random number between 0 and 1, one has a prescription for how to evolve the system in a kinetic rather than an equilibrium manner. While this provides the means to get to long time scales, it does have one major drawback. One must specify in advance the allowed elementary processes; in so doing, one may actually constrain the behavior of the material under study. Further, deterministic dynamics have been sacrificed at the altar of stochastic behavior in order to gain the increase in time scales. However, first principles-derived KMC has been used successfully to study kinetic behavior for a number of different materials by our group[247] and others[248–252] (not all of which are listed here).

Recently, Voter proposed a method known as hyperdynamics, or hyper-MD, designed to accelerate molecular-dynamics simulations.[253,254] This acceleration is achieved by including a bias potential that raises the potential energy in the valley regions of the potential energy surface but leaves the potential in transition state regions unaffected. Because the bias potential does not affect transition state regions, even though the Transition State Theory (TST) escape rate is enhanced, the ratio of TST escape rates from a given state to adjacent states is preserved. As a result, the system should advance sequentially at an accelerated pace while preserving the relative probabilities of exact dynamics. Provided that the assumptions inherent in TST are obeyed for the processes of interest (i.e. no saddle point recrossings and no correlated events), this is an attractive method for reaching time scales orders of magnitude longer than those achieved by conventional molecular dynamics. Voter has also proposed a method to effectively parallelize MD simulations.[255] This technique and the hyper-MD are beneficial in decreasing simulation time necessary for the observation of infrequent events.

Kinetic Monte Carlo and hyperdynamics methods have yet to be applied to processes involved in thermal barrier coating failure or even simpler model metal–ceramic or ceramic–ceramic interface degradation as a function of time. A hindrance to their application is lack of a clear consensus on how to describe the interatomic interactions by an analytic potential function. If instead, for lack of an analytic potential, one must resort to full-blown density functional theory to calculate the interatomic forces, this will become the bottleneck that will limit the size and complexity of systems one may examine, even with multiscale methods.

8. Phenomenological Approaches

Phenomenological approaches have been very successful in some areas, e.g. Miedema theory[256] for predicting many quantities in metallurgy. The essential task in phenomenological theories is to identify a suitable set of physically meaningful variables, which are "linearly independent", to characterize the materials. Experimental data are then correlated against this set of variables and functional relations are fitted.

Li[257] explored such possibilities for M/C interfaces. He found that the interface cohesion correlated well with average metal electron density in bond regions, the free energy of formation, the band gap and the conductivity of the ceramic. This study mainly focused on transition metals, but the correlation with electron density seemed also to encompass simple metals and semiconductors as interface partners. Chatain *et al.*[258] stressed the local chemical aspect by correlating interface cohesion with a linear combination of the solution enthalpy of the ceramic anion and cation, respectively, in the bulk metal, plus a van der Waals term. For simple M/C interfaces, Bordier and Noguera's model tight binding study[259] indicates that the Fermi level of the metal and the degree of ionicity of the ceramic should be important variables for correlating experimental data. Of course, the diversity of the approaches and variable sets above illustrates that there is no unique strategy for understanding M/C cohesion.

9. Conclusions and Outlook

Understanding the behavior of the interfaces and bulk materials involved in thermal barrier coating failure due to the extreme environment created in aircraft engines is still in its infancy. This is primarily because the system involves complex interfacial chemistry and the materials issues span large length and time scales. In this review, we have focused on the atomic level characterization. Once that is specified, it will be imperative to draw links between the atomic and the microstructural scales in order to understand the materials failure mechanisms completely.

Ultimately, an important goal of the structural studies described above is to provide a foundation for experiment and theory to investigate the dynamic behavior of M/C interfaces, especially under conditions similar in severity to those found in the combustion region of a gas turbine engine. However, as mentioned above, theoretical dynamical studies of this type would be highly nontrivial and to date are nonexistent. In fact, constructing techniques to make these dynamical studies feasible constitutes an active area of our own research program. Why have such simulations not been performed? It is because of the — at best — murky understanding of M/C bonding. In other words, the lack of dynamics simulations on such systems is indicative of a much deeper, unanswered question; namely, what is the true nature of the interaction between metal atoms and the atoms present in a ceramic? It is a much more difficult and complex system to characterize than either a metal itself or an isolated ceramic, each of which can be described independently with much simpler models (e.g. glue or embedded-atom models for metals, ionic or shell models for ceramics, etc.). It is the interface interaction which eludes understanding to date. Thus, the emphasis in this review on static rather than dynamic properties of these materials points out that theorists are not yet in a position to examine these systems dynamically on the large length and time scales which may prove necessary to fully simulate thermal barrier coating failure.

That said, we can still glean some insight from the static investigations, enough to even make some tentative conclusions about how to improve TBC's. It is clear that the ideal coating should consist of a stress-tolerant ceramic and an oxidation-resistant bond coating with nearly the same co-efficient of thermal expansion and elastic modulus as both the ceramic and the Ni-based superalloy. The cost savings that such an ideal coating would produce is nontrivial: it has been estimated at 10^7 gallons of fuel per year for a fleet of 250 airplanes.

Use of nanostructured pure ZrO_2 might help create such an ideal coat-ing, as recent experiments have shown that one can stabilize the tetragonal phase of undoped ZrO_2 up to a critical grain radius of 6 nm.[49] We have provided an explanation for the special stability of t-ZrO_2 in nanocrys-tals, based on surface energy arguments. Thus, appropriately synthe-sized nanocrystals of ZrO_2 might allow the use of pure rather than doped zirconia, which would eliminate possible sources of materials failure due to dopant diffusion. Furthermore, use of t-ZrO_2 would provide the stress tolerance required via the transformation toughening mechanism observed for t-ZrO_2.[48]

It may be that the life expectancy of a TBC is even more sensitive to the bond coating composition than to the ceramic top coat composition.[260] Oxidation of the bond coating to Al_2O_3 may also be a significant source of TBC spallation. We suggest that reducing the amount of Al in the bond coat may be necessary to avoid spallation, based on our first principles calculations.

Based on the theoretical work done to date, it appears that rare earth or transition metals with open d- or f-shells (the less noble the better) show the greatest adhesion to ceramics, presumably because there can be a degree of covalent bonding at the interface. On the other hand, one wants to utilize transition metals that are either oxidation-resistant or form oxides that adhere strongly to ZrO_2. This may explain the presence of Y in the bond coating, since oxidation of the Y metal to Y_2O_3 will produce an oxide that forms solid solutions with ZrO_2, i.e. it is strongly interacting. It suggests that one should consider metals that form other cubic-based metal oxides, such as cerium, for addition to the bond coat.

As one can see, exploring the atomic-level properties of thermal barrier coatings involves a complex set of issues. We have attempted to provide some background on methods available for preparing and characterizing TBC's experimentally, as well as presenting the hierarchy of theoretical methods available in computational materials science, including the trade-offs involved in the use of each one. It is our hope that a microscopic understanding, provided both by ever-improving theory and experiment, of the basic mechanisms for metal–ceramic adhesion, ion and metal atom diffusion, and ceramic phase transformations eventually will lead to substantive improvements in macroscopic properties of TBC's, resulting in extended survival in the extreme environment present in an aircraft engine.

Acknowledgment

This work was supported by the Air Force Office of Scientific Research.

References

1. S. M. Meier, D. K. Gupta and K. D. Sheffler, *J. Miner. Metals Mat. Soc.* **43**, 50 (1991).
2. F. C. Toriz, A. B. Thakker and A. K. Gupta, *Surf. Coat. Tech.* **39/40**, 161 (1989).
3. D. J. Wortman, B. A. Nagaraj and E. C. Duderstadt, *Mat. Sci. Eng.* **A121**, 433 (1989).
4. F. Ernst, *Mat. Sci. Eng.* **R14**, 97 (1995).
5. R. Brydson, H. Müllejans, J. Bruley, P. A. Trusty, X. Sun, J. A. Yeomans and M. Rühle, *J. Microscopy* **177**, 369 (1995).
6. S. Musikant, *What Every Engineer Should Know about Ceramics* (Marcel Dekker, New York, 1991).
7. E. Ryshkewitch and D.W. Richerson, *Oxide Ceramics* (Academic Press, Orlando, 1985).
8. R. Taylor, J. R. Brandon and P. Morrell, *Surf. Coat. Tech.* **50**, 141 (1992).
9. F. H. Stott, D. J. de Wet and R. Taylor, *MRS Bull.* 46 (October 1994).
10. T. K. Gupta, J. H. Bectold, R. C. Kuznicki, L. H. Cadoff and B. R. Rossing, *J. Mat. Sci.* **12**, 2421 (1977).
11. A. Dominguez-Rodriguez and A. H. Heuer, in *Surfaces and Interfaces of Ceramic Materials*, Eds. L. C. Dufour *et al.* (Kluwer, 1989) pp. 761–776.

12. A. E. Hughes, in *Science of Ceramic Interfaces II*, Ed. J. Nowotny (Elsevier, Amsterdam, 1994).
13. I. Nettleship and R. Stevens, *Int. J. High Tech. Ceram.* **6**, 1 (1987).
14. T. Gupta, J. Bechtold and R. Kvznicki, *J. Mat. Sci.* **12**, 2421 (1977).
15. R. Garvie, *J. Am. Ceram. Soc.* **20**, 23 (1984).
16. J. Brandon and R. Taylor, *Surf. Coat. Tech.* **39–40**, 143 (1989).
17. H. Frank, H. Brown and P. Duwez, *J. Am. Ceram. Soc.* **37**, 129 (1954).
18. V. Pandolfelli, I. Nettleship and R. Stevens, *Proc. Br. Ceram. Soc.* **42**, 139 (1989).
19. W. Pyda, R. Haberko and M. Bucko, *Ceram. Int.* **18**, 321 (1992).
20. W. Pyda, R. Haberko and M. Bucko, in *Science and Technology of Zirconia V*, Ed. S. Badwal *et al.* (Technomic, PA, 1993).
21. P. Diaz, M. J. Edirisinghe and B. Ralph, *Surf. Coat. Tech.* **82**, 284 (1996).
22. P. Diaz, M. J. Edirisinghe and B. Ralph, *J. Mat. Sci. Lett.* **13**, 1595 (1994).
23. J. A. Haynes, E. D. Rigney, M. K. Ferber and W. D. Porter, *Surf. Coat. Tech.* **86–87**, 102 (1996).
24. Dongming Zhu and R. A. Miller, in *Fundamental Aspects of High Temperature Corrosion*, Ed. D. A. Shores *et al.* (Electrochemical Society Proceedings, Pennington, 1997), pp. 288–307.
25. B. A. Pint, I. G. Wright, W. Y. Lee, Y. Zhang, K. Prüßner and K. B. Alexander, *Mat. Sci. Eng.* **A245**, 201 (1998).
26. D. R. Clarke, R. J. Christensen and V. Tolpygo, *Surf. Coat. Tech.* **94–95**, 89 (1997).
27. Kh. G. Schmitt-Thomas, H. Haindl and D. Fu, *Surf. Coat. Tech.* **94–95**, 149 (1997).
28. P. Harmsworth and R. Stevens, *J. Mat. Sci.* **27**, 616 (1992).
29. W. Y. Lee and D. P. Stinton, *J. Am. Ceram. Soc.* **79**, 3003 (1996).
30. U. Leushake, T. Krell, U. Schulz, M. Peters, W. A. Kaysser and B. H. Rabin, *Surf. Coat. Tech.* **94–95**, 131 (1997).
31. C. R. Aita, U.S. Patent 5.472.795 (1995).
32. C. M. Scanlan, M. Gajdardziska-Josifovska and C. R. Aita, *Appl. Phys. Lett.* **64**, 3548 (1994).
33. J. Yuan, V. Gupta and M. Kim, *Acta Metall. Mat.* **43**, 769 (1995).
34. V. Gupta, J. Wu and A. N. Pronin, *J. Am. Ceram. Soc.* **80**, 3172 (1997).
35. C. Funke, J. C. Mailand, B. Siebert, R. Vaßen and D. Stöver, *Surf. Coat. Tech.* **94–95**, 106 (1997).
36. R. E. Taylor and K. D. Maglic, in *Compendium of Thermophysical Property Measurement Methods 1*, Ed. K. D. Maglic (Plenum Press, New York, 1984) p. 336.
37. R. E. Taylor, *Mat. Sci. Eng.* **A245**, 160 (1998).
38. R. E. Taylor, in *Compendium of Thermophysical Property Measurement Methods 1*, Ed. K. D. Maglic (Plenum Press, New York, 1984), p. 125.
39. S. Alaruri, L. Bianchini and A. Brewington, *Opt. Lasers Eng.* **30**, 77 (1998).
40. B. Ealet, E. Gillet, V. Nehasil and P. J. Møller, *Surf. Sci.* **318**, 151 (1994).

41. C. D. Qin and B. Derby, *J. Mat. Res.* **7**, 1480 (1992).

42. C. D. Qin and B. Derby, *J. Mat. Sci.* **28**, 4366 (1993).

43. T. Wagner, R. Kirchheim and M. Rühle, *Acta Metall. Mat.* **43**, 1053 (1995).

44. K. P. Trumble and M. Rühle, *Acta Metall. Mat.* **39**, 1915 (1991).

45. N. L. Loh, Y. L. Wu and K. A. Khor, *J. Mat. Proc. Tech.* **37**, 711 (1993).

46. Q. Zhong and F. S. Ohuchi, *J. Vac. Sci. Tech.* **A8**, 2107 (1990).

47. C. R. Aita, C. M. Scanlan and M. Gajdardziska-Josifovska, *JOM* **46**, 40 (1994).

48. A. Christensen and E. A. Carter, *Phys. Rev.* **B58**, 8050 (1998).

49. M. A. Shofield, C. R. Aita, P.M . Rice and M. Gajdardziska-Josifovska, *Thin Solid Films* **326**, 106 (1998) *ibid.*, p. 117.

50. M. Gajdardziska-Josifovska and C. R. Aita, *J. Appl. Phys.* **79**, 1316 (1996).

51. G. M. Newaz, S. Q. Nusier and Z. A. Chaudhury, *J. Eng. Mat. Tech.* **120**, 149 (1998).

52. A. G. Evans, G. B. Crumley and R. E. Demaray, *Oxid. Metals* **20**, 193 (1983).

53. H. E. Evans and M. P. Taylor, *Surf. Coat Tech.* **94–95**, 27 (1997).

54. H. E. Evans, *J. Mat. Sci. Eng.* **A120**, 139 (1989).

55. Z. Suo, *J. Mech. Phys. Solids* **43**, 829 (1995).

56. S. M. Meier and D. K. Gupta, *Trans. ASME* **23**, 245 (1993).

57. T.S. Oh, R. M. Cannon and R. O. Ritchie, *J. Am. Ceram. Soc.* **70**, C352 (1987).

58. I. E. Reimanis, B. J. Dalgleish and A. G. Evans, *Acta Mat.* **39**, 3133 (1991).

59. J. D. Kuenzly and D. L. Douglass, *Oxid. Metals* **8**, 139 (1974).

60. G. S. A. M. Theunissen, A. J. A. Winnubst and A. J. Burggraaf, in *Surfaces and Interfaces of Ceramic Materials*, Ed. L. C. Dufour *et al.* (Kluwer, 1989) pp. 365–372.

61. K. Schindler, D. Schmeisser, U. Vohrer, H. D. Wiemhöffer and W. Göpel, *Sensors and Actuators* **17**, 555 (1989).

62. G. M. Ingo and G. Padeletti, *Surf. Interface Anal.* **21**, 450 (1994).

63. J. Daloe and D. Boone, in Failure Mechanisms of Coating Systems Applied to Advanced Turbine Engine Components, *Proc. 42nd ASME Gas Turbine Aeroeng. Congr.* (Orlando, 1997).

64. B. A. Pint, A. J. Garrat-Reed and L. W. Hobbs, *Mat. High Temp.* **13**, 3 (1995).

65. B. A. Pint, *Oxid. Metals* **45**, 1 (1996).

66. B. A. Pint, A. J. Garratt-Reed and L. W. Hobbs, *J. Am. Ceram. Soc.* **81**, 305 (1997).

67. B. A. Pint, *MRS Bull.* **19**, 26 (1994).

68. P. Y. Hou, K. Prüßner, D. H. Fairbrother and J. G. Roberts, *et al.*, *Scripta Materialia* **40**, 241 (1998).

69. P. Y. Hou and J. Stringer, *Oxid. Metals* **38**, 323 (1992).

70. H. J. Grabke, G. Kurbatov and H. J. Schmutzler, *Oxid. Metals* **43**, 97 (1995).

71. K. Prüßner, E. Schumann and M. Rühle, in *Fundamental Aspects of High Temperature Corrosion VI*, Ed. D. A. Shores *et al.* (Electrochemical Society of Pennington, 1996), pp. 344–356.

72. B. A. Pint, *Oxid. Metals* **48**, 303 (1997).

73. W. P. Allen and N. S. Bornstein, in *High Temperature Coatings I*, Ed. N. Dahotre *et al.* (TMS, Warrendale, PA, 1995) pp. 193–202.

74. G. H. Meier, F. S. Pettit and J. L. Smialek, *Mat. Corrosions* **46**, 232 (1995).

75. M. A. Smith, W. E. Frazier and B. A. Pregger, *Mat. Sci. Eng.* **A203**, 388 (1995).

76. J. C. Schaeffer, W. H. Murphy and J. L. Smialek, *Oxid. Metals* **43**, 1 (1995).

77. J. G. Smeggil, A. W. Funkenbusch and N. S. Bornstein, *Metal Trans.* **A17**, 923 (1986).

78. C. Sarioglu, J. R. Blachre, F. S. Pettit and G. H. Meier, in *Microscopy of Oxidation 3*, Eds. S. B. Newcomb and J. A. Little, Institute of Metals, London, United Kingdom (1997), pp. 41–50.

79. R. J. Christensen, D. M. Lipkin and D. R. Clarke, *Appl. Phys. Lett.* **69**, 3754 (1996).

80. T. Troczynski and J. Camire, *Eng. Fracture Mech.* **51**, 327 (1995).

81. V. Gupta, A. S. Argon and J. A. Cornie, *J. Mat. Sci.* **24**, 41 (1989).

82. A. G. Evans and B. J. Dalgleish, *Acta Metall. Mat.* **40**, S295 (1992).

83. A. G. Evans, M. Rühle, B. J. Dagleigh and P. G. Charalambides, *Metall. Trans.* **A21**, 2419 (1990).

84. M. Arai, T. Sakuma, T. Mizutani, K. Kishimoto and M. Saito, *J. Ceram. Soc. Japan* **106**, 198 (1998).

85. J. Yuan and V. Gupta, *Acta Metall. Mat.* **43**, 781 (1995).

86. A. S. Argon, J. A. Cornie, V. Gupta, L. Lev and D. M. Parks, *Mat. Sci. Eng.* **A273**, 224 (1997).

87. V. Gupta, A. S. Argon, D. M. Parks and J. A. Cornie, *J. Mech. Phys. Solids* **40**, 141 (1992).

88. J. N. Johnson, *J. Appl. Phys.* **52**, 2812 (1981).

89. S. Eliezer, I. Gilath and T. Bar-Noy, *J. Appl. Phys.* **67**, 715 (1991).

90. T. W. Conlon, *Contemp. Phys.* **26**, 521 (1985).

91. M. F. Stroosnijder, in *Application of Particle and Laser Beams in Materials Technology*, Ed. P. Misaelides (1994).

92. M. F. Stroosnijder and G. Macchi, *Nucl. Instrum. Meth. Phys. Res.* **B100**, 155 (1995).

93. L. Tollier, R. Fabbro and E. Bartnicki, *J. Appl. Phys.* **83**, 1224 (1998).

94. M. W. Finnis, *J. Phys.: Condens. Matter* **8**, 5811 (1996).

95. V. E. Heinrich, *J. Cat.* **88**, 519 (1983).

96. K. M. Neyman, S. Ph. Ruzankin and N. Rösch, *Chem. Phys. Lett.* **246**, 546 (1995).

97. X. Xu, H. Nakatsuji, M. Ehara, X. Lü, N. Q. Wang and Q. E. Zang, *Chem. Phys. Lett.* **292**, 282 (1998).

98. F. Rittner, R. Fink, B. Boddenberg and V. Staemmler, *Phys. Rev.* **B57**, 4160 (1998).
99. G. Pacchioni, A. M. Ferrari, A. M. Marquez and F. Illas, *J. Comp. Chem.* **18**, 617 (1997).
100. M. Fernandez-Garcia, J. C. Conesa and F. Illas, *Surf. Sci.* **349**, 207 (1996).
101. Y. Ferro, A. Allouche, F. Corà, C. Pisani and C. Girardet, *Surf. Sci.* **325**, 139 (1995).
102. D. E. Ellis, G. A. Benesh and E. Byrom, *Phys. Rev.* **B16**, 3308 (1977).
103. M. A. Nygren, L. G. M. Petterson, Z. Barandiaran and L. Seijo, *J. Chem. Phys.* **100**, 2010 (1994).
104. N. Lopez, G. Pacchioni, F. Maseras and F. Illas, *Chem. Phys. Lett.* **294**, 611 (1998).
105. P. E. M. Siegbahn and U. Wahlgren, *Int. J. Quant. Chem.* **42**, 1149 (1992).
106. G. te Velde and E. J. Baerends, *Chem. Phys.* **177**, 399 (1993).
107. N. Govind, Y. A. Wang, A. J. R. da Silva and E. A. Carter, *Chem. Phys. Lett.* **295**, 129 (1998); N. Govind, Y. A. Wang and E. A. Carter, *J. Chem. Phys.* **110**, 7677 (1999).
108. J. Neugebauer and M. Scheffler, *Phys. Rev.* **B46**, 16067 (1992).
109. G. Makov and M. C. Payne, *Phys. Rev.* **B51**, 4014 (1995).
110. P. E. Blöchl, *J. Chem. Phys.* **103**, 7422 (1995).
111. H. L. Skriver and N. M. Rosengaard, *Phys. Rev.* **B43**, 9538 (1991).
112. B. C. Bolding and E. A. Carter, *Mol. Simul.* **9**, 269 (1992).
113. N. W. Ashcroft and N. D. Mermin, *Solid State Physics*, 444, Saunders College, Philadelphia, 1976.
114. L. D. Landau and E. M. Lifshitz, *Theory of Elasticity*, 3rd edition, **7** (Butterworth-Heinenann Ltd., Oxford, 1995).
115. M. S. Daw, *Phys. Rev.* **B47**, 10895 (1993).
116. X.-P. Li, R. W. Nunes and D. Vanderbilt, *Phys. Rev.* **B47**, 10891 (1993).
117. R. Baer and M. Head-Gordon, *Phys. Rev. Lett.* **79**, 3962 (1997).
118. K. R. Bates, A. D. Daniels and G. E. Scuseria, *J. Chem. Phys.* **109**, 3308 (1998).
119. D. Sanchez-Portal, P. Ordejon, E. Artacho and J. M. Soler, *Int. J. Quant. Chem.* **65**, 453 (1997).
120. W. Yang, *Phys. Rev. Lett.* **66**, 1438 (1991).
121. G. Galli and M. Parrinello, *Phys. Rev. Lett.* **69**, 3547 (1992).
122. F. Mauri, G. Galli and R. Car, *Phys. Rev.* **B47**, 9973 (1993).
123. P. Ordejon, D. A. Drabold, M. P. Grumbach and R. M. Martin, *Phys. Rev.* **B48**, 14646 (1993).
124. E. B. Stechel, A. R. Williams and P. J. Feibelman, *Phys. Rev.* **B49**, 10088 (1994).
125. R. Baer and M. Head-Gordon, *J. Chem. Phys.* **107**, 10003 (1997).
126. R. Baer and M. Head-Gordon, *Phys. Rev.* **B58**, 15296 (1998).
127. R. Baer and M. Head-Gordon, *J. Chem. Phys.* **109**, 10159 (1998).

128. T. Zhu, W. Pan and W. Yang, *Phys. Rev.* **B53**, 12713 (1996).
129. S. C. Watson and P. A. Madden, *Phys. Chem. Comm.* **1**, 1 (1998).
130. L.-W. Wang and M. P. Teter, *Phys. Rev.* **B45**, 13196 (1992).
131. M. Pearson, E. Smargiassi and P. A. Madden, *J. Phys.: Condens. Matter* **5**, 3321 (1993).
132. F. Perrot, *J. Phys.: Condens. Matter* **6**, 431 (1994).
133. E. Smargiassi and P. A. Madden, *Phys. Rev.* **B49**, 5220 (1994).
134. M. Foley, E. Smargiassi and P. A. Madden, *J. Phys.: Condens. Matter* **6**, 5231 (1994).
135. E. Smargiassi and P. A. Madden, *Phys. Rev.* **B51**, 117 (1995).
136. M. Foley and P.A. Madden, *Phys. Rev.* **B53**, 10589 (1996).
137. B. J. Jesson, M. Foley and P. A. Madden, *Phys. Rev.* **B55**, 4941 (1997).
138. E. Chacón, J. E. Alvarellos and P. Tarazona, *Phys. Rev.* **B32**, 7868 (1985).
139. P. García-González, J. E. Alvarellos and E. Chacón, *Phys. Rev.* **B53**, 9509 (1996).
140. P. García-González, J. E. Alvarellos and E. Chacón, *Phys. Rev.* **A54**, 1897 (1996).
141. P. García-González, J. E. Alvarellos and E. Chacón, *Phys. Rev.* **B57**, 4192 (1998).
142. P. García-González, J. E. Alvarellos and E. Chacón, *Phys. Rev.* **B57**, 4857 (1998).
143. Y. A. Wang, N. Govind and E. A. Carter, unpublished.
144. Y. A. Wang, N. Govind and E. A. Carter, *Phys. Rev.* **B58**, 13465 (1998).
145. N. Lopez and F. Illas, *J. Phys. Chem.* **100**, 16275 (1996).
146. N. Lopez and F. Illas, *J. Chem. Phys.* **107**, 7345 (1997).
147. N. C. Bacalis and A. B. Kunz, *Phys. Rev.* **B32**, 4857 (1985).
148. A. D. Zdetsis and A. B. Kunz, *Phys. Rev.* **B32**, 6358 (1985).
149. J. P. Perdew, *Phys. Rev.* **B33**, 8822 (1986); J. P. Perdew, *ibid.* **34**, 7406(E) (1986).
150. A. D. Becke, *Phys. Rev.* **A38**, 3098 (1988).
151. J. P. Perdew and Y. Wang, *Phys. Rev.* **B43** 13244 (1992); J. P. Perdew, J. A. Chevary, S. H. Vosko, K. A. Jackson, M. R. Pederson, D. J. Singh and C. Fiolhais, *ibid.* **46** 6671 (1992).
152. J. P. Perdew, K. Burke and M. Ernzerhof, *Phys. Rev. Lett.* **77**, 3865 (1997).
153. B. Hammer, L. B. Hansen and J. K. Norskov, *Phys. Rev.* **B59**, 7413 (1999).
154. M. Causá and A. Zupan, *Chem. Phys. Lett.* **220**, 145 (1994).
155. Yu. F. Zhukovskii, M. Alfredsson, K. Hermansson and E. A. Kotomin, *Nucl. Instrum. Meth. Phys. Res.* **B141**, 73 (1998).
156. E. Heifets, E. A. Kotomin and R. Orlando, *J. Phys.: Condens. Matter* **8**, 6577 (1996).
157. A. Trampert, F. Ernst, C. P. Flynn, H. F. Fischmeister and M. Rúhle, *Acta Metall. Mat.* **40**, S227 (1992).
158. R. G. Parr and W. Yang, *Density-Functional Theory of Atoms and Molecules* (Oxford University Press, New York, 1989).

159. A. D. Becke, *J. Chem. Phys.* **98**, 5648 (1993).
160. N. Lopez, F. Illas, N. Rösch and G. Pacchioni, *J. Chem. Phys.* **110**, 4873 (1999).
161. P. Hohenberg and W. Kohn, *Phys. Rev.* **B136**, 864 (1964); W. Kohn and L. Sham, *Phys. Rev.* **A140**, 1133 (1965).
162. R. O. Jones and O. Gunnarsson, *Rev. Mod. Phys.* **61**, 689 (1989).
163. E. G. Moroni, G. Kresse, J. Hafner and J. Furthmüller, *Phys. Rev.* **B56**, 15629 (1997).
164. J. E. Jaffe, Z. Lin and A. C. Hess, *Phys. Rev.* **B57**, 11834 (1998).
165. P. Ziesche, S. Kurth and J.P. Perdew, *Comput. Mat. Sci.* **11**, 122 (1998).
166. Y. M. Juan and E. Kaxiras, *Phys. Rev.* **B48**, 14944 (1993).
167. L. Vitos, A. V. Ruban, H. L. Skriver and J. Kollar, *Surf. Sci.* **411**, 186 (1998).
168. J. Goniakowski, J. M. Holender, L. N. Kantorovich, M. J. Gillan and J. A. White, *Phys. Rev.* **B53**, 957 (1996).
169. V. I. Anisimov, F. Aryasetiawan and A. I. Lichtenstein, *J. Phys.: Condens. Matter* **9**, 767 (1997).
170. J. P. Perdew and A. Zunger, *Phys. Rev.* **B23**, 5048 (1981).
171. D. Vogel, P. Kruger and J. Pollmann, *Phys. Rev.* **B54**, 5495 (1996).
172. V. I. Anisimov, J. Zaanen and O. K. Andersen, *Phys. Rev.* **B44**, 943 (1991); V. I. Anisimov, I. V. Solovyev, M. A. Korotin, M. T. Czyzyk and G. A. Sawatzky, *ibid.* **48**, 16929 (1993); A. I. Lichtenstein, J. Zaanen and V. I. Anisimov, *ibid.* **R52**, 5467 (1995).
173. P. Blöchl, G. P. Das, H. F. Fischmeister and U. Schönberger, in *Metal–Ceramic Interfaces*, Acta-Scripta Metallurgica Proceedings, Series 4, Eds. M. Ruhle, A. G. Evans, M. F. Ashby and J. P. Hirth (Oxford, New York, 1990), p. 9; A. J. Freeman and C. Li, *ibid.* p. 2.
174. C. Li, R. Wu, A. J. Freeman and C. L. Fu, *Phys. Rev.* **B48**, 8317 (1993).
175. I. Yudanov, G. Pacchioni, K. Neyman and N. Rösch, *J. Phys. Chem.* **B101**, 2786 (1997).
176. U. Schönberger, O. K. Andersen and M. Methfessel, *Acta Metall. Mat.* **40**, S1 (1992).
177. T. Hong, J. R. Smith and D. J. Srolovitz, *Acta Metall. Mat.* **43**, 2721 (1995).
178. We used the plane-wave code VASP, which is described in G. Kresse and J. Hafner, *Phys. Rev.* **B47**, 558 (1993); *ibidem*, **49**, 14251 (1994); G. Kresse and J. Furthmüller, *Comput. Mat. Sci.* **6**, 15 (1996); G. Kresse and J. Furthmüller, *Phys. Rev.* **B55**, 11169 (1996); ultrasoft pseudopotentials are described in G. Kresse and J. Hafner, *J. Phys.: Condens. Matter* **6**, 8245 (1994).
179. In each case, a plane-wave cutoff at 340 eV was used, implying that absolute energies were converged to within 6 meV/atom. The ultrasoft pseudopotential augmentation charge cutoff was 554 eV. A k-point sampling density of 0.05 Å^{-1} was used and all LDA and GGA calculations, respectively, were performed in the same unit cells to optimize error cancellation. The GGA

parameterization used is that of Perdew *et al.* [PW91 = *Phys. Rev.* **B46**, 6671 (1991)] Ultrasoft pseudopotentials with outermost pseudization radii $r_c = 1.52$, 1.01, 1.50 Å were used for Mg, O, Ag, respectively. Separate pseudopotential sets were generated for both LDA and GGA, i.e. the pseudization was consistent with the exchange correlation functional applied in the interface calculation. For LDA and GGA, the cubic bulk equilibrium lattice constants for MgO are predicted to be $a_0 = 4.126$ Å and 4.224 Å, respectively, both in good agreement with the experimental value $a_0 = 4.205$ Å. Our MgO(001) slab and vacuum region were each 4 layers thick, and we considered only one monolayer of Ag. Corrections for interslab multipole interactions, although small, were added.

180. R. Benedek, M. Minkoff and L. H. Yang, *Phys. Rev.* **B54**, 7697 (1996).
181. J. Goniakowski, *Phys. Rev.* **B57**, 1935 (1998).
182. B. Hammer, L. B. Hansen and J. K. Nørskov, *Phys. Rev.* **B59**, 7413 (1999).
183. G. Pacchioni and N. Rösch, *J. Chem. Phys.* **104**, 7329 (1996).
184. Y. Li, D. C. Langreth and M. R. Pederson, *Phys. Rev.* **B52**, 6067 (1995).
185. A. V. Matveev, K. M. Neyman, G. Pacchioni and N. Rösch, *Chem. Phys. Lett.* **299**, 603 (1999).
186. V. Musolino, A. Selloni and R. Car, *Surf. Sci.* **402–404**, 413 (1998).
187. K. H. Johnson and S. V. Pepper, *J. Appl. Phys.* **53**, 6634 (1982).
188. M. Gautier, G. Renaud, L. P. Van and B. Villette, *J. Am. Ceram. Soc.* **77**, 323 (1994).
189. P. Guenard, G. Renaud, A. Barbier and M. Gautier-Soyer, *Surf. Rev. Lett.* **5**, 321 (1998).
190. J. Guo, D. E. Ellis and D. J. Lam, *Phys. Rev.* **B45**, 13647 (1992).
191. F. S. Ohuchi and M. Kohyama, *J. Am. Ceram. Soc.* **74**, 1163 (1991).
192. J. A. Haynes, E. D. Rigney, M. K. Ferber and W. D. Porter, *Surf. Coat. Tech.* **86–87**, 102 (1996).
193. We found that the metal's electronic and geometric structures were well-converged with respect to slab thickness for a three layer Ni(111) slab. Accordingly, we used a three layer Ni(111) slab with a hexagonal unit cell of 4.975 Å side lengths and periodic boundary conditions. This corresponded to a fully relaxed bulk Ni lattice constant. The top two layers of the Ni and all of the Al_2O_3 layers were allowed to relax with no symmetry constraints. Over 10 Å of vacuum layer was included above the M/C layers and a $5 \times 5 \times 1$ k-point sampling grid was used. The pseudization radii $r_c = 1.29$ Å for Ni, 1.40 Å for Al, and 1.01 Å for O were used. Fermi surface smearing of 0.10 eV was employed during the relaxation and the kinetic energy cutoff for the basis was 270 eV with a 554 eV cutoff used for the augmentation charge of the ultrasoft pseudopotential. Once we reached the equilibrium structure, a final point calculation was performed with the kinetic energy cutoff increased to 340 eV, and the tetrahedron method with Blöchl corrections [P. E. Blöchl, O. Jepsen and O. K. Andersen, *Phys. Rev* **B49**, 16223 (1994)] was employed instead of Fermi smearing. The nonlinear core correction for exchange and

correlation [S. G. Louie, S. Froyen and M. L. Cohen, *Phys. Rev.* **B26**, 1738 (1982); W. Maysenhölder, S. G. Louie and M. L. Cohen, *Phys. Rev.* **B31**, 1817 (1985)], was employed. These calculations included spin polarization and were performed at the GGA (PW91) level. Our calculated c/a lattice constant ratio for the hexagonal cell was 2.73 which is the same as that found by experiment [H. d'Amour, D. Schiferl, W. Denner, H. Schultz and W. B. Holzapfel, *J. Appl. Phys.* **49**, 4411 (1978)]. Our orthorhombic cell's lattice constant agreed within one percent of experiment (5.19 Å to 5.14 Å respectively). The GGA Ni lattice constant, used for the M/C interface calculations, was 3.51 Å, in close agreement with the experimental value of 3.52 Å.[113]

194. C. Kruse, M. W. Finnis, V. Y. Milman, M. C. Payne, A. de Vita and M. J. Gillan, *J. Am. Ceram. Soc.* **77**, 431 (1994).

195. C. Kruse, M. W. Finnis, J. S. Lin, M. C. Payne, V. Y. Milman, A. de Vita and M. J. Gillan, *Phil. Mag. Lett.* **73**, 377 (1996).

196. We again used a three layer Ni(111) slab. We employed a hexagonal unit cell (6.5 Å × 6.5 Å) with periodic boundary conditions. We fixed the unit cell at the fully relaxed bulk lattice constant of Ni. Again, we allowed only metal atom relaxations of the top two layers of Ni; the ceramic was allowed to relax fully. No symmetry constraints were imposed. We include 10 Å of vacuum above the M/C film couple. Only the valence $5s$ and $4d$ electrons are treated explicitly for Zr; a justification of this is given in our earlier work on ZrO_2.[48] The pseudization radii used were 1.01 for O, 1.62 for Zr, and 1.29 for Ni. Once again, the nonlinear core correction for exchange and correlation was used. Calculations were carried out spin-polarized at both the LDA and GGA levels. A k-point sampling density of 0.05 Å^{-1} (a $3 \times 3 \times 1$ grid) was used. Fermi surface smearing by 0.30 eV was employed to enhance convergence. Kinetic energy cutoffs were 270 eV for the plane wave basis and 554 eV for the augmentation charge of the ultrasoft pseudopotential. Although dipole corrections were calculated for the final structure's total energy, the effect was small. The Ni lattice constant was the same as in the Ni/Al_2O_3 calculations and was used for the interface calculations. The cubic ZrO_2 lattice constant for GGA was 5.10 Å, which agrees with the experimental value extrapolated to T = 0 [H. G. Scott, *J. Mat. Sci.* **10**, 1527 (1975)].

197. W. A. Harrison, *Electronic Structure and the Properties of Solids* (Dover Pub., Inc., New York, 1989) p. 231.

198. P. E. Blöchl, *Phys. Rev.* **B50**, 17953 (1994).

199. We considered a large number of possible interface matchings for cells with surface areas up to 50 Å^2. The lowest lattice misfit of 4 percent for a reasonably sized cell was determined to involve the α-$Al_2O_3(1-102)$ surface bonding to c-$ZrO_2(001)$. This growth orientation is in fact the same as what has been observed experimentally for YSZ grown on the (1–102) plane of sapphire, [F. Konushi, T. Doi, H. Matsunaga, Y. Kakihara,

M. Koba, K. Awane and I. Nakamura in *Layered Structures and Epitaxy Symposium*, Ed. J. M. Gibson *et al.* (Materials Research Society, Pittsburgh, 1986) pp. 259–64; X. D. Wu, R. E. Muenchausen, N. S. Nogar, A. Pique, R. Edwards, B. Wilkens, T. S. Ravi, D. M. Hwang and C. Y. Chen, *Appl. Phys. Lett.* **58**, 304 (1991); and G. Garcia, J. Casado, J. Llibre, A. Figueras, S. Schamm, D. Dorignac and Ch. Grigis *Proc. 13th Int. Conf. Chem. Vapor Deposition*, Ed. T. M. Besmann *et al.* (Electrochemical Society of Pennington, 1996) pp. 699–705], so we are modeling a relevant interface. The alumina substrate was 9 Å thick, enough to emulate a bulk ceramic surface, both with regards to electronic structure and relaxation effects (based on our systematic studies of ZrO_2 and Al_2O_3). The ZrO_2 overlayer is also about 9 Å thick and we also include 10 Å of vacuum in between these thin film couples, which is usually enough to ensure negligible coupling between the thin film couples unless a large dipole is present at the interface. Apart from periodic boundary conditions on the unit cell, no symmetry or constraints were imposed on the electronic density and ionic motion. The interface unit cells (lattice constants and cell angles) were not relaxed, but fixed to the values derived from the fully relaxed bulk unit cell of sapphire. We then performed fully self-consistent ionic relaxation for this alumina/zirconia interface structure. These calculations included 4 k-points (for an equispaced grid) and were performed at the GGA level [J. P. Perdew and A. Zunger, *Phys. Rev.* **B23**, 5048 (1981); and A. D. Becke, *J. Chem. Phys.* **96**, 2155 (1992)].

200. J. Harris, *Phys. Rev.* **B31**, 1770 (1985).
201. W. M. C. Foulkes and R. Haydock, *Phys. Rev.* **B39**, 12520 (1989).
202. M. W. Finnis, *J. Phys.: Condens. Matter* **2**, 331 (1990).
203. N. Chetty, K. W. Jacobsen and J. K. Norskov, *J. Phys.: Condens. Matter* **3**, 5437 (1991).
204. O. F. Sankey and D. J. Niklewski, *Phys. Rev.* **B40**, 3979 (1989).
205. K. Stokbro, N. Chetty, K. W. Jacobsen and J. K. Norskov, *Phys. Rev.* **B50**, 10727 (1994).
206. A. P. Horsfield, *Phys. Rev.* **B56**, 6594 (1997).
207. A. M. Stoneham, *Appl. Surf. Sci.* **14**, 249 (1983).
208. M. W. Finnis, *Acta Metall. Mat.* **40**, S25 (1992).
209. J. R. Smith, T. Hong and D. J. Srolovitz, *Phys. Rev. Lett.* **72**, 4021 (1994).
210. W. A. Harrison, *Electronic Structure and Properties of Solids* (Dover, New York, 1989).
211. D. Pettifor, *Bonding and Structure of Molecules and Solids* (Oxford, New York, 1995).
212. P. W. Atkins, *Phys. Chem.* (W. H. Freeman, New York, 1994).
213. R. Hoffmann, *J. Chem. Phys.* **39**, 1397 (1963).
214. A. B. Anderson, *J. Chem. Phys.* **60**, 2477 (1974); *ibid.* **62**, 1187 (1975).
215. N. Chetty, K. Stokbro, K. W. Jacobsen and J. K. Norskov, *Phys. Rev.* **B46**, 3798 (1992).

216. K. Stokbro, N. Chetty, K. W. Jacobsen and J. K. Norskov, *J. Phys.: Condens. Matter* **6**, 5415 (1994).
217. H. Haas, C. Z. Wang, M. Fähnle, C. Elsässer and K. M. Ho, *Phys. Rev.* **B57**, 1461 (1998).
218. S. Froyen, *Phys. Rev.* **B22**, 3119 (1980).
219. M. Elstner, D. Porezag, G. Jungnickel, J. Elsner, M. Haugk, Th. Frauenheim, S. Suhai and G. Seifert, *Phys. Rev.* **B58**, 7260 (1998).
220. P. Alemany, R. S. Boorse, J. M. Burlitch and R. Hoffmann, *J. Phys. Chem.* **97**, 8464 (1993).
221. K. Nath and A. B. Anderson, *Phys. Rev.* **B39**, 1013 (1989).
222. A. B. Anderson, C. Ravimohan and S. P. Mehandru, *Surf. Sci.* **183**, 438 (1987).
223. T. R. Ward, P. Alemany and R. Hoffmann, *J. Phys. Chem.* **97**, 7691 (1993).
224. S. Y. Hong, A. B. Anderson and J. L. Smialek, *Surf. Sci.* **230**, 175 (1990).
225. A. B. Anderson, S. P. Mehandru and J. L. Smialek, *J. Electrochem. Soc.* **132**, 1695 (1985).
226. M. S. Daw and M. I. Baskes, *Phys. Rev. Lett.* **50**, 1285 (1983).
227. M. W. Finnis and J. E. Sinclair, *Phil. Mag.* **A50**, 45 (1984).
228. F. Ercolessi, E. Tosatti and M. Parrinello, *Phys. Rev. Lett.* **57**, 719 (1986).
229. K. W. Jacobsen, J. K. Norskov and M. J. Puska, *Phys. Rev.* **B35**, 7423 (1987).
230. A. P. Sutton and J. Chen, *Phil. Mag. Lett.* **61**, 139 (1990).
231. J. A. Moriaty, *Phys. Rev.* **B38**, 3199 (1988).
232. A. E. Carlsson, *Phys. Rev.* **B44**, 6590 (1991).
233. M. Aoki, *Phys. Rev. Lett.* **71**, 3842 (1993).
234. B. G. Dick and A. W. Overhauser, *Phys. Rev.* **112**, 603 (1958).
235. M. Wilson and P. A. Madden, *J. Phys.: Condens. Matter* **5**, 2687 (1993).
236. M. Wilson, M. Exner, Y.-M. Huang and M. W. Finnis, *Phys. Rev.* **B54**, R15683 (1996); M. Wilson, U. Schönberger and M. W. Finnis, *ibid.* **54**, 9147 (1996).
237. M. Wilson, P. A. Madden, N. C. Pyper and J. H. Harding, *J. Chem. Phys.* **104**, 8068 (1996).
238. A. J. Rowley, P. Jemmer, M. Wilson and P. A. Madden, *J. Chem. Phys.* **108**, 10209 (1998).
239. A. M. Stoneham and P. W. Tasker, *J. Phys. C: Solid State Phys.* **18**, 543 (1985).
240. D. M. Duffy, J. H. Harding and A. M. Stoneham, *J. Appl. Phys.* **76**, 2791 (1994).
241. J. Purton, S. C. Parker and D. W. Bullett, *J. Phys.: Condens. Matter* **9**, 5709 (1997).
242. S. Kohlhoff, P. Gumbsch and H. F. Fischmeister, *Phil. Mag.* **A64**, 851 (1991).
243. R. Thomson, S. J. Zhou, A. E. Carlsson and V. K. Tewary, *Phys. Rev.* **B46**, 10613 (1992).

244. R. E. Rudd and J. Q. Broughton, *Phys. Rev.* **B58**, R5893 (1998).
245. E. B. Tadmor, M. Ortiz and R. Phillips, *Phil. Mag.* **A73**, 1529 (1996).
246. K. A. Fichthorn and W. H. Weinberg, *J. Chem. Phys.* **95**, 1090 (1991).
247. M. R. Radeke and E. A. Carter, *Phys. Rev.* **B55**, 4649 (1997).
248. P. Ruggerone, A. Kley and M. Scheffler, *Comments Condens. Matter Phys.* **18**, 261 (1998).
249. J. G. LePage, M. Alouani, D. L. Dorsey, J. W. Wilkins and others, *Phys. Rev.* **B58**, 1499 (1998).
250. M. Neurock and E. W. Hansen, *Comp. Chem. Eng.* **22**, S1045 (1998).
251. M. Jiang, Y.-J. Zhao and P.-L. Cao, *Phys. Rev.* **B57**, 10054 (1998).
252. M. J. Caturla, *Comp. Mat. Sci.* **12**, 319 (1998).
253. A. F. Voter, *J. Chem. Phys.* **106**, 4665 (1997).
254. A. F. Voter, *Phys. Rev. Lett.* **78**, 3907 (1997).
255. A. F. Voter, *Phys. Rev.* **B57**, R13985 (1998).
256. A. R. Miedema and R. Boom, *Z. Metallk.* **69**, 183 (1978); F. R. de Boer, R. Boom, W. C. M. Mattens, A. R. Miedema and A. K. Nissen, *Cohesion in Metals: Transition Metal Alloys* (North-Holland, Amsterdam, 1988).
257. J.-G. Li, *J. Am. Ceram. Soc.* **75**, 3118 (1992).
258. D. Chatain, I. Rivollet and N. Eustathopoulos, *J. Chim. Phys. et de Physico-Chim. Biol.* **83**, 561 (1986).
259. G. Bordier and C. Noguera, *Phys. Rev.* **B44**, 6361 (1991).
260. S. Stecura, *Thin Solid Films* **182**, 121 (1989).

CHAPTER 11

MOLECULAR DYNAMICS SIMULATIONS OF DETONATIONS[*]

C. T. White[†,‡,§] and D. R. Swanson[§]

‡ *Department of Physics and Institute for Shock Physics*
Washington State University, Pullman, WA 99164
§ *Naval Research Laboratory, Washington, DC 20375*

D. H. Robertson
Department of Chemistry, IUPUI, 402 N. Blackford St.
Indianapolis, IN 46202

Contents

[*]Supported by ONR through NRL and DoE through the Institute for Shock Physics at WSU.
[†]To whom correspondence should be addressed.

1. Introduction

Detonations race through condensed-phase chemical explosives at supersonic velocities, inducing the chemical reactions that sustain them. Once begun, a detonation rapidly consumes the material, often with catastrophic results. For example, chemical energy can be released at rates that exceed 10^{11} Watts on a 10 cm^2 detonation front — a figure comparable to the electrical generating capacity of the United States.[1]

Immediately ahead of the detonation front the explosive rests quietly in its metastable state, while to the rear the shocked and reacted material flows at several kilometers per second with a pressure of several hundred thousand atmospheres and temperature of several thousand Kelvins.[1,2] The rapid compression and heating of matter to these extreme conditions and the associated high velocity flow are properties of detonations that can be shared by strong shockwaves. However, with detonations the heated and compressed flow is selfsustaining. Typically, detonations are maintained by the exothermic chemistry they induce. Detonations driven by first order phase transitions have been envisioned,[3] but have not yet been observed.

There are two fundamental length scales in even the simplest detonation. They are conveniently introduced by starting from a point of view that ignores them. Imagine the detonation as an inviscid planar flow driven by a steadily moving piston that mimics the work done on the system by expansion of the products. The continuum Euler equations for compressive inviscid fluids, resulting from the conservation of mass, energy, and momentum, allow only for discontinuous selfsimilar steady solutions satisfying the boundary conditions.[4] Ahead of the discontinuity the unreacted material is in its quiescent state while immediately to its rear — prior to expansion — the shocked and reacted material is flowing at high velocities at elevated pressures.[1,2] In this description the detonation front has no width and the detonation profile is a step function. As a result, there is no fundamental length scale.

Real explosives cannot respond instantaneously. Hence, the front cannot be a step function. In particular, relaxing the assumption of inviscid flow while still mimicking the chemistry with a steadily moving piston introduces a first length scale — that associated with the initial response at the detonation front. Even without chemistry, the rapid changes in the flow variables across the shockfront must cause dissipative processes that arise from viscosity and heat conduction, which smear the front. Intuitively, this initial broadening should correspond to several molecular mean free paths. Thus, for a gas at ambient conditions these dissipative mechanisms should smear the front by a hundred nanometers or more, but for liquids and solids the corresponding broadening should be only several nearest neighbor distances and hence only a few nanometers. These conclusions should remain unaltered by the inclusion of slower chemistry. Therefore, within the continuum formulation of the problem the initial rise of condensed phase shockfronts, including detonation fronts, should remain sharp on a scale of ten's of nanometers. These widths correspond to rise times of picoseconds for shock velocities of kilometers per second.

Unsupported detonations are not driven by a piston but rather by the induced exothermic chemistry. This exothermic chemistry introduces a second important length scale — that associated with the time it takes for these reactions to run to completion. Although this chemistry is thought to proceed promptly, estimates for reaction zone widths range from millimeters[5] to microns,[2] which correspond to passage times from microseconds to nanoseconds. Thus, although small, these length and time scales can be much larger than the scales associated with the initial compression. There is no physical reason, however, why the length of the reaction zone cannot be much shorter — even down to a few nanometers for systems driven by sufficiently simple chemical reactions. The reaction zone length is determined by the complexity of the chemical processes driving the detonation and hence will vary from system to system. In any event, the finite length reaction zone must keep pace with the front. Therefore, even if the dissipative processes associated with the initial shock compression were ignored, the detonation front would have a fixed width.

The explicit inclusion of chemistry forms the basis of the ZND theory of detonation.[1,2,6,7] This theory was formulated independently during the war years of the 1940's by Yakov B. Zel'dovich in the Soviet Union, John von Neumann in the United States, and Werner S. Doering in Germany[2] and is named in honor of them. It has continued to evolve over the last sixty years into the modern continuum theory of detonations — a theory that yields accurate predictions for the macroscopic properties of detonations. In spite of these advances, the atomic scale behavior of detonations in solids remains poorly understood. Continuum theories can only address this domain indirectly, while the extremely short time and length scales of the processes involved together with their destructive nature make experiments difficult, if not currently impossible. This state of affairs is unfortunate, because the processes leading to detonations must start at the atomic scale. A better understanding of the nature of these processes could prove crucial in the design of safer explosives.

Processes at detonation fronts occur on such short time scales (sub-picosecond) and length scales (subnanometer) that they are ideal for molecular dynamics simulations. Such simulations, which follow individual atomic trajectories, not only have the potential to probe how detonations are initiated but are also capable of studying the subsequent chemistry at the shockfront. The approach might also clarify how the discrete shock-induced chemistry relates to the continuum theory of detonations.

Molecular dynamics simulations of chemically sustained shockwaves, however require potentials that are capable of simultaneously following the dynamics of thousands of atoms in a rapidly changing environment while including the possibility of exothermic chemical reactions. The reactions should also proceed along chemically reasonable reaction paths from the cold solid state reactants to the hot gas-phase molecular products. For the molecular solids that are typical of energetic materials these potentials must incorporate both strong intramolecular forces for binding atoms into molecules and weak intermolecular forces for binding molecules into solids.

In Sec. 2, we review the reactive empirical bond order potentials that we have developed over the last decade which possess these essential characteristics.[8,9] Our model systems and computational techniques are also described. Simulations using the first of our models to exhibit classic detonation behavior are discussed in Sec. 3, where results involving both detonation and initiation are presented. In Sec. 4, the results of our molecular dynamics simulations are compared in detail with the predictions of continuum theory. Then, in Sec. 5, we review some simulations that raise the possibility of a detonation accompanied by several phase transitions. A summary with conclusions is given in the final section.

2. Models and Methods

Our earliest attempt to model shock-induced chemistry with molecular dynamics[10] was based on LEPS-like potentials, which were used to describe a nitric oxide chain with the NO molecules arranged head to tail and tail to head along the chain (NO, ON, NO, ...). This approach was successful in demonstrating that chemically sustained shockwaves driven by exothermic reactions (in this case the half reactions $N + NO \rightarrow N_2 + O$) could be modeled with molecular dynamics. However, the potential form assumed required a fixed number of nearest neighbors and hence was not readily generalized to higher dimensions.

To overcome this limitation we developed a series of potentials in the late 1980's and early 1990's that have become known as reactive empirical bond order (REBO) potentials.[8,9] These potentials are based on the empirical bond order potential form introduced by Tersoff to describe the static properties of silicon[11] but were tailored by us to incorporate a modicum of chemistry. In Sec. 2.1, after introducing the REBO potential form, we describe our simple models for energetic materials that are based on these potentials. In Sec. 2.2, we provide an overview of the approach taken to implement our simulations of shock-induced chemistry and detonations.

2.1. *Reactive Empirical Bond Order (REBO) Potentials*

Within the REBO approach, the binding energy of a system on N atoms is represented by

$$E_b = \sum_i^N \sum_{j>i}^N \{f_c(r_{ij})[V_R(r_{ij}) - \bar{B}_{ij} V_A(r_{ij})] + V_{vdW}(r_{ij})\}, \qquad (1)$$

where the key functions entering this expression are defined in Table 1, r_{ij} is the scalar distance between atoms i and j, and $f_c(r_{ij})$ is a function that restricts the range of the potential. The molecular bonding portion of the potential consists of a repulsive term, $V_R(r)$, and attractive term, $V_A(r)$, both of which are modeled by exponentials. The weak long ranged van der Waals interaction is described with a Lennard–Jones potential, $V_{vdW}(r)$, that is truncated not only at large distances but also at bonding separations

Table 1. REBO components and parameters used in Eq. (1).

$$V_R(r) = \frac{D_e}{S-1} \exp[-\alpha\sqrt{2S}\,(r - r_e)]$$

$$V_A(r) = \frac{SD_e}{S-1} \exp\left[-\alpha\sqrt{\frac{2}{S}}\,(r - r_e)\right]$$

$$B_{ij} = \left\{1 + G \sum_{k \neq i,j} f_c(r_{ik}) \exp[m(r_{ij} - r_{ik})]\right\}^{-n}$$

$$f_c(r) = \begin{cases} 1 & r < r_e + \delta \\ \frac{1}{2}\left\{1 + \cos[\beta\pi(r - (r_e + \delta))]\right\} & r_e + \delta \leq r < r_e + \delta + \beta^{-1} \\ 0 & r_e + \delta + \beta^{-1} \leq r \end{cases}$$

$Mass_{A,B} = 14.0$ amu; $D_e^{AB} = 2.0$ eV; $D_e^{AA} = D_e^{BB} = 5.0$ eV; $S = 1.8$; $\alpha = 2.7$ Å^{-1}; $G = 5.0$; $m = 2.25$ Å^{-1}; $n = 0.5$.

AB Model I: $r_e = 1.2$ Å, $\delta = 0.467$ Å; $\beta = 1.2$ Å^{-1}.
AB Model II: $r_e = 1.0$ Å, $\delta = 0.400$ Å, $\beta = 1.0$ Å^{-1}.
AB Model III: $r_e = 1.0$ Å, $\delta = 1.000$ Å, $\beta = 1.0$ Å^{-1}.

$V_{vdW}(r)$ is a Lennard–Jones potential splined quickly to zero (Ref. 8). It is not the cause of any significant differences between the models.

to allow for covalent bonding. The bond order function, $\bar{B}_{ij} \equiv (B_{ij} + B_{ji})/2$, introduces many-body effects into the potential by modifying $V_A(r)$ in response to the local bonding environment. These many-body effects arise from electronic degrees of freedom that are not explicitly treated in the model. For an isolated diatomic molecule, $\bar{B}_{ij}$ equals one so that the potential reduces to a generalized Morse function of the type used to describe diatomic bonding.[12] This diatomic bond has a length r_e, a well depth D_e, and a force constant $2D_e\alpha^2$. For more highly coordinated structures, $\bar{B}_{ij}$ is no longer one, but decreases with the increasing number of competing bonds available to atoms i and j. This decrease in $\bar{B}_{ij}$ reflects the finite number of valence electrons that atoms i and j have available for bonding. When the coupling G entering $\bar{B}_{ij}$ is small, the strength of the bond between these two atoms is only slightly reduced by the competing bonds, and the potential will favor highly coordinated metallic systems. However, when G is large, the potential favors low coordinated structures with a few strong bonds such as those in molecular solids. Cases intermediate between these two extremes correspond to the bonding in semiconductors such as silicon.[13] In essence, $\bar{B}_{ij}$ represents an effective valence that can be adjusted to reproduce different bonding behavior.

Together with the ability of closely related approaches to model a wide range of static solid state and dynamic gas-phase phenomena, the simple form of Eq. (1) suggests that the model potential has the flexibility necessary for incorporating large scale chemical reactivity into simulations of solid state detonations. The expression is sufficiently realistic that it can be tailored to reproduce such solid state properties as the changes in bonding that occur in differently coordinated polymorphs of silicon and carbon. It is also easy to fit the potential to the observed bond lengths, binding energies, and vibrational frequencies of the isolated diatomic molecule. It contains enough of the essential chemistry to allow fitting to few-body reactive potential energy surfaces, so that the possibility of chemical reactions can be included.

Rather than attempt to fit the potential to a particular system, we chose the parameters given in Table 1 to yield a simple generic model of an explosive — a molecular solid composed initially of AB molecules which, when shocked, can undergo exothermic chemical reactions to form stable A_2 and B_2 products. The parameters entering the model were chosen so that the bonding pair terms are significantly weakened for atomic coordinations greater than one. This insures that the ground state of the system at zero temperature and pressure is a collection of diatomic molecules and maintains the molecular nature of the system at ambient conditions.

The possibility of exothermic chemistry is incorporated into the model by choosing different well depths for the reactants and products. For the diatomic AB molecular reactants, a well depth of 2 eV is assumed, while a well depth of 5 eV is used for the diatomic A_2 and B_2 molecular products. Neglecting the small van der Waals contributions, these parameter choices imply that the gas-phase half-reactions $A + AB \rightarrow A_2 + B$, and $B + BA \rightarrow B_2 + A$ — taken as identical in the model — are exothermic by 3 eV. This energy release is similar to, but slightly smaller than, that occurring in the exothermic reaction[14] $N + NO \rightarrow N_2 + O$ thought important in the detonation of nitric oxide. The model parameters also yield chemically reasonable potential energy surfaces for these generic half reactions with the transition state occurring at a collinear geometry, as in F + H_2 → FH + H.[15]

Three closely related but different versions of the AB model are described in Table 1. These different versions — termed AB Model I,[8] AB Model II,[16] and AB Model III[9] — differ mainly in the choices of the bond length, r_e, a scale factor, β, and a parameter δ which together with β determines the range of $f_c(r)$. The potential contour map for the collinear exothermic reaction A + AB → A_2 + B for Model I is depicted in the lefthand panel of Fig. 1, with the minimum energy path leading from the reactant to the product valleys shown as a dashed line. As can be seen from Fig. 1, this reaction has an early[15] barrier which is typical of atom–diatom exothermic reactions. The potential energy along the minimum energy pathway, depicted as a solid line in the righthand panel of Fig. 1, shows that this barrier is much less than the AB dissociation energy of 2 eV. This realistically reflects the concerted nature of the reaction at the transition state, which occurs when the original AB bond is broken and the new A_2 bond is formed. The barrier to reaction at the transition state is 0.08 eV, which is close to the barrier for the exothermic reaction F + H_2 → FH + H.[15] Also shown in the righthand panel of Fig. 1 is the potential energy along

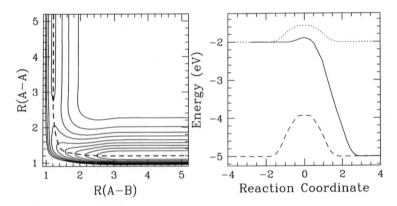

Fig. 1. Left: Potential energy surface for the collinear gas phase reaction A + AB → A_2 + B. The contours range from −4.5 to −0.5 eV with a new contour drawn every 0.5 eV. The minimum energy pathway for the reaction is shown as a dashed line. Right: Potential energy along the minimum energy pathway of the reactions A + AB → A_2 + B (solid line), A + A → A_2 + A (dashed line), and A + BA → AB + A (dotted line).

the reaction paths for the two transfer reactions: $A + A_2 \rightarrow A_2 + A$ and $B + B_2 \rightarrow B_2 + B$. Models II and III yield potential energy surfaces very similar to that for Model I, although they have a slightly shorter bond distance (0.1 nm versus 0.12 nm).

All three models predict reasonable bond lengths and vibrational frequencies for both the reactants and the products. For the parameters given in Table 1, each of the reactant (product) diatomic molecules (A_2 and B_2) has a vibrational frequency of 1064 cm^{-1} (1682 cm^{-1}) with both the reactants and products having identical equilibrium bond distances, 0.12 nm for Model I and 0.10 nm for Models II and III. Finally, inclusion of V_{vdW} causes the AB molecular reactants in all three models to condense into a diatomic AB molecular solid with reasonable crystalline binding energies, reasonable distances of closest approach between atoms, and reasonable solid state sound speeds.[8,9]

One might wonder why three different parametizations were introduced when the properties of the three models seem so similar. The problem is not in the normal properties of these models, which are similar, but in their behavior under the extreme conditions of a detonation.[8,9] The majority of the simulations discussed here were performed using Model I, our first model to yield a behavior consistent with a classic detonation. However our more recent studies (Sec. 4) have been carried out using the parametization corresponding to Model II. Although Models I and II yield qualitatively similar results, this change has been made to facilitate comparison with Model III (Sec. 5) which exhibits qualitatively different behavior at high pressures. When formulated as in Table 1, other than minor and unimportant changes in V_{vdW}, Models II and III differ only in a single parameter, δ.

2.2. *Initial Conditions and Computational Methods*

Most of the simulations reviewed were performed in two dimensions, although some simulations of three-dimensional detonations are discussed in Sec. 3.1.2. Equilibrated molecular crystals for AB Model I in both two and three dimensions are shown in Fig. 2. This model has slightly longer bond lengths than either Model II or III, but otherwise the starting crystal structures are similar for all three models. To model an infinite crystal, almost all detonation simulations were carried out with periodic boundary conditions enforced perpendicular to the direction of the shock propagation.

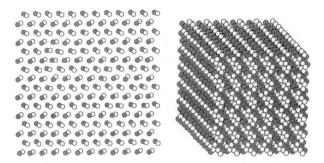

Fig. 2. Equilibrated crystal structures for the AB molecular solid for Model I in two (left) and three (right) dimensions.

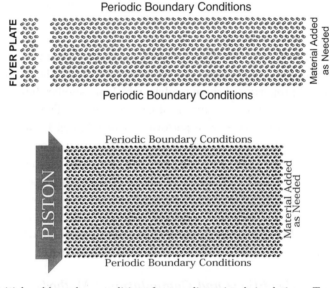

Fig. 3. Initial and boundary conditions for two-dimensional simulations. Top: Simulation begun with a 4-layer flyer plate. Bottom: Simulation begun with a steadily moving piston.

The sole exceptions are discussed in Sec. 3.2, where free boundary conditions perpendicular to the direction of shock propagation were assumed so that critical widths in two dimensions could be studied. Spot checks were made to ensure that increasing the size of the repeated simulation cell did not significantly affect the results obtained with periodic boundary conditions. When the model material was taken as infinite in the direction of the shock propagation, this was modeled by the addition of material in front of the propagating shock. In the detonation simulations, the initial, temperature was fixed at a low but nonzero value to allow the formation of stable molecular crystals while preventing spurious soliton solutions. Shockwaves were introduced into the simulations by either impacting the crystal's free edge with a flyer plate, as depicted at the top of Fig. 3, or by driving the material with a piston moving at constant velocity relative to the undisturbed solid, as depicted at the bottom of Fig. 3. The piston was modeled by a quadratic potential. Once the boundary and initial conditions were established, individual atomic trajectories were followed using a Nordsiek predictor-corrector method to integrate Hamilton's equations of motion.[17]

3. AB Model I: Detonation and Initiation

In this section we first present a series of results (Sec. 3.1) which show that molecular dynamics simulations can be used to directly link atomic scale chemistry to the continuum theory of detonations. We then show (Secs. 3.2 and 3.3) that complex initiation behavior can arise even within the simple AB Model I system. Taken together, the results reviewed in this section demonstrate that simulations using REBO potentials provide a powerful probe of the interplay between the continuum properties of shockwaves and the atomic scale chemistry induced in the initiation and propagation of condensed phase detonations.

3.1. *Detonation*

The continuum based ZDN theory views a detonation as starting with an initial very rapid compression and heating caused by the shockwave.[1] It is assumed that the explosive is brought to its ignition point during the initial compression and heating with no reactions occurring. The reactions then begin accompanied by a gradual drop in pressure and run to completion in the reaction zone. Beyond the reaction zone the system relaxes to satisfy the rear boundary condition. An essential feature of the theory is the von Neumann spike at the front, which occurs because the initial compression is much faster than the ensuing chemistry. In the following, we review molecular dynamic simulations with Model I that demonstrate that detonations in two and three dimensions (Secs. 3.1.1 and 3.1.2) have properties consistent with ZND theory.

3.1.1. *Detonations in Two Dimensions*

Simulations using Model I[8] were initiated by impacting a flyer plate composed of several layers of a nonenergetic AA molecular solid with the edge of the 2-D semiinfinite energetic AB solid. The latter is initially at rest at near zero temperature and pressure. Figure 4 shows a typical snapshot of a chemically sustained shockwave that can result. A distinct shockfront is visible with reactant molecules at the right and product molecules at the far left. After initiation, the shock front rapidly approaches a constant

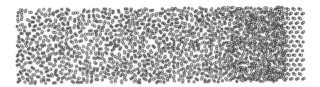

Fig. 4. Snapshot of a detonation 12.5 ps after impact by a 6 km/s flyer plate. The shockwave is propagating from left to right. The two types of atoms are shown as open and solid circles.

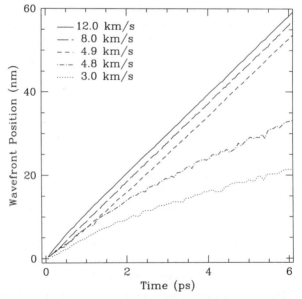

Fig. 5. Wave front position versus time for a series of two-dimensional simulations begun with 4-layer flyer plates. Results are shown for flyer plate velocities of 12.0 (solid), 8.0 (long dash), 4.9 (short dash), 4.8 (dot-dash), and 3.0 (dotted line) km/s.

velocity of about 9.5 km/s, which is within the range of observed detonation velocities.[2] The detonation velocity is independent of the conditions of the initiation, as is demonstrated by the upper three lines in Fig. 5. Although offset from one another because of different initial conditions, the lines rapidly become parallel and hence correspond to the same detonation velocity.

Figure 6, which is for the same simulation as Fig. 4, shows sectional averages of the pressure, particle velocity, and density as a function of position with the dashed lines denoting the initial values in the undisturbed crystal. These quantities peak at the front and then relax during the reaction and expansion behind the shock. The shockfront shape is in accord with ZND continuum theory for unsupported planar detonations,[1,6,7] which predicts a von Neumann peak near the front followed by a reacting flow and a (Taylor) rarefaction wave. The peak pressure around 1.0 eV/Å^2 (Fig. 6, top), which corresponds to an effective pressure of approximately 400 kbar, and maximum particle velocity of 4.8 km/s (Fig. 6, middle) are consistent

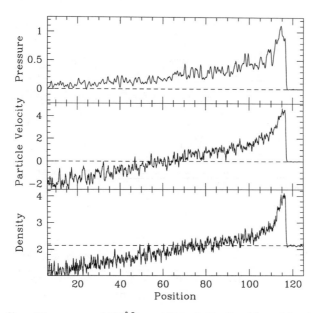

Fig. 6. Profiles of the pressure (eV/Å^2), particle velocity (km/s), and density (amu/Å^2) for the snapshot in Fig. 4. Positions (nm) are measured relative to the initial edge of the unreacted material. Dotted lines represent the initial values of the quantities in the unshocked material.

with a condensed phase detonation, as is the forward propagating region of product material that develops behind the detonation front.[1]

Once a detonation is achieved, the results of the simulations are soon in good quantitative agreement with the Rankine–Hugoniot relations[4,7] of continuum theory. These relations state that the following quantities are conserved across a planar shockfront:

$$\bar{\rho} \equiv \rho u \,, \tag{2a}$$

$$\bar{P} \equiv P + \rho u^2 \,, \tag{2b}$$

$$\bar{E} \equiv E + \frac{P}{\rho} + \frac{1}{2}u^2 \,, \tag{2c}$$

where ρ is the density, u is the local particle velocity, P is the pressure, and E is the internal energy per unit mass. The Rankine–Hugoniot relations

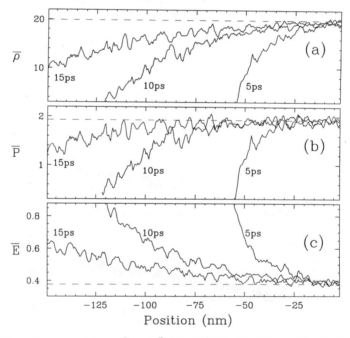

Fig. 7. Plots of the quantities $\bar{\rho}$, $\bar{P}$, and $\bar{E}$ defined in Eq. (2) of the text for the reference frame in which the shock is stationary. The horizontal axes gives position relative to the shockfront at times of 5, 10, and 15 ps. The dashed lines indicate values ahead of the shockfront.

assume steady flow conditions in the reference frame fixed with respect to the shock and require that $\bar{E}$ and $\bar{P}$ be measured sufficiently far enough behind the shockfront that the longitudinal component of the stress tensor is equal to the hydrostatic pressure and that the heat flux has vanished. In Fig. 7, the quantities $\bar{\rho}$, $\bar{P}$, and $\bar{E}$ are plotted relative to the shockfront position and starting about 2 nm behind it for simulation times of 5, 10, and 15 ps. The degree to which the values of these quantities behind the shockfront are equal to their initial values in the unreacted material indicates how close the simulation has approached steady flow with an unchanging shockprofile. The initial values are indicated by the dashed lines. Because there is a rapid relaxation to a constant shockvelocity, the rapid stabilization close to the shockfront seen in Fig. 7 indicates that the Rankine–Hugoniot relations are satisfied over a progressively wider region, with near steady flow conditions being reached 50 nm behind the shock in as little as 15 ps.

The fact that the Rankine–Hugoniot relations are well satisfied from 2 nm to 5 nm behind the front implies that dissipative processes rapidly vanish behind the front, because these relations assume an isotropic stress tensor and a vanishing heat flux. In addition it is easily seen from Fig. 4 that the chemistry is largely complete by 5 nm behind the front. Therefore, the length scales introduced into the problem through the initial response of the material and the ensuing chemistry might no longer be important beyond about 3 to 5 nm behind the front. If this is true, then by this point the flow should become selfsimilar and hence depend only on displacement scaled by time. However, if the flow is selfsimilar and not a squarewave, it will not be steady in the reference frame of the material. Consequently, it would be wrong to conclude from Fig. 7 that, if the simulations were run up to say 500 ps, the Rankine–Hugoniot relations would be satisfied at 50 nm behind the front to the same degree of accuracy that they are at 5 nm behind the front, even at 10 ps. Near the front the flow is steady, but it relaxes once it becomes selfsimilar.

Continuum theory predicts the existence of the special point behind the front known as the Chapman–Jouguet (CJ) point.[7] From the CJ point to the shockfront the flow is steady, while in unsupported detonations this is nowhere true behind the CJ point. The measured steady detonation velocity implies that an approximate CJ point exists in these simulations, while Fig. 4 suggests that it could well be within 5 nm of the front — a point beyond which the chemistry is essentially complete. The fact that the

Rankine–Hugoniot relations are so well obeyed within about 5 nm of the front implies, via a classic argument[7] not repeated here, that the simulated detonation is moving at the minimum velocity consistent with the conservation conditions, as predicted by ZND theory. We return to a discussion of the CJ point and related issues in Sec. 4, where several methods are presented for establishing the existence and position of the CJ point from the simulations.

The results reviewed so far establish that the Model I is capable of producing detonations in two dimensions with properties that are consistent with the ZND continuum theory of unsupported planar detonations.

3.1.2. *Detonations in Three Dimensions*

In this section we review results[18] that show that Model I is also capable of producing detonations in three dimensions with results that, if anything, are closer to experiment. Figure 8 shows results of a three-dimensional simulation of a core of energetic material with periodic boundary that was

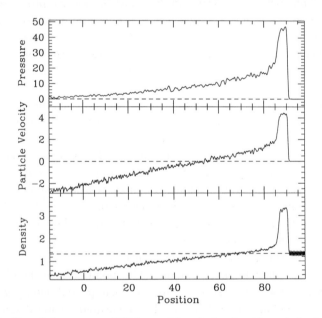

Fig. 8. Profiles of the pressure (GPa), particle velocity (km/s), and density (g/cm^3) for a three-dimensional simulation 12.5 ps after impact. Positions (nm) are measured relative to the initial edge of the unreacted material. The dashed lines are the unshocked values of the properties.

impacted with a flyer plate. Plots are given of the pressure, particle velocity, and density at 12.5 ps after impact. The results are very similar to those of our two-dimensional simulations, but show less fluctuation than Fig. 6. This is because of the greater amount of spatial averaging in three dimensions, which was possible because of the greater number of atoms. The profiles for three dimensions also possess a better defined initial peak.

If one compares the shockfront positions for two and three dimensions in Figs. 6 and 8, it can be seen that the shockfront in the two-dimensional simulation has moved farther during the same time following impact. This is confirmed in Fig. 9 where the shockfront position versus the time since impact is plotted. As indicated by the straight but diverging lines, the shockfronts quickly reach constant but different velocities. The slopes of the lines in Fig. 9 correspond to velocities of about 9.5 and 7.2 km/s for the two- and three-dimensional simulations, respectively.

The data presented above shows that both the two- and three-dimensional simulations yield similar results in terms of shock profiles and in

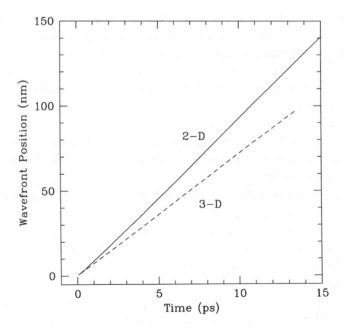

Fig. 9. Shockfront position versus time since impact for both two and three dimensional simulations.

the ability to support chemically sustained shockwaves. The main difference is that in three dimensions, the detonation has a lower velocity that
is closer to experimental values. The difficulty with three-dimensional simulations is that the number of atoms that must be treated is much greater
than in two dimensions. For example, for the two-dimensional simulation
presented above for a 5 nm wide sample required only 10 000 atoms, while
33 000 atoms were used for the narrow 3×3 nm core of material in the three-
dimensional simulations. This difference in the number of atoms needed
plus the added dimension with its larger number of neighbors causes the
three-dimensional simulations to be much more computationally intensive.
For this reason, we focus on the two-dimensional results in the remainder of this review. Two-dimensional simulations using Model I evidently
yield results that are very similar to those that would be obtained from
computationally much more demanding simulations in three dimensions.

3.2. *Critical Widths*

As mentioned in Sec. 3.1.1, the chemical reactions driving the detonation
depicted in Fig. 4, obtained with Model I, are largely complete between
2 nm and 5 nm behind the detonation front. When coupled with continuum
theory,[5,19] this extremely short reaction zone length implies that the model
should exhibit a critical width for detonation that is small enough to be
directly accessible in a molecular dynamics simulation of a two-dimensional
strip. The results of this section show that this is the case.[18] The results
further establish the connection between our simulations and continuum
theory and suggest the possibility of observing nanoscale detonations in
highly reactive ribbons.

Figure 10 depicts a detonation propagating through a 15 nm wide planar
AB strip at 24 ps after initiation. Visible in the snapshot is the unreacted
AB molecular solid, followed by a narrow reaction zone, which is followed by
an expansion region consisting mainly of A_2 and B_2 product molecules. For
this choice of width, the detonation continues to propagate in the ribbon,
but with a curved front that is consistent with expectations from continuum theory.[5,19] The small kink visible in the curved front results from a
shortlived fluctuation that does not quench the detonation. A pie-shaped
region of higher density is seen behind the front. It is caused by rarefaction
waves that propagate into the detonating sheet from the sides and pinch

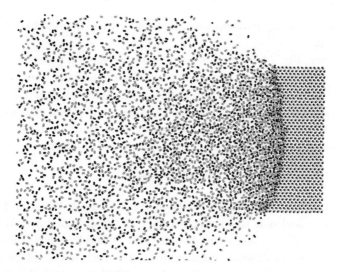

Fig. 10. Snapshot of a 15 nm wide sample 24 ps after impact. The two types of atoms are shown as open and solid circles. The detonation is moving from left to right. The figure has been clipped to fit the page.

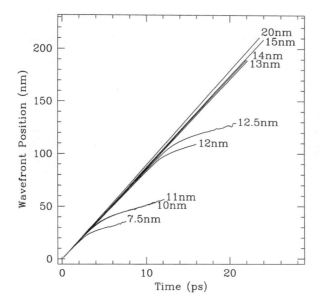

Fig. 11. Shockfront position versus time for a series of finite width 2-D simulations obtained using Model I.

off the high density reaction zone, thereby slowing the detonation but not stopping it.

However, Fig. 11 shows it is possible for rarefaction waves from the side to quench detonations in narrower strips. This quenching leads to a critical strip width below which the detonation does not propagate. The situation is made clearer in Fig. 12 where the measured detonation velocity, D, as a function of the width of the strip is shown. As a guide to the eye, the data points for different widths are fit to a function of the form:

$$D = D_0 \left(1 - \frac{\alpha}{W^3}\right), \qquad (3)$$

where D_0 is the detonation velocity in the 2-D sheet, W is the halfwidth of the strip, and α is the fitting parameter. We obtained $\alpha = 171.7$ nm^3. For values of $2W$ greater than about 20 nm, D rapidly approaches D_0. For smaller widths D depends sensitively on $2W$ with lower velocities corresponding to narrower strips. For values of $2W$ less than about 12.5 nm

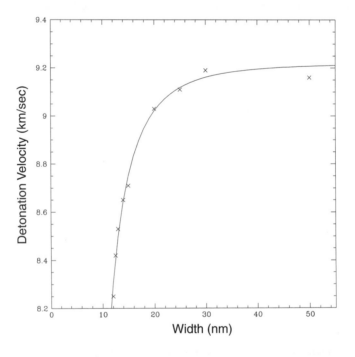

Fig. 12. Detonation velocity versus width of the strip for Model I.

the detonation does not propagate. If anything, this cutoff is approached with decreasing width even more rapidly than suggested by Eq. (3). The presence of a curved front, a value of D that decreases with decreasing $2W$, and a critical width for propagation are all consistent with the predictions of continuum theory.[5,19] However, further comparisons will require tailoring the well developed continuum theory of critical diameters to the specific strip geometry studied here.

The above results establish that Model I leads to a small critical width that can be probed with molecular dynamcs simulations. This small critical width is a consequence of the small reaction zone in the model. Although reaction zone lengths in complex detonating materials are often millimeters or longer, the model suggests that there is no physical reason why they cannot be much smaller — even down to the nanometer scale for simple detonating materials driven by a single exothermic chemical reaction. These small reaction zone lengths imply extremely small critical widths and diameters. Such materials, although potentially dangerous at the macroscale, could ultimately find use at the nanoscale. As shown below, the fact that the model has a very small critical width does not imply that it is extremely sensitive to impact initiation.

3.3. *Initiation*

So far we have concentrated mainly on the propagation characteristics of Model I. We now use this model to study the problem of initiation. As mentioned earlier, a better understanding of the shock to detonation transition is key to designing more reliable and safer explosives. After reviewing results indicating the model is insensitive to initiation in Sec. 3.3.1,[18] we show in Sec. 3.3.2 that the model exhibits delayed initiation around the initiation threshold.[20] In Sec. 3.3.3, it is shown that crystal structure only weakly affects initiation.[21] Finally, we close this section on initiation by reviewing results that indicate that the initiation threshold can be significantly lowered by line defects such as nanocracks.[22] The results show that molecular dynamics can provide significant insights into the complex processes leading to detonation.

3.3.1. *Insensitivity to Initiation*

High purity explosives are known to be very insensitive to the method of initiation.[23] However, the very short reaction zone lengths and the resulting small critical widths suggest that Model I might be quite sensitive to impact initiation. Reviewed in this section are results that show that this is not the case. Figure 5 indicates, at least for very thin flyer plates, that the model is insensitive to initiation as a four-layer plate velocity of approximately 4.9 km/s is required for initiation in two dimensions.

We have further studied the problem of impact initiation in two dimensions by varying both the number of layers in the flyer plate and its impact velocity.[18] Results from these simulations are shown in Fig. 13. In this figure the solid squares denote simulations where the detonation was sustained, while the open squares denote simulations where, if exothermic chemistry began at all, it was insufficient to maintain a detonation. The solid line follows the interface between these two regions. Figure 13 demonstrates that as the number of layers in the flyer plate is increased, the minimum impact velocity for initiation decreases, approaching an asymptotic value

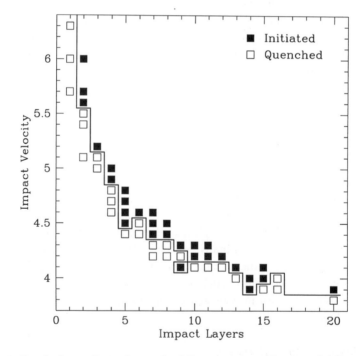

Fig. 13. Results from 2-D simulations for different numbers of layers and different velocities for the impacting flyer plate. The solid and open squares correspond to simulations where detonations were initiated or quenched, respectively.

of about 3.9 km/s. Thus, flyer plates of even macroscopic thickness must move at this high velocity to cause a detonation. This result agrees with the known insensitivity of crystalline explosives to initiation[23] and suggests that defects and hot-spots could have special importance in many actual initiations. Independent studies with this model substantiate this suggestion.[24] In addition, the analysis presented there[24] shows that the results from the model are both consistent with thermal explosion theory with chemically reasonable parameters, as well as with the behavior of real explosives.

3.3.2. *Delayed Initiation*

We saw in the previous section that Model I is insensitive to initiation. In particular, we reviewed results showing that, when the detonation is caused by flyer plate impact, the plate velocity required for initiation decreases to a limiting value of about 3.9 km/s as the mass of the flyer plate is increased. This behavior is consistent with the known insensitivity of crystalline explosives to high velocity impacts. In this section the initiation characteristics of Model I are examined more closely in the vicinity of the threshold for initiation. We find that the detonation begins behind a compaction front that is subsequently overrun by the detonation front.[20] Hence the behavior of the model is qualitatively similar to that observed in experimental studies of homogeneous initiation and delayed detonation associated with low velocity impacts.[25,26]

Figure 14 presents a series of snapshots of a detonation initiated by impacting a massive (on an atomic scale) flyer plate composed of 20 layers of the molecular solid with the edge of a two-dimensional semiinfinite diatomic molecular crystal. In this simulation the initial flyer plate velocity was taken as 3.9 km/s in the reference frame where the reactant crystal was initially at rest near zero temperature and pressure. As can be seen from the 3.0 ps snapshot of Fig. 14 (top), the impact of the flyer plate does not immediately induce chemical reactions. Rather a compaction shock is generated in which the average pressure and density are below that required to initiate reactions but with fluctuations that are sufficient to cause some chemistry behind the compaction front. More specifically, in the 3.0 ps snapshot products from a series of local reactions can be seen between 6 nm and 12 nm in a region that begins about 7 nm behind the compaction front which is located at 19 nm. Although reactions have occurred in this region with accompanying expansion there is not yet a well-defined

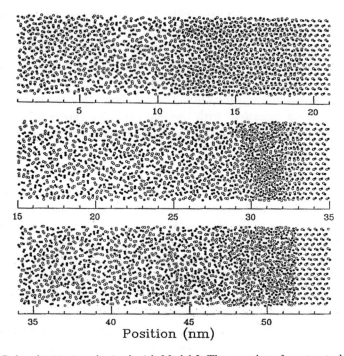

Fig. 14. Delayed initiation obtained with Model I. The snapshots from top to bottom show the atomic positions for times 3.0, 5.0, and 7.0 ps, respectively. The shockwaves are traveling from left to right, and the two types of atoms are shown as open and solid circles.

detonation front. However, by 5.0 ps (middle snapshot) sufficient reactions have occurred to generate a detonation front at about 32 nm, while the compaction front is still visible at about 33 nm. Finally, as observed from the 7.0 ps snapshot of Fig. 14 (bottom), this chemically driven front has overtaken the compaction front so that only the detonation front is visible.

The process of forming a detonation front associated with the onset of chemical reactions that eventually overruns the compaction shockfront is also illustrated in Fig. 15, where the position of these fronts is plotted as a function of time. Initially the compaction front propagates into the material with a velocity of 6.4 km/s, as indicated by the slope of the dashed line. Chemical reactions begin near 2.0 ps, and by 3.0 ps the leading edge of the reaction zone is definable. The position of this detonation front is shown as a solid line in Fig. 15. Because the detonation front is traveling faster

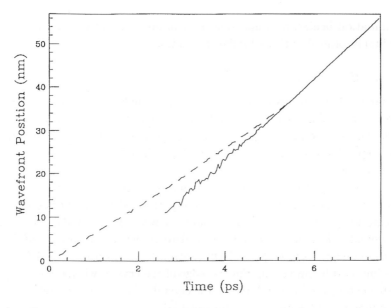

Fig. 15. Positions of the compaction (dashed) and detonation (solid) fronts as functions of time.

than the compaction shock, it overtakes the compaction front at around 5.3 ps. The initial fluctuations in the position of the detonation front in Fig. 15, while this front is still behind the compaction shockfront, indicate that reactions are initiated at several locations before the formation of a sharper better defined shockfront. Although the time and length scales differ, this simulation — showing the generation of a detonation front behind a compaction shock that is overrun by the detonation front — predicts qualitative behavior consistent with experimental studies of homogeneous initiation and of delayed detonation from low velocity impacts.

Among other things, the overall scale of the delayed initiation near threshold is related to the complexity of the reaction sequences that are required to drive the detonation. However, based on the results of Figs. 14 and 15, we expect a delay in the shock to detonation transition to occur in even the simplest detonating crystalline materials when shocked near threshold. In addition, the longer delays anticipated in more complex materials could have the same fundamental origin as the shorter delays expected in simpler materials. These observations stand as a challenge to experiment while suggesting that the fundamental phenomena of fluctuations leading to initiation behind a compaction front in homogeneous explosives can be profitably probed with molecular dynamics.

3.3.3. *Effects of Crystal Orientation*

So far we have discussed initiation with a single orientation of the un-shocked crystalline solid. It is known however that initiation depends on crystal orientation in energetic materials such as PETN, and this orientation dependence has been exploited to gain insight into the steps leading to initiation.[27]

To investigate the possibility that initiation is affected by orientation, a set of simulations with Model I were performed assuming two different orientations of the unshocked crystal.[21] A weak orientational dependence was found. The two orientations studied are shown at the far right of Fig. 16 and are denoted O1 (top) and O2 (bottom). In both cases a detonation was initiated by impacting the free edge of the crystal with a 96-atom flyer plate moving at 6 km/s relative to the crystal. The atomic trajectories were then monitored for an elapsed time of 12.6 ps.

The positions of the detonation fronts as functions of time are shown in Fig. 17 with the upper and lower lines corresponding to orientations O1 and O2, respectively. After about 5 ps the lines become parallel, indicating that the two detonation fronts have essentially the same velocity. This conclusion was verified by least-squares fits to the wavefront versus time profiles between 5 ps and 12.6 ps. The fits yielded detonation velocities

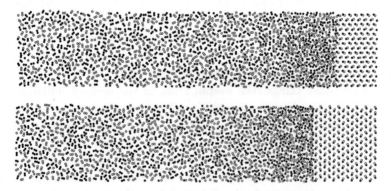

Fig. 16. Snapshots of atoms near the detonation fronts 12.6 ps after impact. The upper snapshot is for crystal orientation O1 and the lower is for O2. The two types of atoms are shown as open and solid circles.

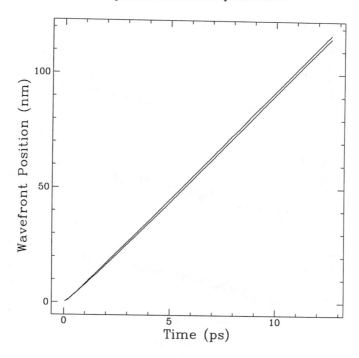

Fig. 17. Detonation front position versus time for orientation O1 (upper line) and O2 (lower line).

of 9.478 km/s for O1 and 9.451 km/s for O2 with a correlation coefficient of 1.0000 for both orientations. Not only are the front velocities virtually identical after about 5 ps, but so are the corresponding profiles for pressure, density, and particle flow velocity behind the fronts, as can be seen in Fig. 18. Thus, once initiation has occurred there is no significant difference in the character of the detonations.

However, the shifted front positions seen in Figs. 16 through 18 are evidence that orientation does affect initiation. Specifically, Fig. 17 shows that these shifts arise predominately in the first 5 ps after impact and thus occur during initiation. Detonation apparently starts earlier in the O1 orientation than in the O2 orientation, a conclusion that is consistent with what one would expect from the interaction of the shockfronts with the undisturbed material for the two orientations. An inspection of Fig. 16 suggests that the advancing shock should cause an initial one-dimensional compression that rotates the molecules so the line between the two atoms

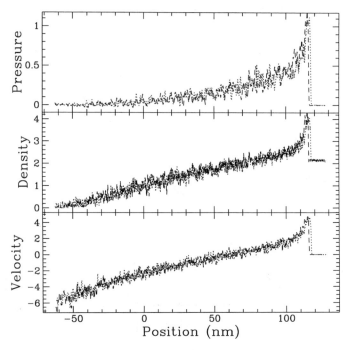

Fig. 18. The pressure (eV/Å^2), density (amu/Å^2), and particle flow velocity (km/s) at 12.6 ps for the O1 (dotted lines) and O2 (dashed lines) orientations.

in a molecule tends to align parallel to the shockfront. For orientation O1 the expected rotations will tend to place the A-atom in each molecule next to A atoms in the neighboring molecules and the B-atom next to B-atoms. Furthermore, this will be true both for neighboring molecules in the direction of propagation and for those in the direction parallel to the shockfront. In contrast, for orientation O2, the expected rotation will place the A-atom in a molecule next to B atoms in neighboring molecules and the B-atom next to A atoms. Because the configuration with A's abutting A's and B's abutting B's is the ideal orientation for exothermic chemical reactions, while A's abutting B's tends to frustrate these chemical reactions, exothermic chemistry is expected to be more easily initiated in orientation O1 than in orientation O2. Nevertheless, this effect is quite weak in the AB model, as the reactants quickly mix independently of the initial orientation.

However, in more complex structures with greater mechanical integrity such effects are expected to be more important. The fact that we see any orientational effects in the AB model strongly suggests that molecular dynamics simulations should be quite useful for studying the effect of molecular orientation on initiation in more complex materials. Indeed, we have seen much more pronounced orientational effects in a preliminary study of a model of ozone with REBO type potentials.[28]

3.3.4. *Effects of Defects*

In materials with defects the transition from a shock to a detonation is thought to involve the formations of local hot spots,[29-34] which are regions within the heterogeneous material that couple efficiently to the shockwave causing high local temperatures that ultimately lead to initiation. A range of defect types have been studied[35-38] as possible candidates for these initiation sites, including structural defects such as voids and shear bands. Nevertheless, the atomic scale mechanism for the transfer of energy from the shockfront to the hot spots remains a key unanswered question in studies of the predetonation process. Both molecular dynamics[34-36] and model quantum approaches have been used to address the question.[37]

In our earliest work on defect related initiation we used molecular dynamics simulations to study the coupling of a strong shockwave to a nanometer diameter cylindrical void oriented at right angles to the direction of propagation of the shockfront.[39] Although these simulations suggested possible atomic scale mechanisms for coupling shockwaves to material defects, they involved potentials that did not allow for chemical reactions,[39] so that the possibility of shock-induced chemistry could not be studied. In later work we carried out simulations of nanometer wide voids interacting with a strong shockwave in our AB energetic material[22] that did allow for chemical reactions to result from the shock-defect interaction. We found from the model that the simple defect studied could actually cause a detonation under strong shock conditions.

In these latter studies, strong shockwaves were produced by driving the free edge of the molecular solid with a steadily moving piston as depicted in the lower part of Fig. 3. Two-dimensional simulations were initially carried out to determine the piston driven shock-to-detonation threshold in the perfect crystal. Once this threshold was determined, a crack such as that depicted at the top of Fig. 19 was introduced. Additional simulations were then performed for a series of piston velocities near, but below, the critical piston velocity, V_p, that is necessary to cause detonation in defect-free

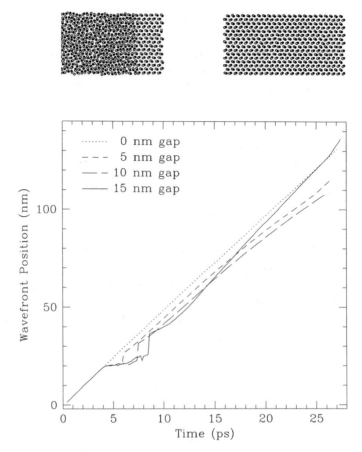

Fig. 19. Top: Illustration of a nanocrack in an otherwise perfect AB crystal. The two types of atoms are shown as open and solid circles. The piston driven shockfront approaches the crack from the left. Periodic boundary conditions are used perpendicular to the direction of shock propagation. Bottom: Front positions versus times for a series of crack widths. Piston velocity is 1.4 km/s in all cases.

material. A series of simulations were performed to study how the crack width would affect the coupling to the piston driven shockwave. In all cases, the crack had infinite length because of periodic boundary conditions.

For strong shockwaves the presence of a single crack such as that depicted in Fig. 19 (top) was found to significantly reduce the shock-to-detonation threshold. This is demonstrated in Fig. 19 (bottom). This figure depicts the shockfront position versus time for a series of simulations each driven by a piston moving with a constant velocity of 1.4 km/s. There is a single crack of constant width in each simulation. The straight line prior to about 5 ps in Fig. 19 indicates that these simulations rapidly approach a constant shockfront (particle flow) velocity prior to contacting the crack. At around 5 ps this steadily moving shockfront contacts the leading edge of the crack, which is reflected in Fig. 19 as the onset of the near flat part of the curves. For a piston velocity of 1.4 km/s we find that the 5 and 10 nm cracks do not cause sufficient perturbation to initiate significant chemical reactions. Hence, the curves of Fig. 19 corresponding to these simulations only wobble as the front leaves the region of the crack on their way back to a constant slope corresponding to the defect free crystal. The situation for the 15 nm simulation is different — not only does the 15 nm crack cause significant chemical reactions but actually causes a detonation which is reflected in the increasing slope of this curve beyond about 12 ps. Because V_p in the model without defects is about 1.8 km/s, the reduction in the shock to detonation threshold caused by the introduction of a single 15 nm wide crack is over 20%.

The results demonstrate that nanometer wide cracks can have severe effects on the shock-to-detonation threshold. It might be tempting to conclude that the chemical reactions caused by these defects result from a velocity doubling as atoms are spalled into the crack. Indeed, we found that the velocities of the leading particles that are spalled into the crack by the shockwave had approximately twice the particle flow velocity in the shockwave, as predicted by the continuum theory. However, we also observed that when these high velocity molecules struck the opposite side of the crack, reactions were not induced immediately. Rather, the complex motions of the many atoms within the crack appears to seed the chemical reactions that ultimately cause detonation. These studies lay the foundation for additional studies with more complex models.

4. AB Model II: Detailed Comparisons with Classic Detonation Theory

In Sec. 3.1, we reviewed results that established that Model I supports detonations with properties consistent with ZDN theory. In this section we make more detailed comparisons between the results of our simulations of detonations and the predictions of the continuum theory. In Sec. 4.1, we define the detonation Hugoniot curve, a concept central to continuum theory, and present several methods for calculating it directly from simulations. The calculated Hugoniot is also used to establish the approximate location of the CJ point. In Sec. 4.2, we show that the simulations are selfsimilar behind the CJ point, as asserted in Sec. 3.1.1. Overall, the molecular dynamics simulations reviewed in this section are in quantitative agreement with the predictions of continuum theory.

The results reported in Secs. 4.1 and 4.2 were obtained with Model II. As mentioned earlier, the normal state and the detonation properties of Model II are qualitatively similar to those of Model I, but the use of Model II facilitates the comparison with Model III, which yields qualitatively different results, as will be discussed in Sec. 5.

4.1. *Detonation Hugoniots and the CJ Point*

The Rankine–Hugoniot relations can be recast in the form[4]:

$$\frac{(D-u)}{V_1} = \frac{D}{V_0}, \tag{4a}$$

$$P_1 = P_0 + \left(\frac{D}{V_0}\right)^2 (V_0 - V_1), \tag{4b}$$

$$E_1(P_1, V_1) - E_0(P_0, V_0) = \frac{1}{2}(P_0 + P_1)(V_0 + V_1). \tag{4c}$$

These equations relate the undisturbed explosive lying at rest with pressure $P_0 = 0$ and specific volume $V_0 \equiv \frac{1}{\rho_0}$ to the state behind the detonation front, which is characterized by a pressure P_1, a specific volume V_1, and a particle flow velocity u. Both u and the detonation velocity, D, are measured in the reference frame of the undisturbed material. Because P_0 and V_0 are known, the Rankine–Hugoniot relations are a set of three equations for the four unknowns, u, D, P_1, and V_1. The first relation determines u in terms of D, P_1, and V_1, which leaves two equations with three unknowns. The first of the remaining equations, Eq. (4b) defines the Rayleigh line while Eq. (4c) defines the Hugoniot curve. The problem is formally determined by selecting the solution of Eqs. (4b) and (4c) that corresponds to the minimum value of D for an unsupported detonation.[40] This additional condition is the Chapman–Jouguet hypothesis, which was put on a firmer foundation by Zel'dovich.[7]

Since Eq. (4c) does not involve D, the Hugoniot curve can be represented by a function of the form $P_1 = H(V_1; V_0, P_0)$, which defines a set $\{P_1, V_1\}$ that satisfies Eq. (4c) assuming that P_0 and V_0 are known. The final state Hugoniot is a central construct of the ZDN theory of detonations and can be determined provided the equation of state, $E(P, V)$, is known. However, the Hugoniot function can be determined quite accurately from detonation simulations without calculating $E(P, V)$. This is done by performing a series of supported simulations with each corresponding to a different piston velocity. In these simulations the flow immediately in front of the piston face soon settles down to a constant profile moving at the piston velocity and characterized by P_1 and V_1, which can be measured accurately from the simulations by averaging the results near the piston face in both space and time. The state of the system in front of the piston is fully reacted and in thermodynamic equilibrium. In addition, if the shock profile is steady from the piston face to the detonation front, then this state will satisfy the Rankine–Hugoniot equations so that P_1 and V_1 will lie on the Hugoniot curve. The set $\{P_1, V_1\}$ defining the Hugoniot curve is then determined by performing a series of simulations with each one corresponding to a different piston velocity.

The Hugoniot curve determined in this way with Model II is shown in Fig. 20. Calculated points on this curve are represented by diamonds connected by solid lines that are guides to the eye. Also shown is the Rayleigh line (dotted) determined by the CJ hypothesis. This hypothesis selects the Rayleigh line with minimum D, that is the minimum slope line that originates at the initial state. As can be seen from Fig. 20, the requirement of minimum slope makes this line tangent to the Hugoniot curve. The point of tangency is known as the CJ point. We find that D determined from the slope of this Rayleigh line agrees with D measured directly from the simulations within less than a percent. This result provides strong confirmation of ZND theory in general and of the CJ hypothesis in particular.

Determining the Hugoniot from supported piston simulations requires that the detonation profile from the piston face to the front be steady for a given piston velocity, for only then will the Rankine–Hugoniot relations be satisfied near the piston face. We have shown directly from the simulations that this is an excellent assumption so long as the detonation is overdriven, that is so long as P_1 and V_1 exceed their corresponding values at the CJ point. This is illustrated in Fig. 21, where we show a series

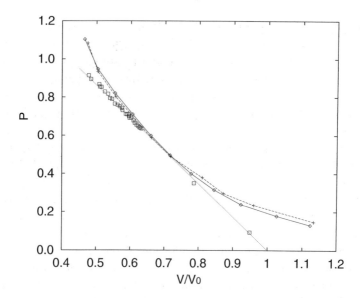

Fig. 20. Hugoniot curve calculated from supported detonation simulations (diamonds) and equation of state simulations (crosses). The solid and dashed lines connecting the points are guides to the eye. The dotted line is the Rayleigh line determined according to the CJ hypothesis. The open squares along the Rayleigh line were determined directly by sampling the unsupported detonation simulation between the CJ point to the shockfront.

of density profiles at different times throughout a simulation with a detonation supported by a piston moving at constant speed. The fact that these averaged profiles all coincide when aligned provides good evidence that the shock profile is indeed everywhere steady. On the other hand, this is no longer true for supported detonations that are underdriven. In this case the detonation still moves at a steady velocity. The flow in front of the piston is also steady and in equilibrium, but there is a region behind the front that is relaxing to satisfy the rear boundary conditions and thus is changing with time. Mass, energy, and momentum accumulate within this region, so that the Rankine–Hugoniot relations are no longer satisfied at the piston face. This failure to satisfy the Rankine–Hugoniot relations with the piston method poses no problem, because simulations of an unsupported detonation should never reach these states.

The above discussion shows that the points calculated from the supported simulations should no longer lie on the equation of state Hugoniot curve below the CJ point. To test this conclusion we have determined the

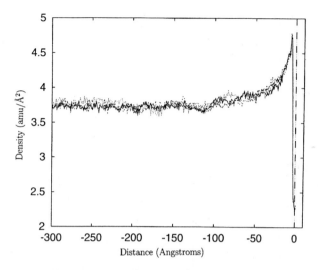

Fig. 21. Density profiles from an over-driven simulation for a piston velocity of 3.75 km/s relative to unreacted material. The results shown were obtained by averaging over 0.02 ps intervals between 10 and 20 ps (solid line), 20 and 30 ps (dashed line), 30 and 40 ps (dash-dotted line) and 40 and 50 ps (dotted line.)

Hugoniot curve directly from Eq. (4c) by carrying out a series of simulations within a rectangle box with periodic boundary conditions enforced at all edges. The width of this rectangle was chosen to equal the width of the periodically repeated strips in the supported two-dimensional simulations and its length was adjusted to give specified specific volumes V_1 assuming 640 particles, 320 of which are A's with the rest B's. The pressure, P_1, was measured from the simulations and the kinetic temperature changed until P_1 and V_1 satisfy Eq. (4c). The procedure was then repeated to determine additional values of P_1 and V_1. The results from these calculations are shown as crosses in Fig. 20 with dashed lines connecting neighboring points. Figure 20 establishes that results for P_1 and V_1 calculated from the piston simulations do lie on the Hugoniot curve obtained with Eq. (4c) above the CJ point but differ below the CJ point. This then provides a method for determining the CJ point without recourse to the Rayleigh line. Regardless of how the CJ point is found, once known, its position behind the front can be located as shown in Fig. 22. These results show the CJ point lies approximately 3 nm behind the front for Model II in two dimensions. Thus, although this model has an extremely short reaction zone, it nevertheless behaves in a manner consistent with continuum theory.

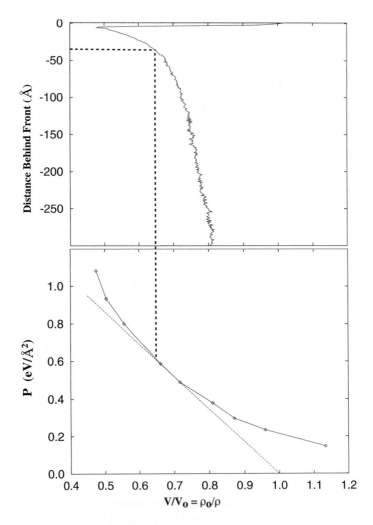

Fig. 22. Determination of position of CJ point behind the detonation front.

The Rayleigh line drawn in Fig. 20 assumes an equilibrium steady flow from the detonation front to the CJ point. However, this cannot be true because, although the flow at the front could be steady, it certainly is not in equilibrium as the stress tensor is not equal the hydrostatic pressure. However, if P in Eq. (4b) is replaced with P_{xx}, the component perpendicular to the shockfront, Eq. (4b) will be valid as long as the flow is steady, even though not in equilibrium. More specifically, to derive the result

$$P_{xx1} = P_{xx0} + \left(\frac{D}{V_0}\right)^2 (V_0 - V_1) \qquad (5)$$

requires only steady flow and conservation of mass and momentum, but to convert this result to Eq. (4b) requires the additional condition that $P = P_{xx}$. Thus, even if the system is not in equilibrium it must move along the line described by Eq. (5) provided the flow is steady. To test whether the flow is indeed steady from the front to the CJ point in unsupported simulations, as required by ZND theory, we have calculated P_1 and V_1 over this range from a simulation. The quantities were obtained with 1 Å wide bins averaged over 10 ps time intervals. The results are shown as open squares in Fig. 20, where the results do lie along a straight line with slope proportional D, as required by steady flow. The fact that the points lie along the Rayleigh line determined by the initial state and tangency to the Hugoniot curve implies that at the CJ point the chemistry is essentially complete and the dissipative processes are inactive. This follows because the Hugoniot curve is obtained assuming equilibrium. Equilibrium at the CJ point in the simulations is also consistent with Fig. 23 where the components of the kinetic temperature at the front are depicted. Although initially $T_{xx} \neq T_{yy}$,

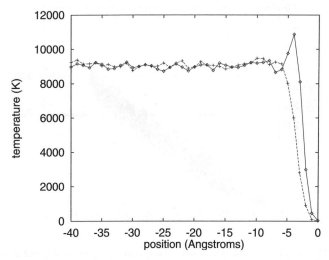

Fig. 23. Components of the kinetic temperature. The diamonds and crosses are for the components parallel and perpendicular to the direction of shock propagation, respectively.

these components become equal prior to the CJ point. Steady flow from the CJ point to the front in the simulations provides another important verification of the assumptions of ZDN theory. These results also indicate that behind the CJ point the simulations should become selfsimilar, since the length and time scales associated with dissipative processes and chemistry have been lost.

4.2. *Selfsimilar Behavior*

As discussed in Secs. 3.1.1 and 4.1, continuum theory predicts that an unsupported detonation is selfsimilar behind the CJ point, since beyond this point the chemistry is complete and the other nonequilibrium processes have vanished. The simulations reviewed in Sec. 4.1 provide strong evidence for the existence of a CJ point at approximately 3 nm behind the detonation front. At 3 nm, the reactions have run to completion and the effects of thermal transport and viscosity are no longer important. Thus, unsupported detonations in Model II should become selfsimilar at about 3 nm behind the front.

This conclusion is tested in Fig. 24, where we plot the density versus x/t at 4 ps (diamonds) and 20 ps (crosses) after initiation for a simulation of

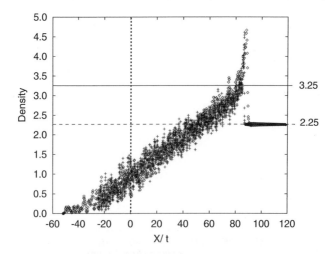

Fig. 24. The density (amu/Å^2) versus x/t (Å/ps) at 4 ps (diamonds) and 20 ps (crosses) after initiation. The horizontal line at 3.25 amu/Å^2 corresponds to the density behind the peak where the simulation first becomes selfsimilar. The horizontal line at 2.25 amu/Å^2 corresponds to the density of the unshocked crystal.

an unsupported detonation. The data was collected in 0.1 nm bins without smoothing or time averaging. Thus, there is considerable scatter in the results. Nevertheless, the results at 4 ps do coincide quite well with those at 20 ps at all points behind the peak where the density is less than 3.25 amu/Å^2. The density where the detonation first becomes selfsimilar should be the CJ density, ρ_{CJ}. Thus, we find directly from Fig. 24 that $\rho_{CJ} \approx 3.25$ amu/Å^2. This value corresponds to a normalized specific volume at the CJ point of $\frac{V_{CJ}}{V_0} = \frac{\rho_0}{\rho_{CJ}} \approx \frac{2.25}{3.25} \approx 0.7$, which is close to the specific volume $\sim$3 nm behind the detonation front found in Fig. 22. Thus, Fig. 24 not only establishes the selfsimilar character of the detonation behind the CJ point, but also determines its properties without recourse to the Hugoniot.

The method presented in the previous paragraph for determining system properties at the CJ point relies on the selfsimilar nature of the relaxation beyond that point, while the methods presented in Sec. 4.1 relied on the steady flow ahead of the CJ point. That these complementary approaches yield similar results provides further strong evidence of the agreement between our simulations to the continuum theory of detonations.

5. AB Model III: Phase Transitions and Detonations

In this section we turn to the last of the models to be studied in detail, Model III. The results reviewed show a detonation accompanied by a dissociative phase transition with concomitant flat-topped, split shockwaves.[9] Both of the results raise the possibility of a novel type of detonation and emphasize the importance of paying close attention to the high pressure characteristics of a model when developing potentials.

Figure 25 depicts a typical snapshot of a detonation that can result from the hypervelocity impact of a flyer plate at the edge of a two-dimensional Model III system. That the character of the detonation obtained with Model III is much different than those obtained with either Model I or II is already evident at only about 5 ps into the simulation. All three models

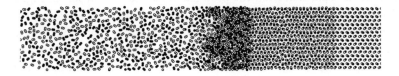

Fig. 25. Snapshot of the split chemically sustained shockwave that developed soon after initiation with Model III.

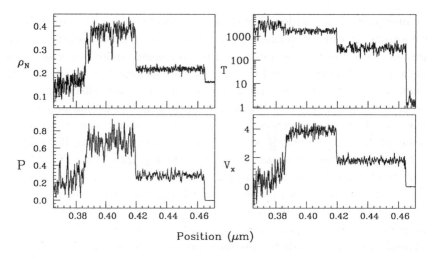

Fig. 26. Profiles of the density ρ_N (atoms/Å^2), two-dimensional pressure P (eV/Å^2), kinetic temperature T (K), and particle velocity in the direction of the shock detonation V_x (km/s), for a simulation begun with an impact plate velocity of 6.0 km/s. The results shown are at an elapsed time of 65 ps.

transform AB reactants into products, but with Model III the transformation is accomplished in a distinctly different fashion. As shown in the previous sections, a single shockfront with a profile consistent with the ZND continuum theory of planar detonations results with Models I and II. In contrast classic ZDN behavior is not obtained with Model III. First, with Model III there is a leading compressional wave — absent with either Model I or II — which heats and compresses the material and starts it flowing but does not induce any chemistry. This compressional region is then followed by a dissociative zone where the molecules are compressed to the point that they lose their molecular identity. Although the density is higher in this compressed region, the average nearest-neighbor interatomic separation has actually increased. In addition, the dissociative region for Model III is separated from the products region at the far left in Fig. 25 by a distinct boundary across which the majority of products are abruptly produced. Figure 26 shows that with Model III a flat-topped, split shockwave structure is obtained, rather than the peak near the front with a subsequent Taylor wave that resulted with Models I and II. Although the three boundaries separating the four distinct regions — like the single boundary obtained with Models I and II — rapidly stabilize to steady velocities, the different values of these velocities can cause the compressional and dissociative regions to enlarge to millimeter size in microseconds.[9] This unsteady behavior contrasts with the steady reaction zone length found with Model I and II.

We have traced the shockwave splitting present with Model III to a first-order phase transition accompanied by a volume collapse. This transition is a dissociative one in which the AB diatomic molecular solid transforms to a close-packed material with a much higher density but a larger average nearest-neighbor interatomic separations. More than 45 years ago the first shock induced polymorphic phase transition was observed.[41] An important property of these phase transitions, which is predicted by continuum theory and is observed experimentally, is the associated compressional shockwave splitting that can occur.[4] If split compressional shockwaves are present, continuum theory predicts that the leading shockfront starts the material flowing and brings it to the point of transition with the transition occurring across the second compressional shockfront.[4] This is exactly the behavior observed with Model III.

Actually, there is not one but two transitions visible in the snapshot of the Model III simulations. This second transition occurs at the interface between the dissociative zone and the rarefaction region as the material transforms from the dissociative phase to A_2 and B_2 molecular products. It is the second transition that produces the required behavior in the shock Hugoniot for a rarefaction shockwave. This product rarefaction shockfront appears to act as a steadily moving piston producing the observed flat-topped shockwave structure.

The distinctly different behavior of Model III is further clarified in Fig. 27, where the Hugoniots calculated from piston simulations for both Models II and III are shown. The Hugoniot for Model II, which has the classic ZDN behavior, is shown in the top panel of Fig. 27. The Hugoniot for Model III in the lower panel shows much different behavior. In this case, the system proceeds from the initial state I shown in Fig. 27 to the dissociative state B via the intermediate state A. It then proceeds from the dissociative phase to a product phase beginning with a C through a rarefaction shock. The position of the point A is determined by the properties of the phase transition. If A had occurred at a somewhat lower pressure, the system would have been able to proceed directly from I to B and the leading compressional shockfront would have been overrun by the dissociative

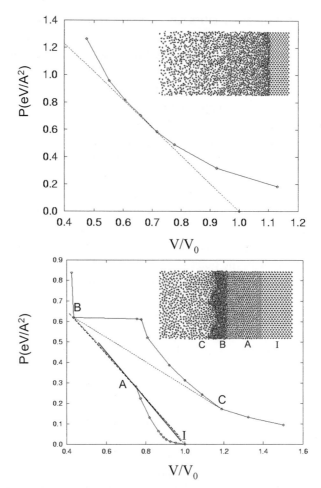

Fig. 27. Hugoniots from piston driven two-dimensional simulations assuming Model II (top) and III (bottom). The inserts show snapshots of the unsupported detonations that result in these models soon after initiation. The straight lines drawn are Rayleigh lines with slopes proportional to the shockfront velocities. The lines from I to A and from B to A in the lower panel were extended to make it easier to see their slopes.

shockfront. If that had occurred, then region A of the insert would have been absent and there would have been two, rather than three shockfronts present in the snapshot. The first shock is a compressional shock that separates the initial crystal from the dissociative phase, while the second shock is a rarefaction shock that separates the dissociative phase from the product phase. Since the velocities of compressional and the rarefaction shocks can differ, the region between B and C can continue to grow with time.

As can be seen from Table 1, Models II and III essentially differ only in the choice of δ, which determines the range of the cutoff function $f_c(r)$. These different choices for δ produce little differences in normal state properties such as bond distances and strengths, vibrational frequencies, crystal structures, binding energies, solid state speeds of sound, and potential energy surfaces for few-body reactions. For example, the bonding part of the REBO potential for gas-phase products within Models II and III are compared in Fig. 28. For these isolated product molecules the differences in δ has no effect on the equilibrium bond distances and strengths and vibrational frequencies. Differences only arise far from equilibrium where the bonding part of the potential is cutoff faster in Model II. However, in the extreme conditions of a detonation the larger range of $f_c(r)$ tends to stabilize the dissociative phase through the many-body part of the potentials and thus lead to detonations accompanied by flat-topped, split shockwaves.

The possibility of a dissociative phase transition induced by a detonation is fascinating because the behavior could result in more efficient conversion of chemical energy to useful work. Although detonations accompanied by phase transitions are perfectly compatible with continuum theory,[3] they have yet to be reported in detonating solids. In any event, the results presented in this section, when compared to those for either

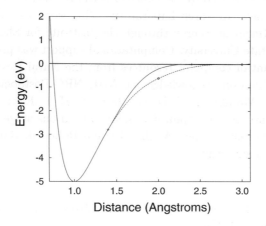

Fig. 28. REBO pair potential for Models II (solid line) and III (dotted line).

Models I or II, demonstrate the importance of carefully investigating the high pressure behavior associated with a model when developing potentials for the simulation of detonations.

6. Summary and Conclusions

The results reviewed demonstrate the utility of molecular dynamics simulations both for studying discrete shock induced chemistry in energetic materials and for relating the associated atomic scale chemistry to the classic ZND continuum theory of detonations. The results also demonstrate the usefulness of molecular dynamics for studying the effects of fluctuations, crystal orientation, and defects on the shock-to-detonation transition, studies that might ultimately aid in the design of safer explosives. Although only the simplest model systems have been studied, REBO potentials have a form that can be extended to more complex energetic materials.[28] Three-dimensional simulations of shocks should be possible in these more complex systems with numerical algorithms that can follow the dynamics of 10^8 to 10^7 atoms on parallel computers.[42] Molecular dynamics simulations with reactive many-body potentials represent a promising tool for probing the relationship between the continuum properties of shockwaves and the associated atomic scale chemistry.

Acknowledgments

This work was supported by the Office of Naval Research both directly and through the Naval Research Laboratory. This work was also supported by the Department of Energy through the Institute for Shock Physics at Washington State University. Computational support was provided in part through a grant of computer resources from the Naval Research Laboratory. D. R. Swanson acknowledges an NRL/NRC Postdoctoral Research Associateship. We also thank D. W. Brenner, M. L. Elert, B. L. Holian, and J. W. Mintmire for helpful discussions at various stages of this work. In addition, we thank Rainer A. Dressler and Robert J. Hardy for careful editing of the manuscript.

References

1. W. Fickett, *Introduction to Detonation Theory* (University of California Press, Berkeley, 1985).
2. W. C. Davis, *Sci. Am.* **256**, 106 (1987).

3. R. L. Rabie, G. R. Fowles and W. Fickett, *Phys. Fluids* **22**, 422 (1979).
4. Ya. B. Zel'dovich and Yu. P. Raizer, *Physics of Shockwaves and High-Temperature Hydrodynamic Phenomena*, Vols. 1 and 2 (Academic Press, New York, 1966, 1967).
5. H. Eyring, R. Powell, G. Duffey and R. Parlin, *Chem. Rev.* **45**, 69 (1949).
6. Ya. B. Zel'dovich and A. S. Kompaneets, *Theory of Detonation* (Academic Press, New York, 1960).
7. L. D. Landau, *Fluid Mechanics* (Elsevier Science, New York, 1959).
8. D. W. Brenner, D. H. Robertson, M. L. Elert and C. T. White, *Phys. Rev. Lett.* **70**, 2174 (1993); *ibid.* **76**, 2202 (1996).
9. C. T. White, D. H. Robertson, M. L. Elert and D. W. Brenner, in *Macroscopic Simulations of Complex Hydrodynamic Phenomena*, Eds. M. Mareschal and B. L. Holian (Plenum Press, New York, 1992), p. 111.
10. M. L. Elert, D. M. Deaven, D. W. Brenner and C. T. White, *Phys. Rev.* **B39**, 1453 (1989).
11. J. Tersoff, *Phys. Rev. Lett.* **56**, 632 (1986); *Phys. Rev.* **B37**, 6991 (1988).
12. J. N. Murrell, S. Carter, S. C. Farantos, P. Huxley, and A. J. C. Varando, *Molecular Potential Energy Functions* (John Wiley & Sons, Chichester, 1984).
13. G. C. Abell, *Phys. Rev.* **B31**, 6184 (1985).
14. S. P. Walch and R. L. Jaffe, *J. Chem. Phys.* **86**, 6946 (1987).
15. I. Shavitt, R. M. Stevens, F. L. Minn and M. Karplus, *J. Chem. Phys.* **48**, 2700 (1968).
16. D. S. Swanson, M. L. Elert and C. T. White, in *Shock Compression of Condensed Matter* — 1999, accepted for publication.
17. C. W. Gear, *Numerical Initial Value Problems in Ordinary Differential Equations* (Prentice-Hall, Englewood Cliffs, 1971).
18. D. H. Robertson, D. W. Brenner, and C. T. White, in *High Pressure Shock Compression of Solids III*, Eds. Lee Davison and Mohsen Shahinpoor (Springer-Verlag, 1998), Chap. 2.
19. W. W. Wood and J. G. Kirkwood, *Chem. Phys.* **22**, 1920 (1954).
20. D. H. Robertson, D. W. Brenner and C. T. White, *Mat. Res. Soc. Proc.* **296**, 183 (1993).
21. D. H. Robertson, D. W. Brenner and C. T. White, in *High Pressure Science and Technology* — 1993, Eds. S. C. Schmidt, J. W. Shaner, G. A. Samar and M. Ross (AIP Press, New York, 1994), p. 1369.
22. C. T. White, J. J. C. Barrett, J. W. Mintmire, M. L. Elert and D. H. Robertson, in *Shock Compression of Condensed Matter* — 1995, Eds. S. C. Schmidt and W. C. Tao, AIP Press, Woodbury, New York, 1996, p. 187.
23. F. P. Bowden and Y. D. Yoffe, *Initiation and Growth of Explosives in Liquids and Solids* (Cambridge University Press, Cambridge, 1985).
24. P. J. Haskins and M. D. Cook, *High Pressure Science and Technology* — 1993, Eds. S. C. Schmidt, J. W. Shaner, G. A. Samara and M. Ross (AIP Press, New York, 1994), p. 134.
25. A. W. Campbell, W. C. Davis and J. R. Travis, *Phys. Fluids* **4**, 498 (1961).

26. L. G. Green, E. James, E. L. Lee, E. S. Chambers, C. M. Tarver, C. West-moreland, A. M. Weston and B. Brown, *Proc. 7th Symp. (Int.) Detonation,* NSWC MP 82-334, Silver Spring, MD, 1981, p. 256.
27. J. J. Dick, *Appl. Phys. Lett.* **44**, 859 (1984).
28. J. J. C. Barrett, D. H. Robertson and C. T. White, *J. Chem. Phys.* (Russian), accepted for publication.
29. J. N. Johnson, P. K. Tang and C. A. Forest, *J. Appl. Phys.* **57**, 4323 (1985).
30. J. N. Johnson, *Proc. Roy. Soc. London, Ser.* **A413**, 329 (1987).
31. S. A. Bordzilovskii, S. M. Karakhanov and V. F. Lobanov, *Fiz. Goreniya Vzryva* **23**, 132 (1987); *Combustion, Explosion, and Shockwaves* (USSR) **23**, 624 (1987).
32. P. K. Tang, *J. Appl. Phys.* **63**, 1041 (1988).
33. F. E. Walker and A. M. Karo, in *Shockwaves in Condensed Matter,* Eds. S. C. Schmidt and N. C. Holmes (Elsevier, Amsterdam, 1988), p. 543.
34. A. M. Karo and J. R. Hardy, *Int. J. Quant. Chem. Symp.* **20**, 763 (1986).
35. A. M. Karo, T. M. Deboni, J. R. Hardy and G. A. Weiss, *Int. J. Quant. Chem. Symp.* **24**, 277 (1990).
36. D. H. Tsai, in *Chemistry and Physics of Energetic Materials,* Ed. S. N. Bulusu, NATO ASI Series C, Vol. 309 (Kluwer, Dordrecht, 1990), p. 195; D. H. Tsai, *J. Chem. Phys.* **95**, 7497 (1991).
37. D. D. Dlott and M. D. Fayer, *J. Chem. Phys.* **92**, 2798 (1990).
38. C. S. Coffey, *J. de Physique Colloq.* **48**, C4-253 (1987).
39. J. W. Mintmire, D. H. Robertson and C. T. White, *Phys. Rev.* **B49**, 14859 (1994).
40. J. J. Erpenbeck, *Phys. Rev.* **A46**, 6406 (1992).
41. D. Bancroft, E. L. Peterson and S. Minshall, *J. Appl. Phys.* **27**, 291 (1956).
42. B. L. Holian and P. S. Lomdahl, *Science* **280**, 2085 (1998).

GLOSSARY

albedo — the ratio of the amount of incident solar radiation reflected from an object to the total amount incident upon it.

annealing — the removal of crystal defects of materials through heating and subsequent cooling.

Auger Electron Spectroscopy (AES) — a keV electron beam is directed at a surface. Core electrons are ejected from atoms of the solid. Electrons with smaller binding energies fall into the core electron holes. The energy thus released can emit a low energy electron. The energy spectrum of Auger electrons is characteristic of the elemental identity of the respective atoms.

brine — water containing large amounts of salt.

dayglow — atmospheric optical emissions in daylight. The emissions are mostly attributable to sunlight-induced fluorescence.

E × B drift: The drift motion accompanying the gyration motion of a charged particle moving in a combination of electric and magnetic fields, **E** and **B**.

equilibrium — the present book differentiates between chemical, thermal and microscopic equilibrium. In chemical equilibrium, forward and backward rates of all chemical reactions of a system are equal ($\Delta G = 0$). Thermal equilibrium refers to the internal energies of all species having equilibrated in a manner that the various internal energy modes (rotational, vibrational and electronic) can be described by Boltzmann distributions of the same temperature. Microscopic equilibrium is the equivalent of the statistical energy partitioning in an isolated molecular system.

geological unit — material type based on its optical reflectance spectra.

homopause — also called the *turbopause*, defines the atmospheric altitude above which molecular transport is governed by molecular diffusion. Below the homopause, molecular transport occurs through bulk Eddy diffusion (turbulent mixing).

ionosphere — the atmospheric layer of a planet which is partially ionized by the penetration of solar ultraviolet. Earth's ionosphere ranges from altitudes of ~ 80 to ~ 1000 km.

Jeans escape — the process associated with the planetary loss of those molecules to space that have velocities exceeding $\sqrt{2gr}$, where g is the planet's gravitational acceleration and r is the distance to the planet's center.

Jovian — of or relating to the planet Jupiter.

LEPS (London–Eyring–Polanyi–Sato) Potential — approximate polyatomic potential surface obtained from diatomic Morse functions and related repulsive functions.

Madelung potential — refers to the electrostatic interaction per ion pair in a crystal. The electrostatic energy per ion pair is the Madelung energy, E_{electro}:

$$E_{\text{electro}} = -(\alpha Z^2 e^2)/d$$

where α is the Madelung constant, the ion pair consists of ions with charges Ze and $-Ze$, and d is the separation.

magnetosphere — a region of a planet's atmosphere that is dominated by the planet's magnetic field so that charged particles are trapped in it.

mesopause — the mesopause is the upper limit of the *mesosphere* and the lower boundary of the *thermosphere*. At ~ 85 km, it is the coldest altitude of Earth's atmosphere.

mesosphere — Earth's atmospheric region between the stratosphere and the thermosphere ranging from 50 to ~ 80 km. The temperature in the mesosphere decreases sharply with increasing altitude.

meteor — the visual phenomenon associated with the entry of interplanetary material into a planetary atmosphere.

meteorite — a stony or metallic object from interplanetary space that impacts a planetary surface.

meteoroid — the interplanetary objects that are the source of meteors.

mixing ratio — also referred to as *volume, molecular and molar mixing ratios* in atmospheric chemistry. The three are identical for ideal gases, and are given by the ratio of the partial pressure or density of an atmospheric consitituent to the atmospheric pressure or density for all atmospheric consitituents. A mixing ratio of 10^{-6} corresponds to 1 part per million (1 ppm).

Molière central potential — an analytical expression to approximate the repulsive potential between two atoms at short range. The universal pairwise potential requires only two parameters — the atomic numbers of the interacting atoms.

nightglow — the atmospheric glow during the night. The emissions are attributable to electronic, atomic and molecular collision processes.

porosity — fraction of the volume of a body that is void space.

shakeup peak — peak in a photoelectron spectrum attributable to a multielectron process.

spinel — a mineral composed of magnesium aluminate, MgA_2O_4.

supercritical fluid — a fluid which is at a temperature greater than its critical temperature (the temperature above which it cannot be liquefied by increasing pressure). Under these conditions, the two fluid phases, liquid and vapor, become indistinguishable.

thermosphere — the outermost shell of the atmosphere, between the mesosphere and outer space; where temperatures increase steadily with altitude.

threshold gain — the optical gain in a laser cavity that leads to amplification of an incident wave.

Wulff construction — a way to obtain the equilibrium shape of a crystal introduced by Wulff in 1901.

X-ray Photoelectron Spectroscopy (XPS) — in XPS, keV photons produce photoelectrons highly specific to the surface. The binding energies of the electrons are characteristic of the surface elements as well as the binding energy to the surface and neighboring atoms.

INDEX